Sustainable Development and Biodiversity

Volume 6

Series editor
Kishan Gopal Ramawat
M.L. Sukhadia University, Botany Department
Udaipur, Rajasthan, India

This book series provides complete, comprehensive and broad subject based reviews about existing biodiversity of different habitats and conservation strategies in the framework of different technologies, ecosystem diversity, and genetic diversity. The ways by which these resources are used with sustainable management and replenishment are also dealt with. The topics of interest include but are not restricted only to sustainable development of various ecosystems and conservation of hotspots, traditional methods and role of local people, threatened and endangered species, global climate change and effect on biodiversity, invasive species, impact of various activities on biodiversity, biodiversity conservation in sustaining livelihoods and reducing poverty, and technologies available and required. The books in this series will be useful to botanists, environmentalists, marine biologists, policy makers, conservationists, and NGOs working for environment protection.

More information about this series at http://www.springer.com/series/11920

Dinesh K. Maheshwari • Meenu Saraf
Editors

Halophiles

Biodiversity and Sustainable Exploitation

Springer

Editors
Dinesh K. Maheshwari
Department of Botany and Microbiology
Gurukula Kangri University
Haridwar (Uttarakhand), India

Meenu Saraf
Department of Microbiology and Biotechnology
University School of Sciences
Gujarat University
Ahmedabad, Gujarat, India

ISSN 2352-474X ISSN 2352-4758 (electronic)
Sustainable Development and Biodiversity
ISBN 978-3-319-14594-5 ISBN 978-3-319-14595-2 (eBook)
DOI 10.1007/978-3-319-14595-2

Library of Congress Control Number: 2015948373

Springer Cham Heidelberg New York Dordrecht London

Printed on acid-free paper

Springer International Publishing AG Switzerland is part of Springer Science+Business Media (www.springer.com)

Preface

There is a great interest on study of extreme microorganisms, particularly halophilic microorganisms which live in saline environments through different adaptation mechanisms and produce metabolites with great potential. Saline environments such as saline habitats are excellent sources for isolation of halophilic microorganisms. Microbial adaptation has played a major role when bacteria in all forms branched out into different environmental niches, arising from the hypersaline conditions of the primordial sea. Halophiles are among the best model microorganisms to study cellular adaptation. Their low nutritional requirements and resistance to high concentrations of salt make them a potent candidate in a wide range of biotechnological applications. Halophilic bacteria are very divergent and more than 150 species in 70 genera of halophilic bacteria are reported. Use of these halophilic and halotolerant microorganisms and their metabolites under harsh industrially conditions is a strikingly important subject for industrial sectors.

This volume deals with the diversity and importance of halophilic bacteria and actinobacteria for biotechnological industries; methods for their isolation and techniques for physiological, taxonomical and molecular characterization have been highlighted. The halophiles as important sources of enzyme production has been discussed. Metabolites and biological functions may resolve the ever-increasing thirst of industry to cope up with a range of issues from environmental pollutions to diseases and world's hunger. The combination of these bio-molecules with various nanomaterials like thin-layers, nanotubes, and nanospheres that results in novel compounds possessing both biological properties of biomolecules and physico-chemical characteristics of nanomaterials has been suitably described. The book presents insights into the main biomolecules produced by both halophilic archaea and bacteria revealing their potential implications in some nanotechnologies. Attempts have been made to outline progress in our understanding of their environmental diversity and biological survivability. The adaptation of enzymes during the course of their evolutionary development and some metabolic differences helped them expand and achieve environmental diversity. One of the most interesting findings in this field of research is spatial and temporal variation in microbial community structure, which was related to variations in salinity of the microenvironment. Many of the isolated halophiles have been artificially augmented and applied directly to saline soils to improve the nutrient status of such soil and contribute to soil fertility and remediation.

Metabolic processes, like osmoregulation in halotolerant cells, dictate the regulation of the bacterial cell membrane. Some of their biomolecules that have been studied and applied in industrial processes include exopolysaccharides, carotenoid pigments, bacteriorhodopsin, hydrolytic enzymes, etc. Various halophilic enzymes in different enzymatically processes, compatible solutes as macromolecules stabilizers, biopolymers, biofertilizers and pharmaceutically active molecules from halophilic bacteria are among the important applications of these group. Quorum sensing (QS) could influence the production of these biomolecules; thus a better understanding of halophilic bacterial communication mechanisms can help to improve the yields of these biotechnological processes.

The highlights of this book which include the distribution of the halobacteria, their adaptation in different stress conditions, and mining of their unique antimicrobial and enzymatic potential shall be useful to senior as well as budding researchers in the emerging fields of microbiology, molecular biology, biochemistry, biotechnology, environmental sciences and nanotechnology.

We would like to express our gratitude to all the subject experts and reviewers for their much needed mutual co-operation of scientific benefits. Their authoritative contribution and up-to-date information could make this project a success. Assistance rendered by our research scholars, Mohit and Shrivardhan, is thankfully acknowledged. We extend our sincere appreciation to Dr. Valeria Rinaudo and Dr. Ineke from the publisher Springer for their valuable support to facilitate the completion of this book.

Haridwar, Uttarakhand, India Dinesh K. Maheshwari
Ahmedabad, Gujarat, India Meenu Saraf

Contents

Contributors

C.N. Aguilar Food Research Department, School of Chemistry, Universidad Autónoma de Coahuila, Coahuila, Mexico

Dina M. Al-Mailem Microbiology Program, Department of Biological Sciences, Kuwait University, Safat, Kuwait

Mohammad Ali Amoozegar Extremophiles Laboratory, Department of Microbiology, School of Biology and Center of Excellence in Phylogeny of Living Organisms, College of Science, University of Tehran, Tehran, Iran

Nicolaus Barbara National Research Council of Italy, Institute of Biomolecular Chemistry, Pozzuoli, Naples, Italy

Shilajit Barua Department of Microbiol, Vijaygarh Jyotish Ray College, Kolkata, India

Kalyan Chakrabarti Institute of Agric Science, University of Calcutta, Kolkata, India

Roxana Cojoc Department of Microbiology, Institute of Biology Bucharest, Romanian Academy, Bucharest, Romania

Kruti G. Dangar Department of Biosciences, Saurashtra University, Rajkot, Gujarat, India

Mona Dehhaghi Department of Microbial Biotechnology, School of Biology and Center of Excellence in Phylogeny of Living Organisms, College of Science, University of Tehran, Tehran, Iran

Microbial Technology and Products Research Center, University of Tehran, Tehran, Iran

M. Delgado-García Food Research Department, School of Chemistry, Universidad Autónoma de Coahuila, Coahuila, Mexico

Toksoy Oner Ebru Department of Bioengineering, Marmara University, Istanbul, Turkey

Mădălin Enache Department of Microbiology, Institute of Biology Bucharest, Romanian Academy, Bucharest, Romania

Sevim Feyza Erdoğmuş Department of Laboratory and Veterinary Health, Technical Vocational School of Higher Education of Bayat, Afyon Kocatepe University, Afyonkarahisar, Turkey

Tommonaro Giuseppina National Research Council of Italy, Institute of Biomolecular Chemistry, Pozzuoli, Naples, Italy

Sangeeta D. Gohel Department of Biosciences, Saurashtra University, Rajkot, Gujarat, India

Javad Hamedi Department of Microbial Biotechnology, School of Biology and Center of Excellence in Phylogeny of Living Organisms, College of Science, University of Tehran, Tehran, Iran

Microbial Technology and Products Research Center, University of Tehran, Tehran, Iran

Anton Hartmann Research Unit Microbe-Plant Interactions, Helmholtz Zentrum Muenchen, German Research Center for Environmental Health (GmbH), Neuherberg, Germany

Wael N. Hozzein Bioproducts Research Chair (BRC), College of Science, King Saud University, Riyadh, Kingdom of Saudi Arabia

Botany and Microbiology Department, Faculty of Science, Beni-Suef University, Beni-Suef, Egypt

Masahiro Kamekura Halophiles Research Institute, Noda, Chiba-ken, Japan

Muhsin Konuk Department of Molecular Biology and Genetics, Faculty of Engineering and Natural Sciences, Üsküdar University, İstanbul, Turkey

Safiye Elif Korcan Biology Department, Faculty of Science and Literature, Afyon Kocatepe University, Afyonkarahisar, Turkey

Meenal Kowshik Department of Biological Sciences, BITS Pilani, Zuarinagar, Goa, India

Fatemeh Mohammadipanah Department of Microbial Biotechnology, School of Biology and Center of Excellence in Phylogeny of Living Organisms, College of Science, University of Tehran, Tehran, Iran

Microbial Technology and Products Research Center, University of Tehran, Tehran, Iran

Elhafid Nabti Laboratory of Renewable Energies, Group of Biomass and Environment, University of Bejaia, FSNV, Bejaïa, Algeria

B. Nicolaus Istituto Di Chimica Biomolecolare, Comprensorio ex Olivetti, Pozzuoli, Naples, Italy

Hamed Kazemi Shariat Panahi Department of Microbial Biotechnology, School of Biology and Center of Excellence in Phylogeny of Living Organisms, College of Science, University of Tehran, Tehran, Iran

Microbial Technology and Products Research Center, University of Tehran, Tehran, Iran

Stuti Patel Department of Microbiology and Biotechnology, University School of Sciences, Gujarat University, Ahmedabad, Gujarat, India

A. Poli Istituto Di Chimica Biomolecolare, Comprensorio ex Olivetti, Pozzuoli, Naples, Italy

Megha K. Purohit Department of Biosciences, Saurashtra University, Rajkot, India

Samir S-A. Radwan Microbiology Program, Department of Biological Sciences, Kuwait University, Safat, Kuwait

Narayanan Rajendran MASC, Kentucky State University, Frankfort, KY, USA

Vikram H. Raval Department of Biosciences, Saurashtra University, Rajkot, India

Abbamondi Gennaro Roberto National Research Council of Italy, Institute of Biomolecular Chemistry, Pozzuoli, Naples, Italy

R. Rodríguez-Herrera Food Research Department, School of Chemistry, Universidad Autónoma de Coahuila, Coahuila, Mexico

Meenu Saraf Department of Microbiology and Biotechnology, University School of Sciences, Gujarat University, Ahmedabad, Gujarat, India

Michael Schmid Research Unit Microbe-Plant Interactions, Helmholtz Zentrum Muenchen, German Research Center for Environmental Health (GmbH), Neuherberg, Germany

Amit K. Sharma Department of Biosciences, Saurashtra University, Rajkot, Gujarat, India

Satya P. Singh Department of Biosciences, Saurashtra University, Rajkot, Gujarat, India

Maryam Siroosi Extremophiles Laboratory, Department of Microbiology, School of Biology and Center of Excellence in Phylogeny of Living Organisms, College of Science, University of Tehran, Tehran, Iran

Pallavee Srivastava Department of Biological Sciences, BITS Pilani, Zuarinagar, Goa, India

Foram J. Thakrar Department of Biosciences, Saurashtra University, Rajkot, Gujarat, India

Sudipta Tripathi Institute of Agric Science, University of Calcutta, Kolkata, India

Chapter 1
Biodiversity of Halophilic and Halotolerant Actinobacteria

Wael N. Hozzein

Abstract The halophiles possess interesting biotechnological capabilities which increase the importance of their exploitation for commercial purposes. The halophilic microorganisms are highly diverse and among those organisms the halophilic actinobacterial group is of special interest for their secondary metabolites. The halophilic actinobacteria have morphological, taxonomical, ecological and physiological diversity. In terms of morphological diversity, they are ranging from coccoid, rod-shaped, fragmented to extensively branched filamentous organisms. In terms of taxonomical diversity, they are distributed among different taxonomical orders, suborders, families and genera. Ecologically, they are found in different ecosystems from hypersaline soils, salt lakes, marine sediments, salted food, and brines to endophytes. The physiological diversity is apparent from the variation in tolerance or response to the salt in their environment. The focus of the following chapter will be the diversity of halophilic and halotolerant actinobacteria from different aspects.

Keywords Biodiversity • Halophilic • Halotolerant • Actinobacteria • Taxonomy • Ecology

1.1 Introduction

Halophilic microorganisms are salt-loving organisms inhabiting diverse environments with the capacity to balance the osmotic pressure of the environment. The world of the halophilic microorganisms is highly diverse, as representatives of the three domains of life, Archaea, Bacteria, and Eucarya belong to this world. There are aerobic as well as anaerobic halophiles, heterotrophic, phototrophic, and chemoautotrophic types (Oren 2002, 2008) (Table 1.1).

W.N. Hozzein (✉)
Bioproducts Research Chair (BRC), College of Science, King Saud University, P.O. Box 2455, Riyadh 11451, Kingdom of Saudi Arabia

Botany and Microbiology Department, Faculty of Science, Beni-Suef University, Beni-Suef, Egypt
e-mail: hozzein29@yahoo.com

D.K. Maheshwari, M. Saraf (eds.), *Halophiles*, Sustainable Development and Biodiversity 6, DOI 10.1007/978-3-319-14595-2_1

Table 1.1 Taxonomic diversity of halophilic actinobacteria

Sub-order	Family	Genus	Species	Reference
Actinopolysporineae	Actinopolysporaceae	*Actinopolyspora*	*A. halophila*,	Gochnauer et al. (1975)
			A. mortivallis,	Yoshida et al. (1991)
			A. xinjiangensis,	Guan et al. (2010)
			A. alba and *A. erythraea*,	Tang et al. (2011b)
			A. algeriensis,	Meklat et al. (2012)
			A. saharensis,	Meklat et al. (2013a)
			A. dayingensis,	Guan et al. (2013b)
			A. righensis,	Meklat et al. (2013b)
			A. lacussalsi	Guan et al. (2013a)
			and *A. mzabensis*	Meklat et al. (2013c)
Corynebacterineae	Corynebacteriaceae	*Corynebacterium*	*C. halotolerans*	Chen et al. (2004)
Glycomycineae	Glycomycetaceae	*Glycomyces*	*Glycomyces halotolerans*	Guan et al. (2011b)
Jiangellineae	Jiangellaceae	*Haloactinopolyspora*	*H. alba*	Tang et al. (2011a)
		Jiangella	*H. alkaliphila*	Zhang et al. (2014)
Micrococcineae	Bogoriellaceae	*Georgenia*	*Georgenia halophile*	Tang et al. (2010a)
	Cellulomonadaceae	*Cellulomonas*	*C. phragmiteti*	Rusznyák et al. (2011)
			C. pakistanensis	Ahmed et al. (2014)
	Intrasporangiaceae	*Serinicoccus*	*S. marinus*	Yi et al. (2004)
			S. chungangensis	Traiwan et al. (2011)
			S. profundi	Xiao et al. (2011)
	Microbacteriaceae	*Microbacterium*	*M. halophilum*	Takeuchi and Hatano (1998)
			M. halotolerans	Li et al. (2005a)
			M. sediminis	Yu et al. (2013)

		Salinibacterium	*S. amurskyense*	Han et al. (2003)
			S. xinjiangense	Zhang et al. (2008)
	Micrococcaceae	*Arthrobacter*	*Arthrobacter halodurans*	Chen et al. (2009a)
		Kocuria	*Kocuria halotolerans*	Tang et al. (2009b)
			Kocuria marina	Kim et al. (2004)
		Nesterenkonia	*Nesterenkonia halobia*	Onishi and Kamekura (1972)
			N. aethiopica	Delgado et al. (2006)
			N. alba	Luo et al. (2009)
			N. flava	Luo et al. (2008)
			N. halophile	Li et al. (2008b)
			N. halotolerans, *N. xinjiangensis*	Li et al. (2004b)
			N. jeotgali	Yoon et al. (2006)
			N. lacusekhoensis	Collins et al. (2002)
			N. lutea, *N. sandarakina*	Li et al. (2005b)
			N. suensis	Govender et al. (2013)
	Promicromonosporaceae	*Isoptericola*	*I. halotolerans*	Zhang et al. (2005)
	Ruaniaceae	*Ruania*	*R. albidiflava*	Gu et al. (2007)
		Haloactinobacterium	*H. album*	Tang et al. (2010b)
Micromonosporineae	Micromonosporaceae	*Micromonospora*	*M. halophytica*	Weinstein et al. (1968)
			M. halotolerans	Carro et al. (2013)
Propionibacterineae	Nocardioidaceae	*Aeromicrobium*	*A. halocynthiae*	Kim et al. (2010)
		Nocardioides	*N. halotolerans*	Dastager et al. (2008a)
			N. tritolerans	Dastager et al. (2008b)
			N. daedukensis	Yoon et al. (2010)

(continued)

Table 1.1 (continued)

Sub-order	Family	Genus	Species	Reference
Pseudonocardineae	Pseudonocardiaceae	*Amycolatopsis*	*A. halotolerans*	Lee (2006)
			A. halophile	Tang et al. (2010a)
			A. salitolerans	Guan et al. (2012)
			A. cihanbeyliensis	Tatar et al. (2013)
		Haloechinothrix	*H. alba*	Tang et al. (2010d)
		Prauserella	*P. marina*	Wang et al. (2010)
			P. muralis	Schäfer et al. (2010)
			P. rugosa	Kim and Goodfellow (1999)
			P. alba, *P. halophile*	Li et al. (2003c)
			P. aidingensis, *P. flava*, *P. salsuginis* and *P. sedimina*	Li et al. (2009)
		Saccharomonospora	*S. halophile*	Al-Zarban et al. (2002b)
			S. paurometabolica	Li et al. (2003d)
			S. saliphila	Syed et al. (2008)
		Saccharopolyspora	*S. halophila*	Tang et al. (2009a)
			S. qijiaojingensis	Tang et al. (2009d)
			S. lacisalsi	Guan et al. 2011a
			S. dendranthemae	Zhang et al. (2013a, b)
			S. jiangxiensis	Zhang et al. (2009)

Streptomycineae	Streptomycetaceae	*Streptomyces*	*S. albiaxialis*	Kuznetsov et al. (1992)
			S. chilikensis	Ray et al. (2013)
Streptosporangineae	Nocardiopsaceae	*Haloactinospora*	*H. alba*	Tang et al. (2008)
		Nocardiopsis	*N. baichengensis*, *N. chromatogenes*, *N. composta*, *N. gilva*, *N. halotolerans*, *N. rhodophaea*, and *N. rosea*	Al-Zarban et al. (2002a), Kämpfer et al. (2002), and Li et al. (2006)
			N. halophila, *N. kunsanensis*, *N. xinjiangensis*, *N. salina*, *N. litoralis* and *N. terrae*	Al-Tai and Ruan (1994), Chun et al. (2000), Li et al. (2003a, 2004c), and Chen et al. (2009b, 2010b)
		Salinactinospora	*S. qingdaonensis*	Chang et al. (2012)
		Streptomonospora	*S. salina*	Cui et al. (2001)
			S. alba	Li et al. (2003b)
			S. halophile	Cai et al. (2008)
			S. amylolytica and *S. flavalba*	Cai et al. (2009)
			S. Arabica	Hozzein and Goodfellow (2008)
			S. nanhaiensis and *S. sediminis*	Zhang et al. (2013a, b)
		Thermobifida	*T. halotolerans*	Yang et al. (2008)

The unusual properties of the halophilic microorganisms make them valuable resources in the development of novel biotechnological processes and industrial applications. Therefore, halophiles have wide range of biotechnological potential in industry, e.g. biosurfactant production, biopolymers in oil recovery, proteases and amylases in detergent industry, poly-beta hydroxyalkanoate as biodegradable plastic, exopolysaccharide and bioremediation of contaminated hypersaline brines etc. (Kanekar et al. 2012).

Among the halophilic microorganisms, the halophilic actinobacteria are of special interest for their amazing metabolic diversity, biological activities and biotechnological applications. It is well known that members of the actinobacteria group are the richest source of natural products amongst the prokaryotes (Berdy 2005; Olano et al. 2009) and the filamentous actinobacteria account for about 45 % of all microbial bioactive secondary metabolites (Berdy 2005). On the same regard, there is also evidence that actinobacteria isolated from the extremobiosphere will be a rich source of novel natural products (Bull 2011).

Actinobacteria are Gram-positive or Gram-variable microorganisms with high G+C content which have a rigid cell wall that contains muramic acid. Most are chemoorganotrophs and some of them are halophiles. Recently, members of actinobacteria were raised to the taxonomic rank of a phylum which is one of the major phyla in the domain Bacteria, as inferred from its branching pattern in the 16S rRNA gene tree (Garrity and Holt 2001; Ludwig and Klenk 2005). The phylum actinobacteria includes phenotypically diverse microorganisms which show diverse morphological properties that range from cocci to highly differentiated mycelia (Goodfellow 2012). The halophilic actinobacteria are widely distributed in saline, hypersaline terrestrial and aquatic habitats. Isolation of them does not demand special enrichment techniques (Kanekar et al. 2012), supplementing the isolation media with salt would be enough. Taxonomically, they belong to different taxa.

In this chapter, the biodiversity of halotolerant and halophilic actinobacteria from different aspects will be reviewed. The focus will be on their taxonomy and ecological distribution in different habitats as well as a short description of the halophilic genera and species of the actinobacteria group.

1.2 The Halophilic and Halotolerant Actinobacteria

Microorganisms show quite different responses to salt and different classification schemes have been proposed by many authors (Trüper and Galinski 1986; Vreeland 1987; Ramos-Cormenzana 1989). The most widely used classification by most scientists was proposed by Kushner and Kamekura (1988). This classification is based on the optimal growth of microorganisms with respect to the concentration of NaCl. On the basis of this classification, halophilic microorganisms are divided into the following categories: (1) extreme halophiles, able to grow optimally in media with 15–30 % (2.5–5.2 M) NaCl, (2) moderate halophiles, growing optimally in media with 3–15 % (0.5–2.5 M) NaCl, and (3) slight halophiles, able to grow optimally between 1 and 3 % (0.2–0.5 M) NaCl. In contrast, non-halophilic microorganisms are those organisms that grow optimally in media with less than 1 % (0.2 M) NaCl. However, bacteria able to grow in the absence as well as in the presence of salt which can tolerate relatively high NaCl concentrations are designated halotolerant or extremely tolerant if tolerance extends above 15 % (2.5 M) NaCl. There are representatives of actinobacteria in all of the above mentioned categories.

An important point that should be mentioned here is that the salt requirement and tolerance of many species may vary according to growth conditions such as medium composition and temperature. Therefore, the growth temperature should be specified in the definition of the salt range for growth. An example for this effect is *Marinococcus halophilus* which grows in presence of low concentrations starting from 0.01 M NaCl at 20 °C, but at least 0.5 M is required at 25 °C (Novitsky and Kushner 1976) for optimum growth.

1.3 Taxonomical Diversity of Halophilic and Halotolerant Actinobacteria

In contrast to the halophilic Archaea, the halophilic and halotolerant Bacteria are included in many different phylogenetic branches (phyla). The domain Bacteria contains many halophilic and halotolerant organisms spread in different bacterial taxa and true halophiles are well known within the phyla Cyanobacteria, Proteobacteria, Firmicutes, Actinobacteria, Spirochaetes and Bacteroidetes (Oren 2011). The halophilic and halotolerant actinobacteria belong to phylum Actinobacteria. This taxon is one of the major phyla in the domain Bacteria, as inferred from its branching position in the 16S rRNA phylogenetic tree (Garrity and Holt 2001; Ludwig and Klenk 2005) and taxon-specific 16S rRNA signatures (Zhi et al. 2009). Moreover, some conserved indels in 23S rRNA and protein sequences support the distinctness of members of actinobacteria from all other bacteria (Gao and Gupta 2005).

In the most recent edition of the Bergey's Manual of Systematic Bacteriology (Goodfellow 2012), the phylum Actinobacteria encompasses 5 classes, 19 orders, 50 families, and 221 genera. The constituent classes are Acidimicrobiia class. nov., Actinobacteria (Stackebrandt et al. 1997), Coriobacteriia class. nov., Rubrobacteria class. nov., and Thermoleophilia class. nov. However, many new taxa were discovered after the publication of the Manual. To date, the halophilic and halotolerant actinobacteria have been accommodated only in the subclass Actinobacteridae of class Actinobacteria, and of the three orders Actinomycetales, Bifidobacteriales, and Nitriliruptorales of the subclass Actinobacteridae, only the Actinomycetales comprises halophilic and halotolerant actinobacteria. Nowadays, the order Actinomycetales (Buchanan 1917; Stackebrandt et al. 1997; Zhi et al. 2009) is composed of 23 suborders. It was reported that the halophilic and halotolerant actinobacteria are included in the following suborders of the order Actinomycetales: Actinopolysporineae, Corynebacterineae, Glycomycineae, Jiangellineae, Micrococcineae, Micromonosporineae, Propionibacterineae, Pseudonocardineae, Streptomycineae and Streptosporangineae.

These taxa that comprises halophilic and halotolerant actinobacteria do not necessarily consist solely of halophiles. The opposite is true: there are only few phylogenetically consistent groups that are composed entirely of halophiles. In most cases halophiles and non-halophilic relatives are found together in the phylogenetic tree, and many genera, families and orders have representatives with greatly different salt requirement and tolerance patterns (Oren 2011). However, in most cases the halophilic species will be grouped in close clades in the phylogenetic tree and not scattered within their non-halophilic relatives in the tree.

The halophilic and halotolerant actinobacteria as accommodated in different taxa, they show a remarkable range of different morphologies from organisms that form cocci (e.g. *Ruania* and *Serinicoccus*), short rods (e.g. *Cellulomonas*), irregular rods (e.g. *Microbacterium* and *Salinibacterium*), rods and cocci (e.g. *Arthrobacter*), and mycelia that fragment into coccoid and rod-like elements (e.g. *Nocardiopsis*). Others show more extensive morphological differentiation ranging from those which produce extensively branched substrate hyphae that bear spores (e.g. *Micromonospora*) to those that form a stable branched mycelium that carries aerial hyphae which differentiate into short or long chains of spores (e.g. *Actinopolyspora* and *Streptomyces*). In general, spores are nonmotile.

Currently, the number of species names that have been validly published as halophilic and halotolerant actinobacteria is very large and is growing exponentially due to the recent efforts to characterize microorganisms from hypersaline environments. In the following section, the halophilic and halotolerant actinobacterial species will be reviewed with their classification and main characteristics. The focus will be only on those species names that have been validly described. More detailed data about their characteristics can be found on the original descriptions or in the most recent edition of the Bergey's Manual of Systematic Bacteriology. It should be clear here that besides the halophilic and halotolerant actinobacterial species included here whose names have been validly published, there are a large number of halophilic and halotolerant actinobacterial organisms that have been isolated and studied from other points of view, such as for their biotechnological applications or other reasons, which have not been properly characterized.

Another point of interest is that some species have been described as halophilic or halotolerant without a scientific basis as their response to NaCl has not been studied in detail or their NaCl concentration for optimal growth has not been reported on their taxonomic descriptions.

1.3.1 Suborder Actinopolysporineae

1.3.1.1 Family Actinopolysporaceae and Genus Actinopolyspora

Family Actinopolysporaceae is the only family of order *Actinopolysporales*. It is also a monogeneric family that contains the genus *Actinopolyspora* (Gochnauer et al. 1975; Zhi et al. 2009). Genus *Actinopolyspora* is very interesting as all the validly described species in the genus are halophilic actinobacteria and some of them can be considered as extreme halophiles.

Members of the genus are aerobic and Gram-positive actinomycetes that form an extensively branched substrate mycelium. Fragmentation of the substrate mycelium is occasionally observed, but substrate hyphae are mostly non-fragmented. Sporophores containing ten or more smooth-walled spores which are produced basipetally on aerial hyphae are formed. Spores are not observed on the substrate mycelium (Gochnauer et al. 1975; Tang et al. 2011b).

The genus *Actinopolyspora* was originally established by Gochnauer et al. (1975) and the description was later emended by Tang et al. (2011b). The genus currently contains 11 species with validly published names: *A. halophila* (Gochnauer et al. 1975), *A. mortivallis* (Yoshida et al. 1991), *A. xinjiangensis* (Guan et al. 2010), *A. alba* and *A. erythraea* (Tang et al. 2011b), *A. algeriensis* (Meklat et al. 2012), *A. saharensis* (Meklat et al. 2013a), *A. dayingensis* (Guan et al. 2013b), *A. righensis* (Meklat et al. 2013b), *A. lacussalsi* (Guan et al. 2013a) and *A. mzabensis* (Meklat et al. 2013c).

The strain *A. iraqiensis* (Ruan et al. 1994) was misclassified in the genus and has been transferred as a later heterotypic synonym of *Saccharomonospora halophila* (Tang et al. 2011b). A thermophilic species, *A. thermovinacea* was described by Lu and Yan (1983), but the name has not been validly published. Also, the species *A. egyptensis* was described (Hozzein and Goodfellow 2011) and has not yet been validated.

Strains of genus *Actinopolyspora* require salt for growth and most are extreme halophiles. On solid media, growth occurs in the presence of 5–30 % NaCl, with optimal growth in media containing 10–20 % NaCl.

1.3.2 Suborder Corynebacterineae

1.3.2.1 Family Corynebacteriaceae and Genus *Corynebacterium*

The family Corynebacteriaceae (Lehmann and Neumann 1896; Stackebrandt et al. 1997; Zhi et al. 2009) contains a single genus, *Corynebacterium*. The genus *Corynebacterium* represents a large group of Gram-positive, asporogenous, rod-shaped bacteria (Liebl et al. 1991). The majority of the *Corynebacterium* species were isolated from clinical samples or animals, but *Corynebacterium halotolerans* is the only halotolerant species of the genus with optimal growth occurs at 10 % NaCl or KCl (Chen et al. 2004).

1.3.3 Suborder Glycomycineae

1.3.3.1 The Family Glycomycetaceae

The family Glycomycetaceae (Stackebrandt et al. 1997; Labeda and Kroppenstedt 2005) contains three genera, *Glycomyces*, *Haloglycomyces* and *Stackebrandtia*, but the genus *Stackebrandtia* (Labeda and Kroppenstedt 2005) does not contain halophilic or halotolerant species.

Genus *Glycomyces* Members of the genus *Glycomyces* (Labeda et al. 1985; Labeda and Kroppenstedt 2004) are aerobic with branching vegetative mycelia and aerial mycelium that may be produced on certain growth media. Spores may be formed on the vegetative hyphae in some species and they are oval, spherical or rod-like. Chains of square-ended conidia may be produced on aerial hyphae. All the previously described species were not reported to be halophilic or halotolerant except the recently described *Glycomyces halotolerans* (Guan et al. 2011b).

Genus *Haloglycomyces* Closely related to genus Glycomyces is the monospecific genus *Haloglycomyces*, with *Haloglycomyces albus* which is a moderately halophilic species that has optimum NaCl concentration for growth of 8–12 % (Guan et al. 2009). It was described as an aerobic actinobacterium with well-developed aerial mycelium on most media and branched substrate mycelium that fragments into short or elongated rods.

1.3.4 Suborder Jiangellineae

1.3.4.1 Family Jiangellaceae and Genus *Haloactinopolyspora*

The family Jiangellaceae comprises two genera, *Jiangella* and *Haloactinopolyspora*. The genus *Haloactinopolyspora* contains the type species *H. alba* (Tang et al. 2011a) and the recently described species *H. alkaliphila* (Zhang et al. 2014). Both have substrate mycelium that fragments into short or elongated rods, aerial mycelium that differentiates well into long spore chains and described as moderately halophiles. They are strictly aerobic showing growth at 7–23 % NaCl with optimal growth at 10–15 % NaCl, and no growth occurs in the absence of NaCl.

1.3.5 Suborder Micrococcineae

Among all the families included in the suborder Micrococcineae, seven of them comprise halophilic species: Bogoriellaceae, Cellulomonadaceae, Intrasporangiaceae, Microbacteriaceae, Micrococcaceae, Promicromono-sporaceae and Ruaniaceae.

1.3.5.1 Family Bogoriellaceae and Genus *Georgenia*

The genus *Georgenia* (Altenburger et al. 2002; Li et al. 2007), belonging to the family Bogoriellaceae is currently comprises eight species. Only *Georgenia halophila* (Tang et al. 2010b) of the genus was described as a moderately halophilic actinomycete with optimal growth at 5–10 % NaCl. Cells of the species are facultatively anaerobic, nonmotile, non-endospore-forming and short rods.

1.3.5.2 Family Cellulomonadaceae and Genus *Cellulomonas*

Recently two halotolerant species were described in the genus *Cellulomonas*, namely, *C. phragmiteti* (Rusznyák et al. 2011) and *C. pakistanensis* (Ahmed et al. 2014). Cells of the species are facultatively anaerobic and motile rods. They are moderately halotolerant actinobacteria showing growth in the presence of 2–7 % NaCl with optimum growth at 5 % NaCl.

1.3.5.3 Family Intrasporangiaceae and Genus *Serinicoccus*

All the validly described three species of genus *Serinicoccus* are moderately halophilic actinobacteria; *S. marinus* (Yi et al. 2004), *S. chungangensis* (Traiwan et al. 2011), and *S. profundi* (Xiao et al. 2011). They are strictly aerobic, non sporeforming and non-motile cocci. They can grow in the presence of 0–15 % NaCl with optimum growth between 2 and 5 % NaCl, and interestingly at 13 % NaCl in the case of *S. chungangensis*.

1.3.5.4 Family Microbacteriaceae

Of the many genera in the family *Microbacteriaceae*, only *Microbacterium* and *Salinibacterium* contains halotolerant actinobacteria.

Genus *Microbacterium* Genus *Microbacterium* currently includes 86 species and only *M. halophilum* (Takeuchi and Hatano 1998), *M. halotolerans* (Li et al. 2005a) and *M. sediminis* (Yu et al. 2013) are halotolerant among them. Cells of the organisms are nonmotile and irregular short rods that may form V shapes but no branching or rod-coccus cycle are observed. They are tolerant to salt and show growth in the presence of 0–15 % NaCl with optimal growth at 2–6 % NaCl.

Genus *Salinibacterium* *Salinibacterium* is currently contains two species, *S. amurskyense* (Han et al. 2003) and *S. xinjiangense* (Zhang et al. 2008). Both species have nonmotile, aerobic, non-sporeforming, irregular rods. They are halotolerant showing growth in the presence of 0–14 % NaCl with optimal growth at 1–3 %.

1.3.5.5 Family Micrococcaceae

The family *Micrococcaceae* includes several genera and some of them include halophilic or halotolerant species.

Genus *Arthrobacter* The majority of species in the genus *Arthrobacter* (Conn and Dimmick 1947; Koch et al. 1995) exhibit a marked rod-coccus growth cycle when grown in complex media. The stationary-phase cultures (generally after 2–7 days) are composed entirely or largely of coccoid cells and some species are showing only spherical cells throughout the growth cycle. *Arthrobacter halodurans* (Chen et al. 2009a) is the only species reported to be halotolerant in the genus with growth occurs in the presence of 0–12 % NaCl.

Genus *Kocuria* Genus *Kocuria* contains only two halotolerant actinobacteria; *Kocuria halotolerans* (Tang et al. 2009b) and *Kocuria marina* (Kim et al. 2004). Cells of the species are coccoid, non-motile and non-endospore-forming. They grow in the presence of 0–10 % NaCl and optimally with 5 % NaCl.

Genus *Nesterenkonia* The genus *Nesterenkonia* was first proposed by Stackebrandt et al. (1995) with the reclassification of *Micrococcus halobius* (Onishi and Kamekura 1972) as *Nesterenkonia halobia*. The description of the genus was later emended by Collins et al. (2002) and Li et al. (2005b). Members of the genus are coccoid or short rods that sometimes showing branching, non-spore-forming, chemo-organotrophic with strictly respiratory metabolism. Species of the genus are moderately halophilic or halotolerant and some species are alkaliphilic or alkalitolerant (Stackebrandt et al. 1995; Collins et al. 2002; Li et al. 2005b).

Recently, the genus comprises 12 species, the type species *N. halobia* (Onishi and Kamekura 1972), *N. aethiopica* (Delgado et al. 2006), *N. alba* (Luo et al. 2009), *N. flava* (Luo et al. 2008), *N. halophila* (Li et al. 2008b), *N. halotolerans*, *N. xinjiangensis* (Li et al. 2004b), *N. jeotgali* (Yoon et al. 2006), *N. lacusekhoensis* (Collins et al. 2002), *N. lutea*, *N. sandarakina* (Li et al. 2005b), and *N. suensis* (Govender et al. 2013). They are showing optimal growth at 2.5–10 % NaCl.

Genus *Yaniella* The type species of the genus is *Yaniella halotolerans* (Li et al. 2004a, 2008a) which is a halotolerant species (illegitimate homotypic synonym of *Yania halotolerans* (Li et al. 2004a)). Cells of members of the genus are nonmotile, aerobic, non-spore-forming, coccoid or oval, and occur singly or in clusters.

Genus *Zhihengliuella* The five validly described species of genus *Zhihengliuella* are either halotolerant or moderately halophilic with *Z. halotolerans* (Zhang et al. 2007) as the type species. Cells of members of the genus are aerobic, short rods, non-motile and non-spore-forming (Zhang et al. 2007; Tang et al. 2009c; Hamada et al. 2013). Growth occurs in the presence of 0–15 % NaCl with optimal growth at 5–10 % NaCl.

1.3.5.6 Family Promicromonosporaceae and Genus *Isoptericola*

Of all the genera included in the family Promicromonosporaceae, only the genus *Isoptericola* includes a halophilic species. The genus *Isoptericola* (Stackebrandt et al. 2004) currently comprises seven species of which, only *Isoptericola halotolerans* is a halotolerant actinobacterium with optimal growth at 10 % NaCl (Zhang et al. 2005). Cells of the species are coccoid or rod-shaped, nonmotile, and do not form spores. Primary mycelium is formed.

1.3.5.7 Family Ruaniaceae

The family Ruaniaceae contains the genera *Ruania* and *Haloactinobacterium* and the two genera are monospecific.

Genus *Ruania* *Ruania albidiflava*, the type and only species of genus *Ruania*, is aerobic, moderately halotolerant, nonmotile, non-spore-forming cocci (Gu et al. 2007). The strain can tolerate up to 10 % NaCl.

Genus *Haloactinobacterium* Cells of *Haloactinobacterium album*, the type and only species of genus *Haloactinobacterium*, are short rods, nonmotile, moderately halophilic with optimal growth occurs at 7–10 % NaCl (Tang et al. 2010b).

1.3.6 Suborder Micromonosporineae

1.3.6.1 Family Micromonosporaceae and Genus *Micromonospora*

Of the several genera of family *Micromonosporaceae*, only genus *Micromonospora* has described halophiles. The first species reported to be halophilic actinomycete in the genus was *M. halophytica* (Weinstein et al. 1968) and then the recently described species *M. halotolerans* (Carro et al. 2013). Both species have typical characteristics of the genus *Micromonospora* as they produce spherical to ellipsoidal spores randomly along branching mycelium on short or long sporophores and abundant dark colored spores occur in older cultures. *M. halotolerans* was observed to grow optimally in the presence of 5 % NaCl.

1.3.7 Suborder Propionibacterineae

1.3.7.1 Family Nocardioidaceae

Family Nocardioidaceae currently contain nine genera of which the genera *Aeromicrobium* and *Nocardioides* were reported to have halotolerant validly published species.

Genus *Aeromicrobium* Cells of the genus are typically aerobic, non-spore-forming, and non-motile irregular rods; straight, curved and rudimentarily branched filaments can be produced. *Aeromicrobium marinum* was reported to grow well in the range 0.63–10.7 % NaCl with a maximal growth at 5.35 %; but no growth was observed in the absence of salts (Bruns et al. 2003). Although, *A. halocynthiae* showed good growth with 0–6 % NaCl, but NaCl was required for robust growth (Kim et al. 2010).

Genus *Nocardioides* The type genus of family *Nocardioidaceae* has three halotolerant species, all of them were recently described; *Nocardioides halotolerans* (Dastager et al. 2008a), *Nocardioides tritolerans* (Dastager et al. 2008b), and *Nocardioides daedukensis* (Yoon et al. 2010). They can tolerate up to 10 % NaCl and exhibit optimal growth at 0–3 %. Their cells are aerobic, non-spore-forming, non-motile rods or cocci and neither substrate nor aerial mycelium is formed.

1.3.8 Suborder Pseudonocardineae

1.3.8.1 Family Pseudonocardiaceae

The family Pseudonocardiaceae (Embley et al. 1988; Labeda et al. 2011) currently includes 31 genera, only 5 of them have halophilic species, namely: *Amycolaptosis*, *Haloechinothrix*, *Praseurella*, *Saccharomonospora* and *Saccharopolyspora*.

Genus *Amycolatopsis* The genus *Amycolatopsis* (Lechevalier et al. 1986; Tang et al. 2010c) comprises four species with validly published names that were described as halotolerant or moderately halophilic. The four species are *A. halotolerans* (Lee 2006), *A. halophila* (Tang et al. 2010c), *A. salitolerans* (Guan et al. 2012) and *A. cihanbeyliensis* (Tatar et al. 2013). Members of these species are aerobic, nonmotile and filamentous actinomycetes which forms well-developed, branched aerial and substrate mycelia that fragment into rod-shaped elements. Growth occurs in the presence of 1–15 % NaCl with optimal growth at 5 % NaCl.

Genus *Haloechinothrix* The type and only species of genus *Haloechinothrix*, *H. alba* was described as a moderately halophilic actinobacterium. It is strictly aerobic and filamentous actinomycete with spiny aerial mycelium and substrate mycelium that fragments into rod-like elements and does not form chains of spores at maturity. Growth occurs in presence of 9–23 % NaCl with optimum concentration for growth at 15 % NaCl (Tang et al. 2010d).

Genus *Prauserella* Genus *Prauserella* (Kim and Goodfellow 1999; Li et al. 2003c) comprises nine species and with the exception of *P. marina* (Wang et al. 2010) and *P. muralis* (Schäfer et al. 2010), all of them are moderately halophilic. These species are *P. rugosa* (the type species) (Kim and Goodfellow 1999), *P. alba* and *P. halophila* (Li et al. 2003c), *P. aidingensis*, *P. flava*, *P. salsuginis* and *P. sedimina* (Li et al. 2009). Members of the genus are aerobic, non-motile actinobacteria which form an extensively branched substrate mycelium that fragments into irregular rod-shaped elements. Their aerial hyphae differentiate into branched short or, at maturity, long, straight to flexuous spore chains. Optimal growth occurs on agar media supplemented with 8–15 % NaCl.

Genus *Saccharomonospora* Of the several species with validly published names included in the genus *Saccharomonospora* (Nonomura and Ohara 1971), only three are moderately to extreme halophilic actinomycetes: *S. halophila* (Al-Zarban et al.

2002b), *S. paurometabolica* (Li et al. 2003d), and *S. saliphila* (Syed et al. 2008). All of them show optimal growth at 10 % NaCl, though *S. halophila* and *S. paurometabolica* require NaCl for growth, but it is not essential for growth of *S. saliphila*. Members of the three species are aerobic organisms that produce well-developed aerial and substrate mycelia. They produce single or pairs of spores with smooth or wrinkled surfaces on either the aerial or substrate mycelium, short chains on vegetative or aerial hyphae are occasionally present.

Genus *Saccharopolyspora* Most members of the genus *Saccharopolyspora* (Lacey and Goodfellow 1975; Korn-Wendisch et al. 1989) grow well in media supplemented with salt. Of the validly described species, *S. halophila* (Tang et al. 2009a), *S. qijiaojingensis* (Tang et al. 2009d), *S. lacisalsi* (Guan et al. 2011a) and *S. dendranthemae* (Zhang et al. 2013b) are halotolerant or halophilic actinomycetes. Their members form a well-developed substrate mycelium that may be fragment into rod-shaped elements and aerial hyphae that differentiate into long straight chains of oval or spherical spores which have smooth surfaces. They show NaCl tolerance range between 3 and 22 % with optimal growth occurs at 10–15 % NaCl. Other than the above mentioned species, *S. jiangxiensis* (Zhang et al. 2009) can tolerate up to 11 % NaCl and the type strain of *S. cebuensis* has a strict requirement for salt which suggests that it is an obligate marine actinomycete (Pimentel-Elardo et al. 2008).

1.3.9 Suborder Streptomycineae

1.3.9.1 Family Streptomycetaceae and Genus *Streptomyces*

It is very interesting to know that in the biggest actinomycete genus with currently 655 validly published species there are only 2 described as halotolerant or halophilic species. The two species are *Streptomyces albiaxialis* (Kuznetsov et al. 1992) and *Streptomyces chilikensis* (Ray et al. 2013) with typical morphological features of members of the genus. *S. albiaxialis* was reported to be a halotolerant streptomycete and *S. chilikensis* was reported as a halophilic streptomycete.

1.3.10 Suborder Streptosporangineae

1.3.10.1 Family Nocardiopsaceae

The family Nocardiopsaceae currently comprises eight genera and in five of them halophilic filamentous actinomycetes are found.

Genus *Haloactinospora* The only and type species of the genus is *Haloactinospora alba* which is a moderately halophilic actinomycete that is able to grow between 9 and 21 % NaCl and showing optimal growth at 15 % NaCl (Tang et al. 2008). The organism is aerobic which form well-developed and non-fragmented hyphae.

Morphologically, long chains of non-motile spores with smooth surfaces are formed on the aerial mycelium and spore chains with wrinkled surfaces and terminal pseudosporangia are formed on the substrate mycelium.

Genus *Nocardiopsis* The genus *Nocardiopsis* (Meyer 1976) currently contains 45 species with validly published names. Many *Nocardiopsis* strains can tolerate NaCl concentrations up to 20 % (Hozzein and Trujillo 2012). Halotolerant species include *N. baichengensis*, *N. chromatogenes*, *N. composta*, *N. gilva*, *N. halotolerans*, *N. rhodophaea*, and *N. rosea*, which can grow in media supplemented with 15–18 % NaCl (Al-Zarban et al. 2002a; Kämpfer et al. 2002; Li et al. 2006). On the other hand, *N. halophila*, *N. kunsanensis*, *N. xinjiangensis*, *N. salina*, *N. litoralis* and *N. terrae* (Al-Tai and Ruan 1994; Chun et al. 2000; Li et al. 2003a, 2004c; Chen et al. 2009b, 2010b) are considered as true halophilic species as NaCl (at least 3 %) is necessary for growth, with an optimal concentration of 10–15 %.

Members of the genus *Nocardiopsis* are typically aerobic, chemo-organotrophic, non-motile filamentous actinomycetes with well-developed substrate mycelia and long and densely branched hyphae. They are characterized by forming well-developed aerial mycelia which are sparse to abundant and aerial hyphae which are long, branched, straight to flexuous, or irregularly zig-zagged and completely fragmenting into smooth surfaced spores of various lengths. Fragmentation into coccoid and bacillary elements may also occur in some species (Hozzein and Trujillo 2012).

Genus *Salinactinospora* The monospecific genus *Salinactinospora* was recently described (Chang et al. 2012). The type and only species, *Salinactinospora qingdaonensis*, is aerobic and moderately halophilic that forms aerial mycelium with long chains of rod-shaped spores with smooth surfaces and extensively branched substrate mycelium with non-fragmenting hyphae. The organism grows with 1–23 % NaCl with optimum growth at 9–12 % NaCl.

Genus *Streptomonospora* The genus *Streptomonospora* was first established by Cui et al. (2001) and currently comprises eight species all of them are strictly halophilic filamentous actinomycetes. The type species of the genus is *S. salina* (Cui et al. 2001), subsequently, the following species were described: *S. alba* (Li et al. 2003b), *S. halophila* (Cai et al. 2008), *S. amylolytica* and *S. flavalba* (Cai et al. 2009). The most recent species included in the genus are *S. arabica*, after the transfer of *Nocardiopsis arabia* (Hozzein and Goodfellow 2008) to the genus, *S. nanhaiensis* and *S. sediminis* (Zhang et al. 2013a).

Members of the genus are aerobic organisms with branching hyphae and aerial mycelium, at maturity, forms short chains of non-motile spores, which may be oval to rod-shaped with wrinkled surfaces. Their substrate mycelium is extensively branched with non-fragmenting hyphae and single, non-motile, oval to round spores are borne on sporophores or dichotomously branched sporophores of the substrate hyphae. Optimum growth occurs in media supplemented with NaCl at a concentration of 10–15 %.

Genus *Thermobifida* The genus *Thermobifida* contains only one species among the validly published four species which is halotolerant. *Thermobifida halotolerans* was described by Yang et al. (2008) as an aerobic filamentous actinomycete that forms an extensively branched substrate mycelium with single, smooth-ridged spores formed at the tips of dichotomously branched sporophores borne on aerial hyphae. It can tolerate up to 10 % NaCl for growth.

1.4 Ecological Diversity and Distribution

Halophilic and halotolerant microorganisms are present usually where high salt concentrations are found. Compared to the extensive literature on the ecology of the halophilic archaea, the halophilic bacteria have been relatively poorly studied and often neglected during the early studies on the hypersaline environments.

Actinobacteria are widely distributed in aquatic and terrestrial habitats, including extreme habitats, such as the hypersaline environments which are widely distributed on our planet (Ventosa et al. 2008). The halophilic and halotolerant actinobacteria usually referred to be inhabitants of the hypersaline environments. The hypersaline environments are defined as those environments with salt concentrations above that of sea water (3.3 % total dissolved salts) (Oren 2002). These environments spread over a wide variety of aquatic and terrestrial ecosystems. In such hypersaline environments the occurrence of gradients in salinity as a result of the evaporation of sea water is a common phenomenon (Ventosa et al. 1998).

The halophilic and halotolerant actinobacteria exhibited interesting diversity in terms of distribution in different habitats. There are several reports on inhabitance of them in diverse hypersaline environments such as hypersaline soils, saline salt lakes, solar salterns, salt mines, salted food products, and in some unexpected places as brines deep in the sea, on plants that excrete salts from their leaves, and on ancient wall paintings (Chen et al. 2007; Xiang et al. 2008; Sabet et al. 2009; Guan et al. 2011a).

Culture-dependent approaches using synthetic media supplemented with salt or sea water to enrich and isolate the halophilic and halotolerant actinobacteria from the environmental samples were often used. One of the reasons for the limited value of using the conventional isolation procedures is that the cultivation conditions used to isolate organisms do not reflect usually the conditions in the environment (Zengler et al. 2002; Stevenson et al. 2004).

Nowadays, it is known that the majority of microorganisms in the environment cannot be cultured by using conventional plate culture methods hence the vast majority of the microbiota remain undiscovered (Vartoukian et al. 2010). This limitation has promoted the use of culture-independent approaches recently to detect previously uncultured microorganisms and evaluate the microbial diversity in natural habitats. Thus, in a study of the microbial communities in the chemocline of Urania Basin, a hypersaline deep-sea basin in the Mediterranean Sea, bacterial numbers estimated by most probably number methods using dilutions in 11 different growth media were only between 0.006 and 4.3 % of the total cell count (Sass et al. 2001).

Advances in techniques for studying environmental microbiology during the last decades and application of molecular biology methods has revealed an astonishing level of microbial diversity in the hypersaline environments, only a small proportion of which has been assessed through cultivation (DeLong and Pace 2001; Rappè and Giovannoni 2003). In this regard, in a recent review, Hamedi et al. (2013) described that actinomycetes form a stable, metabolically active and persistent population in various marine hypersaline ecosystems. The occurrence of actinomycetes in highly saline environments and the tolerance of these organisms to high salt concentrations were first described by Tresner et al. (1968). Recently, there is a range of procedures available for the selective isolation of halophilic actinobacterial communities present in different habitats (Goodfellow 2010).

It is interesting to mention here that the first halophilic species of the genus *Actinopolyspora*, *A. halophila* (Gochnauer et al. 1975) was isolated as a contaminant of a culture medium containing 25 % NaCl. Earlier, *Nesterenkonia halobia*, previously described as *Micrococcus halobia*, was isolated from unrefined solar salt of unknown origin obtained from Noda, Japan (Onishi and Kamekura 1972).

Then, most of the known and validly described halophilic actinobacteria species were isolated from saline soils. The examples include *Actinospolyspora mortivallis* (Yoshida et al. 1991) which is a moderately halophilic actinomycete isolated from salty soil obtained from the Death Valley, California (USA) and *Actinopolyspora iraqiensis*, later transferred to the genus *Saccharomonospora* as *Saccharomonospora halophila*, which was obtained concomitantly with *Nocardiopsis halophila* from a saline soil from Iraq (Ruan et al. 1994; Al-Tai and Ruan 1994). In the same regard, the moderately halophilic species *Nesterenkonia lutea* was isolated from a saline soil from China (Li et al. 2005b), *Yaniella flava* was isolated from a saline soil sample originating from Qinghai Province in North-West China (Li et al. 2008a) and *Amycolatopsis salitolerans* which was isolated from a hypersaline soil (Guan et al. 2012).

Nocardiopsis species has been reported to predominate in saline soils (Hozzein and Trujillo 2012). Some examples of the moderately halophilic species of the genus *Nocardiopsis* which were isolated from saline soil samples in China are *N. salina* (Li et al. 2004c), *N. quinghaiensis* (Chen et al. 2008) and *N. terrae* (Chen et al. 2010b).

It is worth mentioning here that the studies on halophilic actinomycetes in hypersaline soils of Xinjiang Province, North-West China, led to the isolation of several novel species. Some of them are *Haloglycomyces albus* (Guan et al. 2009), *Nesterenkonia halotolerans*, *Nesterenkonia xinjiangensis* (Li et al. 2004b), *Nesterenkonia halophila* (Li et al. 2008b), *Nocardiopsis xinjiangensis* (Li et al. 2003a), *Nocardiopsis salina* (Li et al. 2004c), *Nocardiopsis gilva*, *Nocardiopsis rosea*, *Nocardiopsis rhodophaea*, *Nocardiopsis chromatogenes*, and *Nocardiopsis baichengensis* (Li et al. 2006), *Prausella alba*, *Prauserella halophila* (Li et al. 2003c), *Saccharomonospora paurometabolica* (Li et al. 2003d), and *Yaniella halotolerans* (Li et al. 2004a). Moreover, *Saccharomonospora halophila* was isolated from a salt marsh soil sample in Kuwait (Al-Zarban et al. 2002b), while

Saccharomonospora saliphila (Syed et al. 2008) was isolated from a muddy soil from Gulbarga, Karnataka, India, and interestingly, *Nesterenkonia sandarakina* was isolated from a soil sample collected from the Eastern desert of Egypt (Li et al. 2005b).

Apart from the terrestrial saline sources, many halophilic and halotolerant species were isolated from saline water bodies, mainly the saline salt lakes. The following species are examples: *Streptomonospora amylolytica* and *Streptomonospora flavalba* were isolated from a salt lake in the North-West of China (Cai et al. 2009), *Saccharopolyspora halophila* and *Saccharopolyspora qijiaojingensis* were isolated from saline salt lakes (Tang et al. 2009a, d), *Saccharopolyspora lacisalsi* was isolated from a salt lake in Xinjiang, China (Guan et al. 2011a), and *Haloactinobacterium album* was isolated from a soil sample collected from Qijiaojing Lake, which is a salt lake in Xinjiang, China (Tang et al. 2010b). On the other hand, *Nesterenkonia lacusekhoensis* was isolated from a 23 m-deep water sample of Ekho Lake (a hypersaline, meromictic, and heliothermal lake in the ice-free Vestfold Hills, East Antarctica) (Collins et al. 2002).

Other than the salt lakes, many actinobacteria species were derived from the marine sources. Of them, *Nocardiopsis litoralis* is a halophilic marine actinomycete isolated from a sea anemone collected from a tidal flat in the South China Sea (Chen et al. 2009b), *Microbacterium sediminis* is a halotolerant actinomycete isolated from deep-sea sediment (Yu et al. 2013), *Arthrobacter halodurans* is a halotolerant actinobacterium isolated from sea water (Chen et al. 2009a), and *Kocuria marina* is a halotolerant actinobacterium isolated from marine sediment taken from the Troitsa Bay of the Gulf of Peter the Great, East Siberian Sea (Kim et al. 2004).

The increased interest in the last decade in isolating and identifying new halophilic and halotolerant actinobacteria led to the discovery of many novel taxa in other unexplored habitats. In this regard, *Zhihengliuella salsuginis*, a moderately halophilic actinobacterium, was isolated from a subterranean brine sample collected from a salt mine in Hunan Province, China (Chen et al. 2010a). Also, *Amycolatopsis cihanbeyliensis*, a halotolerant actinomycete, was isolated from a salt mine (Tatar et al. 2013). More interestingly, *Saccharopolyspora dendranthemae*, a halotolerant endophytic actinomycete, was isolated from a coastal salt marsh plant in Jiangsu, China (Zhang et al. 2013a, b).

1.5 Conclusion

Keeping in view of their importance economically, halophilic microorganisms have many advantages. First, most of them can grow at high salt concentrations, minimizing the risk of contamination. Second, they are easy to grow, and their nutritional requirements are simple: the majority can use a large range of compounds as their sole carbon and energy source (Vreeland 1987). It is well known that actinomycetes are the most economically and biotechnologically valuable prokaryotes. They are responsible for the production of about half of the discovered bioactive secondary metabolites (Berdy 2005). Therefore, the halophilic and halotolerant actinobacteria have gained significance in recent years as they are believed to produce novel metabolites of unique properties.

What we know about halophiles is far less than what exists in the nature and that our picture of the true diversity of halophiles is still incomplete. With new approaches derived from genomics, proteomics and other modern sciences, it is likely that new knowledge would be generated on biodiversity of halophilic actinobacteria (Kanekar et al. 2012). This hypothesis opens up a new horizon to microbiologists interested in biodiversity of halophilic actinobacteria to do more serious work. It is believed that the halophilic and halotolerant actinobacteria hold a prominent position due to their biodiversity and potentiality to produce novel compounds. Therefore, investigation of new ecosystems for the isolation of them is crucial for the discovery of novel taxa and subsequently for natural product-based drug discovery (Ventosa et al. 2008). To date, few ecosystems have been explored for the isolation of this interesting group of halophiles and a number of saline environments on earth are yet to be explored for their isolation.

The author would like to suggest here that the non-saline environments should be also explored for the isolation of halophilic and halotolerant actinobacteria. This is because of the presence of saline microenvironments in the non-saline environments. Also, this could yield many novel taxa, especially the halotolerant actinobacteria hidden in the environment. Along with investigating new ecosystems for isolation of new halophilic and halotolerant actinobacteria, new isolation methodologies and media need to be developed. Although, several researchers have described various enrichment methods and media that they used for isolation of halophilic actinobacteria from different saline environments, there is a need to develop more versatile media and cultivation conditions for the isolation process. This will enhance our understanding of the diversity of those bacteria in nature and increases the possibility of discovery of novel taxa.

Although a number of compounds have been recently identified and reported from halophilic and halotolerant actinobacteria, only few of them have been produced. Also, isolation of new halophilic and halotolerant actinobacteria may lead to novel molecules that could be used for applications in different fields. Many ideas and biotechnological applications are yet to be exploited for the halophilic actinobacteria. Therefore, addition of novel taxa of them will open our way for new biotechnological applications, e.g. synthesis of nanomaterials, bioremediation of saline waste waters and construction of molecular sensors, for the benefit of human kind.

References

Ahmed I, Kudo T, Abbas S, Ehsan M, Iino T, Fujiwara T, Ohkuma M (2014) *Cellulomonas pakistanensis* sp. nov., a moderately halotolerant *Actinobacteria*. Int J Syst Evol Microbiol 64:2305–2311

Al-Tai AM, Ruan JS (1994) *Nocardiopsis halophila* sp. nov., a new halophilic actinomycete isolated from soil. Int J Syst Bacteriol 44:474–478

Altenburger P, Kämpfer P, Schumann P, Vybiral D, Lubitz W, Busse H-J (2002) *Georgenia muralis* gen. nov., sp. nov., a novel actinobacterium isolated from a medieval wall painting. Int J Syst Evol Microbiol 52:875–881

Al-Zarban SS, Abbas I, Al-Musallam AA, Steiner U, Stackebrandt E, Kroppenstedt RM (2002a) *Nocardiopsis halotolerans* sp. nov., isolated from salt marsh soil in Kuwait. Int J Syst Evol Microbiol 52:525–529

Al-Zarban SS, Al-Musallam AA, Abbas I, Stackebrandt E, Kroppenstedt RM (2002b) *Saccharomonospora halophila* sp. nov., a novel halophilic actinomycete isolated from marsh soil in Kuwait. Int J Syst Evol Microbiol 52:555–558

Berdy J (2005) Bioactive microbial metabolites. J Antibiot (Tokyo) 58:1–26

Bruns A, Philipp H, Cypionka H, Brinkhoff T (2003) *Aeromicrobium marinum* sp. nov., an abundant pelagic bacterium isolated from the German Wadden Sea. Int J Syst Evol Microbiol 53:1917–1923

Buchanan RE (1917) Studies in the nomenclature and classification of the bacteria: II. The primary subdivisions of the *Schizomycetes*. J Bacteriol 2:155–164

Bull AT (2011) Actinobacteria of the extremobiosphere. In: Horikoshi K, Antranikian G, Bull AT, Robb FT, Stetter KO (eds) Extremophiles handbook. Springer Japan, Tokyo, pp 1203–1240

Cai M, Zhi XY, Tang SK, Zhang YQ, Xu LH, Li WJ (2008) *Streptomonospora halophila* sp. nov., a halophilic actinomycete isolated from a hypersaline soil. Int J Syst Evol Microbiol 58:1556–1560

Cai M, Tang SK, Chen YG, Li Y, Zhang YQ, Li WJ (2009) *Streptomonospora amylolytica* sp. nov. and *Streptomonospora flavalba* sp. nov., two novel halophilic actinomycetes isolated from a salt lake. Int J Syst Evol Microbiol 59:2471–2475

Carro L, Pukall R, Spröer C, Kroppenstedt RM, Trujillo ME (2013) *Micromonospora halotolerans* sp. nov., isolated from the rhizosphere of a *Pisum sativum* plant. Antonie Van Leeuwenhoek 103:1245–1254

Chang X, Liu W, Zhang XH (2012) *Salinactinospora qingdaonensis* gen. nov., sp. nov., a halophilic actinomycete isolated from a salt pond. Int J Syst Evol Microbiol 62:954–959

Chen HH, Li WJ, Tang SK, Kroppenstedt RM, Stackebrandt E, Xu LH, Jiang CL (2004) *Corynebacterium halotolerans* sp. nov., isolated from saline soil in the west of China. Int J Syst Evol Microbiol 54:779–782

Chen YG, Cui XL, Pukall R, Li HM, Yang YL, Xu LH, Wen ML, Peng Q, Jiang CL (2007) *Salinicoccus kunmingensis* sp. nov., a moderately halophilic bacterium isolated from a salt mine in Yunnan, south-west China. Int J Syst Evol Microbiol 57:2327–2332

Chen YG, Cui XL, Kroppenstedt RM, Stackebrandt E, Wen ML, Xu LH, Jiang CL (2008) *Nocardiopsis quinghaiensis* sp. nov., isolated from saline soil in China. Int J Syst Evol Microbiol 58:699–705

Chen YG, Tang SK, Zhang YQ, Li ZY, Yi LB, Wang YX, Li WJ, Cui XL (2009a) *Arthrobacter halodurans* sp. nov., a new halotolerant bacterium isolated from sea water. Antonie Van Leeuwenhoek 96:63–70

Chen YG, Wang YX, Zhang YQ, Tang SK, Liu ZX, Xiao HD, Xu LH, Cui XL, Li WJ (2009b) *Nocardiopsis litoralis* sp. nov., a halophilic marine actinomycete isolated from a sea anemone. Int J Syst Evol Microbiol 59:2708–2713

Chen YG, Tang SK, Zhang YQ, Liu ZX, Chen QH, He JW, Cui XL, Li WJ (2010a) Zhihengliuella salsuginis sp. nov., a moderately halophilic actinobacterium from a subterranean brine. Extremophiles 14:397–402

Chen YG, Zhang YQ, Tang SK, Liu ZX, Xu LH, Zhang LX, Li WJ (2010b) *Nocardiopsis terrae* sp. nov., a halophilic actinomycete isolated from saline soil. Antonie Van Leeuwenhoek 98:31–38

Chun J, Bae KS, Moon EY, Jung SO, Lee HK, Kim SJ (2000) *Nocardiopsis kunsanensis* sp. nov., a moderately halophilic actinomycete isolated from a saltern. Int J Syst Evol Microbiol 50:1909–1913

Collins MD, Lawson PA, Labrenz M, Tindall BJ, Weiss N, Hirsch P (2002) *Nesterenkonia lacusekhoensis* sp. nov., isolated from hypersaline Ekho Lake, East Antarctica, and emended description of the genus *Nesterenkonia*. Int J Syst Evol Microbiol 52:1145–1150

Conn HJ, Dimmick I (1947) Soil bacteria similar in morphology to *Mycobacterium* and *Corynebacterium*. J Bacteriol 54:291–303

Cui XL, Mao PH, Zeng M, Li WJ, Zhang LP, Xu LH, Jiang CL (2001) *Streptimonospora salina* gen. nov., sp. nov., a new member of the family *Nocardiopsaceae*. Int J Syst Evol Microbiol 51:357–363

Dastager SG, Lee JC, Ju YJ, Park DJ, Kim CJ (2008a) *Nocardioides halotolerans* sp. nov., isolated from soil on Bigeum Island, Korea. Syst Appl Microbiol 31:24–29

Dastager SG, Lee JC, Ju YJ, Park DJ, Kim CJ (2008b) *Nocardioides tritolerans* sp. nov., isolated from soil in Bigeum Island, Korea. J Microbiol Biotechnol 18:1203–1206

Delgado O, Quillaguaman J, Bakhtiar S, Mattiasson B, Gessesse A, Hatti-Kaul R (2006) *Nesterenkonia aethiopica* sp. nov., an alkaliphilic, moderate halophile isolated from an Ethiopian soda lake. Int J Syst Evol Microbiol 56:1229–1232

DeLong EF, Pace NR (2001) Environmental diversity of bacteria and archaea. Syst Biol 50:470–478

Embley MT, Smida J, Stackebrandt E (1988) The phylogeny of mycolate-less wall chemotype IV Actinomycetes and description of *Pseudonocardiaceae* fam. nov. Syst Appl Microbiol 11:44–52

Gao B, Gupta RS (2005) Conserved indels in protein sequences that are characteristic of the phylum *Actinobacteria*. Int J Syst Evol Microbiol 55:2401–2412

Garrity GM, Holt JG (2001) The road map to the manual. In: Boone C (ed) Bergey's manual of systematic bacteriology, vol 1, 2nd edn. Springer, New York, pp 119–155

Gochnauer MB, Leppard GG, Komaratat P, Kates M, Novitsky T, Kushner DJ (1975) Isolation and characterization of *Actinopolyspora halophila*, gen. sp. nov., an extremely halophilic actinomycete. Can J Microbiol 2:1500–1511

Goodfellow M (2010) Selective isolation of actinobacteria. In: Bull AT, Davies JE (eds) Manual of industrial microbiology and biotechnology, 3rd edn, vol 1, section 1. Isolation and screening of secondary metabolites and enzymes (Baltz RH, Davies JE, Demain AL, eds). ASM Press, Washington, DC, pp 13–27

Goodfellow M (2012) Phylum XXVI. Actinobacteria phyl. nov. In: Whitman WB, Goodfellow M, Kämpfer P, Busse HJ, Trujillo ME, Ludwig W, Suzuki KI, Parte A (eds) Bergey's manual of systematic bacteriology, vol 5, 2nd edn, The Actinobacteria, Part B. Springer, New York, pp 33–34

Govender L, Naidoo L, Setati ME (2013) *Nesterenkonia suensis* sp. nov., a haloalkaliphilic actinobacterium isolated from a salt pan. Int J Syst Evol Microbiol 63:41–46

Gu Q, Pasciak M, Luo H, Gamian A, Liu Z, Huang Y (2007) *Ruania albidiflava* gen. nov., sp. nov., a novel member of the suborder *Micrococcineae*. Int J Syst Evol Microbiol 57:809–814

Guan TW, Tang SK, Wu JY, Zhi XY, Xu LH, Zhang LL, Li WJ (2009) *Haloglycomyces albus* gen. nov., sp. nov., a halophilic, filamentous actinomycete of the family *Glycomycetaceae*. Int J Syst Evol Microbiol 59:1297–1301

Guan TW, Liu Y, Zhao K, Xia ZF, Zhang XP, Zhang LL (2010) *Actinopolyspora xinjiangensis* sp. nov., a novel extremely halophilic actinomycete isolated from a salt lake in Xinjiang, China. Antonie Van Leeuwenhoek 98:447–453

Guan TW, Wu N, Xia ZF, Ruan JS, Zhang XP, Huang Y, Zhang LL (2011a) *Saccharopolyspora lacisalsi* sp. nov., a novel halophilic actinomycete isolated from a salt lake in Xinjiang, China. Extremophiles 15:373–378

Guan TW, Xia ZF, Xiao J, Wu N, Chen ZJ, Zhang LL, Zhang XP (2011b) *Glycomyces halotolerans* sp. nov., a novel actinomycete isolated from a hypersaline habitat in Xinjiang, China. Antonie Van Leeuwenhoek 100:137–143

Guan TW, Xia ZF, Tang SK, Wu N, Chen ZJ, Huang Y, Ruan JS, Li WJ, Zhang LL (2012) *Amycolatopsis salitolerans* sp. nov., a filamentous actinomycete isolated from a hypersaline habitat. Int J Syst Evol Microbiol 62:23–27

Guan TW, Wei B, Zhang Y, Xia ZF, Che ZM, Chen XG, Zhang LL (2013a) *Actinopolyspora lacussalsi* sp. nov., an extremely halophilic actinomycete isolated from a salt lake. Int J Syst Evol Microbiol 63:3009–3013

Guan TW, Zhao HP, Che ZM, Zhang XP, Zhang LL (2013b) *Actinopolyspora dayingensis* sp. nov., a novel halophilic actinomycete isolated from a hypersaline lake. Antonie Van Leeuwenhoek 104:787–792

Hamada M, Shibata C, Tamura T, Suzuki K (2013) *Zhihengliuella flava* sp. nov., an actinobacterium isolated from sea sediment, and emended description of the genus *Zhihengliuella*. Int J Syst Evol Microbiol 63:4760–4764

Hamedi J, Mohammadipanah F, Ventosa A (2013) Systematic and biotechnological aspects of halophilic and halotolerant actinomycetes. Extremophiles 17:1–13

Han SK, Nedashkovskaya OI, Mikhailov VV, Kim SB, Bae KS (2003) *Salinibacterium amurskyense* gen. nov., sp. nov., a novel genus of the family *Microbacteriaceae* from the marine environment. Int J Syst Evol Microbiol 53:2061–2066

Hozzein WN, Goodfellow M (2008) *Nocardiopsis arabia* sp. nov., a halotolerant actinomycete isolated from a sand-dune soil. Int J Syst Evol Microbiol 58:2520–2524

Hozzein WN, Goodfellow M (2011) *Actinopolyspora egyptensis* sp. nov., a new halophilic actinomycete. Afr J Microbiol Res 5:100–105

Hozzein WN, Trujillo M (2012) Genus *Nocardiopsis*. In: Whitman WB, Goodfellow M, Kämpfer P, Busse HJ, Trujillo ME, Ludwig W, Suzuki KI, Parte A (eds) Bergey's manual of systematic bacteriology, vol 5, 2nd edn, The Actinobacteria, Part B. Springer, New York, pp 1891–1928

Kämpfer P, Busse H-J, Rainey FA (2002) *Nocardiopsis compostus* sp. nov., from the atmosphere of a composting facility. Int J Syst Evol Microbiol 52:621–627

Kanekar PP, Kanekar SP, Kelkar AS, Dhakephalkar PK (2012) Halophiles–Taxonomy, diversity, physiology and applications. In: Satyanarayana T, Johri BN, Prakash A (eds) Microorganisms in environmental management: microbes and environment. Springer, Dordrecht, pp 1–34

Kim SB, Goodfellow M (1999) Reclassification of *Amycolatopsis rugosa* Lechevalier et al. 1986 as *Prauserella rugosa* gen. nov., comb. nov. Int J Syst Bacteriol 49:507–512

Kim SB, Nedashkovskaya OI, Mikhailov VV, Han SK, Kim KO, Rhee MS, Bae KS (2004) *Kocuria marina* sp. nov., a novel actinobacterium isolated from marine sediment. Int J Syst Evol Microbiol 54:1617–1620

Kim SH, Yang HO, Sohn YC, Kwon HC (2010) Aeromicrobium halocynthiae sp. nov., a taurocholic acidproducing bacterium isolated from the marine ascidian Halocynthia roretzi. Int J Syst Evol Microbiol 60:2793–2798

Koch C, Schumann P, Stackebrandt E (1995) Reclassification of *Micrococcus agilis* (Ali-Cohen 1889) to the genus *Arthrobacter* as *Arthrobacter agilis* comb. nov. and emendation of the genus *Arthrobacter*. Int J Syst Bacteriol 45:837–839

Korn-Wendisch F, Kempf A, Grund E, Kroppenstedt RM, Kutzner HJ (1989) Transfer of *Faenia rectivirgula* Kurup and Agre 1983 to the genus *Saccharopolyspora* Lacey and Goodfellow 1975, elevation of *Saccharopolyspora hirsuta* subsp. *taberi* Labeda 1987 to species level, and emended description of the genus *Saccharopolyspora*. Int J Syst Bacteriol 39:430–441

Kushner DJ, Kamekura M (1988) Physiology of halophilic eubacteria. In: Rodriguez-Valera F (ed) Halophilic bacteria, vol 1. CRC Press, Boca Raton, pp 109–140

Kuznetsov VD, Zajtseva TA, Vakulenko LV, Flippova SN (1992) *Streptomyces albiaxialis* sp. nov. a new oil hydrocarbon degrading species of thermo- and halotolerant streptomyces. Mikrobiologiya 61:84–91

Labeda DP, Kroppenstedt RM (2004) Emended description of the genus *Glycomyces* and description of *Glycomyces algeriensis* sp. nov., *Glycomyces arizonensis* sp. nov. and *Glycomyces lechevalierae* sp. nov. Int J Syst Evol Microbiol 54:2343–2346

Labeda DP, Kroppenstedt RM (2005) *Stackebrandtia nassauensis* gen. nov., sp. nov. and emended description of the family *Glycomycetaceae*. Int J Syst Evol Microbiol 55:1687–1691

Labeda DP, Testa RT, Lechevalier MP, Lechevalier HA (1985) *Glycomyces*, a new genus of the *Actinomycetales*. Int J Syst Bacteriol 35:417–421

Labeda DP, Goodfellow M, Chun J, Zhi XY, Li WJ (2011) Reassessment of the systematics within the suborder *Pseudonocardineae*: transfer of the genera within the family *Actinosynnemataceae* Labeda and Kroppenstedt 2000 emend. Zhi et al. 2009 into an emended family *Pseudonocardiaceae* Embley et al. 1989 emend. Zhi et al. 2009. Int J Syst Evol Microbiol 61:1259–1264

Lacey J, Goodfellow M (1975) Novel actinomycete from sugarcane Bagasse *Saccharopolyspora hirsuta* gen. et sp. nov. J Gen Microbiol 88:75–85

Lechevalier MP, Prauser H, Labeda DP, Ruan JS (1986) Two new genera of nocardioform actinomycetes, *Amycolata* gen. nov. and *Amycolatopsis* gen. nov. Int J Syst Bacteriol 36:29–37

Lee SD (2006) *Amycolatopsis jejuensis* sp. nov. and *Amycolatopsis halotolerans* sp. nov., novel actinomycetes isolated from a natural cave. Int J Syst Evol Microbiol 56:549–553

Lehmann KB, Neumann R (1896) Atlas und Grundriss der Bakteriologie und Lehrbuch der speciellen bakteriologischen Diagnostik. J.F. Lehmann, Munich

Li MG, Li WJ, Xu P, Cui XL, Xu LH, Jiang CL (2003a) *Nocardiopsis xinjiangensis* sp. nov., a halophilic actinomycete isolated from a saline soil sample in China. Int J Syst Evol Microbiol 53:317–321

Li WJ, Xu P, Zhang LP, Tang SK, Cui XL, Mao PH, Xu LH, Schumann P, Stackebrandt E, Jiang CL (2003b) *Streptomonospora alba* sp. nov., a novel halophilic actinomycete, and emended description of the genus *Streptomonospora* Cui et al. 2001. Int J Syst Evol Microbiol 53:1421–1425

Li WJ, Xu P, Tang SK, Xu LH, Kroppenstedt RM, Stackebrandt E, Jiang CL (2003c) *Prauserella halophila* sp. nov. and *Prauserella alba* sp. nov., moderately halophilic actinomycetes from saline soil. Int J Syst Evol Microbiol 53:1545–1549

Li WJ, Tang SK, Stackebrandt E, Kroppenstedt RM, Schumann P, Xu LH, Jiang CL (2003d) *Saccharomonospora paurometabolica* sp. nov., a moderately halophilic actinomycete isolated from soil in China. Int J Syst Evol Microbiol 53:1591–1594

Li WJ, Chen HH, Xu P, Zhang YQ, Schumann P, Tang SK, Xu LH, Jiang CL (2004a) *Yania halotolerans* gen. nov., sp. nov., a novel member of the suborder *Micrococcineae* from saline soil in China. Int J Syst Evol Microbiol 54:525–531

Li WJ, Chen HH, Zhang YQ, Schumann P, Stackebrandt E, Xu LH, Jiang CL (2004b) *Nesterenkonia halotolerans* sp. nov. and *Nesterenkonia xinjiangensis* sp. nov., actinobacteria from saline soils in the west of China. Int J Syst Evol Microbiol 54:837–841

Li WJ, Park DJ, Tang SK, Wang D, Lee JC, Xu LH, Kim CJ, Jiang CL (2004c) *Nocardiopsis salina* sp. nov., a novel halophilic actinomycete isolated from saline soil in China. Int J Syst Evol Microbiol 54:1805–1809

Li WJ, Chen HH, Kim CJ, Park DJ, Tang SK, Lee JC, Xu LH, Jiang CL (2005a) *Microbacterium halotolerans* sp. nov., isolated from a saline soil in the west of China. Int J Syst Evol Microbiol 55:67–70

Li WJ, Chen HH, Kim CJ, Zhang YQ, Park DJ, Lee JC, Xu LH, Jiang CL (2005b) *Nesterenkonia sandarakina* sp. nov. and *Nesterenkonia lutea* sp. nov., novel actinobacteria, and emended description of the genus *Nesterenkonia*. Int J Syst Evol Microbiol 55:463–466

Li WJ, Kroppenstedt RM, Wang D, Tang SK, Lee JC, Park DJ, Kim CJ, Xu LH, Jiang CL (2006) Five novel species of the genus *Nocardiopsis* isolated from hypersaline soils and emended description of *Nocardiopsis salina* Li et al. 2004. Int J Syst Evol Microbiol 56:1089–1096

Li WJ, Xu P, Schumann P, Zhang YQ, Pukall R, Xu LH, Stackebrandt E, Jiang CL (2007) *Georgenia ruanii* sp. nov., a novel actinobacterium isolated from forest soil in Yunnan (China), and emended description of the genus *Georgenia*. Int J Syst Evol Microbiol 57:1424–1428

Li WJ, Zhi XY, Euzeby JP (2008a) Proposal of *Yaniellaceae* fam. nov., *Yaniella* gen. nov. and *Sinobaca* gen. nov. as replacements for the illegitimate prokaryotic names *Yaniaceae* Li et al. 2005, *Yania* Li et al. 2004, emend. Li et al. 2005, and *Sinococcus* Li et al. 2006, respectively. Int J Syst Evol Microbiol 58:525–527

Li WJ, Zhang YQ, Schumann P, Liu HY, Yu LY, Zhang YQ, Stackebrandt E, Xu LH, Jiang CL (2008b) *Nesterenkonia halophila* sp. nov., a moderately halophilic, alkalitolerant actinobacterium isolated from a saline soil. Int J Syst Evol Microbiol 58:1359–1363

Li Y, Tang SK, Chen YG, Wu JY, Zhi XY, Zhang YQ, Li WJ (2009) *Prauserella salsuginis* sp. nov., *Prauserella flava* sp. nov., *Prauserella aidingensis* sp. nov. and *Prauserella sediminis* sp. nov., isolated from a salt lake. Int J Syst Evol Microbiol 59:2923–2928

Liebl W, Ehrmann M, Ludwig W, Schleifer KH (1991) Transfer of *Brevibacterium divaricatum* DSM 20297T, "*Brevibacterium flavum*" DSM 20411, "*Brevibacterium lactofermentum*" DSM

20412 and DSM 1412, and *Corynebacterium glutamicum* and their distinction by rRNA gene restriction patterns. Int J Syst Bacteriol 41:255–260

Lu Y, Yan X (1983) Classification of thermophilic actinomycetes. 4. Determination of thermophilic members of *Nocardiaceae*. Acta Microbiol Sin 23:220–228

Ludwig W, Klenk HP (2005) Overview: a phylogenetic backbone and taxonomic framework for procaryotic systematics. In: Brenner DJ, Krieg NR, Staley JR, Garrity G (eds) Bergey's manual of systematic bacteriology, vol 2, 2nd edn, The proteobacteria, Part A, Introductory essays. Springer, New York, pp 49–65

Luo HY, Miao LH, Fang C, Yang PL, Wang YR, Shi PJ, Yao B, Fan YL (2008) *Nesterenkonia flava* sp. nov., isolated from paper-mill effluent. Int J Syst Evol Microbiol 58:1927–1930

Luo HY, Wang YR, Miao LH, Yang PL, Shi PJ, Fang CX, Yao B, Fan YL (2009) *Nesterenkonia alba* sp. nov., an alkaliphilic actinobacterium isolated from the black liquor treatment system of a cotton pulp mill. Int J Syst Evol Microbiol 59:863–868

Meklat A, Bouras N, Zitouni A, Mathieu F, Lebrihi A, Schumann P, Spröer C, Klenk HP, Sabaou N (2012) *Actinopolyspora algeriensis* sp. nov., a novel halophilic actinomycete isolated from a Saharan soil. Extremophiles 16:771–776

Meklat A, Bouras N, Zitouni A, Mathieu F, Lebrihi A, Schumann P, Spröer C, Klenk HP, Sabaou N (2013a) *Actinopolyspora saharensis* sp. nov., a novel halophilic actinomycete isolated from a Saharan soil of Algeria. Antonie Van Leeuwenhoek 103:771–776

Meklat A, Bouras N, Zitouni A, Mathieu F, Lebrihi A, Schumann P, Spröer C, Klenk HP, Sabaou N (2013b) *Actinopolyspora righensis* sp. nov., a novel halophilic actinomycete isolated from Saharan soil in Algeria. Antonie Van Leeuwenhoek 104:301–307

Meklat A, Bouras N, Zitouni A, Mathieu F, Lebrihi A, Schumann P, Spröer C, Klenk HP, Sabaou N (2013c) *Actinopolyspora mzabensis* sp. nov., a halophilic actinomycete isolated from an Algerian Saharan soil. Int J Syst Evol Microbiol 63:3787–3792

Meyer J (1976) *Nocardiopsis*, a new genus of order *Actinomycetales*. Int J Syst Bacteriol 26:487–493

Nonomura H, Ohara Y (1971) Distribution of actinomycetes in soil. X. New genus and species of monosporic actinomycetes in soil. J Ferment Technol 49:895–903

Novitsky TJ, Kushner DJ (1976) *Planococcus halophilus* sp. nov., a facultatively halophilic coccus. Int J Syst Bacteriol 26:53–57

Olano C, Mendez C, Salas JA (2009) Antitumor compounds from marine actinomycetes. Mar Drugs 7:210–248

Onishi H, Kamekura M (1972) *Micrococcus halobius* sp. nov. Int J Syst Bacteriol 22:233–236

Oren A (2002) Halophilic microorganisms and their environments. Kluwer, Dordrecht

Oren A (2008) Microbial life at high salt concentrations: phylogenetic and metabolic diversity. Saline Syst 4:2

Oren A (2011) Diversity of halophiles. In: Horikoshi K, Antranikian G, Bull AT, Robb FT, Stetter KO (eds) Extremophiles handbook. Springer Japan, Tokyo, pp 309–325

Pimentel-Elardo SM, Tiro LP, Grozdanov L, Hentschel U (2008) *Saccharopolyspora cebuensis* sp. nov., a novel actinomycete isolated from a Philippine sponge (*Porifera*). Int J Syst Evol Microbiol 58:628–632

Ramos-Cormenzana A (1989) Ecological distribution and biotechnological potential of halophilic microorganisms. In: Da Costa MS, Duarte JC, Williams RAD (eds) Microbiology of extreme environments and its potential for biotechnology. Elsevier Applied Science, London, pp 289–309

Rappè MS, Giovannoni SJ (2003) The uncultured microbial majority. Annu Rev Microbiol 57:369–394

Ray L, Suar M, Pattnaik AK, Raina V (2013) *Streptomyces chilikensis* sp. nov., a halophilic streptomycete isolated from brackish water sediment. Int J Syst Evol Microbiol 63:2757–2764

Ruan JS, Al-Tai AM, Zhou ZH, Qu LH (1994) *Actinopolyspora iraqiensis* sp. nov., a new halophilic actinomycete isolated from soil. Int J Syst Bacteriol 44:759–763

Rusznyák A, Tóth EM, Schumann P, Spröer C, Makk J, Szabó G, Vladár P, Márialigeti K, Borsodi AK (2011) *Cellulomonas phragmiteti* sp. nov., a cellulolytic bacterium isolated from reed

(*Phragmites australis*) periphyton in a shallow soda pond. Int J Syst Evol Microbiol 61:1662–1666
Sabet S, Diallo L, Hays L, Jung W, Dillon JG (2009) Characterization of halophiles isolated from solar salterns in Baja California, Mexico. Extremophiles 13:643–656
Sass AM, Sass H, Coolen MJL, Cypionka H, Overmann J (2001) Microbial communities in the chemocline of a hypersaline deep-sea basin (Urania Basin, Mediterranean Sea). Appl Environ Microbiol 67:5392–5402
Schäfer J, Martin K, Kämpfer P (2010) *Prauserella muralis* sp. nov., from the indoor environment. Int J Syst Evol Microbiol 60:287–290
Stackebrandt E, Koch C, Gvozdiak O, Schumann P (1995) Taxonomic dissection of the genus *Micrococcus*: *Kocuria* gen. nov., *Nesterenkonia* gen. nov., *Kytococcus* gen. nov., *Dermacoccus* gen. nov., and *Micrococcus* Cohn 1872 gen. emend. Int J Syst Bacteriol 45:682–692
Stackebrandt E, Rainey FA, Ward-Rainey NL (1997) Proposal for a new hierarchic classification system, *Actinobacteria* classis nov. Int J Syst Bacteriol 47:479–491
Stackebrandt E, Schumann P, Cui XL (2004) Reclassification of *Cellulosimicrobium variabile* Bakalidou et al. 2002 as *Isoptericola variabilis* gen. nov., comb. nov. Int J Syst Evol Microbiol 54:685–688
Stevenson BS, Eichorst SA, Wertz JT, Schmidt TM, Breznak JA (2004) New strategies for cultivation and detection of previously uncultured microbes. Appl Environ Microbiol 70:4748–4755
Syed DG, Tang SK, Cai M, Zhi XY, Agasar D, Lee JC, Kim CJ, Jiang CL, Xu LH, Li WJ (2008) *Saccharomonospora saliphila* sp. nov., a halophilic actinomycete from an Indian soil. Int J Syst Evol Microbiol 58:570–573
Takeuchi M, Hatano K (1998) Proposal of six new species in the genus *Microbacterium* and transfer of *Flavobacterium marinotypicum* ZoBell and Upham to the genus *Microbacterium* as *Microbacterium maritypicum* comb. nov. Int J Syst Bacteriol 48:973–982
Tang SK, Tian XP, Zhi XY, Cai M, Wu JY, Yang LL, Xu LH, Li WJ (2008) *Haloactinospora alba* gen. nov., sp. nov., a halophilic filamentous actinomycete of the family *Nocardiopsaceae*. Int J Syst Evol Microbiol 58:2075–2080
Tang SK, Wang Y, Cai M, Zhi XY, Lou K, Xu LH, Jiang CL, Li WJ (2009a) *Saccharopolyspora halophila* sp. nov., a novel halophilic actinomycete isolated from a saline lake in China. Int J Syst Evol Microbiol 59:555–558
Tang SK, Wang Y, Lou K, Mao PH, Xu LH, Jiang CL, Kim CJ, Li WJ (2009b) *Kocuria halotolerans* sp. nov., an actinobacterium isolated from a saline soil in China. Int J Syst Evol Microbiol 59:1316–1320
Tang SK, Wang Y, Chen Y, Lou K, Cao LL, Xu LH, Li WJ (2009c) *Zhihengliuella alba* sp. nov., and emended description of the genus *Zhihengliuella*. Int J Syst Evol Microbiol 59:2025–2031
Tang SK, Wang Y, Wu JY, Cao LL, Lou K, Xu LH, Jiang CL, Li WJ (2009d) *Saccharopolyspora qijiaojingensis* sp. nov., a halophilic actinomycete isolated from a salt lake. Int J Syst Evol Microbiol 59:2166–2170
Tang SK, Wang Y, Lee JC, Lou K, Park DJ, Kim CJ, Li WJ (2010a) *Georgenia halophila* sp. nov., a halophilic actinobacterium isolated from a salt lake. Int J Syst Evol Microbiol 60:1317–1321
Tang SK, Zhi XY, Wang Y, Wu JY, Lee JC, Kim CJ, Lou K, Xu LH, Li WJ (2010b) *Haloactinobacterium album* gen. nov., sp. nov., a halophilic actinobacterium, and proposal of *Ruaniaceae* fam. nov. Int J Syst Evol Microbiol 60:2113–2119
Tang SK, Wang Y, Zhang H, Lee JC, Lou K, Kim CJ, Li WJ (2010d) Haloechinothrix alba gen. nov., sp. nov., a halophilic, filamentous actinomycete of the suborder Pseudonocardineae. Int J Syst Evol Microbiol 60:2154–2158
Tang SK, Zhi XY, Wang Y, Shi R, Lou K, Xu LH, Li WJ (2011a) *Haloactinopolyspora alba* gen. nov., sp. nov., a halophilic filamentous actinomycete isolated from a salt lake, with proposal of *Jiangellaceae* fam. nov. and *Jiangellineae* subord. nov. Int J Syst Evol Microbiol 61:194–200

Tang SK, Wang Y, Klenk HP, Shi R, Lou K, Zhang YJ, Chen C, Ruan JS, Li WJ (2011b) *Actinopolyspora alba* sp. nov. and *Actinopolyspora erythraea* sp. nov., isolated froma salt field, and reclassification of *Actinopolyspora iraqiensis* Ruan et al. 1994 as a heterotypic synonym of *Saccharomonospora halophila*. Int J Syst Evol Microbiol 61:1693–1698

Tatar D, Sazak A, Guven K, Cetin D, Sahin N (2013) *Amycolatopsis cihanbeyliensis* sp. nov., a halotolerant actinomycete isolated from a salt mine. Int J Syst Evol Microbiol 63:3739–3743

Traiwan J, Park MH, Kim W (2011) *Serinicoccus chungangensis* sp. nov., isolated from tidal flat sediment, and emended description of the genus *Serinicoccus*. Int J Syst Evol Microbiol 61:1299–1303

Tresner HD, Hayes JS, Backus EJ (1968) Differential tolerance of streptomycetes to sodium chloride as a taxonomic tool. Appl Microbiol 16:1134–1136

Trüper HG, Galinski EA (1986) Concentrated brines as habitats for microorganisms. Experientia 42:1182–1187

Vartoukian SR, Palmer RM, Wade WG (2010) Strategies for culture of 'unculturable' bacteria. FEMS Microbiol Lett 309:1–7

Ventosa A, Nieto JJ, Oren A (1998) Biology of moderately halophilic aerobic bacteria. Microbiol Mol Biol Rev 62:504–544

Ventosa A, Mellado E, Sanchez-Porro C, Marquez MC (2008) Halophilic and halotolerant microorganisms from soils. In: Dion P, Nautiyal CS (eds) Microbiology of extreme soils, vol 13, Soil biology. Springer, Berlin, pp 87–115

Vreeland RH (1987) Mechanisms of halotolerance in microorganisms. Crit Rev Microbiol 14:311–356

Wang J, Li Y, Bian J, Tang SK, Ren B, Chen M, Li WJ, Zhang LX (2010) *Prauserella marina* sp. nov., isolated from ocean sediment of the South China Sea. Int J Syst Evol Microbiol 60:985–989

Weinstein MJ, Luedemann GM, Oden EM, Wagman GH (1968) Halomicin, a new *Micromonospora-produced* antibiotic. Antimicrob Agents Chemother (Bethesda) 1967:435–441

Xiang W, Guo J, Feng W, Huang M, Chen H, Zhao J, Zhang J, Yang Z, Sun Q (2008) Community of extremely halophilic bacteria in historic Dagong Brine Well in south western China. World J Microbiol Biotechnol 24:2297–2305

Xiao J, Luo Y, Xie S, Xu J (2011) *Serinicoccus profundi* sp. nov., an actinomycete isolated from deep-sea sediment, and emended description of the genus *Serinicoccus*. Int J Syst Evol Microbiol 61:16–19

Yang LL, Tang SK, Zhang YQ, Zhi XY, Wang D, Xu LH, Li WJ (2008) *Thermobifida halotolerans* sp. nov., isolated from a salt mine sample, and emended description of the genus *Thermobifida*. Int J Syst Evol Microbiol 58:1821–1825

Yi H, Schumann P, Sohn K, Chun J (2004) *Serinicoccus marinus* gen. nov., sp. nov., a novel actinomycete with l-ornithine and l-serine in the peptidoglycan. Int J Syst Evol Microbiol 54:1585–1589

Yoon JH, Jung SY, Kim W, Nam SW, Oh TK (2006) *Nesterenkonia jeotgali* sp. nov., isolated from jeotgal, a traditional Korean fermented seafood. Int J Syst Evol Microbiol 56:2587–2592

Yoon JH, Park S, Kang SJ, Lee JS, Lee KC, Oh TK (2010) *Nocardioides daedukensis* sp. nov., a halotolerant bacterium isolated from soil. Int J Syst Bacteriol 60:1334–1338

Yoshida M, Matsubara K, Kudo T, Horikoshi K (1991) *Actinopolyspora mortivallis* sp. nov., a moderately halophilic actinomycete. Int J Syst Bacteriol 41:15–20

Yu L, Lai Q, Yi Z, Zhang L, Huang Y, Gu L, Tang X (2013) *Microbacterium sediminis* sp. nov., a psychrotolerant, thermotolerant, halotolerant and alkalitolerant actinomycete isolated from deep-sea sediment. Int J Syst Evol Microbiol 63:25–30

Zengler K, Toledo G, Rappe M, Elkins J, Mathur EJ, Short JM, Keller M (2002) Cultivating the uncultured. Proc Natl Acad Sci U S A 99:15681–15686

Zhang YQ, Schumann P, Li WJ, Chen GZ, Tian XP, Stackebrandt E, Xu LH, Jiang CL (2005) *Isoptericola halotolerans* sp. nov., a novel actinobacterium isolated from saline soil from Qinghai Province, north-west China. Int J Syst Evol Microbiol 55:1867–1870

Zhang YQ, Schumann P, Yu LY, Liu HY, Zhang YQ, Xu LH, Stackebrandt E, Jiang CL, Li WJ (2007) *Zhihengliuella halotolerans* gen. nov., sp. nov., a novel member of the family *Micrococcaceae*. Int J Syst Evol Microbiol 57:1018–1023

Zhang DC, Liu HC, Xin YH, Yu Y, Zhou PJ, Zhou YG (2008) *Salinibacterium xinjiangense* sp. nov., a psychrophilic bacterium isolated from the China No. 1 glacier. Int J Syst Evol Microbiol 58:2739–2742

Zhang J, Wu D, Liu Z (2009) *Saccharopolyspora jiangxiensis* sp. nov., isolated from grass-field soil. Int J Syst Evol Microbiol 59:1076–1081

Zhang DF, Pan HQ, He J, Zhang XM, Zhang YG, Klenk HP, Hu JC, Li WJ (2013a) Description of *Streptomonospora sediminis* sp. nov. and *Streptomonospora nanhaiensis* sp. nov., and reclassification of *Nocardiopsis arabia* Hozzein & Goodfellow 2008 as *Streptomonospora arabica* comb. nov. and emended description of the genus *Streptomonospora*. Int J Syst Evol Microbiol 63:4447–4455

Zhang YJ, Zhang WD, Qin S, Bian GK, Xing K, Li YF, Cao CL, Jiang JH (2013b) *Saccharopolyspora dendranthemae* sp. nov. a halotolerant endophytic actinomycete isolated from a coastal salt marsh plant in Jiangsu, China. Antonie Van Leeuwenhoek 103:1369–1376

Zhang YG, Liu Q, Wang HF, Zhang DF, Zhang YM, Park DJ, Kim CJ, Li WJ (2014) *Haloactinopolyspora alkaliphila* sp. nov., and emended description of the genus *Haloactinopolyspora*. Int J Syst Evol Microbiol 64:1945–1951

Zhi XY, Li WJ, Stackebrandt E (2009) An update of the structure and 16S rRNA gene sequence-based definition of higher ranks of the class *Actinobacteria*, with the proposal of two new suborders and four new families and emended descriptions of the existing higher taxa. Int J Syst Evol Microbiol 59:589–608

Chapter 2
Antimicrobial and Biocatalytic Potential of Haloalkaliphilic Actinobacteria

Sangeeta D. Gohel, Amit K. Sharma, Kruti G. Dangar, Foram J. Thakrar, and Satya P. Singh

Abstract Actinobacteria have the genetic potential for the production of wide variety of yet-to-discover secondary metabolites. They are known for their metabolic versatility that allows them to survive even under extreme environmental conditions. Further, considerable increase in the interest pertaining to the natural products originating from the saline and alkaline environments has brought haloalkaliphilic actinobacteria into the focus of intensive research. These organisms harbour huge potential for the discovery of new antibiotics and enzymes with novel properties. Besides, they have vast biosynthetic potential that remains unexplored. In the present chapter, we have evaluated the present state of research on the haloalkaliphilic actinobacteria and their unique antimicrobial properties. The highlights include the distribution of the haloalkaliphilic actinobacteria, their adaptation in different stress conditions, and mining of their unique antimicrobial and enzymatic potential.

Keywords Haloalkaliphilic actinobacteria • Salt stress • Antimicrobial activity • Novel enzymes

2.1 Introduction

Actinobacteria for long have been explored as a potential source of bioactive compounds, with the ability to produce diverse secondary metabolites (Ghorbel et al. 2014; Malviya et al. 2014; Das et al. 2008). These organisms having high guanosine-cytosine (G+C) contents (65–75 %) belong to the phylum Actinobacteria, one of the largest groups in the domain Bacteria (Miao and Davies 2010), dwelling in varied habitats: soils, the rhizosphere, marine and extreme arid environments (Li and Liu 2006; Bull 2011). Actinobacteria represents one of the largest taxonomic units among the 18 major lineages currently recognized within the bacteria domain,

S.D. Gohel • A.K. Sharma • K.G. Dangar • F.J. Thakrar • S.P. Singh (✉)
Department of Biosciences, Saurashtra University, Rajkot, 360 005 Gujarat, India
e-mail: satyapsingh@yahoo.com; satyapsingh125@gmail.com

D.K. Maheshwari, M. Saraf (eds.), *Halophiles*, Sustainable Development and Biodiversity 6, DOI 10.1007/978-3-319-14595-2_2

including five subclasses and 14 suborders (Stackebrandt 2000). Recently, potent actinomycetes were reported to have approximately 7,000 of the metabolites in the dictionary of the natural products (Manivasagan et al. 2013).

Actinobacteria from the extreme habitats are largely unexplored for the discovery of novel bioactive secondary metabolites (Hamedi and Mohammadipanah 2013). In fact, efforts have been made to isolate actinobacteria from the saline environments and most of them appeared to have potential biotechnological applications (Vasavada et al. 2006; Thumar and Singh 2009; Thumar et al. 2010; Gulder and Moore 2010; Vijayakumar et al. 2012; Gohel and Singh 2012a, b, 2013; Hamedi and Mohammadipanah 2013). Recently, phylogenetically diverse actinomycetes are reported from an Indian coastal solar saltern (Jose and Jebakumar 2012). These organisms hold prominent positions and have proven abilities to produce novel metabolites, industrially important enzymes and other molecules of pharmaceutical importance (Ellaiah et al. 2004; Singh et al. 2013a, b; Gohel and Singh 2015). A large number of actinomycetes has been reported to produce antibiotic compounds and halo-tolerant enzymes from the coastal areas and solar salterns (Gohel and Singh 2013; Zafrilla et al. 2010; Thumar et al. 2010; Jose et al. 2011; Thumar and Singh 2009; Vasavada et al. 2006). A user-friendly database has been created and published on the diversity, biocatalytic potential and phylogeny of the salt-tolerant alkaliphilic actinomycetes (Sharma 2011; Sharma et al. 2012).

Among the different genera of the actinomycetes, the genus *Streptomyces* is represented by the largest number of species and varieties with the ability to produce a large number of antibiotics and bioactive secondary metabolites (Devi et al. 2006). In fact, 80 % of the recognized antibiotics have been produced from this genus (Procopio et al. 2012), and many representatives of this group produce substances of high commercial value and are being extensively screened for novel bioactive compounds (Anderson and Wellington 2001; Vijayakumar et al. 2007). Exploring new habitats is one of the most promising ways to isolate new strains of actinomycetes with antimicrobial activities (Zitouni et al. 2005; Khanna et al. 2011; Wadetwar and Patil 2013). Further, actinobacteria have a wide range of enzyme activities. They have been commercially used in harsh industrial processes such as food processing, detergent formulation, biosynthetic processes and production of pharmaceuticals, neutraceuticals, enzymes, antitumor agents, and enzyme inhibitors (Ventosa et al. 2005; Remya and Vijaykumar 2008). The alkaliphilic actinobacteria such as *Streptomyces* and *Nocardiopsis* have been reported to produce range of hydrolytic enzyme and other value based molecules for the industrial applications. Although there are several reports on the physiology, energetics and structure and function relationship of the enzymes of the alkaliphilic bacteria (Purohit and Singh 2014; Raval et al. 2014; Kumar and Kumar 2014; Pandey and Singh 2012; Ventosa et al. 2005; Yoshimune et al. 2003), only few reports are available on the alkaliphilic salt-tolerant actinomycetes (Gohel and Singh 2013, 2012a, b; Thumar et al. 2010).

The unique characteristics of the microbial enzymes include their capability and appreciable activity under the unconventional conditions. New physical characteristics and unique properties, such as high specificity or specific activity, large productivity, extreme thermo-alkali stability, low cost of production, and tolerance to solvents and inhibitors add to the robustness and application prospects of the enzymes (Prakash et al. 2013 Badoei et al. 2014; Moreira et al. 2014; Panwar et al. 2014; Siroosi et al. 2014). Halophilic proteins compete effectively with salts for hydration, a property that may result in resistance to conditions of low water activity and maintain stability under non-aqueous conditions (Kerkar 2004; Thumar and Singh 2009). The members of the actinomycetes, such as *Streptomycetes*, *Nocardiopsis*, *Thermomonospora*, and *Thermoactinomyces* have been explored for the production of various industrially important enzymes, prominently being proteases, amylases, cellulases, chitinases and lipases (Bakhtiar et al. 2003; Ramesh and Mathivanan 2009). Usually, these enzymes possess excess of the acidic residues over the basic residues and have a low content of hydrophobic residues at their surface (Mevarech et al. 2000). Further, the increasing tendency of the multi-drug resistance in the pathogenic bacteria has created a pressing demand for more rational approach and strategies to search for broad spectrum antibiotics. Therefore, exploration of the antimicrobial and biocatalytic features of the lesser explored haloalkalophilic actinobacteria would be highly significant.

2.2 Distribution of Haloalkaliphilic Actinomycetes in the Saline and Alkaline Habitats

The actinomycetes, in general, have been extensively explored for their secondary metabolites. However, there is a lack of comprehensive effort to understand the diversity, distribution and ecology of the actinomycetes of the extreme habitats. More than 70 species of halotolerant and halophilic actinomycetes of 24 genera have recently been described (Hamedi and Mohammadipanah 2013). During the recent years, halophilic actinomycetes have been described from sea water, saline soils, salt lakes, brines, alkaline saline habitats and other habitats. Considerable number of salt-tolerant and drought-resistant actinomycetes are reported from arid areas such as those of the Coastal Gujarat (India), Arabian Peninsula, Iranian Plateau, Central Asia and Australia coastal regions (Gohel and Singh 2013; Al et al. 2007; Thumar and Singh 2007a). Few groups of scientists in China are actively involved in the exploration of actinomycetes of the saline and alkaline habitats. They have isolated more than 30 novel actinomycetes from the saline and alkaline habitats during the last several years (Chen et al. 2009; Cai et al. 2009; Wu et al. 2009; Yang et al. 2008; Li et al. 2006a, b). Further, actinomycetes are also reported from the Saline lakes, Soda Lake and Salt mines in Egypt, Korea and East Africa (Sabry et al. 2004; Hozzein et al. 2004; Schippers et al. 2002; Chun et al. 2000).

Approaches to culture-dependent diversity are based on the sediment pretreatments, taxon-selective isolation media, preliminary circumscription and replication of isolates. While the culture-independent explorations depend on the actinomycetes specific primers designed on the basis of the current 16S rRNA sequence databases (Stach et al. 2003), dereplication and community profiling with SSCP, DGGE or T-RFLP, and sequencing usually of the representative members of the resulting OTU clusters. The ecological role of actinomycetes in the saline ecosystem is least explored and largely unknown. Hence, it would be interesting to explore vast genetic resources particularly represented by the non-cultivable actinomycetes using the recently developed approaches of the metagenomics. This allows the estimation of the ecological parameters such as richness, diversity and degree of similarity between microbial communities (Schloss and Handelsman 2005, 2006). In India, Syed and co-workers are involved with the exploration of the actinobacterial diversity of Gulbarga, Karnataka. They have characterized few species with respect to their proteolytic potential (Syed et al. 2009; Dastager et al. 2008). Another research group in India, led by Prof. Satya P. Singh at the Saurashtra University, is actively involved with the exploration of the salt tolerant alkaliphilic actinomyctes from varied saline habitats of the Coastal Gujarat (India) (Thumar and Singh 2009, 2010, 2011; Gohel and Singh 2012a, b, 2013, 2015; Singh et al. 2013a, b). The halophilic actinomycetes are listed in Table 2.1.

2.3 Adaptation of Haloalkaliphilic Actinobacteria to Different Stress Conditions

Halophiles are distributed throughout the three domains of life; eukaryotes, prokaryotes and archaebacterial. Among the halophilic bacteria, Cyanobacteria, Proteobacteria, Firmicutes, Actinobacteria, Spirochaetes, and Bacteroidetes are included. Adaptation to environmental changes is essential for the survival and proliferation of the microbes. Adaptation is an evolutionary process through which an organism develops ability to live in its habitat (Dobzhansky et al. 1968). Environmental changes shift the balance of the complex microbial communities by favoring some populations and restricting others, through mechanisms such as microbial competition for nutrients, antibiosis and by selecting the most suitable organism to environmental stress. Thus, majority of the work in microbial ecology has focused on the evolution of microbial populations subjected to the changes in the environmental conditions (Amoroso et al. 1998; Baath et al. 1998; Baek and Kim 2009; Baek et al. 2009; Carrasco et al. 2005; Belanger et al. 2011).

Haloalkaliphilic actinomycetes face a succession of environmental challenges as they grow and develop. Thus they develop adaptive mechanisms, such as the phenotype traits in response to the obligatory conditions. Haloalkaliphilic actinomycetes survive and grow under the extreme environments of high pH and salt concentrations (Vasavada et al. 2006; Gohel and Singh 2012a, b). They may also face high pressure, high/low temperature, nutrient scarcity, high levels of radiation, harmful heavy metals and toxic compounds (Satyanarayana et al. 2005; Ogino et al. 2007). These organisms are of considerable practical interest, as they provide sources of highly stable enzymes, least affected by the variation in environmental conditions (Thumar and Singh 2009; Gohel and Singh 2012b; Chakraborty et al. 2012).

Table 2.1 Distribution of haloalkaliphilic actinomycetes

No.	Name of isolates	NaCl (w/v)	pH	Site of isolation	References
1.	*Actinopolyspora alba*	10–25 %	7.0–8.0	Salt field in Xinjiang province, North-West China	Tang et al. (2011b)
2.	*Actinopolyspora saharensis* sp. nov.	15–25 %	6.0–7.0	Saharan soil sample collected in El-Oued province, south Algeria	Meklat et al. (2013)
3.	Actinopolyspora sp.	8–15 %	7.9	Marine sediment. Alibag coast, Maharashtra	Kokare et al. (2004)
4.	*Amycolatopsis halophila* sp. nov.	1–15 %	6.0–8.0	Salt lake in Xinjiang Province, north-west China	Tang et al. (2010)
5.	*Blastococcus jejuensis* sp. nov.	0–1 %	6.1–10.1	Sand sediment of a beach in Jeju, Korea	Lee (2006)
6.	*Bogoriella caseilytica* gen. nov., sp. nov.	2–8 %	9.0–10.0	Soda soil (Lake Bogoria, Kenya)	Groth et al. (1997)
7.	*Haloactinopolyspora alba* gen. nov., sp. nov.	10–15 %	7.0–8.0	Salt lake in Xinjiang province, North-West China	Tang et al. (2011a)
8.	*Ilumatobacter fluminis* gen. nov., sp. nov.	0–5 %	7.0–11.0	Sediment sample collected at the mouth of the Kuiragawa River, Iriomote, Okinawa Prefecture, Japan	Matsumoto et al. (2009)
9.	*Kocuria gwangalliensis* sp.	0–7 %	8.0	Gwangalli coast of Korea	Yong et al. (2009)
10.	*Nocardiopsis alba* OK-5	0–15 %	8.0–11.0	Sandy soil from Okha, Gujarat, India	Gohel and Singh (2012b)
11.	*Nesterenkonia suensis* sp. nov.	2.5 %	9.0	Brine from Sua salt pan in Botswana	Govender et al. (2013)
12.	*Nocardiopsis arabia* sp. nov.	0–15 %	7.2	Sand-dune soil collected from Borg Al-Arab, Egypt	Hozzein and Goodfellow (2008)
13.	*Nocardiopsis arvandica* sp. nov.	0–15 %	6.0–11.0	Sandy soil from the banks of the Arvand River, Khoramshahr, Iran	Hamedi et al. (2011)
14.	*Nocardiopsis fildesensis* sp. nov.	0–12 %	9.0–11.0	Soil sample collected from the fields of Peninsula, King George Island, West Antarctica	Xu et al. (2014)
15.	*Nocardiopsis kunsanensis* sp. nov.	10 %	9.0	Saltern in Kunsan, Republic of Korea,	Chun et al. (2000)
16.	*Nocardiopsis litoralis* sp. nov.	1–15 %	6.0–10.5	Sea anemone collected from a tidal flat in the South China Sea	Chen et al. (2009)
17.	*Nocardiopsis quinghaiensis* sp. nov.	0–10 %	6.0–8.0	Saline soil collected from the Qaidam Basin in Qinghai Province, North-West China	Chen et al. (2008)

(continued)

Table 2.1 (continued)

No.	Name of isolates	NaCl (w/v)	pH	Site of isolation	References
18.	*Nocardiopsis sinuspersici* sp. nov.	0–15 %	5.0–12.0	Sandy rhizospheric soil in Sarbandar, Persian Gulf, Iran	Hamedi et al. (2010)
19.	*Nocardiopsis* sp. *VITSVK 5*	2 %	7.0–8.0	Marine sediment samples collected at the Puducherry coast of India	Vimal et al. (2009)
20.	*Nocardiopsis terrae* sp. nov.	1–15 %	6.0–10.5	Saline soil collected from the Qaidam Basin, North-west China	Chen (2010)
21.	*Nocardiopsis valliformis* sp. nov.	1–5 %	9.5–13.0	Alkali lake soil in China	Yang et al. (2008)
22.	*Brachystreptospora Xinjiangensis*	0–10 %	8.0–10.0	Powdery soft soil, from Okha Madhi, Gujarat, India	Gohel and Singh (2012b)
23.	*Prauserella halophila And Prauserella alba*	10–15 %	7.0	Xinjiang Province in the west of China,	Li et al. (2003)
		10 %			
24.	*Prauserella marina* sp. nov.	0–10 %	6.0–9.0	Ocean sediment sample collected from the South China Sea	Wang et al. (2010)
25.	*Prauserella salsuginis*	5–15 %	6.0–9.0	Salt lake in Xinjiang province, North-West China	Li et al. (2009)
	Prauserella flava	5–15 %			
	Prauserella aidingensis	5–15 %			
	Prauserella sediminis	5–20 %			
26.	*Sciscionella marina* sp. nov.	0–13 %	6.0–8.0	Marine sand sediment at a depth form Northern South China Sea,	Tian et al. (2009)
27.	*Streptomyces clavuligerous*	0–15 %	7.0–10.0	Saline soil of Mithapur, Gujarat, India	Vasavada et al. (2006)
28.	*Streptomyces deccanensis* sp. nov.	7 %	7.0–12.0	Gulbarga, Karnataka Province, India	Dastager et al. (2008)
29.	*Streptomyces oceani* sp. nov.	2.5–12.5 %	6.0–8.0	Deep-sea sample of seep authigenic carbonate nodule in South China Sea	Tian et al. (2012)
30.	*Verrucosispora maris* sp. nov.	2.5 %	5.0–10.0	Sediment sample collected from the Sea of Japan	Goodfellow et al. (2012)
31.	*Verrucosispora sediminis* sp. nov.	0–6 %	5.0–7.5	Deep-sea sediment sample of the South China Sea	Dai et al. (2010)
32.	*Zhihengliuella salsuginis* JSM 071043T	0.5–20 %	6.5–10.5	Subterranean brine sample collected from a salt mine in Hunan Province, China	Chen et al. (2010a, b)

2.3.1 Osmotic Adaptation

The occurrence of actinomycetes in saline environments and their tolerance to high salt concentrations were first described more than 40 years ago (Tresner et al. 1968; Gottlieb 1973; Hamedi and Mohammadipanah 2013). Relatively few organisms are adapted to the salinity and the upper salinity limit of the dissimilatory process is correlated with the amount of energy generated and the energetic cost of the osmotic adaptation. There are two main approaches adapted by the microorganisms to cope with the osmotic stress of high salt concentrations. First, bacterial cells may maintain high intracellular salt concentrations, as a minimum osmotic condition to the external concentrations (the "salt-in" strategy). For this to happen all intracellular systems should be adapted to high salt concentrations. Alternatively, as a second option, the cells maintain low salt concentrations within their cytoplasm (the "compatible-solute" strategy). The osmotic pressure of the medium is then balanced by organic compatible solutes without necessitating any special adaptation of the intracellular systems (Oren 2002). Most halophilic and all halo-tolerant organisms expend energy to eliminate salt from their cytoplasm to avoid protein aggregation ('salting out'). The osmoadaptation takes place by the accumulation of K^+ ions and one or more of a restricted range of low molecular mass organic solutes (osmoprotectants), called compatible solutes (Welsh 2000). They can be synthesized within the cell or accumulated from the environment to high intracellular concentrations, in order to balance the osmotic pressure of the growth medium and maintain cell turgor pressure (Santos and Costa 2002). Most common compatible solutes are neutral or zwitterionic and include amino acids, sugars, polyols, betaines and ectoines and byproducts of some of these compounds.

2.3.2 Compatible Solutes

Compatible solutes act as stress protectants and can stabilize biomacromolecules against harsh environments such as high temperature, desiccation and freezing (Roberts 2005; Nieto and Vargas 2002). Usually, amino acids and polyols, e.g., glycine betaine, ectoine, sucrose, trehalose and glycerol, accumulate in halophiles as compatible solutes. Some of these molecules especially glycine, betaine and ectoines have considerable biotechnological importance (Ventosa et al. 1998).

Halophilic actinomycetes, however, are only scarcely examined for the presence and distribution of organic solutes. The compatible solutes or their biosynthetic genes of halophilic actinomycetes appear to be an important source of stress protectants. Improved osmotic tolerance was observed when the genes encoding the pathway of the compatible solutes of *A. halophila* were cloned and expressed in *E. coli* (Nyyssola and Leisola 2001). Actinomycetes from saline soils include *Actinopolyspora halophila* which is one of the few heterotrophic bacteria that can de novo synthesize the compatible solutes, glycine and betaine, whereas *Nocardiopsis halophila* uses a hydroxy derivative of ectoine and β-glutamate as compatible solutes (DasSarma and Arora 2001). *Streptomyces coelicolor* A3 (2) synthesizes ectoine and 5-hydroxyectoine when subjected to either salt (0.5 M NaCl) or heat stress (39 °C). The cells maintained highest cellular levels of these compatible solutes when both stress conditions were simultaneously imposed. The gene cluster (*ect-ABCD*) encoding the enzymes for ectoine and 5-hydroxyectoine biosynthesis was identified in the genomes of *S. coelicolor* A3(2), *S. avermitilis*, *S. griseus*, *S. scabiei*, and *S. chrysomallus*, suggesting an important role of these compatible solutes as stress protectants in the genus *Streptomyces* (Bursy et al. 2008). Further, a thermophilic and halo-tolerant *Rubrobacter xylanophilus* of the Phylum Actinobacteria accumulates trehalose and mannosylglycerate (MG) (Empadinhas et al. 2007).

Halophilic and halotolerant microorganisms withstand large osmotic pressure exerted by their highly saline medium. As biological membranes are permeable to water, all microorganisms have to keep their cytoplasm at least iso-osmotic with their environment. When a turgor pressure is to be maintained, the cytoplasm should even be slightly hyperosmotic. In all cases examined, sodium ions are excluded from the cytoplasm as much as possible. However, the nature of the damage caused by the sodium ions is still unknown. Majority of the halophilic microorganisms contain potent transport mechanisms, generally based on Na^+/H^+ antiporters, to expel sodium ions from the interior of the cell (Oren 1999).

Another strategy to accumulate K^+ and Cl^- ions to maintain osmotic balance is used by a limited number of halophiles. The aerobic halophilic archaea of the order Halobacteriales accumulate KCl at concentrations at least as high as the NaCl concentration in their surrounding medium. Extremely halophilic bacteria are represented by 11 genera by Oren (2002). Among the actinomycetes, *Actinopolyspora halophile*, able to grow in saturated NaCl, was first isolated as a contaminant of culture medium containing 25 % NaCl (Johnson et al. 1986). The halotolerant actinomycetes have extensive adaptability to Na^+, K^+ and Mg^{2+}, while a few strains can grow even in low concentrations of $CaCl_2$ (Tang et al. 2003). The haloalkaliphilic actinomycetes have evolved several structural and chemical adaptations, which allow them to survive and grow in extreme environments. The enzymes of these microbes, which function in extreme environments, have several biotechnological applications. Some halophilic actinomycetes and their enzymes are used in waste transformation and degradation (e.g. hypersaline waste brines contaminated with a wide range of organics) (Fathepure 2014). Some compatible solutes produced by halophilic actinomycetes are used as protein and cell protectants in a variety of industrial applications, where freezing and thawing are involved.

2.4 Antimicrobial Activities of Haloalkaliphilic Actinomycetes

Actinobacteria are well known to produce novel bioactive metabolites (Lam 2006). Further, a great metabolic diversity and biotechnological potential has been identified in haloalkaliphilic actinobacteria. Therefore, actinobacteria are valuable prokaryotes to explore antimicrobial activities and other biotechnological avenues.

Although, actinomycetes from various aquatic and terrestrial habitats have been reported for novel bioactive compounds (Williams et al. 2005; Zhang et al. 2005; Pathom-aree et al. 2006; Parungawo et al. 2007; Glen et al. 2008; Sibanda et al. 2010), only few reports on the diversity and activity of the actinomycetes from saline and alkaline habitats exist (Mitsuiki et al. 2002; Tsujibo et al. 2003; Mehta et al. 2006; Thumar and Singh 2007a, b; Thumar and Singh 2009). Therefore, exploration of extreme actinomycetes with novel and unique antimicrobial properties is a fairly recent area of research.

Actinobacteria are biotechnologically valuable and the most economically prokaryotes able to produce novel bioactive compounds, such as antibiotics, antitumor agents, immunosuppressive agents and enzymes. Several biotechnology companies and academic institutions are working on the strategies for the pharmaceutical applications of new compounds produced by extremophilic actinomycetes (Adinarayana et al. 2006; Lombo et al. 2006; Murphy et al. 2010; Mahyudin et al. 2012). It is established that among 23,000 antibiotics discovered from microorganisms, approximately 10,000 of them have been isolated from actinobacteria. Each strain of actinobacteria is likely to have the genetic potential for the production of 10–20 secondary metabolites (Bentley et al. 2002; Lam 2006). These metabolites have antibacterial, antifungal, anticancer, antialgal, antimalarial and anti-inflammatory activities. Many of these secondary metabolites are potent antibiotics, and in this context, *Streptomycetes* have emerged as the primary antibiotic-producing organisms exploited by the pharmaceutical industries.

2.4.1 Antibacterial Activity

The antibacterial substances either inhibit the growth or kill bacteria. Frequency of resistance in microbial pathogens continues to grow at an alarming rate throughout the world. Therefore, the development of effective newer drugs without any side effects has always been an urgent need. Significant antimicrobial activity against gram positive bacteria as compared to the gram negative bacteria is reported in actinomycetes (Kokare et al. 2004). *Streptomyces* species have significant antibacterial activity against *Staphylococcus aureus* and *Pseudomonas aeruginosa. Micromonospora globosa* produces antibacterial substances only in the presence of seawater (Imada et al. 2007). Similarly, a halophilic and alkaliphilic *Streptomyces sannanensis* strain RJT-1 isolated from an alkaline soil was reported as potent antibiotic producer against gram positive bacteria (Vasavada et al. 2006). *Streptomyces* species have significant antimicrobial activity against *Staphylococcus aureus*, *Pseudomonas aeruginosa* and *Vibrio cholera* (Devi et al. 2006). Further, *Bacillus subtilis* strain was reported as highly sensitive against the antimicrobial agents produced by the salt-tolerant alkaliphilic actinomycetes (Thumar et al. 2010). Majority of the antibiotic compounds exhibit exclusive activities against gram-positive bacteria and only fraction of the compounds act against Gram-negative bacteria (Berdy 2005). A large number of strains of actinomycetes have sensitivity against *Candida albicans* (Sundaram et al. 2010). It's quite apparent that actinomycetes from the saline and alkaline habitats have novel antimicrobial compounds, signifying the role of the biologically competitive environment with unique conditions of pH, temperature, pressure, oxygen, light, nutrients and salinity in antibiotic production. Studies have revealed that halophilic actinomycetes from saline habitats are potent sources of the bioactive compounds (Kokare et al. 2004; Li et al. 2005; Manam et al. 2005; Vasavada et al. 2006; Sarkar et al. 2008; Suthindhiran and Kannabiran 2009; Thumar et al. 2010). A halophilic *Saccharopolyspora salina* VITSDK4 produced extracellular compound that inhibited the growth of tumor cells and microbial cells (Suthindhiran and Kannabiran 2009), while a halotolerant alkaliphilic, *Streptomyces aburaviensis* Kut-8 reported to produce effective antibiotics (Thumar et al. 2010).

Streptomyces sp. secreting a broad spectrum antibiotic against some pathogenic bacteria, fungi and *Candida* sp. has emerged as the third most usually present isolates in hemocultures in developed countries. Continuous use of antifungal drugs may also produce resistant strains of *C. albicans* and many infections due to *Candida* species are refractory to antifungal therapy (Cowen 2008; Sangamwar et al. 2008). Although a long list exists on the currently available antibiotics, there are only limited reports on anticandidal activities (Moosa et al. 2004; Khalesi and Bonjarghs 2006; Susithra et al. 2009). Consequently; newer range of the effective antimycotic agents are required, particularly in view of the opportunistic capabilities of candida and yeast in patients suffering from terminal diseases.

Nesterenkonia alba sp. nov., an alkaliphilic actinobacterium, optimally grows at pH 9–10 (Luo et al. 2009). A novel alkaliphilic *Streptomyces* strain secretes pyrocoll, an antimicrobial compound, under alkaline conditions (Dietera et al. 2003). Many *Streptomyces* species are recorded to secrete antibiotics against bacteria, fungi and yeasts at higher salinity and alkaline pH (Basilio et al. 2003).

2.4.2 Antifungal, Antiviral and Anticancer Compounds

Extremophilic actinobacteria have potential as a useful biological tool for the production of antifungal, antiviral and anticancer substances against fungi (Rauert et al. 2001; Phoebe et al. 2001; Wu et al. 2009). While large number of antibiotics has been obtained from different microorganisms, the search for novel antibiotics effective against pathogenic fungi is still going on. Several biotechnology companies, such as Diversa, Cubist and Proteus as well as academic institutions are currently working on strategies for searching new compounds from the marine and other extremophilic microorganisms including actinomycetes. Antifungal and anticancer compounds have been detected in *Streptomyces* sp. (Sivakumar et al. 2007). Similarly, N – (2-hydroxyphenyl) – 2 – phenazinamine (NHP), an anticancer and antifungus compound from *Nocardia dassonvillei* is reported (Gao et al. 2012). Further, antiviral activity of halo-tolerant actinomycetes is recorded against tobacco mosaic tobamovirus and potato Y potyvirus (Sonya and Galal 2005). A new cytotoxic compound, 1-Hydroxy-1- norresistomycin from a marine actinomycete,

Streptomyces chibaensis AUBN1/7 has been reported (Gorajana et al. 2005; Kock et al. 2005). Similarly, Daryamides AC, a weakly cytotoxic polyketides from *Streptomyces* strain CNQ-085 (Asolkar et al. 2006) and Caboxamycin of the benzoxazole family from *Streptomyces* sp. NTK 937 (Hohmann et al. 2009) are known.

2.5 Actinobacterial Extremozymes

Actinobacteria display a diverse range of enzymatic activities (Zhang and Kim 2010). Different commercially viable enzymes, such as α-amylase (Stamford et al. 2001; Kundu et al. 2006), proteases (Dixit and Pant 2000; Thumar and Singh 2007a, b, 2009; Singh et al. 2010; Gohel and Singh 2012a), cellulases (Techapun et al. 2003; Murugan et al. 2007), chitinase (Miyashita et al. 1991), keratinase (Mabrouk 2008), xylanase (Bode and Huber 2005), L-glutaminase (Sivakumar et al. 2006) and α-galactosidase (Anisha and Prema 2006) are known from the actinobacteria. Studies on the alkaliphiles and haloalkaliphiles have led to the discovery of variety of enzymes that exhibit interesting properties (Horikoshi 1999).

2.5.1 Stability of the Haloalkalophilic Enzymes

Enzymes are derived from plants, animals, and microorganisms; however, a major fraction of the commercially available enzymes are of microbial origin. The ease of growth, nutritional requirements, and easy downstream processing of the microbial systems have made them most favorable sources of the enzymes. The halophilic microorganisms, a group of the extremophilic organisms, optimally growing at high salt concentrations are explored from marine salterns, hypersaline lakes and Dead Sea (Oren 2008). The halophilic microorganisms, as discussed in previous sections of this chapter, have their cytoplasm at least isoosmotic with their surrounding medium. At high salt concentrations, proteins are destabilized due to enhanced hydrophobic interactions (Hippel and Schleich 1969).

The activity and stability of the enzymes depend not only on the concentrations but also on the nature of the salt. For some halophilic enzymes, activity in KCl is significantly higher than in NaCl (Madern et al. 2000). The activity of a malate dehydrogenase from *Haloarcula marismortui* activity decreases with increasing salt concentration (Mevarech et al. 2000). Interestingly, temperature optima of a protease from salt tolerant alkaliphilic actinomycete shifts from 60 to 80 °C in the presence of 4M NaCl and 30 % Na-glutamate with above twofold enhancement in the activity (Gohel and Singh 2012a, 2013). The higher stability in 30 % Na-glutamate and 4M NaCl may attribute to decrease in unfavorable electrostatic repulsion. Besides, halophilic enzymes have highly negative surface charge with hydrated carboxyl groups shifted by high salt content, which protects against unfolding of the proteins (Joo and Kim 2005; Akolkar 2009). Most of the halophilic enzymes are inactivated when NaCl or KCl concentrations decrease to less than 2M (Madern et al. 2000).

2.5.2 Extracellular Proteases

The demand for novel proteases and their formulations for a wide range of industries, such as food, pharmaceutical, leather, detergents, animal feed, and breweries is evident from several decades. Alkaline proteases are among the most widely and largest studied groups of hydrolytic enzymes of the microbial origin (Gupta et al. 2002; Rahman et al. 2003; Patel et al. 2006; Mukherjee et al. 2008; Joshi et al. 2008; Singh et al. 2012). Most of the proteases from *Streptomyces* sp. are alkali-tolerant. However, some proteases are also salt tolerant and produced by genera other than the genus *Streptomyces* and *Nocardiopsis* (Horikoshi 1999; Gohel and Singh 2012a). Proteases from *Nocardiopsis* sp. are used in detergent additives (Moreira et al. 2002) and applications of proteases from *Streptomyces* sp. in dehairing of goat skin is economical and environmentally safe (Mitra and Chakrabartty 2005). Alkaline proteases are optimally active under alkaline pH and contain a serine residue at their active site (Gupta et al. 2002). Alkaline proteases are capable to function under alkaline pH, high salt, high temperature and in the presence of organic solvents and other inhibitory compounds (Johnvesly and Naik 2001; Gupta et al. 2002; Thumar and Singh 2009; Gohel and Singh 2012a). In this context, protein engineering will offer possibilities of evolving proteases with new combinations of the features. Hence, although microbial alkaline proteases produced from bacteria and actinobacteria play an important role in several industries such as: washing powders, tannery, food-industry, leather processing, pharmaceuticals and peptide synthesis, their potential is much greater and yet to be explored (Pandey et al. 1999; Kumar and Takagi 1999).

2.5.3 Extracellular Amylases

Amylases are another important group of enzymes employed in starch processing industry for the conversion of starch to high fructose syrups (Ammar et al. 2002; Kikani et al. 2010). The occurrence of amylases is commonly observed in actinomycetes, particularly *Nocardia* and *Streptomyces* (Vigal et al. 1991). A *Streptomyces* strain produces moderately halophilic, detergent and surfactant stable α-amylase (Chakraborty et al. 2009). A surfactant, detergent and oxidant stable α-amylase from a marine haloalkaliphilic *Saccharopolyspora* sp. A9 with the optimum activity at pH 11.0 (range 8.0–12.0) and stability in 11–17 % salt has been reported (Chakraborty et al. 2011). On the other hand, Zhang and Zeng (2008) studied a cold-adapted α-Amylase of *Nocardiopsis* sp. 7326, isolated from Prydz Bay, Antarctic. The enzyme exhibited sigmoidal growth even at 0 °C with the stability at pH 5–10 after 24 h at 4 °C. The optimal pH for enzymatic activity was around 8.0. Thermophilic and acidophilic amylases which can find applications in bakery, brewing, and alcohol industries have been studied and reported from *Streptomyces erumpens* (Kar and Ray 2008). Looking into the commercial significance of this enzyme, the focus has been on the discovery and engineering of new enzymes that are more robust with respect to their pH and temperature kinetics. Thermostable amylases with important applications in bakery and paper industries are reported from *Nocardiopsis* sp. (Stamford et al. 2001). The wide applications of these enzymes have been established in liquefaction of starch, paper, food, sugar and pharmaceutical industries. In food industry, amylolytic enzymes play important roles, such as the production of syrups (glucose, maltose and high fructose corn), reduction of viscosity of the sugar syrups, reduction of turbidity of the clarified fruit juice for longer shelf-life, and saccharification of starch in the brewing industry (Pandey et al. 2000).

2.5.4 Extracellular Cellulases

Cellulases have been investigated mainly with respect to their industrial applications for the conversion of agricultural biomass into useful products. Recently, cellulases have been successfully used as additives for laundry detergents. According to the modern concepts, most of the cellulolytic enzymes comprise modular multi-domain proteins containing at least three separate structural elements of different functions, i.e. catalytic domain (CD), cellulose binding domain (CBD), and interdomain linker. Cellulases convert cellulose to fermentable sugars suitable for human consumption and the largest known producers are from the genus *Streptomyces* (Jang and Chang 2005). The cellulose hydrolyzing thermophilic enzymes are used for the production of bio-ethanol and value-added organic compounds from renewable agricultural residues (Hardiman et al. 2010). Bui (2014) recently examined and isolated cellulolytic bacteria, including actinomycetes; from the coffee exocarps in the coffee-producing areas in Vietnam. The reported isolates belong to the Genus *Streptomyces*, *Clostridium* and *Bacillus* which exhibit optimum activity at pH 7.0. Besides *Streptomyces*, several other genera, such as *Thermobifida* and *Micromonospora* produce recombinant cellulases that can be commercially exploited (Zhang et al. 2005). George et al. (2001) reported a carboxymethyl cellulase produced by a novel alkalothermophilic actinomycetes having optimum growth at pH 9 and 50 °C. The enzyme has an application potential in detergents to clean, soften, and restore the color of the fabrics. The applications of the enzyme in the treatment of textiles, processing of paper and pulp, and as an animal feed additive have been described (Jones et al. 2004). A recombinant cellulase with thermal and pH stability is reported from *Streptomyces thermoviolaceus*. This enzyme retains its activity in the presence of commercial detergents highlighting its superiority to the existing commercial cellulases (Jones et al. 2003). These enzymes not only hold a biotechnological promise but can be economical due to their low cost of production. The cloning and expression of an endo-cellulase gene from a novel alkaliphilic *Streptomycete* isolated from an East African Soda Lake adds to the molecular aspects of the cellulases from the marine actinomycets (Van et al. 2001).

2.5.5 Extracellular Chitinases

Chitin is an abundant renewable natural resource obtained from marine invertebrates, insects, fungi and algae. Chitin is an insoluble linear β-1,4-linked polymer of N-acetylglucosamine (GlcNAc), being the second most abundant polymer in nature. Chitinases are glycosyl hydrolases a class of hydrolases that catalyze the degradation of chitin. This enzyme has gained tremendous importance in the past two decades. This chitin is present in the cell walls of fungi and in the exoskeletons of insects and shells of crustaceans (Bhattacharya et al. 2007). Chitinases occur in several actinomycetes and possess unique properties of thermostability and catalysis in wide pH range (Nawani et al. 2002; Bhattacharya et al. 2007). Chitin oligosaccharides (COS) have anticoagulant, antimicrobial, anticholesteremic, anticancer, wound-healing, antitumor, and antioxidant activities which make them prospective candidates for biomedical applications. A chitinase of *Streptomyces* sp. M-20 was purified and characterized (Kim et al. 2003). An extracellular chitinase with the apparent molecular weight of 55 kDa was characterized from *Streptomyces halstedii* AJ-7 (Joo and Chang 2005). Chitinase from *Microbispora* sp. was used to recover chitobiose, a potential antioxidant used as a food additive and in other biomedical applications. A thermostable chitinase (Chi 40) from *Streptomyces thermoviolaceus* OPC-520 was overexpressed, purified and characterized. The understanding of biochemistry of chitinolytic enzymes will make them more useful in a variety of processes in near future (Patil et al. 2000). The success in using chitinases for different applications depends on the supply of highly active preparation at a reasonable cost. Most of the suppliers use either natural microbial biodiversity or genetically engineered chitinase overproducing microbial strains to obtain efficient preparations. Chitinases are also important in biocontrol of plant pathogens, and developing transgenic plants.

2.5.6 Extracellular Xylanases

Hydrolysis of xylan is undoubtedly an important step toward the utilization of abundantly available lignocellulosic material in nature. Xylan the second most renewable biopolymer is the major component of the hemicelluloses in angiosperm cell walls (Priya et al. 2012). While there are many studies on the xylanases from bacteria and fungi, the actinomycetes are least explored for the xylanases (Ninawe and Kuhad 2005; Ninawe et al. 2006). Genetic diversity of the alkaliphilic actinomycetes of the genera *Streptomyces*, potent producers of alkaline xylanase, has been studied (Ninawe et al. 2006). *Streptomyces* sp. produces high levels of xylanase utilizing untreated rice straw (Rifaat et al. 2006). Similarly, *Streptomyces* sp. hydrolyzes various agricultural residues, such as oil cake and straw waste leading to enhanced biogas production (Priya et al. 2012). A xylanase (XynK) from an alkaliphilic actinomycete (*Kocuria* sp. 3-3) was overexpressed in *Escherichia coli*

(Wang et al. 2014). The enzyme exhibited the optimum catalytic temperature of 55 °C and the cellulose-free xylanase exhibits high activity and stability at pH 7.0–11.0. Similarly, fused xylanases from fungi and actinomycetes have been employed in paper and pulp industries, due to high thermal and pH stability (Fagerstrom et al. 2008). Currently, the focus is on the discovery of enzymes that are effectively stable and active under the extreme conditions with respect to broader range of temperatures, alkaline conditions and high salt concentrations.

2.5.7 Lipases

Lipases (triacylglycerol acylhydrolases E.C.3.1.1.3) catalyze esterification, trans-esterification and aminolysis, and have considerable physiological significance and industrial potential (Babu et al. 2008). Lipases have emerged as one of the leading biocatalysts with proven potential for contributing to the multibillion dollar under-exploited lipid industry. They have been used in-situ lipid metabolism and ex-situ multifaceted industrial application(s) (Sharma and Kanwar 2012). Microbial lipases have been studied in a wide range of species and have been the subject of many important studies in the past few years as they catalyze large number of reactions. The bacterial lipases display highest versatility, reactivity and stability in the organic solvents (Haba et al. 2000; Gao et al. 2000). The lipolytic enzymes are extensively studied from *Pseudomonas* sp. and other oil degrading bacteria (Deb et al. 2006), while the actinomycetes are scarcely attended in this context. However, there are certain studies on the lipases from *Streptomyces* (Zhou et al. 2000; Jain et al. 2003). Lescic and coworkers (2004) structurally characterized a lipase from *Streptomyces rimosus* and confirmed the disulfide bridge pattern using the mass spectroscopy. A lipase gene was over expressed from *Streptomyces rimosus* into a heterologous host, *S. lividans* TK23 to establish the structure-stability relationship (Vujaklija et al. 2003). An endogenous lipase producing actinomycete, *Rhodococcus* UKMP-5M, from Peninsular Malaysia exhibits significant lipase activity at pH 5.0–9.0 (optimum at 5.0) (Nagarajan et al. 2014). Similarly, a cold-adapted, organic solvent stable lipase from *Staphylococcus epidermidis* AT2 has been recently reported (Kamarudin et al. 2014).

Other potent enzymes; lignin peroxidases, ligninase, laccases, lipases, esterases, agarases and tyrosinases with significant industrial applications are also known from actinobacteria (Heumann et al. 2006). Esterases and amidases from *Nocardia* sp. have been used to increase the hydrophilicity of polyethylene terephthalate and polyamide fibers.

On the whole, the recent advances in the discovery of novel antibiotics and stable enzymes from the extremophilic actinomycetes have resulted their applications in organic synthesis and the production of specialty chemicals, pharmaceutical intermediates, and agrochemicals (Demirjian et al. 2001; Sinha and Khare 2014; Siroosi et al. 2014; Schreck and Grunden 2014).

Various enzymes and other value added molecules from the extremophilic actinobacteria described in this chapter have established their prospective applications in bio-industries (Nigam 2013). The acceleration of enzyme discovery from this diverse class of organisms has facilitated the development of new industrial processes. Furthermore, discovery of new antibiotics and bioactive compounds from halophilic and alkaliphilic actinobacteria make them potentially valuable candidates for biotechnology.

2.6 Conclusion

In this chapter, the significance of the haloalkaliphilic actinobacteria for the production of novel antibiotics and other bioactive metabolite is described. These organisms have evolved many structural and chemical adaptations, which allow them to survive and grow under extreme conditions of alkaline pH and salinity. It's revealed that the actinobacteria from the saline and alkaline habitats could provide potential source of novel antibiotics and enzymes. Antibiotics, compatible solutes and other compounds from these microbes have range of applications.

References

Adinarayana G, Venkateshan MR, Bapiraju VV, Sujatha P, Premkumar J, Ellaiah P, Zeeck A (2006) Cytotoxic compounds from the marine actinobacterium. Bioorg Khim 32:328–334

Akolkar AV (2009) Isolation and characterization of Halophilic Archaea: production, characterization and application of extracellular protease from *Halobacterium* sp. SP1. Ph.D. thesis, Maharaja Sayajirao University of Baroda

Al MR, Al DM, Al IS, Patzelt H (2007) Hydrocarbon degradation at high salinity by a novel extremely halophilic actinomycete. Environ Chem 4:5–7

Ammar YB, Matsubara T, Ito K, Iizuka M, Limpaseni T, Pongsawasdi P, Minamiura (2002) Characterization of a thermostable levansucrase from *Bacillus* sp. TH4-2 capable of producing high molecular weight levan at high temperature. J Biotechnol 99:111–119

Amoroso MJ, Castro GR, Carlino FJ, Romero NC, Hill RT, Oliver G (1998) Screening of heavy metal-tolerant actinomycetes isolated from the Sali River. J Gen Appl Microbiol 44:129–132

Anderson AS, Wellington MHE (2001) The taxonomy of streptomyces and related genera. Int J Syst Evol Microbiol 51:797–814

Anisha GS, Prema P (2006) Selection of optical growth medium for the synthesis of a galactosidase from mangrove actinomycetes. Indian J Biotechnol 5:376–379

Asolkar RN, Jensen PR, Kauffman CA, Fenical W (2006) Daryamides AC, weakly cytotoxic polyketides from a marine-derived actinomycete of the genus Streptomyces strain CNQ-085. J Nat Prod 69(12):1756–1759

Baath E, Diaz-Ravina M, Frostegard A, Campbell CD (1998) Effect of metal-rich sludge amendments on the soil microbial community. Appl Environ Microbiol 64:238–245

Babu J, Pramod WR, George T (2008) Cold active microbial lipases: Some hot issues and recent developments. Biotechnol Adv 26:457–470

Badoei-Dalfard A, Karami Z, Ravan H (2014) Purification and characterization of a thermo- and organic solvent-tolerant alkaline protease from *Bacillus* sp. JER02. Prep Biochem Biotechnol. doi:10.1080/10826068.2014.907176

Baek KH, Kim HS (2009) Microbial community structure in hexadecane- and naphthalene-enriched gas station soil. J Microbiol Biotechnol 19:651–657

Baek KH, Yoon BD, Cho DH, Kim BH, Oh HM, Kim HS (2009) Monitoring bacterial population dynamics using real-time PCR during the bioremediation of crude-oil-contaminated soil. J Microbiol Biotechnol 19:339–345

Bakhtiar SJ, Vevodova R, Hatti-Kaul Su XD (2003) Crystallization and preliminary X-ray analysis of an alkaline serine protease from *Nesterenkonia* sp. Acta Crystallogr 59:529–531

Basilio A, Gonzalez I, Vicente MF, Gorrochategui J, Cabello A, Gonzalez A, Genilloud O (2003) Patterns of antimicrobial activities from soil actinomycetes isolated under different conditions of pH and salinity. J Appl Microbiol 95:814

Belanger PA, Beaudin J, Roy S (2011) High-throughput screening of microbial adaptation to environmental stress. J Microbiol Methods 85(2):92–97

Bentley DR et al (2002) Complete genome sequence of the model actinomycete *Streptomyces coelicolor* A3(2). Nature 417:141–147

Berdy J (2005) Bioactive microbial metabolites. A personal view. J Antibiot 58:1–26

Bhattacharya D, Nagpure A, Gupta RK (2007) Bacterial chitinases: properties and potential. Crit Rev Biotechnol 27(1):21–28

Bode W, Huber R (2005) Natural protein proteinase inhibitors and their interaction with proteinases. Eur J Biochem 204(2):433–451

Bui HB (2014) Isolation of cellulolytic bacteria, including actinomycetes, from coffee exocarps in coffee-producing areas in Vietnam. Int J Recycl Org Waste Agric 3:48

Bull AT (2011) Actinobacteria of the extremobiosphere. In: Horikoshi K (ed) Extremophiles handbook. Springer, Japan, pp 1203–1240. doi:10.1007/978-4-431-53898-1-12.1

Bursy J, Kuhlmann UA, Pittelkow M, Hartmann H, Jebbar M, Pierik AJ, Bremer E (2008) Synthesis and uptake of the compatible solutes ectoine and 5-hydroxyectoine by *Streptomyces coelicolor* A3(2) in response to salt and heat stresses. Appl Environ Microbiol 74:7286–7296

Cai M, Tang SK, Chen YG, Li Y, Zhang YQ, Li WJ (2009) *Streptomonospora amylolytica* sp. nov. and *Streptomonospora flavalba* sp. nov., two novel halophilic actinomycetes isolated from a salt lake. Int J Syst Evol Microbiol 59:2471–2475

Carrasco JA, Armario P, Pajuelo E, Burgos A, Caviedes MA, Lopez R, Chamber MA, Palomares AJ (2005) Isolation and characterization of symbiotically effective *Rhizobium* resistant to arsenic and heavy metals after the toxic spill at the Aznalcollar pyrite mine. Soil Biol Biochem 37:1131–1140

Chakraborty S, Khopade A, Kokarea C, Mahadik K, Chopade B (2009) Isolation and characterization of novel α-amylase from marine *Streptomyces* sp. D1. J Mol Catal B Enzym 58:17–23

Chakraborty S, Khopade A, Biao R, Jian W, Yang-Liu X, Mahadik K, Chopade B, Zhang L, Kokare C (2011) Characterization and stability studies on surfactant, detergent and oxidant stable α-amylase from marine haloalkaliphilic *Saccharopolyspora* sp. A9. J Mol Catal B Enzym 68(1):52–58

Chakraborty S, Raut G, Khopade A, Mahadik K, Kokare C (2012) Study on calcium ion independent α- amylase from haloalkaliphilic marine Streptomyces strain A3. Indian J Biotechnol 11:427–437

Chen YG, Cui XL, Kroppenstedt RM, Stackebrandt E, Wen ML, Xu LH, Jiang CL (2008) *Nocardiopsis quinghaiensis* sp. nov., isolated from saline soil in China. Int J Syst Evol Microbiol 58:699–705

Chen YG, Wang YX, Zhang YQ, Tang SK, Liu ZX, Xiao HD, Xu LH, Cui XL, Li WJ (2009) *Nocardiopsis litoralis* sp. nov., a halophilic marine actinomycete isolated from a sea anemone. Int J Syst Evol Microbiol 59(11):2708–2713

Chen YG, Tang SK, Zhang YQ, Liu ZX, Chen QH, He JW, Cui XL, Li WJ (2010a) *Zhihengliuella salsuginis* sp. nov., a moderately halophilic actinobacterium from a subterranean brine. Extremophiles 14(4):397–402

Chen YG, Zhang YQ, Tang SK, Liu ZX, Xu LH, Zhang LX, Li WJ (2010b) *Nocardiopsis terrae* sp. nov., a halophilic actinomycetes isolated from saline soil. Antonie Van Leeuwenhoek 98(1):31–38

Chun J, Bae KS, Moon EY, Jung SO, Lee HK, Kim SJ (2000) *Nocardiopsis kunsanensis* sp. nov., a moderately halophilic actinomycete isolated from a saltern. Int J Syst Evol Microbiol 5:1909–1913

Cowen LE (2008) The evolution of fungal drug resistance: modulating the trajectory from genotype to phenotype. Nat Rev Microbiol 6:187–198

Dai HQ, Wang J, Xin YH, Pei G, Tang SK, Ren B, Ward A, Ruan JS, Li WJ, Zhang LX (2010) *Verrucosispora sediminis* sp. nov., a cyclodipeptide-producing actinomycete from deep-sea sediment. Int J Syst Evol Microbiol 60:1807–1812

Das S, Lyla PS, Ajmal Khan S (2008) Distribution and generic composition of culturable marine actinomycetes from the sediments of Indian continental slope of Bay of Bengal. Chin J Oceanol Limnol 26:166–177

DasSarma S, Arora P (2001) Halophiles. In: Encyclopedia of life sciences. Nature Publishing Group, London, pp 1–9

Dastager SG, Kim CJ, Lee JC, Agasar D, Park DJ, Li WJ (2008) *Streptomyces deccanensis* sp. nov., an alkaliphilic species isolated from soil. Int J Syst Evol Microbiol 58:1089–1093

Deb C, Daniel J, Sirakova TD, Abomoelak B, Dubey VS, Kolattukudy PE (2006) A novel lipase belonging to the hormone-sensitive lipase family induced under starvation to utilize stored triacylglycerol in *Mycobacterium tuberculosis*. J Biol Chem 281(7):3866–3875

Demirjian DC, Morís-Varas F, Cassidy CS (2001) Enzymes from extremophiles. Curr Opin Chem Biol 5:144–151

Devi A, Jeyarani M, Balakrishnan K (2006) Isolation and identification of marine actinomycetes and their potential in antimicrobial activity. Pak J Biol Sci 9(3):470–472

Dietera A, Hamm A, Fiedler HP, Goodfellow M, Muller WE, Brun R, Bringmann G (2003) Pyrocoll, an antibiotic, antiparasitic and antitumor compound produced by a novel alkaliphilic *Streptomyces* strain. J Antibiot 56:639–646

Dixit V, Pant A (2000) Comparative characterization of two serine endopeptidases from *Nocardiopsis* sp. NCIM 5124. Biochim Biophys Acta 1523(2):261–268

Dobzhansky T, Hecht MK, Steere WC (1968) On some fundamental concepts of evolutionary biology. Evol Biol 2:1–34

Ellaiah P, Ramana T, Raju K, Sujatha P, Sankar A (2004) Investigations on marine actinomycetes from bay of Bengal near Kakinada coast of Andhra Pradesh. Asian J Microbiol Biotechnol Environ 6:53–56

Empadinhas N, Mendes V, Simoes C, Santos MS, Mingote A, Lamosa P, Santos H, da Costa MS (2007) Organic solutes in *Rubrobacter xylanophilus*: the first example of di-myo-inositol-phosphate in a thermophile. Extremophiles 11:667–673

Fagerstrom R, Lahtinen T, Lantto R et al (2008) Production and secretion of actinomycete xylanases in a filamentous *Trichoderma* fungus. EP 0876494 B1

Fathepure BZ (2014) Recent studies in microbial degradation of petroleum hydrocarbons in hypersaline environments. Front Microbiol 173(5):1–16

Gao XG, Cao SG, Zhang K (2000) Production, properties and application to non aqueous enzymatic catalysis of lipase from a newly isolated *Pseudomonas* strain. Enzym Microb Technol 27:74–82

Gao X, Lu Y, Xing Y, Ma Y, Lu J, Bao W et al (2012) A novel anticancer and antifungus phenazine derivative from a marine actinomycete BM-17. Microbiol Res 167:616–622

George SP, Ahmad A, Rao MB (2001) Studies on carboxymethyl cellulase produced by alkalothermophilic actinomycetes. Biores Technol 77:171–175

Ghorbel S, Kammoun M, Soltana H, Nasri M, Hmidet N (2014) *Streptomyces flavogriseus* HS1: isolation and characterization of extracellular proteases and their compatibility with laundry detergents. Bio Med Res Int. doi:10.1155/2014/345980

Glen P, Eccleston PR, Brooks D (2008) The occurrence of bioactive *Micromonosporae* in aquatic habitats of the sunshine coast in Australia. Mar Drugs 6:243–261

Gohel SD, Singh SP (2012a) Cloning and expression of alkaline protease genes from two salt-tolerant alkaliphilic actinomycetes in *E. coli*. IJBIOMAC 50:664–671

Gohel SD, Singh SP (2012b) Purification strategies, characteristics and thermodynamic analysis of highly thermostable alkaline protease from a salt tolerant alkaliphilic actinomycete, *Nocardiopsis alba* OK-5. J Chromatogr B 889–890:61–68
Gohel SD, Singh SP (2013) Characteristics and thermodynamics of thermostable protease from salt-tolerant alkaliphilic actinomycetes. Int J Biol Macromol 56:20–22
Gohel SD, Singh SP (2015) Thermodynamics of a Ca^{2+} dependent highly thermostable alkaline protease from a haloalkliphilic actinomycete. Int J Biol Macromol 72:421–429
Goodfellow M, Stach JE, Brown R, Bonda AN, Jones AL, Mexson J, Fiedler HP, Zucchi TD, Bull AT (2012) *Verrucosispora maris* sp. nov., a novel deep-sea actinomycete isolated from a marine sediment which produces abyssomicins. Antonie Van Leeuwenhoek 101:185–193
Gorajana A, Kurada BV, Peela S, Jangam P, Vinjamuri S, Poluri E et al (2005) 1-Hydroxy-1-norresistomycin, a new cytotoxic compound from a marine actinomycete, *Streptomyces chibaensis* AUBN1/7. J Antibiot 58(8):526–529
Gottlieb D (1973) General considerations and implications of the actinomycetes. In: Sykes G, Skinner FA (eds) Actinomycetales: characteristics and practical importance. Academic, London, pp 1–10
Govender L, Naidoo L, Setati ME (2013) *Nesterenkonia suensis* sp. nov., a haloalkaliphilic actinobacterium isolated from a salt pan. *Nesterenkonia suensis* sp. nov., a haloalkaliphilic actinobacterium isolated from a salt pan. Int J Syst Evol Microbiol 63:41–46
Groth I, Schumann P, Rajney FA, Martin K, Schuetze B, Augsten K (1997) *Bogoriella caseilytica* gen. nov., sp. nov., a new alkaliphilic actinomycete from a soda lake in Africa. Int J Syst Bacteriol 47(3):788–794
Gulder TAM, Moore BS (2010) Salinospora midenatural products: potent 20 sproteasome inhibitors as promising cancer chemotherapeutics. Angew Chem Int Ed Engl 49:9346–9367
Gupta R, Beg QK, Lorenz P (2002) Bacterial alkaline proteases: molecular approaches and industrial applications. Appl Microbiol Biotechnol 59:15–32
Haba E, Bresco D, Ferror C, Margues A, Busquets M, Manresa A (2000) Isolation of lipase-secreting bacteria by deploying used frying oil as selective substrate. Enzym Microb Technol 26:40–44
Hamedi J, Mohammadipanah F (2013) Antonio ventosa systematic and biotechnological aspects of halophilic and halotolerant actinomycetes. Extremophiles 17:1–13
Hamedi J, Mohammadipanah F, Von JM, Pötter G, Schumann P, Spröer C, Klenk HP, Kroppenstedt RM (2010) *Nocardiopsis sinuspersici* sp. nov., isolated from sandy rhizospheric soil. Int J Syst Evol Microbiol 60:2346–2352
Hamedi J, Mohammadipanah F, Pötter G, Spröer C, Schumann P, Goker M, Klenk HP (2011) *Nocardiopsis arvandica* sp. nov., isolated from sandy soil. Int J Syst Evol Microbiol 61:1189–1194
Hardiman E, Gibbs M, Reeves R, Bergquist P (2010) Directed evolution of a thermophilic beta-glucosidase for cellulosic bioethanol production. Appl Biochem Biotechnol 161:301–312
Heumann S, Eberl A, Pobeheim H (2006) New model substrates for enzymes hydrolyzing polyethyleneterephthalate and polyamide fibres. J Biochem Biophys Methods 69(1–2):89–99
Hippel PHV, Schleich T (1969) In: Timasheff SN, Fasman GD (eds) Structure and stability of biological macromolecules, vol 2. Marcel Dekker, New York, p 417
Hohmann C, Schneider K, Bruntner C, Irran E, Nicholson G, Bull AT, Jones AL, Brown R, Stach EM, Goodfellow M, Beil W, Kramer M, Imhoff JF, Sussmuth RD, Fiedle HP (2009) Caboxamycin, a new antibiotic of the benzoxazole family produced by the deep-sea strain *Streptomyces* sp. NTK 937. J Antibiot 62:99–104
Horikoshi K (1999) Alkaliphiles: some applications of their products for biotechnology. Microbiol Mol Biol Rev 63(4):735–750
Hozzein WN, Goodfellow M (2008) *Nocardiopsis arabia* sp. nov., a halotolerant actinomycete isolated from a sand-dune soil. Int J Syst Evol Microbiol 58(11):2520–2524

Hozzein WN, Li WJ, Ali MI, Hammouda O, Mousa AS, Xu LH, Jiang CL (2004) *Nocardiopsis alkaliphila* sp. nov., a novel alkaliphilic actinomycete isolated from desert soil in Egypt. Int J Syst Evol Microbiol 54:247–252

Imada C, Koseki N, Kamata M, Kobayashi T, Hamada-Sato N (2007) Isolation and characterization of antibacterial substances produced by marine actinomycetes in the presence of seawater. Actinomycetologica 21:27–31

Jain PK, Jain R, Jain PC (2003) Production of industrially important enzymes by some actinomycetes producing antifungal compounds. Hindustan Antibiot Bull 46(1–4):29–33

Jang HD, Chang KS (2005) Thermostable cellulases from *Streptomyces* sp.: scale-up production in a 50-l fermenter. Biotechnol Lett 27(4):239–242

Johnson KG, Lanthier PH, Gochnauer MB (1986) Studies of two strains of *Actinopolyspora halophila*, an extremely halophilic actinomycete. Arch Microbiol 143(4):370–378

Johnvesly B, Naik GK (2001) Studies on production of thermostable alkaline protease from thermophilic and alkaliphilic *Bacillus* sp. JB-99 in a chemical defined medium. Process Biochem 37:139–144

Jones BE, Wilhelmus AHK, Solingen PV, Weyler W (2003) Cellulase producing actinomycetes, cellulase produced there from and method of producing same. US Patent 6566112

Jones BE, Wilhelmus AHK, Solingen PV, Weyler W (2004) Cellulase producing actinomycetes, cellulase produced there from and method of producing same EP 1408108 B1

Joo HS, Chang CS (2005) Oxidant and SDS stable alkaline protease from a halo-tolerant *Bacillus clausii* I-52: enhanced production and simple purification. J Appl Microbiol 98(2):491–497

Joo WA, Kim CW (2005) Proteomics of Halophilic archaea. J Chromatogr B 815:237–250

Jose PA, Jebakumar SRD (2012) Phylogenetic diversity of actinomycetes cultured from coastal multipond solar saltern in Tuticorin, India. Aquat Biosyst 8:23

Jose PA, Santhi VS, Jebakumar SRD (2011) Phylogenetic affiliation, antimicrobial potential and PKS gene sequence analysis of moderately halophilic Streptomyces sp. inhabiting an Indian saltpan. J Basic Microbiol 51:348–356

Joshi RH, Dodia MS, Singh SP (2008) Optimization of culture parameters for production of commercially valuable alkaline protease from a haloalkaliphilic bacterium isolated from sea water. Biotechnol Bioprocess Eng 13:552–559

Kamarudin NHA, Rahman RNZRA, Ali MSM, Leow TC, Basri M, Salleh AB (2014) A New cold-adapted, organic solvent stable lipase from mesophilic *Staphylococcus epidermidis* AT2. Protein J 33:296–307

Kar S, Ray RC (2008) Statistical optimization of α-amylase production by *Streptomyces erumpens* MTCC 7317 cells in calcium alginate beads using response surface methodology. Pol J Microbiol 57(1):49–57

Kerkar S (2004) Studies on bacteria of the dissimilatory reductive processes of the sulphur cycle from the salt pans of Goa. Ph.D. Thesis, Goa University, Goa

Khalesi E, Bonjarghs AS (2006) Anti yeast activity *Streptomyces olivaceus* strain 115 against *Candida albicans*. J Appl Sci 6:524–526

Khanna M, Solanki R, Lal R (2011) Selective isolation of rare actinomycetes producing novel antimicrobial compounds. Int J Adv Biotechnol Res 2:357–375

Kikani BA, Shukla RJ Singh SP (2010) Biocatalytic potential of thermophilic bacteria and actinomycetes. In: Mendez-Vilas A (ed) Current research, technology and education topics in applied microbiology and microbial biotechnology. Formatex Research Center, Badajoz, pp 1000–1007

Kim KJ, Yang YJ, Kim JG (2003) Purification and characterization of chitinase from *Streptomyces* sp. M-20. J Biochem Mol Biol 36(2):185–189

Kock I, Maskey RP, Biabani MAF, Helmke E, Laatsch H (2005) 1-Hydroxy-1-norresistomycin and resistoflavin methyl ether: new antibiotics from marine-derived streptomycetes. J Antibiot 58(8):530–535

Kokare CR, Kadam SS, Mahadik KR, Chopade BA (2004) Isolation, characterization and antimicrobial activity of marine halophilic *Actinopolyspora* species AH1 from West Coast of India. Curr Sci 86(4):593–597

Kumar R, Kumar S (2014) Bio-prospecting of actinomycetes and its diversity in various environments: a review. Curr Discov 3(1):16–32

Kumar CG, Takagi H (1999) Microbial alkaline proteases: from a bioindustrial viewpoint. Biotechnol Adv 17:561–594

Kundu S, Sahu MK, Sivakumar K, Kannan L (2006) Occurrence of antagonistically active extracellular enzyme producing actinomycetes in the alimentary canal of estuarine fishes. Asian J Microbiol Biotechnol Environ 8:707–710

Lam KS (2006) Discovery of novel metabolites from marine actinomycetes. Curr Opin Microbiol 9:245–251

Lee SD (2006) *Blastococcus jejuensis* sp. nov., an actinomycetes from beach sediment, and emended description of the genus *Blastococcus Ahrens* and *Moll* 1970. Int J Syst Evol Microbiol 56:2391–2396

Lescic I, Zehl M, Muller R, Vukelic B, Abramic M, Pigac J, Allmaier G, Kojic-Prodic B (2004) Structural characterization of extracellular lipase from *Streptomyces rimosus*: assignment of disulfide bridge pattern by mass spectrometry. Biol Chem 385(12):1147–1156

Li ZY, Liu Y (2006) Marine sponge craniella austrialiensis-associated bacterial diversity revelation based on 16S rDNA library and biologically active actinomycetes screening, phylogenetic analysis. Lett Appl Microbiol 43(4):410–416

Li WJ, Xu P, Tang SK, Xu LH, Kroppenstedt RM, Stackebrandt E, Jiang CL (2003) *Prauserella halophila* sp. nov. and *Prauserella alba* sp. nov., moderately halophilic actinomycetes from saline soil. Int J Syst Evol Microbiol 53(5):1545–1549

Li F, Maskey RP, Qin S, Sattler I, Fiebig HH, Maier A, Zeeck A, Laatsch H (2005) Chinikomycins A and B: isolation, structure elucidation, and biological activity of novel antibiotics from a marine *Streptomyces* sp. Isolate M045. J Nat Prod 68:349–353

Li WJ, Kroppenstedt RM, Wang D, Tang SK, Lee JC, Park DJ, Kim CJ, Xu LH, Jiang CL (2006) Five novel species of the genus nocardiopsis isolated from hypersaline soils and emended description of *Nocardiopsis salina* Li et al. 2004. Int J Syst Evol Microbiol 56:1089–1096

Li Y, Tang SK, Chen YG, Wu JY, Zhi XY, Zhang YQ, Li WJ (2009) *Prauserella salsuginis* sp. nov., *Prauserella flava* sp. nov., *Prauserella aidingensis* sp. nov. and *Prauserella sediminis* sp. nov., isolated from a salt lake. Int J Syst Evol Microbiol 59:2923–2928

Lombo F, Velasco A, dela Calle F, Brana AF, Sanchez- Pulles JM, Mendez C, Salas JA (2006) Deciphering the biosynthesis of the antitumor thiocoraline from a marine actinomycete and its expression in two streptomyces species. Chembiochem 7:366–376

Luo HY, Wang YR, Miao LH, Yang PL, Shi PJ, Fang CX, Yao B, Fan YL (2009) *Nesterenkonia alba* sp. nov., an alkaliphilic actinobacterium isolated from the black liquor treatment system of a cotton pulp mill. Int J Syst Evol Microbiol 59:863–868

Mabrouk MEM (2008) Feather degradation by a new keratinolytic *Streptomyces* sp. MS-2. World J Microbiol Biotechnol 24(10):2331–2338

Madern D, Ebel C, Zaccai G (2000) Halophilic adaptation of enzymes. Extremophiles 4:91–98

Mahyudin NA, Blunt JW, Cole ALJ, Munro MHG (2012) The isolation of a new S-methyl benzothioate compound from a marine-derived *Streptomyces* sp. J Biomed Biotechnol, article ID 894708. doi:10.1155/2012/894708

Malviya N, Yadav AK, Solanki MK, Arora DK (2014) Isolation and characterization of novel alkali-halophilic actinomycetes from the Chilika brackish water lake. India Ann Microbiol. doi:10.1007/s13213-014-0831-1

Manam RR, Teisan S, White DJ, Nicholson B, Grodberg J, Neuteboom STC, Lam KS, Mosca DA, Lloyd GK, Potts BCM (2005) Lajollamycin, a nitro-tetraene spiro-b-lactone-c-lactam antibiotic from the marine actinomycete *Streptomyces nodosus*. J Nat Prod 68(2):240–243

Manivasagan P, Venkatesan J, Sivakumar K, Kim S-K (2013) Actinobacterial melanins: current status and perspective for the future. World J Microbiol Biotechnol 29:1737–1750

Matsumoto A, Kasai H, Matsuo Y, Omura S, Shizuri Y, Takahashi Y (2009) *Ilumatobacter fluminis* gen. nov., sp. nov., a novel actinobacterium isolated from the sediment of an estuary. J Gen Appl Microbiol 55(3):201–205

Mehta VJ, Thumar JT, Singh SP (2006) Production of alkaline protease from an alkaliphilic actinomycete. Bioresour Technol 97(14):1650–1654
Meklat A, Bouras N, Zitouni A, Mathieu F, Lebrihi A, Schumann P, Spröer C, Klenk HP, Sabaou N (2013) *Actinopolyspora saharensis* sp. nov., a novel halophilic actinomycete isolated from Saharan soil of Algeria. Antonie Van Leeuwenhoek 103(4):771–776
Mevarech M, Frolow F, Gloss LM (2000) Halophilic enzymes: proteins with a grain of salt. Biophys Chem 86:155–164
Miao V, Davies J (2010) Actinobacteria: the good, the bad, and the ugly. Antonie Van Leeuwenhoek 98:143–150
Mitra P, Chakrabartty PK (2005) An extracellular protease with depilation activity from *Streptomyces nogalator*. J Sci Ind Res 64(12):978–983
Mitsuiki S, Sakai M, Moriyama Y, Goto M, Furukawa K (2002) Purification and some properties of a keratinolytic enzyme from an alkaliphilic *Nocardiopsis* sp. TOA-1. Biosci Biotechnol Biochem 66(1):164–167
Miyashita K, Fujii T, Sawada Y (1991) Molecular cloning and characterization of chitinase genes from *Streptomyces lividans* 66. J Gen Microbiol 137(9):2065–2072
Moosa MY, Sobel JD, Elhalis H, Du W, Akins RA (2004) Fungicidal activity of fluconazole against *Candida albicans* in a synthetic vagina-simulative medium. Antimicrob Agents Chemother 48:161–167
Moreira KA, Albuquerque BF, Teixeira MFS, Porto ALF, Lima FJL (2002) Application of protease from *Nocardiopsis* sp. as a laundry detergent additive. World J Microbiol Biotechnol 18(4):307–312
Moreira AC, Ferreira D, de Almeida FG, Rodrigues-Filho E, Fernandes JB, Silva MFGF, Vieira PC, Pagnocca FC, Souza DHF (2014) Molecular and kinetic characterization of two extracellular xylanases isolated from *Leucoagaricus gongylophorus*. Appl Biochem Biotechnol 173:694–704
Mukherjee AK, Adhikari H, Rai SK (2008) Production of alkaline protease by a thermophilic *Bacillus subtilis* under solid-state fermentation (SSF) condition using imperata cylindrical grass and potato peel as low-cost medium: characterization and application of enzyme in detergent formulation. J Biochem Eng 39:353–361
Murphy BT, Narender T, Kauffman CA, Woolery M, Jensen PR, Fenical W (2010) Saliniquinones AF, new members of the highly cytotoxic anthraquinone-c-pyrones from the marine actinomycete Salinispora arenicola. Aust J Chem 63(6):929–934
Murugan M, Srinivasan M, Sivakumar K, Sahu MK, Kannan L (2007) Characterization of an actinomycete isolated from the estuarine finfish, *Mugil cephalus* Lin. (1758) and its optimization for cellulase production. J Sci Ind Res 6:388–393
Nagarajan J, Nawawi NM, Ibrahim AL (2014) *Rhodococcus* UKMP-5M, an endogenous lipase producing actinomycete from Peninsular Malaysia. Biologia 69(2):123–132
Nawani NN, Kapadnis BP, Das AD, Rao AS, Mahajan SK (2002) Purification and characterization of a thermophilic and acidophilic chitinase from *Microbispora* sp. V2. J Appl Microbiol 93(6):965–975
Nieto JJ, Vargas C (2002) Synthesis of osmoprotectants by moderately halophilic bacteria: genetic and applied aspects. In: Pandalai SG (ed) Recent research developments in microbiology. Research Signpost, Kerala, pp 403–418
Nigam PS (2013) Microbial enzymes with special characteristics for biotechnological applications. Biomolecules 3:597–611
Ninawe S, Kuhad RC (2005) Use of xylan-rich cost effective agro-residues in the production of xylanase by *Streptomyces cyaneus* SN32. J Appl Microbiol 99(5):1141–1148
Ninawe S, Lal R, Kuhad RC (2006) Isolation of three xylanase-producing strains of actinomycetes and their identification using molecular methods. Curr Microbiol 53:178–182
Nyyssola A, Leisola M (2001) *Actinopolyspora halophila* has two separate pathways for betaine synthesis. Arch Microbiol 176(4):294–300
Ogino H, Gemba Y, Yutori Y, Doukyu N, Ishimi K, Ishikawa H (2007) Stabilities and conformational transitions of various proteases in the presence of an organic solvent. Biotechnol Prog 23(1):155–161

Oren A (1999) Bioenergetic aspects of halophilism. Microbiol Mol Biol 63:334–348

Oren A (2002) Diversity of halophilic microorganisms: environments, phylogeny, physiology, and applications. J Ind Microbiol Biotechnol 28:56–63

Oren A (2008) Microbial life at high salt concentrations: phylogenetic and metabolic diversity. Saline Syst 4:2

Pandey S, Singh SP (2012) Organic solvent tolerance of α-amylase from haloalkaliphilic bacteria as a function of pH, temperature and salt concentration. Appl Biochem Biotechnol 166:1747–1757

Pandey A, Selvakumar P, Soccol CR, Nigam P (1999) Solid-state fermentation for the production of industrial enzymes. Curr Sci 77:149–162

Pandey A, Nigam P, Soccol CR, Soccol VT, Singh D, Mohan R (2000) Advances in microbial amylases. Biotechnol Appl Biochem 31:135–152

Panwar D, Srivastava PK, Kapoor M (2014) Production, extraction and characterization of alkaline xylanase from *Bacillus* sp. PKD-9 with potential for poultry feed. Biocatal Agric Biotechnol 3:118–125

Parungawo MM, Maceda EBG, Villano MAF (2007) Screening of antibiotic producing actinomycetes from marine, brackish and terrestrial sediments of Samal Islands, Philippines. J Sci Res 4:29–38

Patel RK, Dodia MS, Singh SP (2006) Purification and characterization of alkaline protease from a newly isolated Haloalkaliphilic *Bacillus* sp. Process Biochem 41(9):2002–2009

Pathom-aree W, Stach JEM et al (2006) Diversity of actinomycetes isolated from challenger deep sediment (10,898 m) from Marina Trench. Extremophiles 10:181–189

Patil RS, Ghormade V, Deshpande MV (2000) Chitinolytic enzymes: an exploration. Enzym Microb Technol 26:473–483

Phoebe CH, Combie J, Albert FG, Tran KV, Cabrera J, Correira HJ, Guo Y, Lindermuth J (2001) Extremophilic organisms as an unexplored source of antifungal compounds. J Antibiot 54:56–65

Prakash D, Nawani N, Prakash M, Bodas M, Mandal A (2013) Actinomycetes: a repertory of green catalysts with a potential revenue resource. BioMed Res Int. doi:10.1155/2013/264020

Priya BS, Stalin T, Selvam K (2012) Efficient utilization of xylanase and lipase producing thermophilic marine actinomycetes (*Streptomyces albus* and *Streptomyces hygroscopicus*) in the production of ecofriendly alternative energy from waste. Afr J Biotechnol 11(78):14320–14325

Procopio RE, Silva IR, Martins MK, Azevedo JL, Araújo JM (2012) Antibiotics produced by *Streptomyces*. Braz J Infect Dis 16(5):466–471

Purohit MK, Singh SP (2014) Cloning, over expression and functional attributes of serine proteases from *Oceanobacillus iheyensis* O.M.$A_1$8 and haloalkaliphilic bacterium O.M.$E_1$2. Process Biochem 49:61–68

Rahman RNZRA, Basri M, Salleh AB (2003) Thermostable alkaline protease from *Bacillus stearothermophilus* F1; nutritional factors affecting protease production. Ann Microbiol 53:199–210

Ramesh S, Mathivanan N (2009) Screening of marine actinomycetes isolated from the Bay of Bengal, India for antimicrobial activity and industrial enzymes. World J Microbiol Biotechnol 25(12):2103–2111

Rauert N, Galbraith W, Selitrennikoff CP (2001) Extremophilic organisms as an unexplored source of antifungal compounds. J Antibiot 54:56–65

Raval VH, Pillai S, Rawal CM, Singh SP (2014) Biochemical and structural characterization of a detergent-stable serine alkaline protease from seawater haloalkaliphilic bacteria. Process Biochem 49:955–962

Remya M, Vijayakumar R (2008) Isolation and characterization of marine antagonistic actinomycetes from west coast of India. Med Biol 15(1):13–19

Rifaat HM, Nagieb ZA, Ahmed YM (2006) Production of xylanases by *Streptomyces* species and their bleaching effect on rice straw pulp. Appl Ecol Environ Res 4(1):151–160

Roberts MF (2005) Organic compatible solutes of halotolerant and halophilic microorganisms. Saline Syst 1:5

Sabry SA, Ghanem NB, Abu-Ella GA, Schumann P, Stackebrandt E, Kroppenstedt RM (2004) *Nocardiopsis aegyptia* sp. nov., isolated from marine sediment. Int J Syst Evol Microbiol 54:453–456

Sangamwar AT, Deshpande UD, Pekamwar SS (2008) Antifungals: need to search for a new molecular target. Indian J Pharm Sci 70:423–430

Santos H, Costa MS (2002) Compatible solutes of organisms that live in hot saline environments. Environ Microbiol 4:501–509

Sarkar S, Saha M, Roy D, Jaisankar P, Das S, Roy GL, Gachhui R, Sen T, Mukherjee J (2008) Enhanced production of antimicrobial compounds by three salt-tolerant actinobacterial strains isolated from the sundarbans in a niche-mimic bioreactor. Mar Biotechnol 10:518–526

Satyanarayana T, Raghukumar C, Shivaji S (2005) Extremophilic microbes: diversity and perspectives. Curr Sci 89(1):78–90

Schippers A, Bosecker K, Willscher S, Spröer C, Schumann P, Kroppenstedt RM (2002) *Nocardiopsis metallicus* sp. nov., a metal-leaching actinomycete isolated from an alkaline slag dump. Int J Syst Evol Microbiol 52:2291–2295

Schloss PD, Handelsman J (2005) Introducing DOTUR, a computer program for defining operational taxonomic units and species richness. Appl Environ Microbiol 71(3):1501–1506

Schloss PD, Handelsman J (2006) Introducing SONS, A tool for operational taxonomic unit-based comparisons of microbial community memberships and structures. Appl Environ Microbiol 72(10):6773–6779

Schreck SD, Grunden AM (2014) Biotechnological applications of halophilic lipases and thioesterases. Appl Microbiol Biotechnol 98:1011–1021

Sharma AK (2011) Assessment of molecular diversity, phylogeny and biocatalytic potential of salt tolerant alkaliphilic actinomycetes using bioinformatics approaches. M. Phil. thesis, Saurashtra University Rajkot

Sharma CK, Kanwar SK (2012) Purification of a novel thermophilic lipase from *B. licheniformis* MTCC-10498. Int Res J Biol Sci 1:43–48

Sharma AK, Gohel S, Singh SP (2012) Actinobase: database on molecular diversity, phylogeny and biocatalytic potential of salt tolerant alkaliphilic actinomycetes. Bioinformation 8(11):535–538

Sibanda T, Mabinya LV, Mazomba N, Akinpelu DA, Bernard K, Olaniran A, Okoh AI (2010) Antibiotic producing potentials of three freshwater actinomycetes isolated from the eastern cape province of south Africa. Int J Mol Sci 11:2612–2623

Singh SP, Thumar JT, Gohel SD, Purohit MK (2010) Molecular diversity and enzymatic potential of salt-tolerant alkaliphilic actinomycetes. In: Mendez-Vilas (ed) Current research, technology and education topics in applied microbiology and microbial biotechnology. Formatex Research Center, Badajoz, pp 280–286

Singh SP, Raval VH, Purohit MK, Pandey S, Thumar JT, Gohel SD, Akbari VG, Rawal CM (2012) Haloalkaliphilc bacteria and actinobacteria from the saline habitats: new opportunities for biocatalysis and bioremediation. In: Satyanarayana T, Johri BN, Prakash A (eds) Microorganisms in environmental management: microbes and environment. Springer (Library of Congress Control Number: 2011944027), pp 415–429. doi:10.1007/978-94-007-2229-3_19

Singh SP, Thumar JT, Gohel S, Kikani BA, Shukla R, Sharma A, Dangar K (2013a) Actinomycetes from marine habitats and their enzymatic potential. In: Trincome A (ed) Marine enzymes for bicatalysis. Woodhead Publishing Series in Biomedicine (Oxford) Ltd, Oxford (Library Congress Number: 2013948122), pp 191–214

Singh SP, Thumar JT, Gohel SD, Kikani B, Shukla R, Sharma A, Dangar K (2013b) Actinomycetes from marine habitats and their enzymatic potential. In: Trincone A (ed) Marine enzymes for biocatalysis: sources, biocatalytic characteristics and bioprocesses of marine enzymes. Woodhead Publishing Series in Biomedicine, Elsevier, pp 191–214

Sinha R, Khare SK (2014) Protective role of salt in catalysis and maintaining structure of halophilic proteins against denaturation. Front Microbiol. doi:10.3389/fmicb.2014.00165

Siroosi M, Amoozegar MA KK, Fazeli M, Rezaei MH (2014) Purification and characterization of a novel extracellular halophilic and organic solvent-tolerant amylopullulanase from the haloarchaeon, *Halorubrum* sp. strain Ha25. Extremophiles 18:25–33

Sivakumar K, Sahu MK, Manivel PR, Kannan L (2006) Studies on L-glutaminase producing actinomycetes strain LG-10 from the estuarine fish, *Chanos chanos* (Forskal, 1775). Indian J Exp Biol 44:256–258

Sivakumar K, Sahu MK, Thangaradjou T, Kannan L (2007) Research on marine actinobacteria in India. Indian J Microbiol 47(3):186–196

Sonya HM, Galal AM (2005) Identification and antiviral activities of some halotolerant streptomycetes isolated from Qaroon Lake. Int J Agric Biol 7(5):747–753

Stach JE, Maldonado LA, Masson DG, Ward AC, Goodfellow M, Bull AT (2003) Statistical approaches for estimating actinobacterial diversity in marine sediments. Appl Environ Microbiol 10:6189–6200

Stackebrandt E (2000) Defining taxonomic ranks. In: Dworking M, Falkow S, Rosenberg E, Schleifer K-H, Stackebrandt E (eds) The prokaryotes: an evolving electronic resource for the microbiological community, 3rd edn. Springer-Verlag, New York

Stamford TLM, Stamford NP, Coelho LCBB, Ara'ujo JM (2001) Production and characterization of a thermostable α-amylase from *Nocardiopsis* sp. endophyte of yam bean. Bioresour Technol 76(2):137–141

Sundaram R, Samikan K, Samuel JI, Murugesan G (2010) Antagonistic activity of marine actinomycetes from Arabian Sea coast. Arch Appl Sci Res 2(6):273–280

Susithra MP, Thenmozhi M, Kannabiran K (2009) Anticandidal activity of *Streptomyces paraguyensis* isolated from marine sediment samples collected at the Puducherry coast, Bay of Bengal, India. Pharmacology 2:527–537

Suthindhiran K, Kannabiran K (2009) Cytotoxic and antimicrobial potential of actinomycete species *Saccharopolyspora salina* VITSDK4 isolated from the Bay of Bengal Coast of India. Am J Infect Dis 5(2):90–98

Syed DG, Agasar D, Pandey A (2009) Production and partial purification of alpha-amylase from a novel isolate *Streptomyces gulbargensis*. J Ind Microbiol Biotechnol 36(2):189–194

Tang SK, Li WJ, Dong W, Zhang YG, Xu LH (2003) Studies of the biological characteristics of some halophilic and halotolerant actinomycetes isolated from saline and alkaline soils. Actinomycetologica 17:06–10

Tang SK, Wang Y, Guan TW, Lee JC, Kim CJ, Li WJ (2010) *Amycolatopsis halophila* sp. nov., a halophilic actinomycetes isolated from a salt lake. Int J Syst Evol Microbiol 60:1073–1078

Tang SK, Wang Y, Klenk HP, Shi R, Lou K, Zhang YJ, Chen C, Ruan JS, Li WJ (2011a) *Actinopolyspora alba* sp. nov. and *Actinop-olyspora erythraea* sp. nov., isolated from a salt field, and reclassification of *Actinopolyspora iraqiensis* Ruan et al. 1994 as a heterotypic synonym of *Saccharomonospora halophile*. Int J Syst Evol Microbiol 61:1693–1698

Tang SK, Zhi XY, Wang Y, Shi R, Lou K, Xu LH, Li WJ (2011b) *Haloactinopolyspora alba* gen. nov., sp. nov., a novel halophilic filamentous actinomycete isolated from a salt lake in China, with proposal of *Jiangellaceae fam.* nov. and *Jiangellineae subord* nov. Int J Syst Evol Microbiol 61:194–200

Techapun C, Poosaran N, Watanabe M, Sasaki K (2003) Thermostable and alkaline-tolerant microbial cellulase free xylanases produced from agricultural wastes and the properties required for use in pulp bleaching bioprocess: a review. Process Biochem 38(1):1327–1340

Thumar JT, Singh SP (2007a) Secretion of an alkaline protease from salt-tolerant and alkaliphilic, *Streptomyces clavuligerus* strain Mit-1. Braz J Microbiol 38:1–9

Thumar JT, Singh SP (2007b) Two – step purification of a highly thermostable alkaline protease from salt-tolerant alkaliphilic *Streptomyces clavuligerus* strain Mit-1. J Chromatogr 854:198–203

Thumar JT, Singh SP (2009) Organic solvent tolerance of an alkaline protease from salt-tolerant alkaliphilic *Streptomyces clavuligerus* strain Mit-1. J Ind Microbiol Biotechnol 36:211–218

Thumar JT, Singh SP (2011) Repression of alkaline protease in salt-tolerant alkaliphilic *Streptomyces clavuligerus* strain Mit-1 under the influence of amino acids in minimal medium. Biotechnol Bioproc Eng 16:1180–1186

Thumar JT, Dhulia K, Singh SP (2010) Isolation and partial purification of an antimicrobial agent from halotolerant alkaliphilic *Streptomyces aburaviensis* strain Kut-8. World J Microbiol Biotechnol 26:2081–2087

Tian XP, Zhi XY, Qiu YQ, Zhang YQ, Tang SK, Xu LH, Zhang S, Li WJ (2009) *Sciscionella marina* gen. nov., sp. nov., a marine actinomycete isolated from a sediment in the northern South China sea. Int J Syst Evol Microbiol 59:222–228

Tian XP, Xu Y, Zhang J, Li J, Chen Z, Kim CJ, Li WJ, Zhang CS, Zhang S (2012) *Streptomyces oceani* sp. nov., a new obligate marine actinomycete isolated from a deep-sea sample of seep authigenic carbonate nodule in South China sea. Antonie Van Leeuwenhoek 102(2):335–343

Tresner HD, Hayes JS, Backus EJ (1968) Differential tolerance of streptomycetes to sodium chloride as a taxonomic tool. Appl Microbiol 16:1134–1136

Tsujibo H, Kubota T, Yamamoto M, Miyamoto K, Inamori Y (2003) Characteristics of chitinase genes from an alkaliphilic actinomycete, *Nocardiopsis prasina* OPC-131. Appl Environ Microbiol 69:894–900

Van SP, Meijer D, Van der Kleij WA, Barnett C, Bolle R, Power SD, Jones BE (2001) Cloning and expression of an endocellulase gene from a novel *Streptomycete* isolated from an East African soda lake. Extremophiles 5(5):333–341

Vasavada S, Thumar J, Singh SP (2006) Secretion of a potent antibiotic by salt-tolerant and alkaliphilic actinomycete *Streptomyces sannanensis* strain RJT-1. Curr Sci 91(10):1393–1397

Ventosa A, Nieto JJ, Oren A (1998) Biology of aerobic moderately halophilic bacteria. Microbiol Mol Biol Rev 62:504–544

Ventosa A, Sanchez-Porro C, Martı'n S, Mellado E (2005) Halophilic archaea and bacteria as a source of extracellular hydrolytic enzymes. In: Gunde-Cimerman A, Oren A, Plemenitas A (eds) Adaptation of life at high salt concentrations in archaea, bacteria and eukarya. Springer, Heidelberg, pp 337–354

Vigal T, Gil JF, Daza A, Garcia-Gonzalez MD, Martin JF (1991) Cloning characterization and expression of an alpha amylase gene from *Streptomyces griseus* IMRU 3570. Mol Gen Genomics 225:278–288

Vijayakumar R, Muthukumar C, Thajuddin N, Panneerselvam A, Saravanamuthu R (2007) Studies on the diversity of actinomycetes in the Palk Strait region of Bay of Bengal, India. Actinomycetologica 21:59–65

Vijayakumar R, Selvam KP, Muthukumar C, Thajuddin N, Panneer selvam A, Saravanamuthu R (2012) Antimicrobial potentiality of a halophilic strain of *Streptomyces* sp. VPTSA18 isolated from the salt pan environment of Vedaranyam, India. Ann Microbiol 62:1039–1047

Vimal V, Rajan BM, Kannabira K (2009) Antimicrobial activity of marine actinomycete, *Nocardiopsis* sp. VITSVK 5 (FJ973467) Asian. J Med Sci 1:57–63

Vujaklija D, Abramic M, Lescic I, Marsic T, Pigac J (2003) *Streptomyces rimosus* GDS(L) lipase: production, heterologous overexpression and structure-stability relationship. Food Technol Biotechnol 41(1):89–93

Wadetwar RN, Patil AT (2013) Isolation and characterization of bioactive actinomycetes from soil in and around Nagpur. Int J Pharm Sci Res 4(4):1428–1433

Wang J, Li Y, Bian J, Tang SK, Ren B, Chen M, Li WJ, Zhang LX (2010) *Prauserella marina* sp. nov., isolated from ocean sediment of the South China sea. Int J Syst Evol Microbiol 60:985–989

Wang W, Wang Z, Cheng B, Zhang J, Li C, Liu X, Yang C (2014) High secretory production of an alkaliphilic actinomycete xylanase and functional roles of some important residues. World J Microbiol Biotechnol. doi:10.1007/s11274-014-1630-3

Welsh DT (2000) Ecological significance of compatible solute accumulation by micro-organisms: from single cells to global climate. FEMS Microbiol Rev 24:263–290

Williams PG, Buchanan GO, Feling RH, Kauffman CA, Jensen PR, Fenical W (2005) New cytotoxic salinosporamides from marine actinomycete *Salinispora tropica*. J Org Chem 70:6196–6203

Wu J, Guan T, Jiang H, Zhi X, Tang S, Dong H, Zhang L, Li W (2009) Diversity of actinobacterial community in saline sediments from Yunnan and Xinjiang, China. Extremophiles 13(4):623–632

Xu S, Yan L, Zhang X, Wang C, Feng G, Li J (2014) *Nocardiopsis fildesensis* sp. nov., an actinomycete isolated from soil. Int J Syst Evol Microbiol 64:174–179

Yang R, Zhang LP, Guo LG, Shi N, Lu Z, Zhang X (2008) *Nocardiopsis valliformis* sp. nov, an alkaliphilic actinomycete isolated from alkali lake soil in China. Int J Syst Evol Microbiol 8:1542–1546
Yong BS, Dong-EK GDK, Hyun WK, Soo WN, Young TK, Jae HL (2009) *Kocuria gwangalliensis* sp. nov., an actinobacterium isolated from seawater. Int J Syst Evol Microbiol 59:2769–2772
Yoshimune K, Wakayama M, Moriguchi M, Nishikawa K (2003) Unique amino acid composition of proteins in halophilic bacteria. J Mol Biol 327(2):347–357
Zafrilla B, Martinez-Espinosa RM, Alonso MA, Bonete MJ (2010) Biodiversity of Archaea and floral of two inland saltern ecosystems in the AltoVinalopó Valley, Spain. Saline Syst 6:10
Zhang C, Kim SK (2010) Research and application of marine microbial enzymes: status and prospects. Mar Drugs 8:1920–1934
Zhang JW, Zeng RY (2008) Purification and characterization of a cold-adapted α-amylase produced by *Nocardiopsis* sp. 7326 isolated from Prydz Bay, Antarctic. Mar Biotechnol 10:75–82
Zhang L, An R, Wang J, Sun N, Zhung S, Hu J (2005) Exploring novel bioactive compounds from marine microbes. Curr Opin Microbiol 8:276–281
Zhou X, Huang J, Ou Z, Wang H, Wang R (2000) Conditions of enzyme production and properties of alkaline lipase by *Streptomyces* Z94-2. Wei Sheng Wu Xue Bao 40:75–79
Zitouni A, Boudjella H, Lamari L, Badji B, Mathieu F, Lebrihi A, Sabaou N (2005) *Nocardiopsis* and *Saccharothrix* genera in Saharan soils in Algeria: isolation, biological activities and partial characterization of antibiotics. Res Microbiol 156:984–993

Chapter 3
Biotechnological Exploitation of Actinobacterial Members

Javad Hamedi, Fatemeh Mohammadipanah, and Hamed Kazemi Shariat Panahi

Abstract Microbial-derived products, while do not have many environmental side effects of their synthetic counterparts, have proved more efficient than those synthetically obtained. The nature has provided a treasure of microorganisms with capabilities to produce vast variety of novel compounds. This ability has been arised during long last of evolution and adaptation to diverse chemical and physical microenviroments. Microbial-derived metabolites have made their own space in industries and therefore human life. Among these microorganisms, halophilic and halotolerant actinobacteria are recently gaining much attentions. Metabolites and biological functions from halophilic or halotolerant members of this phylum of bacteria may resolving the ever-increasing thirst of industry for metabolites with salt-tolerancy to cope a range of issues from environmental pollution to diseases and world's hunger. In the current chapter, it has been tried to introduce the less dealt group of halophilic and halotolerant actinobacteria, and shed light on their potential to be exploited in various industry sectors.

Keywords Actinobacteria • Actinomycetes • Bioremediation • Compatible solutes • Enzymes • Pigments • Secondary metabolites

J. Hamedi (✉) • F. Mohammadipanah • H.K.S. Panahi
Department of Microbial Biotechnology, School of Biology and Center of Excellence in Phylogeny of Living Organisms, College of Science, University of Tehran, Tehran 14155-6455, Iran

Microbial Technology and Products Research Center, University of Tehran, Tehran 14155-6455, Iran
e-mail: jhamedi@ut.ac.ir

D.K. Maheshwari, M. Saraf (eds.), *Halophiles*, Sustainable Development and Biodiversity 6, DOI 10.1007/978-3-319-14595-2_3

3.1 Taxonomy and Phylogeny of Halophile/Halotolerant Actinobacteria

A taxonomic investigation of halophilic and halotolerant actinobacteria clearly demonstrate the relatively heterogeneous nature of energy generation modes and physiological group, associated with different genera. Polyphasic approach including molecular, chemotaxonomic, genotypic and phenotypic investigations has been suggested and/or employed for complete identification of Actinobacteria. Amplified rDNA restriction analysis (ARDRA) or 16S rRNA analyses have been applied for establishment of their nomenclature without cultivation that has increased the knowledge of their diversity. In Table 3.1, halophilic and halotolerant Actinobacterial species, capable of tolerating at least 6 % NaCl (w/v) has been described.

These 272 valid halotolerant or halophilic species from 66 genera, are reported in Table 3.1, belonged to 28 families, including Actinopolysporaceae, Bogoriellaceae, Brevibacteriaceae, Cellulomonadaceae, Corynebacteriaceae, Demequinaceae, Dermabacteraceae, Dermacoccaceae, Dietziaceae, Euzebyaceae, Geodermatophilaceae, Glycomycetaceae, Intrasporangiaceae, Microbacteriaceae, Micrococcaceae, Micromonosporaceae, Nitriliruptoraceae, Nocardiaceae, Nocardioidaceae, Nocardiopsaceae, Promicromonosporaceae, Propionibacteriaceae, Pseudonocardiaceae, Ruaniaceae, Rubrobacteraceae, Streptosporangiaceae, Streptomycetaceae and Thermomonosporaceae and 11 suborders, including Actinopolysporaceae, Micrococcineae, Corynebacterineae, Frankineae, Glycomycineae, Micromonosporineae, Propionibacterineae, Streptosporangineae, Pseudonocardineae, Rubrobacterineae and Streptomycineae, as well as two unclassified families (Euzebyaceae and Nitriliruptoraceae). Therefore, it can be deduced that both halophilic and non-halophilic relatives exist together within phylogenetic clades, however, their tolerances or requirements for salts greatly varied (Oren 2008). The most abundantly described genus with all members reported as halophilic was *Actinopolyspora* (13 species reported) while the largest halophilic and halotolerant genus was *Nocardiopsis* (31 species reported out of which 16 species were halophilic). The most tolerated species were *Actinopolyspora algeriensis* and *Actinopolyspora mzabensis*, could grow in presence of up to 32 % NaCl (w/v), whereas, *Actinopolyspora halophila*, *Actinopolyspora egyptensis*, *Actinopolyspora righensis* and *Actinopolyspora saharensis* showed highest NaCl requirement for optimum growth in 15–20 % NaCl (w/v), followed by *Actinopolyspora mzabensis* which grow optimally at 10–28 % NaCl (w/v). The most flexible species tolerating the widest range for NaCl was *Rubrobacter bracarensis* that could tolerate 3–30 % (w/v) salt concentration.

The commonly believed characteristic of actinobacteria regarding their DNA, containing high load of Guanidine and Cytosine (G+C) subunit bases, and their terrestrial origin are not valid anymore as actinobacteria with low G+C content (Ghai et al. 2012) as well as water harboring actinobacteria have been described. The phylum "Actinobacteria" is well supported by analyses of the 16S and 23S rRNA genes, presence of conserved insertions and deletions in certain proteins, and characteristic gene rearrangements. Class Actinobacteria, including 15 orders, 57 families

Table 3.1 List of 272 valid halophilic and halotolerant actinobacteria species from 26 genera requiring or tolerating at least 6 % NaCl (w/v) up to September 2014

Species name	Type	Optimum grow	NaCl tolerancy	NaCl independency	Reference	Species name	Type	Optimum grow	NaCl tolerance	NaCl independency	Reference
Actinomadura sediminis	–	NR	0–7	Yes	He et al. (2012)	*Nesterenkonia sandarakina*	MH	5	1–15	No	Li et al. (2005c)
Actinopolymorpha alba	HT	NR	0–7	Yes	Cao et al. (2009)	*Nesterenkonia lutea*	MH	5–10	0–20	No	Li et al. (2005c)
Actinopolymorpha singaporensis	HT	NR	0–15	Yes	Wang et al. (2001)	*Nesterenkonia suensis*	HT	2.5	0–18	Yes	Govender et al. (2013)
Actinopolyspora alba	EH	15	10–25	No	Tang et al. (2011a)	*Nesterenkonia halophila*	MH	10	0.5–30	No	Li et al. (2008b)
Actinopolyspora algeriensis	MH	NR	7–32	No	Meklat et al. (2012)	*Nesterenkonia halotolerans*	HT	NR	0–25	Yes	Li et al. (2004b)
Actinopolyspora erythraea	EH	15	20–25	No	Tang et al. (2011a)	*Nesterenkonia xinjiangensis*	HT	NR	0–25	Yes	Li et al. (2004b)
Actinopolyspora halophila	EH	15–20	10–30	No	Gochnauer et al. (1975)	*Nesterenkonia alba*	MH	3	0–6	Yes	Luo et al. (2009)
Actinopolyspora mortivallis	EH	10–15	5–30	No	Yoshida et al. (1991)	*Nesterenkonia aethiopica*	MH	3	3–12	No	Delgado et al. (2006)
Actinopolyspora egyptensis	EH	15–20	5–25	No	Hozzein and Goodfellow (2011)	*Nesterenkonia lacusekhoensis*	MH	6–8	0–15	Yes	Collins et al. (2002)
Actinopolyspora righensis	EH	15–25	10–30	No	Meklat et al. (2013b)	*Nesterenkonia halobia*	MH	5.8–11.7	5.8–23.4	No	Stackebrandt et al. (1995)
Actinopolyspora lacussalsi	MH	12	10–20	No	Guan et al. (2013b)	*Nesterenkonia jeotgali*	MH	2–5	0–16	Yes	Yoon et al. (2006)
Actinopolyspora mzabensis	EH	10–28	7–32	No	Meklat et al. (2013a)	*Nitriliruptor alkaliphilus*	HT	1.2–1.8	0.6–11.7	No	Sorokin et al. (2009)

(continued)

Table 3.1 (continued)

Species name	Type	Optimum grow	NaCl tolerancy	NaCl independency	Reference	Species name	Type	Optimum grow	NaCl tolerance	NaCl independency	Reference
Actinopolyspora dayingensis	MH	13	10–20	No	Guan et al. (2013d)	*Nocardioides salarius*	MH	3	1–10	No	Kim et al. (2008)
Actinopolyspora iraqiensis	MH	10–15	5–20	No	Ruan et al. (1994)	*Nocardioides marinus*	HT	1–3	0.5–8	No	Choi et al. (2007)
Actinopolyspora xinjiangensis	MH	10–15	8–25	No	Guan et al. (2010a)	*Nocardioides dokdonensis*	HT	0–3	0–7	Yes	Park et al. (2008)
Actinopolyspora saharensis	EH	15–20	10–30	No	Meklat et al. (2013c)	*Nocardioides ganghwensis*	HT	0–1	0–8	Yes	Yi and Chun (2004b)
Actinotalea ferrariae	MH	3	0–7	Yes	Li et al. (2013)	*Nocardioides aestuarii*	HT	0–2	0–8	Yes	Yi and Chun (2004a)
Aeromicrobium alkaliterrae	HT	0	0–9	Yes	Yoon et al. (2005)	*Nocardioides albertanoniae*	HT	0–4	0–10	Yes	Alias-Villegas et al. (2013)
Aeromicrobium halocynthiae	–	NR	0–6	Yes	Kim et al. (2010)	*Nocardiopsis xinjiangensis*	MH	10	3–20	No	Li et al. (2003a)
Aeromicrobium ponti	–	NR	0–10	Yes	Lee and Lee (2008)	*Nocardiopsis chromatogenes*	MH	5–8	0–18	Yes	Li et al. (2006b)
Allonocardiopsis opalescens	HT	0	0–10	Yes	Du et al. (2012)	*Nocardiopsis baichengensis*	MH	5–8	0–18	Yes	Li et al. (2006b)
Amycolatopsis salitolerans	MH	5	0–13	Yes	Guan et al. (2012)	*Nocardiopsis gilva*	MH	5–8	0–18	Yes	Li et al. (2006b)
Amycolatopsis halotolerans	HT	0	0–7	Yes	Lee (2006a)	*Nocardiopsis litoralis*	MH	5–7	1–15	No	Chen et al. (2009b)
Amycolatopsis umgeniensis	HT	NR	0–7	Yes	Everest et al. (2013)	*Nocardiopsis salina*	MH	10	3–20	No	Li et al. (2004c)
Amycolatopsis halophila	MH	5	1–15	No	Tang et al. (2010a)	*Nocardiopsis sinuspersici*	HT	2.5	0–15	Yes	Hamedi et al. (2010)

Amycolatopsis jiangsuensis	HT	0	0–10	Yes	Xing et al. (2013)	*Nocardiopsis nikkonensis*	–	NR	0–20	Yes	Yamamura et al. (2010b)
Amycolatopsis cihanbeyliensis	HT	0	0–10	Yes	Tatar et al. (2013)	*Nocardiopsis synnemataformans*	–	NR	0–10	Yes	Yassin et al. (1997)
Amycolatopsis palatopharyngis	HT	NR	0.5–10	No	Huang et al. (2004)	*Nocardioides marinisabuli*	–	NR	0–8	Yes	Lee et al. (2007)
Amycolatopsis thailandensis	–	NR	0–7	Yes	Chomchoei et al. (2011)	*Nocardiopsis exhalans*	–	NR	0–10	Yes	Peltola et al. (2001)
Amycolatopsis marina	MH	5	0.5–12	No	Bian et al. (2009)	*Nocardiopsis umidischolae*	–	NR	0–7.5	Yes	Peltola et al. (2001)
Arthrobacter antarcticus	–	NR	0–6	Yes	Pindi et al. (2010)	*Nocardiopsis halophila*	MH	5–15	3–20	No	Al-Tai and Ruan (1994)
Arthrobacter flavus	HT	NR	0–11.5	Yes	Reddy et al. (2000)	*Nocardiopsis compostus*	MH	10	0–15	Yes	Kämpfer et al. (2002)
Bogoriella caseilytica	HT	NR	0–8	Yes	Groth et al. (1997a)	*Nocardiopsis halotolerans*	MH	10	0–15	Yes	Al-Zarban et al. (2002a)
Brachybacterium saurashtrense	MH	8	0–15	Yes	Gontia et al. (2011)	*Nocardiopsis synnematafomans*	HT	NR	0–10	Yes	Yassin et al. (1997)
Brachybacterium paraconglomeratum	HT	NR	0–15	Yes	Takeuchi et al. (1995)	*Nocardiopsis rosea*	MH	5–8	0–18	Yes	Li et al. (2006b)
Branchiibius hedensis	–	0	0–7	Yes	Sugimoto et al. (2011)	*Nocardiopsis lucentensis*	HT	NR	0–10	Yes	Yassin et al. (1993)
Brevibacterium lutescens	–	NR	0–10	Yes	Wauters et al. (2003)	*Nocardiopsis potens*	–	NR	0–12	Yes	Yassin et al. (2009)
Brevibacterium oceani	HT	NR	0–12	Yes	Bhadra et al. (2008)	*Nocardiopsis arvandica*	–	NR	0–17.5	Yes	Hamedi et al. (2011)
Brevibacterium celere	HT	NR	0–15	Yes	Ivanova et al. (2004)	*Nocardiopsis terrae*	MH	3–5	1–15	No	Chen et al. (2010b)

(continued)

Table 3.1 (continued)

Species name	Type	Optimum grow	NaCl tolerancy	NaCl independency	Reference	Species name	Type	Optimum grow	NaCl tolerance	NaCl independency	Reference
Brevibacterium pityocampae	HT	NR	0–10	Yes	Kati et al.	*Nocardiopsis flavescens*	HT	0–3	0–10	Yes	Fang et al. (2011)
Brevibacterium picturae	HT	NR	0–10	Yes	Heyrman et al. (2004)	*Nocardiopsis coralliicola*	MH	3–7	0–18	Yes	Li et al. (2012b)
Brevibacterium sandarakinum	–	NR	1–10	No	Kämpfer et al. (2010)	*Nocardiopsis fildesensis*	MH	2–4	0–12	Yes	Xu et al. (2013)
Brevibacterium linens	HT	NR	0–10	Yes	Collins et al. (1980)	*Nocardiopsis tropica*	–	NR	0–10	Yes	Evtushenko et al. (2000)
Brevibacterium permense	HT	NR	0–18	Yes	Gavrish et al. (2004)	*Nocardiopsis trehalosi*	–	NR	0–10	Yes	Evtushenko et al. (2000)
Brevibacterium aurantiacum	HT	NR	0–15	Yes	Gavrish et al. (2004)	*Nocardiopsis alkaliphila*	HT	2.5	0–10	Yes	Hozzein et al. (2004)
Brevibacterium antiquum	HT	NR	0–18	Yes	Gavrish et al. (2004)	*Nocardiopsis metallicus*	–	NR	0–10	Yes	Schippers et al. (2002)
Brevibacterium samyangense	HT	NR	0–10	Yes	Lee (2006b)	*Nocardiopsis quinghaiensis*	MH	3	0–10	Yes	Chen et al. (2008)
Brevibacterium marinum	HT	NR	0–10	Yes	Lee (2008)	*Nocardiopsis kunsanensis*	MH	10	3–25	No	Chun et al. (2000)
Brevibacterium jeotgali	MH	5	2–14	No	Choi et al. (2013)	*Nocardiopsis rhodophaea*	MH	5–8	0–18	Yes	Li et al. (2006b)
Brevibacterium yomogidense	HT	1–2	0–17	Yes	Tonouchi et al. (2013)	*Nonomuraea rhizophila*	–	NR	0–7	Yes	Zhao et al. (2011b)
Brevibacterium ammoniilyticum	HT	NA	0–11	Yes	Kim et al. (2013)	*Oceanitalea nanhaiensis*	–	NR	1–10	No	Fu et al. (2012b)
Brevibacterium casei	HT	NR	0–15	Yes	Collins et al. (1983)	*Paraoerskovia sediminicola*	–	NR	0–10	Yes	Hamada et al. (2013b)

Brevibacterium epidermis	HT	NR	0–15	Yes	Collins et al. (1983)	*Paraoerskovia marina*	–	NR	0–8	Yes	Khan et al. (2009)
Brevibacterium salitolerans	MH	3–8	0–18	Yes	Guan et al. (2010b)	*Promicromonospora iranensis*	HT	NR	0–8	Yes	Mohammadipanah et al. (2014)
Brevibacterium album	HT	0–5	0–10	Yes	Tang et al. (2008b)	*Prauserella halophila*	MH	10–15	5–25	No	Li et al. (2003c)
Cellulomonas carbonis	HT	NR	0–7	Yes	Shi et al. (2012)	*Prauserella alba*	MH	10–15	0–25	No	Li et al. (2003c)
Cellulomonas bogoriensis	–	NR	0–8	Yes	Jones et al. (2005)	*Prauserella marina*	HT	0	0–10	Yes	Wang et al. (2010b)
Cellulomonas pakistanensis	HT	1–2	0–12	Yes	Ahmed et al. (2014)	*Prauserella salsuginis*	MH	8–10	5–15	No	Li et al. (2009)
Corynebacterium halotolerans	MH	10	0–25	Yes	Chen et al. (2004)	*Prauserella flava*	MH	8–10	5–15	No	Li et al. (2009)
Corynebacterium humireducens	MH	10	0–13	Yes	Wu et al. (2011)	*Prauserella aidingensis*	MH	8–10	5–15	No	Li et al. (2009)
Corynebacterium maris	HT	0.5–4	0–10	Yes	Ben-Dov et al. (2009)	*Prauserella sediminis*	MH	10	5–20	No	Li et al. (2009)
Corynebacterium marinum	HT	1	0–8	Yes	Du et al. (2010)	*Prauserella rugosa*	MH	5–10	0–20	Yes	Lechevalier et al. (1986), Kim and Goodfellow (1999) and Li et al. (2003c)
Corynebacterium efficiens	HT	NR	0–10	Yes	Fudou et al. (2002)	*Rhodococcus Kroppenstedtii*	HT	NR	0–10	Yes	Mayilraj et al. (2006)
Demequina flava	HT	0	0–10	Yes	Hamada et al. (2013c)	*Rhodococcus fascians*	HT	NR	0–7	Yes	Holt et al. (1994) and Gesheva et al. (2010)

(continued)

Table 3.1 (continued)

Species name	Type	Optimum grow	NaCl tolerancy	NaCl independency	Reference	Species name	Type	Optimum grow	NaCl tolerance	NaCl independency	Reference
Demequina sediminicola	HT	0	0–10	Yes	Hamada et al. (2013c)	*Pseudonocardia antitumoralis*	MH	3	0–15	Yes	Tian et al. (2013b)
Demequina globuliformis	–	NR	0–15	Yes	Ue et al. (2011)	*Pseudonocardia ammonioxydans*	MH	3.5	0–8	Yes	Liu et al. (2006)
Demequina oxidasica	–	NR	0–15	Yes	Ue et al. (2011)	*Pseudonocardia kongjuensis*	–	NR	0–7	Yes	Lee et al. (2001)
Demequina aurantiaca	–	NR	0–15	Yes	Ue et al. (2011)	*Pseudonocardia nantongensis*	MH	3	0–15	Yes	Xing et al. (2012)
Demequina salsinemoris	HT	NR	0–8	Yes	Matsumoto et al. (2010)	*Pseudonocardia kunmingensis*	HT	1–3	0–7	Yes	Zhao et al. (2011a)
Demequina aestuarii	MH	2–4	0–12	Yes	Yi et al. (2007)	*Pseudonocardia alni*	–	NR	0–7	Yes	Evtushenko et al. (1989) and Warwick et al. (1994)
Demetria terragena	–	NR	0–12	Yes	Groth et al. (1997b)	*Pseudonocardia Antarctica*	–	NR	0–8	Yes	Prabahar et al. (2004)
Dermacoccus abyssi	–	NR	0–7.5	Yes	Pathom-Aree et al. (2006)	*Ruania albidiflava*	HT	NR	0–10	Yes	Gu et al. (2007)
Dietzia natronolimnaios	HT	0	0–10	Yes	Duckworth et al. (1998)	*Rubrobacter bracarensis*	MH	3–10	3–30	No	Jurado et al. (2012)
Dietzia timorensis	–	NR	0–7	Yes	Yamamura et al. (2010a)	*Saccharomonospora paurometabolica*	MH	10	5–20	No	Li et al. (2003b)
Dietzia maris	–	NR	0–7	Yes	Rainey et al. (1995) and Nesterenko et al. (1982)	*Saccharomonospora amisosensis*	HT	0	0–10	Yes	Veyisoglu et al. (2013)
Dietzia alimentaria	HT	2	0–10	Yes	Kim et al. (2011)	*Saccharomonospora saliphila*	MH	10	0–20	Yes	Syed et al. (2008)

Dietzia psychralcaliphila	–	NR	0–10	Yes	Yumoto et al. (2002)	*Saccharomonospora halophila*	–	NR	10–30	No	Al-Zarban et al. (2002b)
Dietzia schimae	HT	NR	0–15	Yes	Li et al. (2008a)	*Saccharomonospora azurea*	–	NR	0–7	Yes	Runmao (1987)
Dietzia cercidiphylli	HT	NR	0–10	Yes	Li et al. (2008a)	*Saccharopolyspora lacisalsi*	EH	15	5–25	No	Guan et al. (2011a)
Euzebya tangerina	–	NR	0.5–12	No	Kurahashi et al. (2010)	*Saccharopolyspora cebuensis*	MH	NR	2.5–12.5	No	Pimentel-Elardo et al. (2008)
Geodermatophilus africanus	HT	NR	0–8	Yes	del Carmen Montero-Calasanz et al. (2013)	*Saccharopolyspora qijiaojingensis*	EH	10–15	6–22	No	Tang et al. (2009d)
Georgenia halophila	MH	5–10	1–15	No	Tang et al. (2010b)	*Saccharopolyspora dendranthemae*	MH	3	0–17	Yes	Zhang et al. (2013c)
Georgenia muralis	HT	0–5	0–7	Yes	Altenburger et al. (2002)	*Saccharopolyspora spinosa*	HT	NR	0–11	Yes	Mertz and Yao (1990)
Georgenia thermotolerans	MH	0–5	0–7	Yes	Hamada et al. (2009)	*Saccharopolyspora gloriosae*	HT	0–5	0–11	Yes	Qin et al. (2010)
Georgenia daeguensis	–	NR	1–9	No	Woo et al. (2012)	*Saccharopolyspora antimicrobica*	–	NR	0–7	Yes	Yuan et al. (2008)
Glycomyces halotolerans	MH	4–5	0–11	Yes	Guan et al. (2011b)	*Saccharopolyspora halophila*	EH	10–15	3–20	No	Tang et al. (2009a)
Glycomyces fuscus	MH	5	3–12	No	Han et al. (2014)	*Salinactinospora qingdaonensis*	MH	9–12	1–23	No	Chang et al. (2012)
Glycomyces albus	MH	5–7	0–13	Yes	Han et al. (2014)	*Salinibacterium xinjiangense*	–	NR	0–14	Yes	Zhang et al. (2008)
Gordonia paraffinivorans	–	NR	0.5–7	No	Xue et al. (2003)	*Salinibacterium amurskyense*	HT	1–3	0–10	Yes	Han et al. (2003)

(continued)

Table 3.1 (continued)

Species name	Type	Optimum grow	NaCl tolerancy	NaCl independency	Reference	Species name	Type	Optimum grow	NaCl tolerance	NaCl independency	Reference
Haloactinobacterium album	MH	7–10	2–16	No	Tang et al. (2010e)	*Salinisphaera halophila*	EH	14–19	6–29	No	Zhang et al. (2012c)
Haloactinopolyspora alba	EH	10–15	7–23	No	Tang et al. (2011b)	*Salinispora arenicola*	–	NR	NR	No	Maldonado et al. (2005)
Haloactinopolyspora alkaliphila	MH	2.5–5	0–12.5	Yes	Zhang et al. (2014b)	*Salinispora tropica*	–	NR	NR	No	Maldonado et al. (2005)
Haloactinospora alba	EH	15	9–21	No	Tang et al. (2008a)	*Salinispora pacifica*	–	NR	NR	No	Ahmed et al. (2013)
Haloechinothrix alba	EH	15	9–23	No	Tang et al. (2010c)	*Sciscionella marina*	MH	3–5	0–13	Yes	Tian et al. (2009)
Haloglycomyces albus	MH	8–12	3–18	No	Guan et al. (2009)	*Serinicoccus profundi*	MH	3–5	0–14	Yes	Xiao et al. (2011)
Isoptericola halotolerans	MH	10	0–25	Yes	Zhang et al. (2005)	*Serinicoccus chungangensis*	EH	13	0–15	Yes	Traiwan et al. (2011)
Isoptericola chiayiensis	HT	NR	0–12	Yes	Tseng et al. (2011)	*Serinicoccus marinus*	HT	2–3	0–14	Yes	Yi et al. (2004)
Isoptericola salitolerans	MH	9–11	0–16	Yes	Guan et al. (2013a)	*Spinactinospora alkalitolerans*	MH	3–8	1–15	No	Chang et al. (2011)
Isoptericola hypogeus	HT	NR	0–10	Yes	Groth et al. (2005)	*Streptoalloteichus hindustanus*	HT	NR	0–7	Yes	Tomita et al. (1987)
Isoptericola nanjingensis	MH	3	0.5–10	No	Huang et al. (2012)	*Streptomonospora salina*	EH	15	5–20	No	Cui et al. (2001)
Janibacter corallicola	–	NR	0–7	Yes	Kageyama et al. (2007)	*Streptomonospora Arabica*	MH	5	0–15	Yes	Hozzein and Goodfellow (2008) and Zhang et al. (2013a)

Janibacter cremeus	HT	0	0–10	Yes	Hamada et al. (2013a)	*Streptomonospora flavalba*	MH	10	5–25	No	Cai et al. (2009)
Janibacter alkaliphilus	HT	NR	0–17	Yes	Li et al. (2012a)	*Streptomonospora amylolytica*	MH	10	5–20	No	Cai et al. (2009)
Janibacter terrae	–	NR	0–8	Yes	Yoon et al. (2000)	*Streptomonospora alba*	EH	10–15	5–25	No	Li et al. (2003d)
Kocuria turfanensis	–	NR	0–15	Yes	Zhou et al. (2008)	*Streptomonospora halophila*	MH	10	5–20	No	Cai et al. (2008)
Kocuria flava	–	NR	0–10	Yes	Zhou et al. (2008)	*Streptomonospora algeriensis*	EH	10–15	7–20	No	Meklat et al. (2014)
Kocuria rosea	–	NR	0–7.5	Yes	Stackebrandt et al. (1995)	*Streptomonospora nanhaiensis*	–	0–7	0–20	Yes	Zhang et al. (2013a)
Kocuria kristinae	–	NR	0–10	Yes	Stackebrandt et al. (1995)	*Streptomonospora sediminis*	–	0–7	0–20	Yes	Zhang et al. (2013a)
Kocuria varians	–	NR	0–7.5	Yes	Stackebrandt et al. (1995)	*Streptomyces oceani*	MH	3–5	2.5–12.5	No	Tian et al. (2012b)
Kocuria rhizophila	–	NR	0–15	Yes	Kovács et al. (1999)	*Streptomyces fukangensis*	MH	2.5–5	0–7.5	Yes	Zhang et al. (2013b)
Kocuria marina	–	NR	0–15	Yes	Kim et al. (2004)	*Streptomyces pharmamarensis*	HT	2	0–9	Yes	Carro et al. (2012)
Kocuria halotolerans	MH	5	0–10	Yes	Tang et al. (2009c)	*Streptomycesalbiaxialis*	HT	NR	3–30	No	Kuznetsov et al. (1992)
Koreibacter algae	HT	1–4	0–10	Yes	Lee and Lee (2010)	*Streptomyces glycovorans*	HT	0–3	0–7	Yes	Xu et al. (2012)
Kribbella lupini	–	NR	0–7	Yes	Trujillo et al. (2006)	*Streptomyces xishensis*	HT	0–3	0–7	Yes	Xu et al. (2012)

(continued)

Table 3.1 (continued)

Species name	Type	Optimum grow	NaCl tolerancy	NaCl independency	Reference	Species name	Type	Optimum grow	NaCl tolerance	NaCl independency	Reference
Kytococcus sedentarius	–	NR	0–10	Yes	Stackebrandt et al. (1995)	*Streptomyces abyssalis*	HT	0–3	0–6	Yes	Xu et al. (2012)
Kytococcus schroeteri	–	NR	0–12	Yes	Becker et al. (2002)	*Streptomyces iconiensis*	HT	NR	0–10	Yes	Tatar et al. (2014)
Leucobacter aridicollis	HT	0	0–9	Yes	Morais et al. (2004)	*Streptomyces smyrnaeus*	HT	NR	0–15	Yes	Tatar et al. (2014)
Leucobacter chromiireducens	HT	0	0–9	Yes	Morais et al. (2004)	*Streptomyces chilikensis*	MH	1–8	1–12	No	Ray et al. (2013)
Leucobacter luti	HT	0	0–8	Yes	Morais et al. (2006)	*Streptomyces rochei*	HT	2	0–6	Yes	Reddy et al. (2011)
Leucobacter alluvii	HT	0	0–8	Yes	Morais et al. (2006)	*Streptomyces nanhaiensis*	MH	3	0–7.5	Yes	Tian et al. (2012a)
Leucobacter chironomi	HT	0–1	0–7	Yes	Halpern et al. (2009)	*Streptomyces gulbargensis*	–	NR	0–7	Yes	Dastager et al. (2007)
Longimycelium tulufanense	HT	2	1–8	No	Xia et al. (2013)	*Streptomyces haliclonae*	MH	2–7	1–10	No	Khan et al. (2010)
Lysinimicrobium mangrovi	HT	0	0–7	Yes	Hamada et al. (2012)	*Streptomyces tateyamensis*	MH	2–7	1–10	No	Khan et al. (2010)
Mariniluteicoccus flavus	HT	0–1	0–6	Yes	Zhang et al. (2014a)	*Streptomyces haliclonae*	MH	2–7	1–10	No	Khan et al. (2010)
Microbacterium sediminis	HT	0	0–8	Yes	Yu et al. (2013)	*Thermobifida halotolerans*	HT	0	0–10	Yes	Yang et al. (2008)
Microbacterium halotolerans	MH	5	0–15	Yes	Li et al. (2005b)	*Verrucosispora lutea*	HT	1	0–7	Yes	Liao et al. (2009)
Microbacterium profundi	HT	0–1	0–7.5	Yes	Wu et al. (2008)	*Verrucosispora sediminis*	MH	3.5	0–6	Yes	Dai et al. (2010)

Micrococcus yunnanensis	HT	NR	0–15	Yes	Zhao et al. (2009)	*Verrucosispora qiuiae*	HT	0–5	0–10	Yes	Xi et al. (2012)
Micrococcus luteus	HT	NR	0–10	Yes	Wieser et al. (2002)	*Yania flava*	MH	10–15	0.5–25	No	Li et al. (2005d)
Micrococcus endophyticus	HT	NR	0–10	Yes	Chen et al. (2009a)	*Yania halotolerans*	MH	10KCl	0–25	Yes	Li et al. (2004a)
Micromonospora endolithica	HT	2	0–7	Yes	Hirsch et al. (2004)	*Yimella lutea*	HT	0–1	0–8	Yes	Tang et al. (2010d)
Myceligenerans halotolerans	HT	0–5	0–10	Yes	Wang et al. (2011b)	*Zhihengliuella halotolerans*	MH	10	0–25	Yes	Zhang et al. (2007)
Myceligenerans salitolerans	MH	5	0–16	Yes	Guan et al. (2013c)	*Zhihengliuella salsuginis*	MH	5–10	0.5–20	Yes	Chen et al. (2010a)
Myceligenerans xiligouense	MH	2–7	2–17.5	No	Cui et al. (2004)	*Zhihengliuella alba*	MH	5	0–15	Yes	Tang et al. (2009b)
Nesterenkonia flava	–	NR	0–10	Yes	Luo et al. (2008)						

EH extreme halophile, *MH* moderate halophile, *HT* halotolerant, *NR* not reported and more than 300 genera, is one of the largest taxonomic units within the domain Bacteria. Among the widely diverse microbial taxa, actinobacteria are the most prolific source for production of bioactive metabolites.

3.2 Ecology of Halophile/Halotolerant Actinobacteria

Ubiquity of hypersaline environments with common phenomenon of gradients in salinity are characterized in vast geographical area including the Dead Sea, saltern crystallizer ponds, natural inland salt lakes, brines, alkaline saline habitats, salt-contaminated soils, salt flats, evaporated ponds, subsurface salt formations, deep-sea hypersaline basins, alkaline saline habitats, salted foods, cold saline environments, and decayed monuments which may be the consequence of seawater evaporation. Saline environments are classified based on pH and cations valence into thalassohaline environments, containing dominating sodium and chloride ions with neutral or slightly alkaline pH, and athalassohaline hypersaline environment containing divalent cations such as Mg^{2+} and Ca^{2+} have more concentration of monovalent cations with pH of about 6.0 (Oren 2002; Hamedi et al. 2013; Sass et al. 2001). Moreover, sea and ocean waters cover 70 % of earth's surface while accounting for more than 90 % of the volume earth's crust making an unlimited source of new microorganisms as well as novel natural products (Satpute et al. 2010). Salt-induction stress can divide microorganisms to categories of extreme halophiles preferring 2.5–5.2 M (14.6–30.4 %) salt, borderline extreme halophiles preferring 1.5–4.9 M (8.8–28.6 %) salt, moderate halophiles preferring 0.5–2.5 M (2.9–14.6 %) salt, and halotolerants that grow independently of salt but may resist at least 1.7 M (10 %) salt, depending on the quantity of salt they require or tolerate (Kushner 1978). Many halophilic, halotolerent and/or drought-resistant actinobacteria have also been reported from arid area (Hamedi et al. 2013). Despite the discovery of halophilic actinobacteria more than four decades ago, the knowledge of their biology and potential industrial application are still not fully investigated. Therefore a comprehensive enlightenment demanded for systematic understanding of the compelling functions of halophilic and halotolerant actinobacteria in human beings life to widen the view point which have not been noted fully and found less fascinating than almost saturated field of mesophilic actinobacteria.

3.3 Clinically Significant Natural Products-Derived from Halophilic and Halotolerant Actinobacteria

The chemical space occupied by natural products, including drugs, is more diverse than combinatorial chemicals (Feher and Schmidt 2003) and natural products provide unprecedented skeletons which could be efficiently used for drug development technology (Newman and Cragg 2007). Halophilic and halotolerant microorganisms have proved a reliable and prosperous source of unprecedented natural products.

The marine environment alone has provided around 30,000 natural products through the already exploited secondary metabolites from its harbored organisms (Sagar et al. 2013). In total, 45 % of the drugs have been discovered from actinobacteria, 38 % by fungi and 17 % by unicellular bacteria. Today, it has been accepted that marine actinobacteria have an unparalleled ability to produce bioactive products with wide spectrum of antibiotic activities.

Some novel natural products with no biological activity are reported from halophilic or halotolerant actinobacteria. For instance, in 2005 investigation of mangrove plant *Aegicerus comiculatum* leaves led to isolation of an endophytic *Streptomyces* sp. GT-20026114. The strain produced four new derivatives of cyclopentenone including (5*R*) 3-amino-5-hydroxy-5-vinyl-2-cyclopentent-1-one, (5*R*) 5-hydroxy-3-[(methoxy-carbonyl)amino]-5-vinyl-2-cyclopenten-1-one, (5*R*) 5-hydroxy-3-[[2-(4-hydroxyphenyl)ethyl]amino]-5-vinyl-2-cyclopenten-1-one, and 3-isobutylpropanamide-2-cyclopenten-1-one (Lin et al. 2005). Actinofuranones A and B, two new rare polyketides containing a 3-furanone ring system with a C-2 hemiketal and a C-5 unsaturated alkyl chain were isolated from *Streptomyces* MAR4 strain CNQ766 (Cho et al. 2006). In the same year, a new bioactive natural metabolite (5*S*,8*S*,9*R*,10*S*)-selina-4(14),7(11)-diene-8,9-diol, a novel sesquiterpene belonged to terpenes class of antibiotics screened from culture medium of marine-derived *Streptomyces* sp. QD518 (Wu et al. 2006). Further, a marine-derived *Streptomyces* sp. M491 was isolated which produced a range of natural products including three new natural products, 10*a*,11-dihydroxyamorph-4-ene, 10*a*,15-dihydroxyamorph-4-en-3-one, and 5*a*-10*a*,.11-trihydroxyamorphan-3-one, as well as already known compound, 10*a*-hydroxyamorph-4-en-3-one. All these four compounds were muurolane sesquiterpenes. Finally, two new secondary metabolites, erythronolides H and I which are analogues of clinically significant antibiotic drug erythromycin A, were identified from a novel obligate halophilic actinobacteria, *Actinopolyspora* sp. YIM90600 in addition to erythromycin C. Erythronolide H possesses the C-14 hydroxy moiety and the C-6/C-18-epoxide whereas erythronolide I contains spiroketal moiety and belong to macrolactone class and C-21 polyketides class of antibiotics, respectively (Huang et al. 2009).

Natural products from halophilic or halotolerant actinobacteria with antimicrobial activities and/or cytotoxic activities have been mentioned further in Sects. 3.1 and 3.2, in addition to Table 3.2.

3.3.1 Production of Antimicrobial Compounds Produced from Halotolerant and Halophilic Actinobacteria

During the last eight decades, bioactive compounds derived from microbial origin have always made a noteworthy contribution for introducing novel antibiotics through drug discovery and development programs (Demain and Sanchez 2009). Numerous bioactive natural compounds including some commercially significant antibiotics such as erythromycin, vancomycin, streptomycin, tetracycline, amphotericin, rifamycin and gentamycin have been provided by phylum of actinobacteria

Table 3.2 A number of novel bioactive compounds showing both antimicrobial and anticancer activities isolated from halophilic or halotolerant actinobacteria up to September 2014

Compound	Producing strain	Biological activity	Compound class	Reference
1-hydroxy-1-norresistomycin	*Streptomyces chibaensis* $AUBN_1/7$	Antibacterial; anticancer	Resistomycin	Gorajana et al. (2005)
Carboxamycin	*Streptomyces* sp. NTK 937	Antibacterial; anticancer	Benzoxazole	Hohmann et al. (2009)
Chemomicin A	*Nocardia mediterranei* subsp. *kanglensis* 1747-64	Antibacterial; anticancer	Angucyclinone antibiotics	Sun et al. (2007)
Chlorinated dihydroquinones	Actinomycete strain CNQ-525	Antibacterial; anticancer	Napyradiomycin	Soria-Mercado et al. (2005)
Lajollamycin	*Streptomyces nodosus*	Antibacterial; anticancer	Nitro-tetraene spiro-β-lactone-γ-lactam	Manam et al. (2005)
Lobophorins E and F	*Streptomyces* sp. SCSIO 01127	Antibacterial; anticancer	Polyketide	Niu et al. (2011)
Marinomycins A–D	*Marinispora* strain CNQ-140	Antibacterial; anticancer (except marinomycin D)	Polyketide	Kwon et al. (2006)
Parimycin	*Streptomyces* sp. isolate B8652	Antibacterial; anticancer	–	Maskey et al. (2002)
Pseudonocardians A–C and Deoxynyboquinone	*Pseudonocardia* sp. SCSIO 01299	Antibacterial (except Pseudonocardian C); anticancer	Diazaanthraquinone	Li et al. (2011)
(*R*)-10-methyl-6-undecanolide and (6*R*,10*S*)-10-methyl-6-dodecanolide	*Streptomyces* sp. isolate B6007	Antibacterial; anticancer	Caprolactones of lactones family	Stritzke et al. (2004)
Terpenoid and 2-allyloxyphenol	*Streptomyces* MS 1/7	Antimicrobial; anticancer	–	Saha et al. (2006) and Arumugam et al. (2010)
Trioxacarcins A–D and gutingimycin	*Streptomyces* sp. isolate B8652	Antimicrobial (except trioxacarcin F); anticancer (except trioxacarcins E and F)	–	Maskey et al. (2004)
VIcd	*Amycolatopsis alba* var. nov. DVR D4	Antibacterial; anticancer	Pyridinium	Dasari et al. (2012)

alone (Jose and Jebakumar 2014), and it is predicted that more novel bioactive products with chemotherapeutic applications would be discovered in future from these phylum of bacteria (Baltz 2007). Numerous bioactive natural compounds including some commercially significant antibiotics such as erythromycin, vancomycin, streptomycin, tetracycline, amphotericin, rifamycin and gentamycin have been provided through phylum of actinobacteria alone (Jose and Jebakumar 2014), and it is predicted that more novel bioactive products with chemotherapeutic applications would be discovered in future from these phylum of bacteria (Baltz 2007).

Sophisticated mechanisms has been recruited by pathogenic microorganisms to undo biological activity of antibiotics which reflect a prompt requirement for novel antibiotics that would restore their effects and in turn target the evolved multidrug-resistancy (Butler et al. 2013). Reorientation of current isolation scheme of actinobacteria is therefore, required due to re-isolation of already discovered compounds from conventional environments. In this respect, consideration of new, less or unexplored, uncommon and extreme niches including saline terrestrial and marine environments have to be taken into account (Hamedi et al. 2013; Jose and Jebakumar 2012). As novel chemistry is a factor of environmental conditions (Genilloud et al. 2011), these poorly explored actinobacteria promise a prosperous source of novel and potential natural products, exhibiting antimicrobial, antiviral, antitumor, and anticoagulant and cardioprotective properties (Austin 1989), which can be used further as a raw material for drug discovery against a number of targets (Arul Jose et al. 2011; Subramani and Aalbersberg 2012; Hamedi et al. 2013). This statement is supported by existence of considerable number of chemotherapeutic pharmaceuticals on the sale that are based on scaffolds of actinobacteria-derived natural metabolites (Newman and Cragg 2012; Butler 2005; Demain and Sanchez 2009). Antibiotics derived from neglected resources such as halophilic actinobacteria may display more efficiency at overcoming infections due to lack of establishment of a mechanism for resistance against them (Donia and Hamann 2003).

Two novel aromatic amides belonging to new class of tri-alkyl-substituted benzenes were isolated from a marine actinobacterium (MST-MA190) and named as lorneamide A, active against *Bacillus subtilis*, and lorneamide B. Despite weak antibiotic activity, these compounds possess novel natural carbon skeletons (Capon et al. 2000).

Chalcomycin B, a novel macrolide antibiotic, is produced by the marine isolate *Streptomyces* sp. B7064 and displayed antibacterial activity against Gram-positive bacteria with especial potency on *Staphylococcus aureus* (Asolkar et al. 2002).

Bonactin, a new antimicrobial ester, is produced by marine-sediment-derived *Streptomyces* sp. BD21-2 which can render relatively weak antimicrobial activity against both Gram-positive and Gram-negative bacteria as well as fungi including *Bacillus megaterium*, *Micrococcus luteus*, *Klebsiella pneumoniae*, *Staphylococcus aureus*, *Alcaligenes faecalis*, *Escherichia coli*, and *Saccharomyces cerevisiae*. Bonactin is the first acyclic ester related to the nonactins that possess antimicrobial effects (Schumacher et al. 2003).

A new antibiotics alkaloid containing exceptionally rare dibenzodiazepine core linked to farnesyl side chain with modest antimicrobial activity against Gram-positive bacteria was isolated from marine *Micromonospora* sp. DPJ12 and designated as diazepinomicin (Charan et al. 2004). A novel angucyclinone antibiotic with antibacterial activities against Gram-positive bacteria was identified from marine *Streptomyces griseus* strain NTK 97 isolated from Antarctica and named as frigocyclinose. The molecular weight of frigocyclinose was determined as 463 Da and neither Gram-negative bacteria nor yeasts were found sensitive to it (Bruntner et al. 2005).

Ongoing screening led to isolation of the first triazolopyrimidine antibiotic, named as essramycin, from a marine *Streptomyces* sp. isolate Merv8102. Essramycin is an active antibiotic compound against both Gram-positive and Gram-negative bacteria including; *Escherichia coli*, *Pseudomonas aeruginosa*, *Staphylococcus aureus*, and *Micrococcus luteus*. However, no antifungal activity was displayed. Triazolopyrimidines are smooth muscle cell growth inhibitors, inflammatory agents, with potential in alleviation and prevention of cardiovascular diseases. Before isolation of essramycin, the only way of obtaining triazolopyrimidines was through synthetic production (El-Gendy et al. 2008).

A new thiopeptide antibiotic with rare aminoacetone moiety and 592 Da m.w, is produced by obligate marine *Nocardiopsis* sp. isolate TFS65-07 and showed growth inhibition activity against some *Staphylococcus* and *Streptococcus* species, as vancomycin-resistant bacterial strains including *Enterococcus faecalis* and *Enterococcus faecium* (Engelhardt et al. 2010).

A new 3-hydroxyl derivative of the already identified *Streptomyces* metabolite ikarugamycin, named as butremycin, is produced by novel actinobacterium strain *Micromonospora* sp. K310 isolated from a mangrove river sediment. This compound with molecular weight of 485 Da belongs to polycyclic tetramic acid macrolactams (PTM) family and has been demonstrated for first time in *Micromonospora* sp. It was weakly active against *Staphylococcus aureus*, methicillin-resistant *Staphylococcus aureus*, and *Escherichia coli* (Kyeremeh et al. 2014).

Micromonospora lomaivitiensis strain LL-37I366, a marine-invertebrate-associated halophilic actinobacteria, was found to produce two potent diazobenzofluorene glycosides antitumor natural bioactive products, lomaiviticins A and B, with antibacterial activity against *Staphylococcus aureus* and *Enterococcus faecium*. Lomaiviticin A was also active against a series of cancer cell lines through DNA-damaging mechanism by cleaving double stranded DNA under reducing conditions (He et al. 2001).

3.3.2 Production of Antitumor Compounds

The potential of secondary metabolites is more highlighting by underlining that secondary metabolites-originated drugs have almost doubled the average human beings lifespan (Verdine 1996). Despite great achievement in the treatment of cancer via the application of chemotherapeutic drugs, the requirement of new chemical entities with potential antitumor properties is yet to be explored which emerges from intense mortality rates of many cancer types (Jemal et al. 2005).

Marine-derived actinobacteria are efficient cytotoxic polyketide-producing microorganisms; a new marine-derived *Streptomyces* sp. KS3 was found to be capable of producing a novel neuritogenetic anthracycline, komodoquinone A, and its derivative, komodoquinone B with neuronal cell differentiation induction in the neuroblastoma cell line (Neuro 2A) during solid state fermentation. Upon further structure investigation of komodoquinone A, it was deduced that in these novel natural products, a new amino sugar is bonded to anthracyclinone skeleton D-ring whereas, komodoquinone B is aglycone of komodoquinone A. Komodoquinone B is the first anthracyclines antibiotics exhibiting neuritogenic activity leading to morphological changes of the neuroblastoma, Neuro 2A cell line (Itoh et al. 2003). Later,(*7S*9R*10R**)-pyrromycin,(*7R*9R*10R**)-pyrromycin,1-hydroxyauramycin T, and 1-hydroxysulfumycin T, four new derivatives of anthracycline with good cytotoxicity against the P388 murine leukaemia cell line were obtained from *Streptomyces* sp. CANU Fox 21-2-6 (Phipps et al. 2004). Thereafter, a noticeable cytotoxicity against seven human cancer cell lines including MDA-MB-231, human colon carcinoma (HCT-15), human prostate cancer (PC-3), human lung cancer (NCI-H23), human renal cancer (ACHN), human skin cancer (LOX-IMVI), and human leukemia (K-562) was displayed by new cytotoxic compound streptokordin derived from marine *Streptomyces* sp. KORDI-3238. The streptokordin with 151 Da had no activity against fungi and bacteria. This compound belongs to methylpyridine class of cytotoxic compounds. In addition to streptokordin, four known compounds of nonactic acid, an ionophore macrolide antibiotic nonactic with inhibitory activity against multidrug-resistant erythroleukemia cell line (K-562), dilactone, trilactone and both antibacterial and antifungal activities were identified (Jeong et al. 2006). It was also able to produce a benzyl pyrrolidine derivative, streptopyrrolidine, with significant angiogenesis inhibition activity (Shin et al. 2008). Early control of angiogenesis is generally considered as a promising therapeutic strategy for alleviation of malignant angiogenesis since invasion and metastasis as well as the growth of solid tumors is dependent mainly on angiogenic inducers (Bouïs et al. 2006). The marine-derived actinobacterium, *Salinispora arenicola*, produces two bicyclic polyketides, saliniketals A and B. Further structure elucidation demonstrated that saliniketals possess a new 1,4-dimethyl-2,8-dioxabicyclo[3.2.1]octan-3-yl ring. The compounds proved to be effective in chemoprevention of cancer through activity of ornithine decarboxylase induction-inhibiting, an enzyme having a role in cell growth which is present in wide array of eukaryotic as well as prokaryotic species, at concentration beyond 1 mM (Williams et al. 2007a). Screening of fermentation broth of the obligate marine *Salinispora arenicola* strain CNR-005 resulted in isolation of arenicolides A–C, three new oxygenated bioactive natural products belonging to macrolide polyketides. Arenicolide A showed only a moderate cytotoxic activity towards human colon adenocarcinoma cell line (HTC-116) and no activity was observed against tested bacteria (Williams et al. 2007b). More recently, six novel indoxamycins, A–F, were isolated from sea-water based culture medium of actinobacterium strain NPS-643 that belonged to polyketides class of antibiotics and containing a novel tricyclic ring system. Applying indoxymycins A and F to the human colon adenocarcinoma cell line (HT-29) rendered noticeable antiproliferative activity, however, indoxymycins B, C, D, and E were not significant growth inhibitors possibly due to absence of a lipophilic core (Sato et al. 2009). Following to them, new erythronolides H and I, analogues of clinically significant antibiotic drug erythromycin A, were identified from a novel obligate halophilic actinobacterium, *Actinopolyspora* sp. YIM90600 in addition to erythromycin C. Erythronolide H possesses the C-14 hydroxy moiety and the C-6/C-18-epoxide whereas, erythronolide I contains spiroketal moiety and belongs to macrolactone class and C-21 polyketides class of antibiotics, respectively (Huang et al. 2009). Further, the first three new γ-pyrones metabolites containing a methylhexyl side chain were delivered from marine-derived *Marinactinospora thermotolerans* SCSIO 00606 and named Marinactinones A–C. These polyketides-belonged class of antibiotics displayed moderate cytotoxic activity against pancreatic cancer cell line (SW1990), human hepatocellular liver carcinoma cell line (HepG2) and a hepatoma cell line (SMCC-7721), in addition, marinactinones B was also weak DNA topoisomerase II inhibitor (Wang et al. 2010a). The cell extract of marine *Streptomyces albus* POR-04-15-053 produces four new antitumor actinopyranones named PM05011 and PM050463 with strong cytotoxity, and PM060054 and PM060431 with moderate cytotoxicity against breast (MDA-MB-231), colon (HT29), and lung (A549) cell lines. These compounds contain an α-methoxy-γ-*pyrone* ring with a highly substituted tetraene side chain, therefore belong to polyketides class of antibiotics (Schleissner et al. 2011). Two mechercharmycins A and B, with cyclo-peptide like structure containing four oxazoles and a thiazole, and a linear congener of mechercharmycin A were reported from marine *Thermoactinomyces* sp. YM3-251. Relatively strong cytotoxicity activities against human lung cancer cell line (A549) and human leukemia cell line (Jurkat cells) were exhibited by mechercharmycin A whereas type B displayed no cytotoxic activities (Kanoh et al. 2005).

Salinosporamide A with unusual fused γ-lactam-β-lactone ring structure was isolated from *Salinospora tropica* CNB-392 and showed potent cytotoxic activity against a number of cell-line including HCT-116 human colon carcinoma, NCI's 60-cell-line panel, NCI-H226 non-small cell lung cancer, SF-539 CNS cancer, SK-MEL-28 melanoma, and MDA-MB-435 breast cancer. Further investigation revealed that the unique functionalization of its core bicycle ring structure consequently caused a significantly potent molecule of proteasome inhibitor (Feling et al. 2003). The peculiarity of salinosporamide A was raised by its strong inhibitory effect against the human malaria parasite during the erythrocytic stages through inhibition of 20S proteasome both in vitro and in vivo. These findings led salinosporamide A to advance to phase I trials for evaluation of its safety profile in application against malaria (Prudhomme et al. 2008). Two more new related γ-lactams cytotoxic structural analogues of salinosporamide A named as salinosporamides B and C were identified from the mentioned strain. Salinosporamide B was 500 times less cytotoxic against human colon carcinoma (HCT-116) than compound A. This finding proposes that the key pharmacophore is β-lactone along with significance of chloroethyl group in the cytotoxic activity of compound A. Finally, none of the mentioned compounds showed neither antiviral nor antimicrobial activities against test microorganisms while no virtually cytotoxicity for compound C was yet reported (Williams et al. 2005). Further exploration into the chemistry of this unique marine actinobacterial strain resulted in discovery of two novel halogenated macrolides, sporolides A and B. However, no biological activity was observed from this class of compounds until now (Buchanan et al. 2005). Later, two unprecedented natural products, named lobophorins C and D, belonging to macrolides class of antibiotics were produced from *Streptomyces carnosus* strain AZS17 associated with marine-derived sponges *Hymeniacidin* sp. Lobophorins C and D were derivatives of kijanimicin and showed significant selective cytotoxicity against human liver cancer cell line 7402 and human breast cancer cell (MDA-MB 435), respectively (Wei et al. 2011). The structure of these compounds were very similar to lobophorin B and A produced by actinobacterium strain CNC-837 isolated from the surface of brown alga *Lobophora variegate* (Jiang et al. 1999). Despite relatively identical structure of lobophorins A and B with C and D, they were only anti-inflammatory.

An endophytic actinobacterial strain *Nocardiopsis* sp. A00203 was capable of producing three new 2-pyranone derivatives named norcardiatones A, B, and C during fermentation. Weak cytotoxic activity against Helacyton gartleri (HeLa) cells was proved by norcardiatones in MTT assay (Lin et al. 2010).

Marine *Streptomyces* sp. designated YM14-060 was capable of producing two novel piericidins C_7 and C_8, having 472 and 486 Da m.w. (Hayakawa et al. 2007a). These two members of family piericidin displayed cytotoxicity against rat glial cells transformed with the adenovirus E1A gene (RG-E1A-7), and cytostatic effect on Neuro-2a mouse neuroblastoma cells by inhibiting mitochondrial NADH-ubiquinone oxidoreductase leading to mitochondrial dysfunction and finally cell death (Hayakawa et al. 2007b).

Aureolic acids, chromomycins B, A_2, and A_3, are produced from a marine-derived actinobacterium, *Streptomyces* sp. WBF16, with strong cytotoxic activities against human gastric cancer cell line (SGC7901), human liver hepatocellular carcinoma cell line (HepG2), human lung adenocarcinoma epithelial cell line (A549), human colon cancer cell line (HCT-116), human ovarian cancer cell line (COC1), and human umbilical vein endothelial cells (HUVEC). Despite their limited application due to several side effects, as chromomycin are applied clinically in the alleviation of a number of tumor diseases by binding MTM to the DNA resulting in cross-linkage of two strands which in turn blocks RNA synthesis (Lu et al. 2012). This report along with previous finding about isolation of three chromomycin analogues, chromomycins SA_3, SA_2 and deacetylchromomycin A_3, as well as five more analogues from the marine-derived *Streptomyces* sp. SNB-005 (Hu et al. 2011) suggest that marine actinobacteria are a rich source for aureolic acids.

Three natural products, galvaquinones A, B and C are obtained from marine-derived *Streptomyces spinoverrucosus* SNB-032. Moreover, two already known compounds, anthraquinones lupinacidin A and islandicin, were isolated from fermentation broth of this strain. Among all these six alkylated anthraquinone analogues, only galyaquinones B displayed considerable epigenetic modulatory activity and both galyaquinones B as well as anthraquinones lupinacidin A were moderate cytotoxic agent against non-small-cell lung cancer cell lines (NSCLC) (Calu-3 and H2887).

A cytotoxic bioactive product, 5-(2,4-dimethylbenzyl)pyrrolidin-2-one (DMBPO), against human epithelial type 2 cell line (HEp 2) and liver hepatocellular cell line (HepG2) with a weak membranolytic activity on erythrocytic membrane was identified from marine *Streptomyces* VITSVK5 spp. (Saurav and Kannabiran 2012).

The *Streptomyces aureoverticillatus* NPS001583 was isolated in attempt to discover new bioactive metabolites and a novel macrocyclic lactam and aureoverticillactam with 22-atom ring was identified from that. The isolated compound showed cytotoxic activity against colorectal adenocarcinoma cancer cell line (HT-29), melanoma cancer cell line (B16-F10), and leukemia cancer cell line (Jurkat) (Mitchell et al. 2004).

Four novel natural products, lucentamycins A–D, with new amino acid unit, 3-methyl-4-ethylideneproline were isolated from marine-derived *Nocardiopsis lucentensis* strain CNR-712. These compounds belonged to cytotoxic peptides class of antibiotics. Amongst these four unique peptides, only lucentamycins A and B displayed considerable cytotoxic activity against HCT-116 human colon carcinoma with a greater activity derived from lucentamycin A, containing phenyl and indole rings respectively, which highlighting the key role of the aromatic ring for biological activity in this class of compounds (Cho et al. 2007).

Investigation of fermentation broth of marine *Streptomyces* sp. led to isolation of three promising cytotoxic piperazic acids (piperazimycins A–C) with cyclic hexadepsipeptides structure having rare amino acids of hydroxyacetic acid, *a-methylserine*, γ-hydroxypiperazic acid, and γ-chloropiperazic acid. Moreover, two unique amino acid residues 2-amino-8-methyl-4,6-nonadienoic acid and 2-amino-8-methyl-4,6-decadienoic acid were identified in the structures of piperazimycins A and C, respectively. All piperazimycins, A–C, were significant growth inhibitors against human colon carcinoma cell line (HCT-116). Piperazimycin A was potentially active against all oncologically divers panel of 60 cancer cell line showing three times more cytotoxicity against solid tumors, mostly melanoma and prostate cell lines, than that of leukemia cell lines examined and was active against the colon cancer group, renal cancers, ovarian cancers, non-small cell lung cancers, and breast cancers (Miller et al. 2007).

An investigation of LC-MS analysis of culture broth of a new species of the marine *Salinispora pacifica* strain CNS103 has resulted in the identification of two possible cyclization products of C-1027, a naturally occurring enediyne, which possess 3-keto-pyranohexose sugar and novel carbon skeletons, a cyano- and chloro-substituted cyclopenta[*a*]indene ring system with unique benzyl nitrile moiety in a multiple ring system that named as cyanosporasides A and B. The compounds showed weak cytotoxicity against human colon carcinoma (HCT-116) (Oh et al. 2006).

A novel marine actinobacterium strain CNH-099 was found to be able of biosynthesis of neomarinone, a new metabolite containing a new sesquiterpene and polyketide-derived carbon skeleton as well as three derivatives of the marinone class of naphthoquinone antibiotics, isomarinone, hydroxydebromomarinone, and methoxydebromomarinone. The compounds were moderately active against human colon adenocarcinoma cell line (HCT-116). Neomarinone also rendered cytotoxicity against a panel of 60 cancer cell lines (Hardt et al. 2000).

Chemical investigation of fermentation broth of the marine-derived *Actinoalloteichus cyanogriseus* WH1-2216-6 led to isolation of five new bipyridine alkaloids as well as a novel phenylpyridine alkaloid, caerulomycins F–K, along with five known analogues, caerulomycin A, caerulomycin C, caerulomycinamide, caerulomycinonitrile, and (*Z*)-4-methoxy-2-2′-bipyridine-6-carbaldehyde oxime. All isolated compounds exhibited cytotoxic activity against at least one of the cancer cell lines including the human leukemia (HL-60), K562, KB, and human lung cancer (A549) cell lines except the analogue (*Z*)-4-methoxy-2-2′-bipyridine-6-carbaldehyde oxime which showed no activity. Two analogues, caerulomycin A, caerulomycin C, also displayed antimicrobial effects against *Escherichia coli*, *Aerobacter aerogenes*, *Pseudomonas aeruginosa*, and the fungus *Candida albicans* (Fu et al. 2011).

Proximicin A–C, produced from the marine actinobacterium *Verrucosispora* strain MG-37, belong to aminofuran antibiotics and anticancer compounds. This family of compounds shows no antifungal activity, weak antibacterial activity, but significant cytostatic effect to several human tumor cell lines including; gastric adenocarcinoma (AGS), breast carcinoma (MCF 7), and hepatocellular carcinoma (HepG2) (Fiedler et al. 2008).

Four novel manosouramycin A–D with cytotoxicity activity against 36 tumor cell lines were identified from the marine-derived *Streptomyces* sp. isolate Mei37. A significant cytotoxic selectivity was observed towards non-small cell lung cancer, breast cancer, melanoma, and prostate cancer cells by these compounds. Moreover moderate inhibition activity against bacteria; *Staphylococcus aureus*, *Bacillus subtilis*, and *Escherichia coli*, and strong activity against microalgae; *Chlorella vulgaris*, *Clorella sorokiniana*, and *Scenedesmus subspicatus* were exhibited by Mansouramycin A (Hawas et al. 2009).

Fermentation broth screening of the *Streptomyces chilbaensis* AUBN$_1$/7 yielded a bioactive product, resistoflavin, reported for first time from marine sources with potent cytotoxicity against gastric adenocarcinoma cell line (HMO2) and hepatic carcinoma cell line (HePG2) as well as weak antibacterial effect against both Gram-positive and Gram-negative bacteria. This compound relates to quinone class of antibiotics (Gorajana et al. 2007).

3.4 Biosurfactant-Production Capabilities of Halotolerant and Halophilic Actinobacteria

Surfactants including those originated from microorganisms, called biosurfactants, are populated with increasing environmental awareness (Banat et al. 2000; Desai and Banat 1997). These surface active compounds secreted from microorganisms, have the advantages of ecofriendly and non-toxic nature. Some of these amphiphilic biological compounds produced by microorganisms structurally are glycolipids, rhamnolipids, sophorolipids, trehalolipids, phospholipids fatty acid, lipopeptides and lipoproteins or polymeric biosurfactants (Rahman and Gakpe 2008). Biosurfactants have great potential in facilitated bioremediation owing to their emulsifying, dispersing, or solubilizing activities by (i) catalyzing desorption of highly hydrophobic compounds, such as polyaromatic hydrocarbons (PAH), adsorbed into soil, thus accelerate degradation of these compounds, i.e. increase the compound bioavailability and (ii) avoid adsorption of desired microorganisms for bioremediation into soil of a polluted site and thereby increase the microorganisms mobility (Kitamoto et al. 2002). Biosurfactants are capable of removing heavy metals from contaminated soil by means of adsorption (Mulligan et al. 2001; Banat et al. 2000). Beside the above mentioned applications, some biosurfactants such as glycolipid exhibit biological activity due to their chemical structures. In general, Gram-positive bacteria have more susceptible cell wall to biosurfactants than Gram-negative ones. Moreover, antiviral activity has been demonstrated for some glycolipid biosurfactants (Kitamoto et al. 2002). Biosurfactants have superb dispersing, emulsifying, and foaming activities as well as their effectiveness at extreme conditions make them ideal for number of applications including detergents, personal care, enhanced oil recovery, agrochemicals and bioremediation. Despite various advantages of microbial biosurfactants, their drawback over plant-derived surfactants and synthetic ones is solely their higher expenses; the other demerits of microbial biosurfactants are lack of knowledge for their synthesis regulation and isolation in highly pure form. According to Transparency Market Research, the global demand for microbial biosurfactants was almost 12.7$ million in 2012 and are expected to reach 17.1$ million before 2020 (http://www.transparencymarketresearch.com/microbial-biosurfactants-market.html). It has been pointed out that sophorolipids account for 54 % of biosurfactants world market in 2012, mainly due to their high-yield production with rhamnolipids chasing them which are extensively applied in enhanced oil recovery, biopesticides and bioremediation (http://www.transparencymarketresearch.com/microbial-biosurfactants-market.html).

Although, biosurfactants are mainly accepted as secondary metabolites, some of these compounds vitally function and are required by the producing microorganisms as facilitating transportation of nutrients or microorganism-host interactions as well as presenting antibiotic activity (Kiran et al. 2014). The work addressing actinobacteria-derived biosurfactants are very limited, this is even more adverse in case of halophilic and/or halotolerant actinobacteria capable of biosurfactants production (Khopade et al. 2012b). Among the many, the characteristic driving force behind exploration of halophilic biosurfactants from halophilic and halotolerant actinobacteria is their capability to tolerate and restore activity in saline condition (above 3 % salt concentration), the condition which could efficiently deactivates synthetic surfactant.

A marine-derived *Streptomyces* sp. S1 produces a biosurfactants, containing 82 % protein, 17 % polysaccharide and 1 % reducing sugar, with the yield of 3.8 g/L at optimized fermentation conditions, pH 7, 28 °C and 3 % NaCl (w/v). The compound maintained relatively good stability following storage at 28 °C and at neutral pH. The critical micelle concentration (CMC) of partially purified biosurfactant was calculated as 0.3 mg/mL (Kokare et al. 2007). A halotolerant, up to 26 % NaCl (w/v), actinobacterium *Streptomyces* sp. VITDK3 with ability to produce a lipopeptides class of biosurfactant was reported (Lakshmipathy et al. 2010). No further explanation about the characteristic features of this biosurfactant was provided. The most recent report of genus *Streptomyces* with biosurfactant production ability was a moderately halophilic marine-derived *Streptomyces* sp. B3. The isolated and purified biosurfactant was able to reduce water tension by 29 mN/m and had CMC of 110 mg/L. The partial purified biosurfactant exhibited antimicrobial activity against *Bacillus subtilis*, *Pseudomonas aeruginosa*, *Staphylococcus aureus*, *Escherichia coli*, and *Candida albicans*. The biosurfactant production was observed following of the strain cultivation on non-hydrocarbon substrate with maximum production taking place at 30 °C, pH 7, and 4 % NaCl (w/v). The effect of salinity was investigated in range of 0–9 % NaCl (w/v) and emulsification activity was observed at 9 % salinity tested (Khopade et al. 2012b). This highly active halothermophilic biosurfactants belonged to lipopeptides class of biosurfactants and were stable in alkaline pH 8–12, various temperatures from 30–100 °C, and salt tolerated up to 9 % NaCl (w/v) with an optimum at 6 %. This natural biosurfactant showed higher temperature stability than synthetic surfactants such as sodium dodecyl sulfate which begins to significantly loss its emulsification activity from 70 °C onwards (Khopade et al. 2012b). Therefore, this biosurfactants may found applications in bioremediation sector for degrading spills in saline ecosystem owing to its salinity and alkaline stability (Khopade et al. 2012b). More recently, a glycolipid biosurfactant, comprising lipid 61 % (w/w) and carbohydrate 30 % (w/w), dropping surface tension of water from 73.2 to 32.4 mN/m and CMC of 36 mg/L was identified from a marine-derived actinobacterium *Streptomyces* sp. MAB36. A broad spectrum activity against microorganisms was observed from this glycolipid biosurfactant. Its optimum production conditions were carried out at pH 7, 30 °C, and 1.5 % NaCl (w/v) with higher emulsification activity at presence of 5 % NaCl (w/v). The stability was preserved in temperature of 30–50 °C, pH of 5–9, and salinity up to 1.5 % NaCl (w/v) (Manivasagan et al. 2014).

Rhodofactin, belonging to lipopeptides class of biosurfactants, with a CMC value of 23.7 mg/L and of 801.6 Da m.w and linear structure is derived from marine-derived *Rhodococcus* sp. TW53 which could lower water surface tension to 30.7 mN/m. The producing strain shows 99.3 % similarity to halotolerant *Rhodococcus kroppenstedtii*. The yield of rhodofactin was around 0.2 g/L under un-optimized fermentation conditions (Peng et al. 2008). Also, a seaside soil-derived *Rhodococcus erythropolis* sp. strain 3C-9 produces a biosurfactant consisting of two kinds of glycolipids, glucolipid and a trehalose lipid, and a hydrophobic moiety consists of seven fatty acids. Unfortunately, no further stability or production optimization investigations were done for this newly described biosurfactant (Peng et al. 2007). In 2010, a novel rhamnose-containing glycolipid biosurfactant was delivered from halotolerant *Rhodococcus fascians* strain A-3 with 1.8 g/L yield which could reduce the surface tension of water from 72 mN/m to as low as 27 mN/m and suppress the growth of *Bacillus subtilis*. The bioremediation of cold places such as polar soils may be accomplished using this strain isolated from Antarctica due to prohibition of foreign organisms from Antarctic Treaty (Gesheva et al. 2010).

A potent halotolerant sponge-associated marine actinobacterium strain of *Nocardiopsis alba* was recruited for production of a lipopeptide class of biosurfactant with optimum production conditions at pH 8, temperature 30 °C, and 1 % NaCl (w/v). Constituents of the biosurfactant were reported as carbohydrate 20 μg/0.1 mL, protein 35 μg/0.1 mL, and lipid 573 μg/1 mL with stability in temperature, pH, and salinity of 10–90 °C, 4–9, and 0–2 % NaCl (w/v), respectively. The compound exhibited potent antimicrobial activity against *Enterococcus faecalis* and *Bacillus subtilis* and *Candida albicans* (Gandhimathi et al. 2009). Later, a moderately halophilic marine-derived *Nocardiopsis* sp. B4 was isolated which produced a biosurfactants with emulsification index E24 of 80 % and could reduce the surface tension of water up to 29 mN/m. The maximum production of this biosurfactant occurred at pH 7, 30 °C, in presence of 3 % NaCl (w/v) and the producing strain could retain 80 % of its activity in presence of 12 % NaCl (w/v). The compound was stable in the range of 30–100 °C, pH range of 8–12, and salinity of up to 8 % with optimum of 3 % NaCl (w/v) (Khopade et al. 2012a). Also, a novel glycolipid containing nonanoic acid methyl ester biosurfactant was reported from marine sponge-derived *Nocardiopsis lucentenis* MSA04 following solid-state fermentation using a number of industrial wastes including pretreated sludge from tannery, treated molasses derived from distillery waste, and pretreated molasses. Kerosene was able to enhance production of this newly discovered biosurfactants. The biosurfactant was considerably synthesized at pH 7–9 with optimum at 7, 25 °C, and inevitable NaCl concentration of 0.5–5 % (w/v) with optimum of 2 %. The compound, containing 0.123 mg/mL protein, 0.972 mg/mL carbohydrate, 1.886 mg/mL lipid, and 9.0 mg/mL glycolipid, was capable of reducing surface tension of water from 72.5 mN/m to as low as 16.34 mN/m in pure form. Moreover the compound was highly thermostable even after autoclaving, in pH range of 5–9 and NaCl range of 1–4 % (w/v) with CMC of 16 g/L (Kiran et al. 2010b).

A marine-derived actinobacterium *Brevibacterium aureum* MSA13 was found to provide an unprecedented biosurfactants, containing 0.53 mg/mL protein, 0.012 mg/mL carbohydrate and 4.886 mg/mL lipid, which belongs to lipopeptide class containing an octadecanoic acid methyl ester moiety in addition to short sequence of four amino acids peptide part. Optimization of solid state fermentation of industrial and agro-industrial solid waste residues was done for strain MSA13 leading to three folds increase in the production yield of the biosurfactants. The strain could growth on gelatin, cellulose, chitin and starch with optimum production at pH 7–9, temperature 30 °C, and salinity 2 % NaCl (w/v). The compound was named as brevifactin and proved thermo-stability by resistance to autoclaving. In addition, it was stable at pH and NaCl ranges of 5–9 and 1–5 % (w/v), respectively. Brevifactin can efficiently reduce surface tension of water from 72.54 to 28.56 mN/m and had CMC of 18 g/L as well as broad antimicrobial activity with the more potency against *Candida albicans* and *Klebsiella pneumonia* (Seghal Kiran et al. 2010). In the same year, a glycolipid biosurfactant production from a marine sponge-derived halotolerant actinobacterium *Brevibacterium casei* MSA19 under optimized submerge fermentation conditions, pH 7, temperature 30 °C, and salinity 2 % NaCl (w/v), with in vitro antibiofilm activity against pathogenic biofilms as well as wide range of bacteriostatic activity was reported (Kiran et al. 2010a).

Beside the above mentioned actinobacteria genera, a sponge-derived marine *Brachybacterium paraconglomeratum* MSA21 was applied in a solid-state fermentation to convert agro-industrial wastes, containing 27 % protein and 53 % carbohydrate, along with pre-treated tannery effluent to glycolipid biosurfactants. Although this species is halotolerant (up to 15 % NaCl), salinity-biosurfactant production relationship was not investigated above 3 % salinity (Kiran et al. 2014).

3.5 Production of Industrially Important Enzymes by Halophilic and Halotolerant Actinobacteria

Enzymes have applications in many aspects of human life, including pharmaceuticals, clinical analysis, detergents, food production, organic synthesis and bioremediation. These applications are attracting more and more attention since identification of novel microbial enzymes from different niches such as saline environments will introduce new approaches for industry as well as solution for environmental problems.

3.5.1 Amylases

Starch, an important renewable biological resources, degrading extracellular endo-enzymes, amylases (endo-1,4-*a*-D-glucanohydrolases EC 3.2.1.1), are the first commercially marketed enzymes which have occupied the vending of up to 25 % in enzyme market and add up to 30 % of all the enzyme production in globe. Amylases are classified into two subgroups; starch-hydrolyzing enzymes and the transglycosylating amylases, and starch-modifying enzymes. Amylases give diverse products ranging from sugar syrups, to the cyclodextrins productions which have pharmaceutical applications. Enzymatic hydrolysis is preferred over acid hydrolysis of starch because of reaction specificity, product stability, lower requirements of energy, and shorter process. The average annual market growth rate of 3.3 % makes microbial production of amylases very appealing which demands low-cost medium and efficient producing microorganism (Sivaramakrishnan et al. 2006). Therefore a strain which have better compatibility with production conditions and the stability of the resulting amylases produced must be taken into account carefully from industrial perspective view. For instance, strains demanding high salinity and temperature may be applied during fermentation in warm geographical environments likewise are the produced amylases and salinity profile which should be matched with the expected applications. This created diversity in applications requires the investigation for unique amylases with new and improved physiological and biochemical properties. The enzymes which could tolerate more cruel conditions of preservation may be preferred as they need less caution and are more stable against losing their activities.

Currently, 90 % of the carbohydrases produced, which occupy 40 % of all enzymes, are utilized by the food and beverage industry in preparation of a various digestive juices, as well as reduction viscosity of sugar syrup, haze formation reduction in juices, starch solubilization and saccharification in brewing industries for alcohol fermentation, and postponing staling in baking industry. Other industry sections utilized amylases include detergent industry, as an additive for starch-based dirt removal; paper industry, starch viscosity reduction for appropriate paper coating; textile industry, textile warp sizing of fiber to reduce breakage of yarn, and in pharmaceutical industry as a digestive aid (Van Der Maarel et al. 2002; Riegal and Bissinger 2003; Gupta et al. 2003).

Fungal and bacterial amylases are ruling on enzyme market (Pandey et al. 2000), however, the feasibility of using actinobacters, especially streptomycetes for production of amylases has just begun (Syed et al. 2009). In Table 3.3, it has been tried to cover all reported amylases from halophilic and halotolerant actinobacteria source at the time of writing.

3.5.2 Proteases

From industrial perspective, diverse applications of peptidase including detergent additives, biotreatment of leather, bioprocessing of used X-ray films for silver recovery and also polyester film base recycling, pharmaceutical applications, peptide synthesis and protein processing within food industry sector, waste treatment management, and chemical industry, have made these enzymes appealing (Kumar and Takagi 1999). Microbial-derived proteases are very significant hydrolytic enzymes and are among the three largest industrial enzymes which have occupied the vending of up to 60 % in enzyme market (Rao et al. 1998). Marine microorganisms have enormous capability for economically and ecofriendly production of industrial-feasible enzymes (Lam 2006) which are active in extreme conditions of alkaline pH and high temperatures. The characteristic features of proteases produced from halophilic or halotolerant actinobacteria, beside economically and ecofriendly industrial-feasible production (Lam 2006), are their resistances and activities in many harsh salt-containing industrial processes as well as various tolerating capabilities for extreme conditions, mainly alkaline pH and high temperatures, which consequently decreases enzyme turn over in these situations. There have been large numbers of studies from protease producers within archea and bacteria, but the investigations on protease-producing actinobacteria are sparse. Table 3.4 exemplifies the typical valid protease-producing actinobacteria strains from saline conditions.

3.5.3 Cellulases

Investigations of halophilic cellulases capable of functioning in adverse conditions are lagging from mesophilic enzymes (Percival Zhang et al. 2006). Browsing the literature proves the lack of knowledge on biochemical properties, substrate

Table 3.3 Halophilic and halotolerant amylase-producing actinobacteria up to September 2014

Producing strain	Type	MW kDa	Specific activity U/mg	Optimum condition for activity			Stability towards		Reference
				pH	NaCl % (w/v)	Temperature °C	pH	NaCl % (w/v)	
Saccharopolyspora sp. A9	NR	66	1,640.80	11	11	55	8–12	11–17	Chakraborty et al. (2011)
Streptomyces gulbargensis DAS 131	HT	55	1,341.3	9	NR	45	8.5–11	NR	Syed et al. (2009)
Nocardiopsis sp. 7326	NR	55	548	8	NR	35	5–10	NR	Liu et al. (2011)
Streptomyces sp. D1	NR	66	113.64	9	7	45	7–11	2–15	Chakraborty et al. (2009)

H halophilic, *MH* moderate halophilic, *HT* halotolerant, *NR* not reported

Table 3.4 Showing halophilic and halotolerant protease-producing actinobacteria up to September 2014

Producing strain	Type	MW kDa	Specific activity at most U/mg	Optimum condition			Stability towards			Reference
				pH	NaCl % (w/v)	Temperature °C	pH	NaCl % (w/v)	Temp. °C	
Streptomyces clavuligerus strain Mit-1	HT alkaliphilic	49–50	6,800	10–11	NR	70	8.5–10	NR	60–80	Thumar and Singh (2007)
Nocardiopsis prasina HA-4	MH alkalithermotolerant	NR	NR	7 and 10	NR	55	6–12	NR	20–65	Ningthoujam et al. (2009)
Nocardiopsis alba OK-5	HT alkaliphilic	20	19,011.15	10–11	30 % Na-glutamate then 23.4	70–80	6–12	0–23.4,30 % Na Glutamate	60–80	Gohel and Singh (2012a)
Actinopolyspora sp. strain VITSDK2	H	22	NR	10	NR	60	4–12	NR	25–80	Suthindhiran et al. (2014)
Actinomycete MA1-1	MH alkaliphilic	NR	7,618	9	NR	50	8–13	NR	35–50	Hameş-Kocabaş and Uzel (2007)
Brachystreptospora xinjiangensis OM-6	HT alkaliphilic	25	22,938	10–11	30 % Na-glutamate then 21.4	70	6–12	0–23.430 % Na Glutamate	37–80	Gohel and Singh (2012b)
Streptomyces fungicidicus MML1614	NR	NR	315.66	9	NR	40	6–11	0–11	28–60	Ramesh et al. (2009)
Nocardiopsis sp. SD5	MH	30 and 60	NR	9	NR	50	NR	NR	NR	Saha et al. (2013)
Streptomyces sp. CW1	HT	NR	NR	NR	NR	NR	NR	NR	NR	Kurzbaum et al. (2010)

H halophilic, *MH* moderate halophilic, *HT* halotolerant, *NR* not reported

specificity, genetic characteristic, and gene products of cellulose decomposition with halophilic nature which are crucial for any biotechnological industry containing high salinity processes (Margesin and Schinner 2001).

Genus *Thermobifida* has strong cellulose degradation potential with distinction from other microorganisms and has a halotolerant member, *T. halotolerans*, isolated from salt mine with salt-tolerance up to 10 % (Yang et al. 2008). Another outstanding cellulose degrading genus under phylum actinobacteria is *Cellulomonas* which provides two halotolerant species namely as *Cellulomonas carbonis* and *C. bogoriensis* isolated from coal mine and sediment of the littoral zone of a lake, respectively. It has been shown that cellulose decomposition is greatly affected by salinity-related factor which merely accounted almost 45 % among all the investigated factors. This factor become even more importance when it comes to know that sea level rises as a result of global warming which in turn impacts coastal wetland by increasing salinity (Mendelssohn et al. 1999). Therefore, when halophilic and halotolerant actinobacteria are applied in agriculture, they increase minerals in saline soil through cellulose-decomposition while have ecological function of carbon cycling and waste remediation in saline marshes or can be even applied for commercial purposes through fermentation conversion of salt-contaminated cellulose to desirable products, and in wood and paper industry.

3.5.4 *Chitinases*

Wastes coming from aquaculture (shrimp and cab shells) have infinite treasure of chitin with as high as 60–80 % of the whole shrimp and crabs. However, only a small fraction of this amount is recycled to produce cheap food of shrimp and crabs powder (Wang et al. 2011a). Motivation for recovering unused chitin wastes could be arisen from the potential of chitin-decomposing microorganisms to make highly value-added products in association with modern biotechnology upon processing such as nanofibers for biomedical application (Jayakumar et al. 2010) and production of enzymes and bioactive compounds (Wang et al. 2011a). Currently, enzymes, deodorants, biofertilizers, antioxidants, biofungicides, hydrogen gas, and carotenoids have been reported from microbial fermentation of chitin-containing biowastes. The ecological significance of microbial degradation of chitin as well as its potential for biotechnology has made it valuable enough to be investigated. Among bacteria, *Firmicutes*, actinobacteria, bacilli and clostridia, provide most active chitinolytics. Many literatures have dealt with the main chitinase-producing *Streptomyces* genus which degrades chitin as primary carbon source, as potential biocontrol agents due to their inhibitory activity against many plant pathogenic fungi. One of the prominent works for screening haloalkaliphilic chitinolytic actinobacteria has been done leading to isolation of seven strains belonged to genera *Nocardiopsis*, *Streptomyces*, *Glycomyces* and group of *Cellulomonas-Isoptericola*. These strains could tolerate salt from 8 % to 10.7 % (w/v) in alkaline pH (Sorokin and Kolganova 2014). In other work *Streptomyces* sp. CW1 was able to efficiently utilize chitin as sole source of carbon while tolerating up to 11 % NaCl (w/v) (Kurzbaum et al. 2010). Also, actinobacterium strain VITDDK2 was isolated from marine soil and exhibited chitinase activity (Deepika and Kannabiran 2010).

3.5.5 Xylanases

Potential applications of xylanase in some industry sectors, including bioconversion of cellulose material to value-added products (Butt et al. 2008), paper pulp delignification (Gübitz et al. 1997), production of biofuel, baking and starch from degradation of lignocellulosic materials (Khandeparker and Numan 2008), have popularized this enzyme but at the same time either extreme conditions or extreme pretreatment requirement for these processes discourages the industrial application of this enzyme (Zhang et al. 2012a). Many xylanase-producing microorganisms have been reported in literature, however, the most potent xylanase coping up with many cruel conditions governing industrial processes are provided by extremophiles. Notwithstanding, a high number of extremophile-derived xylanase, a few have been addressed to halophilic species while discussions about xylanase-producing halophilic actinobacteria are even more rare.

One of the xylanase-producing genera within phylum actinobacteria is *Thermobifida*. Despite, outstanding (hemi) cellulolytic degrading abilities revealed from species of this genus, *T. halotolerans* sp. is the only halotolerant species within this genus. A xylanase named ThxynA identified from this species is significantly important because of its thermophilic and halotolerant capabilities (Zhang et al. 2012a). Moderately halophilic xylanase-producing species from *Streptomonospora* capable of growing in up to 15 % or 20 % NaCl (w/v) was recently reported for first time from this genus (Ren et al. 2013). Both the identified xylanases were suitable for direct enzymatic treatment of alkaline pulp without requirement for pH re-adjustment steps which in turn speeding up the process economically (Mishra and Thakur 2011). The relatively high thermostability of these enzymes make them suitable to be exploited in industry for hydrolysis at elevated temperatures with advantages of speeding reactions, avoiding microbial contamination (Maheshwari et al. 2000), and therefore, avoiding lost and/or reducing maintenance expenses Table 3.5.

3.5.6 Other Enzymes

Esterases Extremophilic esterases are of particular attentions because of functional and configuration stability of their proteins and are repeatedly employed for production of optically pure compounds as biocatalysts in fine chemical applications. Despite, the availability of mature market for esterases, little success in their exploration and production from archea and bacteria has been achieved. High temperature and salt-tolerating capabilities of esterase are exploited in development

Table 3.5 Showing halophilic and halotolerant xylanase-producing actinobacteria up to September 2014

Producing strain	Type	MW kDa	Specific activity at most U/mg	K_m mg/mL	V_m µmol/min/mg	Optimum condition		Stability towards		Reference
						pH	Temperature °C	pH	Temp. °C	
Streptomonospora sp. YIM 90494	MH	50	13.4	19.24	6.1	7.5	55	4–10	NR	Ren et al. (2013)
Thermobifida halotolerans YIM 90462	HT	24	173.1	11.6	434	6	80	6–10	45–70	Zhang et al. (2012a)

MH moderate halophilic, *HT* halotolerant

and/or improvement of enzymes which could resist denaturation in organic solvents. One of the few examples of halotolerant esterase-producing actinobacteria is *Rhodococcus* sp. LKE-028. The resulting esterase was thermoalkaliphilic halotolerant with specific activity of 861.2 U/mg proteins, 38 kDa, as well as k_m and V_m of 525 nM and 1,666.7 U/mg proteins, respectively. The esterase exhibited activity in temperature range of 40–100 °C with optimum at 70 °C and pH range of 7–12 with optimum at 11. The pure form of enzyme preserved its activity with 58.4 % NaCl (w/v). The enzyme proved to be stable in various organic solvents tested and also it was resistance to protease (Kumar et al. 2012).

L-glutaminase Potential industrial and therapeutic applications make this enzyme very important (Kashyap et al. 2002). Apart from various applications in food technology sector, this enzyme is a potent cytotoxic compound against leukemia cancer which induces selective death in glutamine-dependent tumor cells through dispossessing glutamine from these cells (Roberts et al. 1970; Pal and Maity 1992). In addition, L-glutaminase is used in biosensors for monitoring the level of glutamine in mammalian hybridoma cell cultures with no requirement for direct glutamate measurement (Sabu et al. 2000). In food industry, L-glutaminase plays as an enhancing-flavor of food produced in fermentation sector by elevating the glutamic acid content and consequently conferring a pleasant taste. In fermentation of soy sauce, increasing the amount of glutamate in the mash is a significant stage for development of delicious taste. The application of allergic monosodium glutamic acid in Chinese dishes has almost been stopped by introduction of L-glutamate as a flavor-enhancing agent. Glutamate synthesis activity of L-glutaminase makes it a significant additive for processing of Shoyu Koji using enzymatic digestion (Sabu et al. 2000; Kashyap et al. 2002). The L-glutaminase produced by halophilic actinobacteria becomes significant in saline circumstances such as some industrial processes with high salinity conditions. Many L-glutaminase are produced by submerged fermentation, however, the more recent technique is solid-state fermentation which is able to utilize natural product and/or agricultural by-products and may yield higher concentration compared to the submerged fermentation. There are only few reports on production of L-glutaminase from marine-derived actinobacteria.

A marine-derived halotolerant actinobacterium, *Streptomyces* sp. SBU1 was identified as an L-glutaminase-producing strain and the yield was enhanced after optimizing the production at pH 9.0; temperature 30 °C; and NaCl 2 % (w/v). The produced enzyme L-glutaminase had maximum activity of 18 U/mL and production efficiency of the strain decreased corresponding with increasing the concentration of salt up to 5 % NaCl (w/v) (Krishnakumar et al. 2011). Further, *Streptomyces olivochromogens* strain P2 could tolerate up to 5 % NaCl (w/v) and optimally produces L-glutaminase at pH 7.0, temperature 30 °C, and NaCl 3.5 % (w/v).

Nitrate Reductase Effluents containing high concentration of salt and nitrate or nitrite are generated as a consequence of some chemicals production including pesticides, herbicides, explosive, and dyes leading to concentration-elevation of salt and nitrite in soil and ground water. The application of bacteria for bioremediation of these compounds are challenging due to their sensitivity to even low amount of nitrate (Carr and Ferguson 1990; Martínez-Luque et al. 1991), nitric as a product of nitrate reduction followed by nitric transformation to genotoxic nitrosamines (Van Maanen et al. 1996). The low water activity and the presence of complex mixtures of xenobiotic compounds reduce the number of potential bacteria extremely (Ventosa et al. 1998; Moreno-Vivián et al. 1999). The salinity of this effluent made halophilic or halotolerant microorganisms as potential candidate among many other nitrate reductase-producing microorganisms since salinity simply can undo nitrate assimilation by blocking nitrate uptake in non halophilic or non halotolerant microorganisms. Potential nitrate-reducing genera within phylum actinobacteria are *Rhodococcus*, *Actinopolyspora* and *Nocardiopsis*.

A new moderately halotolerant actinobacterium *Rhodococcus* sp. RB1 is capable to grow aerobically in the presence of up to 0.9 M nitrate and 60 mM nitrite, consuming both of the mentioned compounds, with high ability to degrade xenobiotic compounds including nitrophenols. The nitrate reductase had molecular weight of 142 kDa, independent to salt availability and was able to use two electron donors of NADH and reduced bromophenol blue. The regulation of assimilatory NADH-nitrate reductase was a dual control, inducing and repressing by availability of nitrate and ammonium/glutamine, respectively. This enzyme showed *in vitro* activity up to 17.5 % NaCl (w/v) with optimum at 2.9 % NaCl (w/v). Optimum temperature for activity was increased from 12 to 32 °C in presence of 11.7 % NaCl (w/v) but optimal pH 10.3 was unaffected (Blasco et al. 2001).

Inulinase This abundant and cheap plant-originated polysaccharide can be applied for generation of a numerous products including fructose syrup (as sweetener), inulo-oligosaccharide (as prebiotics) as well as bioethanol which are highly demanded in various industrial sectors such as pharmaceutical, food, beverage and bioenergy with some advantages including low cariogenicity, iron adsorption elevation and stronger sweetening ability than sucrose as well as beneficial effects in diabetic patients (Yewale et al. 2013; Kango and Jain 2011; Lu et al. 2014). Two classes of inulinases are endoinulinases (2,1-β-D-fructan-fructanohydrolase, EC 3.2.1.7), reduction of long chain of inulin to smaller oligosaccharides, and exoinulinases (β-D-fructan-fructanohydrolase, EC 3.2.1.80), hydrolysis of the terminal fructose on the inulin chains. Microorganisms can produce this industrial demanded enzyme commercially and economically. Unlike many other bacteria and fungi, halophilic inulinase-producing actinobacteria are undervalued. A halotolerant *Nocardiopsis* sp. DN-K15, capable to grow at 12 % NaCl (w/v), produces 25.1 U/mL of a unique halo-alkalitolerant and thermostable inulinase during 60 h fermentation which showed highest inulin-hydrolyzing activity at pH 8.0 and 60 °C, respectively. The enzyme was active in broad ranges of pH 5–11, temperature 40–70 °C, and NaCl 0–15 % (w/v) (Lu et al. 2014). The highest production occurred at pH 8.5, temperature 37 °C, and 1 % NaCl (w/v).

Keratinase Like many other fields, investigation of halophilic actinobacteria for keratinase production is limited. The only valid species reported to render keratinolytic activity is *Nocardiopsis halotolerans* (Al-Zarban et al. 2002a) and *Amycolatopsis keratiniphila* (Al-Musallam et al. 2003), both isolated are from salt marsh soil in Kuwait, *Actinomadura keratinilytica* (Saha et al. 2013) isolated from bovine manure compost, while *Nacardiopsis* sp. (Saha et al. 2013) isolated from a feather waste site. However, the only slightly complete report was about a keratin-degrading *Nocardiopsis* sp. SD5 which was a moderate halotolerant actinobacterium, up to 7.5 % NaCl (w/v) with the ability to decompose the feather. The enzyme produced exhibited noticeable keratinolytic activity which was stable over neutral pH and temperature at 40 °C with optimum activity at pH 9 and 50 °C, respectively (Saha et al. 2013).

3.5.7 Enzyme Inhibitors

An inhibitor of *β*-Glucosidase was first described in the year 2005 from a marine-derived actinobacterium which could tolerate up to 6 % NaCl (w/v) (Imada 2005). Since *β*-Glucosidases function in tumor metastasis and human immunodeficiency virus (HIV), the bioactive metabolites rendering inhibitory activity against this enzyme may find application in medicine. Another example is a halotolerant actinobacterium belonging to genus *Streptomyces* sp. SA-3501 producing two bioactive metabolites pyrostatin A and B, with 4-hydroxy-2-imino-l-methylpyrrolidine-5-carboxylic acid and 2-imino-l-methylpyrrolidine-5-carboxylic acid structures, respectively, with *N*-Acetyl-glucosaminidase competitive inhibitory activity (Aoyama et al. 1995). The potential for exploration of inhibitors for this enzyme is its growing application for causative agent elucidation in patients suffering from diabetes, leukemia or cancer. Similarly, Pyrizinostatin, a non-competitive inhibitor of pyroglutamyl peptidase, was isolated from fermentation broth of marine-derived *Streptomyces* sp. SA2289 showed antibiotic activity against *Proteus vulgaris* OX19 and *Shigella sonnei* as well as no toxicity following intravenously injection to mice (Aoyagi et al. 1992). Terminal amino acids are blocked upon the release of pyroglutamyl residues in many peptides and proteins. This action can be catalyzed by pyroglutamyl peptidase and has application in demonstration of the possible physiological role in the living body, which make search for this type of inhibitor sounds reasonable. One of the few examples of *α*-amylase inhibitors-producing marine microorganisms is *Streptomyces corchorusii* subsp. *rhodomarinus* subsp. nov., is capable of salt-tolerating up to 12 % NaCl (w/v). The enzyme inhibitor isolated from this halotolerant actinobacterium was named as amylostreptin (Imada and Simidu 1988). Inhibitors of *α*-amylase are significantly helpful tools for amylase isozyme activities determination, purification, and suppressing diseases such as diabetes, obesity, and hyperlipemia. More recently, exploration for amylase enzymes inhibitors from marine-derived actinobacteria led to identification of four strains requiring seawater for their growth which showed activity against prokaryotic and eukaryotic amylases (Raja et al. 2010). Further, three γ-pyrones, marinactinones A–C, from

Marinactinospora thermotolerans SCSIO 00606 showed cytotoxic activity (already explained under Sect. 3.3.2) as well as blocking effect against topoisomerase II activity (Wang et al. 2010a). The enzymes responsible for the DNA topology conversion and replication are topoisomerases, therefore, these enzymes are essential for genetic processes in many aspects (Wang 1985). Topoisomerase II inhibitors have potential to be used as antimicrobial and/or cytotoxic drugs. A weak inhibition against the recombinant enzyme sortase B was observed from two newly cyclic peptides, belong to nocardamine class, produced from marine sponge-derived *Streptomyces* sp. strain M1087 (Lee et al. 2005).

3.6 Halophilic and Halotolerant Actinobacteria Capability in Bioremediation of Polluted Area

3.6.1 Hydrocarbon Degradation

The incidents of hypersaline environments contamination with petroleum compounds are globally widespread. Some of them are oil fields, saline industrial effluents, natural saline lakes, sea and oceans. The problem are posed significantly by oilfield specially in regards to number of these sites in globe as well as the salty brackish water during oil and natural gas exploitation in ratio of 1:10 barrel, oil and salty water, respectively (Fathepure 2014). High salt levels (1–250 g/L), oil and grease, many toxic chemicals, naturally occurring radioactive compounds, and heavy metals are the predominant mixtures in effluent water from oilfield (Fathepure 2014). In bioremediation, toxic pollutants are degraded to products including CO_2, H_2O and other inorganic substances through ecologically harmless and cost efficiently processes (Philp et al. 2005). Approximately 25 % of all land contaminated by petroleum hydrocarbons are naturally bioremediated thereby highlighting the significant potential of microorganisms for remediation strategies (Holden et al. 2002). To eliminate detrimental effects of salinity on microorganisms such as cell membrane disruption, enzyme denaturation, oxygen low solubility, hydrocarbons low solubility, and low activity water, the characteristic ability of halophilic or halotolerant microorganisms in tolerating salt are exploited (Pernetti and Palma 2005). Halophilic and halotolerant actinobacteria are among the best possible options for petroleum bioremediation owing to their capabilities to thrive in harsh conditions with occasional incidents of dryness or nutrition run out as well as their abilities of biosurfactants secretions which facilitate uptake of hydrophobic hydrocarbons. Among halophilic and halotolerant hydrocarbon-degrading actinobacteria are reported members of the genera *Streptomyces* (Kuznetsov et al. 1992; Kurzbaum et al. 2010), *Micrococcus* (Ashok et al. 1995), *Arthrobacter* (Plotnikova et al. 2001), *Rhodococcus* (Plotnikova et al. 2001; Borzenkov et al. 2006), *Cellulomonas* (Riis et al. 2003), *Actinopolyspora* (Al-Mueini et al. 2007), *Gordonia* (Xue et al. 2003). Notwithstanding, the relatively high number of reports on hydrocarbon-decomposing microorganisms, little is known about genes and pathways involved in degradation mechanisms. One of the prominent works on this area has been performed on genes and enzymes participated in the initial steps of the benzene decomposition process in halophiles (Dalvi et al. 2012). It was assumed that benzene is first converted to phenol and later catechol is formed due to the action of phenol hydroxylase. In the next step, the ring of produced catechol is cleaved through meta pathway with the support of 2,3-dioxygenase (2,3-CAT) to form 2-hydroxymuconic semialdehyde. For development of alternative and economically more feasible bioremediation techniques, discovery and understanding of more unprecedented genes as well as pathways are vitally important (Dalvi et al. 2012). Table 3.6 presents various halophilic or halotolerant actinobacteria with hydrocarbon-degrading capability.

3.6.2 Heavy Metal Bioremediation and Biomining by Halophilic and Halotolerant Actinobacteria

The expanding inadequacy of mining technology to meet future sustainable development from conventional metal resources causes urgent requirement for investigation of either novel technology-aided, exploitation from unconventional resources or upgrading current mining techniques. One potential example of the former is bacterial dissolution of metals to generate metals from resources including, ores, soils, and solid waste compounds, which their exploitation by conventional mining is not economically feasible (Brandl 2008). This novel technique called biomining (Mishra and Rhee 2014), can be performed directly by physical contact of microorganisms or indirectly through ferric-ferrous cycle mediated by microorganisms. Two processes involved in biomining include; bioleaching process in which sulfides of metal compounds are converted into water-soluble sulfate forms by microorganisms acting as catalyzers and then the elements are extracted from these compounds by filtering water through them, another mechanism is biooxidation process in which the interested elements are concentrated in the solid material while unwanted minerals are oxidized by bacteria. This technique can efficiently cope with recovery of valuable metals from both commercial and environmental viewpoints owing to its relatively inexpensive, versatile, and ecofriendly nature.

Bioleaching activities are prominent in acidophilic and chemolithotrophic bacteria (Brandl 2008). This capability was also reported from alkaliphilic, heterotrophic microorganisms (Willscher and Bosecker 2001), and more recently from some halotolerant actinobacteria such as genera *Nocardiopsis* (Schippers et al. 2002). The growth of acidophilic microorganisms are chloride-dependent and are inhibited in presence of high concentration of this ion (Zammit et al. 2012), thus the halophilic and halotolerant actinobacteria are of special interest due to their ability to thrive in high concentration of NaCl allowing to take advantage of seawater for prospective biomining techniques in drought affected environments.

Table 3.6 Exhibits halophilic and halotolerant hydrocarbon-degrading actinobacteria up to September 2014

Hydrocarbon	Degrader strain	Salinity (%)	Comment/sole carbon source	Reference
Crude oil	*Streptomyces albiaxialis*	3–30	Yes	Kuznetsov et al. (1992)
	Rhodococcus erythropolis	0–10		Zvyagintseva et al. (2001)
	Cellulomonas sp.	>17	Survival for long period of time	Riis et al. (2003)
	Rhodococcus sp.	0–10	Oxidization of *n*-alkane fraction of crude oil	Borzenkov et al. (2006)
	Gordonia sp.	0–15		
	Rhodococcus sp. NCIM 5126	0–10	–	Sharma and Pant (2000, 2001)
	Actinopolyspora sp. DPD1	20	–	Al-Mueini et al. (2007)
Paraffin	*Gordonia paraffinivorans*	0.5–7		Xue et al. (2003)
Aliphatic compounds				
Octane	*Rhodococcus* sp. and *Arthrobacter* sp.	6	Naphthalene and phenanthrene as the sole sources of carbon	Plotnikova et al. (2001, 2011)
n-alkanes and fluorine	*Actinopolyspora* sp. DPD1	25	Degrades 100 % Pentadecane in 4 days, 80 % eicosane in 10 days, and 15 % pentacoase in 14 days. No triacontane was degraded. Novel breakdown pathway for fluorine	Al-Mueini et al. (2007)
n-alkanes	*Dietzia psychralcaliphila*		Yes	Yumoto et al. (2002)
Polycyclic aromatic hydrocarbon				
Naphthalene	*Micrococcus* sp.	7.5	Yes	Ashok et al. (1995)
	Arthrobacter spp. SN17	6–9		Plotnikova et al. (2011)
Anthracene and phenanthrene	*Micrococcus* sp.	5–9	Anthracene as sole carbon source	Ashok et al. (1995)
Fluorene	*Actinopolyspora* sp. DPD1	5–20	–	Al-Mueini et al. (2007)
Biphenyl	*Rhodococcus* sp., and *Arthrobacter* sp.	6–9	Yes	Plotnikova et al. (2001, 2011)

(continued)

Table 3.6 (continued)

Hydrocarbon	Degrader strain	Salinity (%)	Comment/sole carbon source	Reference
Phenol	*Arthrobacter* sp.	6–9	Yes	Plotnikova et al. (2011)
	Streptomyces sp. CW1	0–11	Catechol inhibitory activity maximum phenol concentration tolerated was 50 mg/L	Kurzbaum et al. (2010)
Salicylate, *o*-Phthalate, and gentisate	*Arthrobacter* sp., and *Rhodococcus* sp.	5–9	Salicylate as sole carbon source	Plotnikova et al. (2001, 2011)

The effluents from mining operation, tannery, textile, fabric dye, metallic alloys, petrochemical sectors are rich in salts as well as heavy metal ions including nickel, cadmium, mercury, cobalt, zinc, tellurite, arsenic, and copper (Patterson 1987). Upon exceeding the tolerance levels, these persistent metals exhibit many detrimental effects on biological systems by accumulating in living organisms. Conventional methods including precipitation, reverse osmosis, ion exchange, redox processes, and etc. are expensive and inappropriate when only low concentration of metal is present in relatively large solution volume and high eluent quality is desired (Valdman and Leite 2000). One of the cost-effective techniques for waste water treatment is biosorption in which the sequestering of metals is carried out using biological or industrial biosorbents. Metal ion removal by biosorbents may be through coordination, chelation, ion exchange, complexation, adsorption, and inorganic microprecipitation (Volesky 1990). To recover obtained metals from biosorbents in small, concentrated volume, a desorption mechanism is employed which is identical to that of biosorption involving ion exchange or complexation and biosorbents are eluted with a suitable washing eluent (Vilar et al. 2007). The feasibility of this technique has direct relationship to the metal adsorbent capacity of biosorbents-desorbents and its recycling ability as well as the valid wastewater discharge protocols (Njikam and Schiewer 2012). Halophilic and halotolerant metal-remediating actinobacteria till the time of writing have been presented in Table 3.7.

3.6.3 Humic Acid-Reducing Halotolerant Actinobacteria

A ubiquitous, highly abundant and heterogeneous class of organic compounds on the world is humic substances (Stevenson 1994). These substances can be utilized by microorganism as source of carbon. The microbe-reduced humus, which is stable soils and sediments accumulated organic substances, plays a dynamic role as mediators in the anaerobic oxidation of organic or inorganic pollutants including

Table 3.7 Halophilic and halotolerant metal-remediating actinobacteria up to September 2014

Strain name	Type	NaCl range (optimum) % (w/v)	Heavy metal/comment	Optimum condition	Reference
Actinopolyspora sp. SH-9	H	15–25 (17 and 22)	Mercury chloride (100–150 nmole)	NR	Senthilkuraar (2005)
Streptomyces VITDDK3	HT	NR	Cadmium and lead	NR	Lakshmipathy et al. (2010)
Streptomyces sp. MSI01	NR	NR	Multi-metal resistant	NR	Selvin et al. (2009)
Saccharomonospora sp. MSI36	NR	NR	Multi-metal resistant	NR	Selvin et al. (2009)
Nocardiopsis metallicus	HT alkaliphilic	0–10	Dissolve metals from siliceous	pH 7–10.5 (8.5) temperature 30	Schippers et al. (2002)
Micromonospora sp. MSI28	NR	NR	Multi-metal resistant/ mostly cobalt	NR	Selvin et al. (2009)
Thermoactinomyces sp. QS-2006	MH	2.9–8.8 (1–3)	Tellurite (500 mM) Maximum removal 115 g/L	pH 7–10 (7.5) temperature 30–45 (35) adsorbed to the bacterium wall and inside the cytoplasm	Amoozegar et al. (2012)
Streptomyces sp. MS-2	H	Sea water	Chromium(VI) reduction to Cr(III) (75 mg/L within 12 h)	pH 7 and temperature 37	Mona (2008)

H halophilic, *MH* moderate halophilic, *HT* halotolerant, *NR* not reported

nitroaromatic or halogenated compounds, azo dyes, chlorinated organic contaminants, and iron (III) oxides by transferring electrons in availability of an appropriate electron donor such as sulfide, nitrate and ferric iron (Schwarzenbach et al. 1990; Benz et al. 1998; Cervantes et al. 2000; Hong et al. 2007; Kappler et al. 2004; Wang et al. 2009). Quinone moieties are responsible for humus redox reactions. Many reported humus reducing microorganism are able to transfer electrons to anthraquinone-2,6-disulfonate (AQDS) and subsequently, its reduction to anthrahydroquinone-2,6-disulfonate (AH_2QDS) (Lovley et al. 1996, 1998; Francis et al. 2000). In fact, two halotolerant, alkaliphilic, humic acid-reducing actinobacteria *Corynebacterium humireducens* (Wu et al. 2011), growing at 0–13 % NaCl (w/v), at pH 7–11 and 25–45 °C, and *Kocuria rosea* HN01 (Wu et al. 2014), growing at 0–12 % NaCl (w/v), at pH 6–10 and 30 °C were described. The optimum growth for both actinobacteria was occurred at 10 % NaCl (w/v) and pH 9.0, while temperature optimum for former organism was at 37 °C. These strains were capable of anthraquinone-2,6-disulfonate (AQDS) reduction under anaerobic condition. *Corynebacterium humireducens* were capable of utilizing electron from lactic acid, formic acid, acetic acid, ethanol or sucrose, whereas *Kocuria rosea* HN01 could utilize electron from sucrose, glucose, methanol, glycerol, acetic acid, and ethanol. *Kocuria rosea* HN01 could also reduce iron(III) oxides at pH 9.0 when sucrose was available. The effective dechlorination of *p,p′*-DDT (1,1,1-trichloro-2,2-*bis*(4-chlorophenyl) ethane) was also reported for this strain. To the best of our knowledge, *C. humireducens* was the first reported alkaliphilic bacterium capable of reducing humic acid under alkaline condition, while *K. rosea* HN01 was the second; therefore these two strains increased the upper pH limit for quinone reduction by microorganisms. The higher pH is important as it dissolves humus and in turn increases its availability for humus-reducing bacteria.

3.6.4 Polyester and Bioplastic Degradation

Poly[(*R*)-3-hydroxyalkanoic acids] [poly(HAs)] is a group of polyesters which belongs to class of bacterial storage polymers. This group is synthesized and accumulated intracellularly up to 90 % cellular dry weight in the form of granules by both Gram-positive and Gram-negative bacteria (Jendrossek 2001; Steinbüchel and Hein 2001). The microorganisms capacity to incorporate as many as 60 dissimilar monomer types into their storage compound has resulted in rising a variety of compounds belonging to this family which has been described in the last three decades with poly-3-hydroxybutyrate [poly(3HB)] as well as its copolymer, 3-hydroxyvalerate [poly(3HB-co-3HV)], as the most famous and abundant ones (Steinbüchel 1991). In general, poly(HAs) are classified according to the number of monomer carbon atoms to short-chain-length with crystallinity properties, consisting 3–5 atoms of carbon, and medium-chain-length with elastomeric properties, consisting 6–15 atoms of carbon, denoted as poly(HA_{SCL}) and poly(HA_{MCL}), respectively (Gagnon et al. 1992; Holmes 1988; Anderson and Dawes 1990). This group of polymers has potential commercial applications as thermoplastic elastomers with advantages of ecofriendly and biodegradability to water, methane and/or carbon dioxide as the end products over saturated, non-functionalized carbon skeletons polymers such as poly (ethylene), making poly (HAs) perfect substituents of the latter non-degradable plastics. Poly(HA) as biodegradable plastic is industrially produced and commercialized by Zeneca Bio products (Great Britain) and under trade name BIOPOL®, respectively (Brandi et al. 1995). Poly(HA) can be degraded by various, ubiquitous microorganisms using poly(HA) hydrolyzing enzymes, also known as poly(HA) depolymerases, either extracellularly or intracellularly which is the utilization of an exogenous and active mobilization of an endogenous carbon or energy source, respectively (Jendrossek 2001). The significant merit of poly(HAs) over other biodegradable aliphatic polyesters, including polycaprolactone [poly(LAs)], poly(*β-hydroxybutyrate*) [poly(HBs)], and poly(butylene succinate) [poly(BSs)], is the lack of requirement for specific condition in degradation varying from aerobic to anaerobic situation and, consequently cope up the large space occupied with landfill disposal of polymers.

To the time of writing, bioplastic degradation has been reported for only three halotolerant and halophilic actinobacteria genera including *Stretoalloteichus hindustanus* JCM 3268; degrading poly(LAs) (Jarerat et al. 2002), *Streptomyces* and *Nocardiopsis aegyptia* (Ghanem et al. 2005), and obligate marine *Streptomyces* sp. SNG9 (Mabrouk and Sabry 2001). They were capable of decomposing poly(HBs) and its copolymer poly(3HB-co-3HV).

3.6.5 Reduction Azo-Dye

Industry expansions as the consequences of civilization have made environment more polluted due to discharge of large amount of pollutants to ecosystem. Among these pollutants are dyes effluents which are produced by textile, food, paper, leather, and cosmetic, dyeing and dyestuff industries. These sulphonated waste compounds are colorant, biorecalcitrant, toxic and carcinogenic for organisms (Levine 1991). High biological oxygen demand (BOD), chemical oxygen demand (COD), heat, color, pH and presence of heavy metals make treatments of the textile and dying effluents impractical, highly expensive, and challenging by commonly applicable physical and chemical method (Levine 1991). Bioremediation of these xenobiotics by improving natural degradation capability of indigenous organisms are quite burdensome. In addition the only safe process for the bioremediation of azo dyes is through aerobic treatment since anaerobic treatment leads to production of carcinogenic and mutagenic aromatic amines (Levine 1991). Actinobacteria are capable of azo dye removal through biodegradation, biotransformation or liberalization (Banat et al. 1996; Chung and Stevens 1993). The effectiveness of this treatment depends directly on activities, survivals, and adaptabilities of the interested microorganisms (Cripps et al. 1990; Pasti-Grigsby et al. 1992).

The only described azo dye-reducing halotolerant actinobacterium in literature is *Streptomyces* sp. VITDDK3 with ability to degrade close to 98 % of the azo dye and Reactive Red 5B (RR5B) with initial concentration of 50 mg/L (Lakshmipathy et al. 2010).

3.7 Compatible Solutes from Halophilic and Halotolerant Actinobacteria

Organisms inhabiting in saline water and/or salty environments must be able to cope up injurious osmotic mediated stresses. Halophilic and halotolerant actinobacteria, in addition to ion accumulation like KCl, have the advantage of synthesizing or accumulating compatible solutes or osmolytes for thriving devastating effects of salt. These low molecular densities, highly hydrophilic organic solutes which are compatible with the cellular metabolism and are also responsible for osmotic balance divided in five classes. The first class is disaccharides sugars containing trehalose, mannosucrose and sucrose, the second class is polyols containing glycerol, sorbitol, manitol, mannosyl-glyceramide, mannosyl-glycerol and α-glucosyl-glycerol, the third class is N-acetylated diamino acids compounds such as N-acetylglutaminylglutamine amide, the fourth class is betaines compounds, for example glycine betaine and its derivatives, the fifth class is amino acids including proline, hydroxyproline glutamate and alanine, as well as amino acid derivatives (Pastor et al. 2010), cyclic amino acid ectoine (1,4,5,6-tetrahydro-2-methyl-4-pyrimidine carboxylic acid) (Galinski et al. 1985) and its derivatives (Inbar and Lapidot 1988). It has been accepted that osmotic stress duration, salinity intensity, raw material availability, and surrounding osmolytes or available carbon source in culture media have decisive influence on various compatible solutes synthesized by bacteria (Roberts 2005).

Solutes like ectoine and hydroxyectoine are zwitterions and most abundant, biologically inert and safe for cellular functions which are synthesized and enriched within microorganisms such as *Streptomyces* during stress conditions rendered by high salinity. These amphiphilic compounds function by improving the mobility of the lipid head groups and fluidizing the lipid layer mediated through elevating the surface hydration (Harishchandra et al. 2010). It has been suggested that they inhibit reverse osmotic through increasing the total water content leading to an elevation in the volume of cell cytoplasms (Cayley et al. 1992).

Ectoine and betaine are the most abundant organic compatible solutes found in *Salinisphaera shabanensis*. They accumulate glycerol in prokaryotic cells as was first reported (Antunes et al. 2003). A rare example of extremely halophilic heterotrophic betaine-producing eubacterium is *Actinopolyspora halophila*, uses simple carbon sources for this purpose, showed typical characteristic feature of extreme halotolerant bacteria. In contrast trehalose and other sugars are the main compatible solutes for non-halophilic bacteria (Mackay et al. 1984). It is commonly accepted that high concentration of betaine is more favorably compatible for enzyme activity comparing to sugars such as trehalose (Galinski 1993). In contrast to betaine, trehalose has never reported to be present in molar concentration within cytoplasm (Nyyssölä and Leisola 2001). The synthesis pathway of betaine recruited by bacteria are via threefold methylation of glycine, expending 12 ATPs for each betaine regenerate which are most energy-utilizing mechanisms in nature and costs as equivalent to ATP as for regeneration of each active methyl group from *S*-adenosylmethionine (Atkinson 1977).

3.7.1 Biological and Industrial Applications of Compatible Solutes and Their Productions

Compatible solutes such as ectoine and 5-hydroxyectoine, an ectoine derivative, are osmoprotectant, stabilizer, and rectifier of proteins and provides protection for nucleic acids, enzymes, and other biomolecules as well as cell protection against stress conditions (Harishchandra et al. 2010; Roychoudhury et al. 2012; Sadeghi et al. 2014). Potential therapeutic applications of ectoines have been highlighted for diseases related to misfolding of protein owing to their capabilities in inhibition of misfolding (Barth et al. 2000; Arora et al. 2004). Whole cell damage and viability loss prevention are among the interesting features of ectoines for cosmetics which could lead to production of protective cosmetics products with protective applications for human skin against stresses leading to skin dehydration (Graf et al. 2008).

Two types of microorganisms, genetic engineered non-halophilic microorganisms and wide salt range halophilic or halotolerant microorganisms, are utilized for production of ectoine (Pastor et al. 2010). It has been accomplished by utilization of genes from *Streptomyces crysomallum* to *Chromohalobacter salexigens* (Prabhu et al. 2004). The large scale biosynthesis of ectoine using salt-tolerating actinobacteria has been considered feasible for genus *Brevibacterium*. A final distilled water-ethanol extraction process for ectoine produced by *Brevibacterium epidermis* has also been reported (Onraedt et al. 2005). Earlier, an efficient cyclic system using osmotic downshock through biomass separation by centrifugation and, then re-suspension of obtained pellets on distilled water, known as bacterial milking, has been proposed for ectoine-production with the help of *Brevibacterium* sp. JCM 6894 (Nagata et al. 2008).

Halophilic and halotolerant compatible solutes-producing actinobacteria and those exploited in industrial production of compatible solutes have been mentioned in Tables 3.8 and 3.9, respectively.

Table 3.8 Halophilic and halotolerant compatible solutes-producing actinobacteria up to September 2014

Producing organism	Type	NaCl range of growth (optimum) % (v/w)	NaCl for highest production % (v/w)	Production yield	Determination/comment	Reference
Salinisphaera Shabanensis	MH	1–28 (10)	25	Ectoine and betaine, 16 and 28 mol/g protein, respectively	H- and C-NMR spectroscopy (NaCl investigated range, 5–25 % (w/v))	Antunes et al. (2003)
			5	2.2 glycerol mol/g protein		
Actinopolyspora sp.	NR	NR	20	NR	SDS-PAGE-92 kDa m.w (NaCl investigated range, 15–25 % (w/v))	Kundu et al. (2008)
Brevibacterium sp. JCM6894	H	NR	11.7	Ectoine 800 mg/L culture	HPLC analysis	Nagata and Wang (2001)
Actinopolyspora halophila	EH	NR	24	Betaine 2.8 mmol/gcdw	Betaine produced through choline oxidation to betaine aldehyde and H_2O_2, the betaine aldehyde is oxidized to betaine at the cost of NAD(P)=H2O2-generating and $NAD(P)^+$-reducing HPLC analysis	Nyyssölä and Leisola (2001)
			15	Trehalose 0.28 mmol/gcdw		
Brevibacterium epidermis DSM 20659	HT	0–11.7	5.8	Ectoine 0.14 g/gcdw	HPLC analysis	Onraedt et al. (2004)
Brevibacterium linens CNRZ 211	HT	0–16 (2.9–5.8)	2.9–5.8	Ectoine 825–918 mg/gcdw		Bernard et al. (1993)
Nocardiopsis sp. A5-1	NR	NR	10	Ectoione, hydroxyectoine, trehalose, glutamate and *β*-glutamate	HPLC analysis and NMR spectroscopy	Severin et al. (1992)
Kocuria varians CCM 3316	NR	NR	10	Ectoine, betaine, trehalose, hydroxyectoine	HPLC analysis and NMR spectroscopy	Severin et al. (1992)

Nesterenkonia halobia DSM 20541	NR	NR	10	Ectoine, betaine, trehalose, hydroxyectoine	HPLC analysis and NMR spectroscopy	Severin et al. (1992)
Streptomyces monomycini	HT	3.7–11.8	2.5	Ectoine 2.5 mmol/gcdw	HPLC analysis	Sadeghi et al. (2014)
				Hydroxyectoine 0.6 mmol/gcdw		
Streptomyces rimosus C-2012	HT	0–11.8	2.6	Ectoine 7 mmol/gcdw	HPLC analysis Its *ectABCD* operon is positively affected by salt proved by reverse quantitative PCR (RT-qPCR)	Sadeghi et al. (2014)
				Hydroxyectoine 60 mmol/gcdw		
Streptomyces griseus	NR	3.7–10	2.5	Ectoine 9 mmol/gcdw	HPLC analysis	Sadeghi et al. (2014)
			0	Hydroxyectoine 2.7 mmol/gcdw		
*Streptomyces rimosus*C-2012	NR	4.1–12.8	2.5	Ectoine 15 mmol/gcdw	HPLC analysis	Sadeghi et al. (2014)
				Hydroxyectoine 26 mmol/gcdw		

CDW cell dried weight, *H* halophilic, *MD* moderate halophilic, *HT* halotolerant, *NR* not reported

Table 3.9 Halophilic and halotolerant actinobacteria exploited in industrial production of compatible solutes up to September 2014

Producer	Product (production yield mg/gcdw)	Productivity (g/L day)	Reactor system	Extraction process	NaCl % (w/v)	Carbon source	Reference
Brevibacterium epidermis (DSM 20659)	Ectoine (160)	2	Fed-batch	Bacterial milking followed by ethanol extraction	5	Sodium glutamate, yeast extract	Onraedt et al. (2005)
Brevibacterium sp. JCM 6894	Ectoine (150)	0.34	Batch	Bacterial milking	10.6	Polypeptone yeast extract	Nagata et al. (2008)

3.8 Pigments of Halophilic and Halotolerant Actinobacteria Origin

Significant roles are being played by pigments in various physiology and molecular processes of microorganisms including photosynthesis, adaptation to extreme ecosystems, protectant against solar radiation, and cell-cell communication. Diversity and distribution of pigments in microorganism, especially actinomycetes, has been among the criteria identification and classification for many years (Kuhn and Starr 1960).

The interest for microbial pigments is growing not only because of their abilities to make commodities more appealing and more consumers preferring (Clydesdale 1993) but also due to their production independency to season and geographical conditions. Also microbial-derived pigment can be extracted easily and are commercially utilized (Dharmaraj et al. 2009). Microbial pigments have potentials to be used in textile, cosmetics, animal feeds and pharmaceutical industries (Venil and Lakshmanaperumalsamy 2009; Sutthiwong et al. 2014).

Doxorubicin is a cytotoxic anthracycline pigment antibiotic isolated from cultures of *Streptomyces peucetius* var. *caesius* (Arcamone et al. 1969). It has been approved by FDA and is supplied as a sterile red-orange lyophilized powder in treatment of various cancers (FDA news release 2013). Rubrolone is a water-soluble purple-red pigment produced by *Streptomyces echinoruber*. This pigment has low toxicity and has potential as a food coloring agent (Palleroni et al. 1978). Canthaxanthin is a carotenoid pigment widely distributed in the nature and produced by crustacean, fish and bacteria. It is produced by *Micrococcus roseus*, the actinobacterial member (Cooney et al. 1966). Canthaxanthin is approved in USA and Europe as coloring agent (EFSA 2010; FDA 2014). Indigoidine is non-ribosomal peptide, blue pigment that is produced by various actinobacteria, including *Corynebacterium insidiosum* (Starr 1958) and *Streptomyces aureofaciens* (Novakova et al. 2010). In Table 3.10 a list of halophilic and halotolerant pigment-producing actinobacteria are summarized. Most of these pigments were belonged to below discussed class of pigments.

3.8.1 Phenazine and Prodiginines

These redox-active nitrogen-containing aromatic pigments could render much biological potentials. Their colors vary in intensity. Although, commonly red in color and share a common pyrrolyldipyrromethane structural core. These features have made these class of pigments subjected to various researches during last decade which resulted in identification of them from *Streptoverticillium* (Gerber and Stahly 1975), *Streptomyces* (Pusecker et al. 1997), *Pseudonocardia* and *Actinomadura* (Maskey et al. 2003c), *Streptimonospora* (Liu et al. 2008), and *Nocardiopsis* (Gao et al. 2012).

Table 3.10 Halophilic and halotolerant pigment-producing actinobacteria up to September 2014

Producing strain	Pigment	Appearance	Antimicrobial activity	Cytotoxic activity	Class	Reference
Streptomyces sp. B 8251	5, 10-dihydrophencomycin methyl ester	Brown needles	Weak, *Bacillus subtilis, Escherichia coli*	NR	Phenazine	Pusecker et al. (1997)
Pseudonocardia sp. B6273	Phenazostatin D	Yellow	No	NR	Phenazine	Maskey et al. (2003b)
Actinomadura sp. M048	Iodinin, 1,6-Phenazinediol	Violet and yellow, respectively	*Bacillus subtilis, Staphylococcus aureus, Mucor miehei*	Lung (LXFA 629L and LXFL 529L), breast (MAXF 401NL), kidney (RXF 944L), melanoma (MEXF 462NL), uterus (UXF 1138L) cancer cell lines	Phenazine	Maskey et al. (2003c)
Streptimonospora salina gen. nov., sp. nov. YIM 90002	Phenazine-1-carboxylic acid	Yellow needle	NR	Human renal carcinoma cell line ACHN	Phenazine	Liu et al. (2008)
Nocardiopsis dassonvillei strain BM-17	N-(2-hydroxyphenyl)-2-phenazinamine (NHP)	Golden yellow	*Candida albicans*	Human liver hepatocellular carcinoma cell line (HepG2), human lung adenocarcinoma epithelial cell line (A549), human colon adenocarcinoma cell line (HCT-116), and human ovarian cancer cell line (COC1)	Phenazine	Gao et al. (2012)
Streptoverticillium rubrireticuli	Undecylprodiginine and butylcycloheptyl prodiginine	Red	NR	NR	Prodiginine	Gerber and Stahly (1975)

Streptomyces strain CNQ-085	Daryamides A, B, and C	Yellow to dark yellow	No	Human colon carcinoma cell line (HCT-116)	Manumycins	Asolkar et al. (2006)
Streptomyces sp. Isolate M045	Chinikomycin A	Yellowish brown	No	Mammary cancer cells (MAXF 401NL), melanoma (MEXF 462NL), and renal cancer (RXF 944L)	Manumycins	Li et al. (2005a)
Streptomyces sp. Isolate M045	Chinikomycin B	Red	No	Mammary cancer cells (MAXF 401NL)	Manumycins	Li et al. (2005a)
Actinomadura sp. M048	Chandrananimycins A, B, C	Orange	*Bacillus subtilis*, *Staphylococcus aureus*, and *Mucor miehei*	Colon (CCL HT29), lung (LXFA 526L and LXFA 529L), breast (CNCL SF268, LCL H460, MACL MCF-7), Kidney (PRCL, PC3M, RXF 631L), and melanoma (MEXF 514L) cancer cell lines	Phenoxazinone	Maskey et al. (2003c)
Actinomadura sp. M048	Questiomycin A	Orange	*Bacillus subtilis*, *Mucor miehei*, *Chlorella vulgaris*, *Chlorella sorokiniana*, *Scenedesmus suspicatus*	Colon (CCL HT29), lung (LXFA 526L and LXFA 529L), breast (CNCL SF268, LCL H460, MACL MCF-7), Kidney (PRCL, PC3M, RXF 631L), and melanoma (MEXF 514L) cancer cell lines	Phenoxazinone	Maskey et al. (2003c)
Streptimonospora salinagen. nov., sp. nov. YIM 90002	2-amino-3H-phenoxazin-3-one, 2-methylamino-3H-phenoxazin-3-one, and 2-acetylamino-3H-phenoxazin-3-one	Ranging from orange, red to brown red	NR	Human renal carcinoma cell line ACHN	Phenoxazinone	Liu et al. (2008)

(continued)

Table 3.10 (continued)

Producing strain	Pigment	Appearance	Antimicrobial activity	Cytotoxic activity	Class	Reference
Streptomyces sp. BL-49-58-005	3,6-prenyltryptophol	NR	NR	Panel of 14 different tumor cell lines	Indole	Sánchez López et al. (2003)
Streptomyces sp. 04DH110	Streptochlorin	Yellow	NR	Leukemia cancer cell line (U937)	Indole	Shin et al. (2007)
Streptomyces sp. SCSIO 03032	Spiroindimicins A–D	NR	NR	Spiroindimicin B was active against human T cell lymphoblast-like cell line (CCRF-CEM), mouse melanoma cell line (B16), and H460, spiroindimicin C exhibited growth inhibition against liver hepatocellular cell line (HepG2) and H460 while spiroindimicin D inhibited proliferation of the same cell lines as spiroindimicin C in addition to B16	Bisindole	Zhang et al. (2012b)
Actinomadura sp. 007	ZHD-0501	Pale yellow	NR	Human cancer A549, BEL-7402 and HL60 cells, and mouse leukemia P388 cells	Indolocarbazole	Han et al. (2005)
Streptomyces sp. QD518	N-carboxamido-staurosporine	Yellow	Weak phytotoxic activity and selective activity against *Streptomyces viridochromogenes*	37 cancer cells	Indolocarbazole alkaloids	Wu et al. (2006)

Actinomycete $Z_2$039-2	K252c and arcyriaflavin A	Yellow and orange, respectively	NR	K562 cell line	Indolocarbazole alkaloids	Liu et al. (2007)
Streptomyces sp. FMA	Streptocarbazoles A and B	Yellow	NR	Streptocarbazole A showed cytotoxicity against human leukemia cell line (HL-60) and adenocarcinomic human alveolar basal epithelial cell line (A-549) and Helacyton gartleri cells (HeLa)	Indolocarbazole	Fu et al. (2012a)
Marinispora sp.	Lynamicins A–E	Off-white	*Staphylococci*, *Streptococci*, *Enterococci*, *Haemophili* sp. as well as an *Escherichia coli*	NR	Chlorinated bisindole pyrrole	McArthur et al. (2008)
Streptomyces sp. NPS008187	Glyciapyrroles A, B, and C	Glassy	NR	Colorectal adenocarcinoma (HT-29) and melanoma (B16-F10) tumor cells	Pyrrolosesquiterpenes	Macherla et al. (2005)
Streptomyces strain CNH-287	Chlorizidine A	NR	NR	Human colon cancer cells (HCT-116)	Pyrroloisoindolone	Alvarez-Mico et al. (2013)
Streptoverticillium luteoverticillatum 11014	Butenolides	Yellow	NR	Murine lymphoma P388 and human leukemia K562 cell lines	Butenolide	Li et al. (2006a)
Streptomyces CNQ-418	Marinopyrroles A–F	NR	Marinopyrroles A, B, and C were active against methicillin-resistant *Staphylococcus aureus* (MRSA)		1,3′-bipyrrole	Hughes et al. (2010)

(continued)

Table 3.10 (continued)

Producing strain	Pigment	Appearance	Antimicrobial activity	Cytotoxic activity	Class	Reference
Streptomyces sp. B6921	Himalomycins A and B, and fridamycin D	Yellow to orange	*Bacillus subtilis*, *Streptomyces viridochromogenes*, *Staphylococcus aureus*, and *Escherichia coli*	NR	Quinone	Maskey et al. (2003a)
Streptomyces sp. strain CNH990	Marmycins A and B	Red and pink needle, respectively	NR	Human colon carcinoma cell line (HCT-116) Marmycin A was also active against a panel of 12 human tumor cell lines including breast, prostate, colon, lung, and leukemia	Quinone	Martin et al. (2007)
Streptomyces chilbaensis AUBN$_1$/7	Resistoflavin	Pale yellow	Weak	Gastric adenocarcinoma cell line (HMO2) and hepatic carcinoma cell line (HePG2)	Quinone	Gorajana et al. (2007)
Salinispora arenicola strain CNR-647	Arenimycin	Red	Rifamycin- and methicillin-resistant *Staphylococcus aureus*, coagulase-negative *Staphylococcus*, *Staphylococcus saprophyticus*, *Enterococcus faecalis*, *Enterococcus faesium*, vancomycin-resistant *Enterococcus faesium*, and *Mycobacterium bacilli*	Human colon carcinoma (HCT-116)	Quinone	Asolkar et al. (2010)

Streptomyces strain CNR-698	Ammosamides A and B	Blue and red, respectively	NR	Human colon carcinoma (HCT-116)	Pyrroloiminoquinone	Hughes et al. (2009)
Streptomyces variabilis SNA-020	Ammosamide D	Orange	NR	Pancreatic cancer cell line (MIA PaCa-2)	Pyrroloiminoquinone	Pan et al. (2012)
Arthrobacter flavus	–	Yellow	NR	NR	Carotenoid	Reddy et al. (2000)
Streptomyces strain AQBWWS1	–	Red	NR	NR	Carotenoid	Dharmaraj et al. (2009)
Streptomyces sp.	–	Red	*Escherichia coli*, *Vibrio cholerae*, *Staphylococcus aereus*, *Proteus mirabilus*, *Salmonella paratyphae*, *Klebsiella axytoca*, and *Lactobacillus vulgaris*	NR	Melanin	Vasanthabharathi et al. (2011)

3.8.2 Quinone

Quinone is another group of colored biologically active compounds consists of an aromatic ring structure isolated from saline environment which appeared as yellow to red. This pigment was found to be the secondary metabolite of *Streptomyces* (Gorajana et al. 2007; Martin et al. 2007) (Maskey et al. 2003a; Hughes et al. 2009) (Pan et al. 2012) and *Salinispora* (Asolkar et al. 2010).

3.8.3 Indolocarbazole

Indolocarbazole (ICZ) family of compounds was first described in 1977 and showed many different chemical structures as well as biological activities. A number of arrangements are observed for indole and carbazole to yield different isomeric ring systems among which the most important and the most frequent isolated ICZs from nature are indolo[2,3-α]carbazols (Sánchez et al. 2006). ICZ family of pigments was produced from *Actinomadura* (Han et al. 2005), *Streptomyces* (Wu et al. 2006; Fu et al. 2012a), and unknown specified actinobacteria (Liu et al. 2007).

3.8.4 Manumycins

Manumycins commonly exhibit biological activities and are yellow in appearance. They have been consisted of two separately synthesized polyketide chains. Manumycin-group compounds are usually produced upon submerged cultivation of manumycin-producing organisms with complex nutrient sources. *Streptomyces* (Li et al. 2005a; Asolkar et al. 2006) was reported to produce this class of pigments.

3.8.5 Carotenoids

These compounds consist of a polyene, polyunsaturated hydrocarbons, skeleton with 40 carbon atoms and can be terminated by cyclic or an acyclic end groups (Dharmaraj et al. 2009). These compounds appear in orange, yellow or red-colored pigments. Two different classes of carotenoid are the carotenes and the xanthophylls which are non-oxygen-substituted hydrocarbon carotenoid and oxygen-substituted carotenoid, respectively. In photosynthetic organisms, they are present in light-capturing complexes and function as accessory pigment to absorb light and subsequent transfer of the energy to chlorophyll. These compounds can also provide both phototrophic and non-phototrophic organisms with protection against harmful oxygen radicals (Krinsky 1998; Asker et al. 2012). Human and other animals obtain carotenoid from the diet and are unable to synthesize it themselves. Carotenoid are extensively used as color-enhancing additives to animal feed and also applied as food colorants (Gordon et al. 1982). It is produced by *Arthrobacter* (Reddy et al. 2000) and *Streptomyces* (Dharmaraj et al. 2009).

3.9 Application of Halophilic and Halotolerant Actinobacteria in Agriculture

Todays, food production is a main concern in different parts of world. The setbacks in global agriculture productivity are ecological stresses which can be divided into biotic stresses such as plant diseases (Maheshwari 2013), and abiotic stresses such as salinity, drought, nutrient deficiency, and temperature (Atkinson and Urwin 2012). Conventional ways to control pest, diseases, and abiotic stresses are by means of chemical pesticides in addition to chemical fertilizers. A newer method to chemical pesticides is selection of resistant plants or improving susceptible plants by applying genetically modified engineering technology. The application of many chemical entities, despite their great contribution in sustainability and high agricultural productivities, has been banned following to the confirmation that these chemicals are non-ecofriendly, due to misuse or overuse, and ultimately resulted in the development of resistant pests and microorganisms against chemical pesticides. Also, notwithstanding, great versatility of genetically modified organisms (GMO) including crops, there are a great controversial about possible ecological and health consequences resulted from cultivation and consumption of GMO throughout the world. The situation become harder by the fact that about 5 % of the world's cultivable land (1.5 billion hectares) do not yields efficiently because of high salt level, 20 % (308 million hectares) of the irrigated land is salt-affected, and the area of salt-affected land is increasing (Selvakumar et al. 2014). Therefore, a potential substitute for chemically enhanced agricultural productivity as well as decreasing detrimental effects of salt stress could be through recruitment of agriculturally beneficial microorganisms (Strap 2011).

A number of bacteria referred as plant growth promoting rhizobacteria (PGPR) are found in the rhizoplane and rhizosphere, or as endophytic inside of tissues of plants which could promote plant growth either directly or indirectly (Glick 1995). In indirect method, PGPR decrease or suppress the detrimental effects of single or many phytopathogens through competition for niches and nutrients (Kloepper et al. 1986), secretion of antibiotics, extracellular cell wall hydrolyzing enzymes or siderophore, removal or detoxification of organic and inorganic pollutants, stress tolerance, and synthesizing hydrogen cyanide (Voisard et al. 1989; Nadeem et al. 2013). Secretion of high affinity (Kd=10-20-10-50) siderophores (approximately, 400–1,000 Da) that bind to already sparingly soluble iron, leads to depletion of required iron for assimilation of iron by phytopathogens (Castignetti 1986; Glick 1995), also hydrolyzing detrimental fusaric acid synthesized by *Fusarium* sp. (Toyoda and Utsumi 1991) can promote plants growth. Plants are not commonly susceptible to the localized iron depletion and can grow in much less iron concentrations than microorganisms which makes this biocontrol mechanism specific (Glick 1995). In other hand, increase in the availability of microbial-synthesizing compounds for plant or facilitation of certain nutrients uptake from the surrounding can be fulfilled through direct effect. These include siderophores that provide iron to plants through solubilizing and sequestering it from the soil, production of phytohormones, and mineralization of insoluble sources of minerals (Glick 1995).

The level of soil salinity can drastically affects various physiological processes in plants including seed germination (Raafat et al. 2011), certain enzyme activities (Almansouri et al. 1999), photosynthesis disruption by affecting its bioenergetics processes (Sudhir and Murthy 2004), lengths, and shoot and root weights in term of wet and dry weight (Jamil et al. 2006). The premature death of plants due to salt stress originated from enhanced ethylene production that accelerates petal and leaf dropping (abscission) and tissue aging (senescence) (Mayak et al. 2004). Bacteria with 1-aminocyclopropane-1-carboxylic acid (ACC) deaminase activity could alleviate this salt stress-induced ethylene by decreasing ethylene level through exclusion of ACC, the precursor for ethylene, and further its hydrolyzation to ammonia and α-ketobutyrate (Glick et al. 1998; Siddikee et al. 2010). The most available endogenous auxin is indole-3-acetic acid (IAA) which functions in stem elongation and root growth, is also reported from a number of microorganisms which results in increased amount of this phytohormone in rhizosphere and consequently, these PGPR could promote plant growth through direct mechanisms.

Many halophilic or halotolerant heavy metal-reducing actinobacteria (Sect. 3.6.2) or other bioremediating-capable ones (Sect. 3.6) could promote plant growth under saline conditions through detoxification or degradation of soil pollutants.

Despite, some developments in biological application of agronomically important microorganisms for certain conditions in recent years; currently, market share of these microorganisms is relatively small (Glick 1995). The statistic for halophilic and halotolerant actinobacteria is even worse (Table 3.11).

Table 3.11 Halophilic and halotolerant agriculturally important actinobacteria

Beneficial actinobacteria	Proposed mechanism(s)	Test crop/explanation	Reference
Brevibacterium iodinum *Zhihengliuela alba RS16*	ACC-deaminase activity	Reduction in ethylene production by 53 % and 57 %, respectively, in salt stress (9 % NaCl (w/v)) red pepper seedlings *Capsicum annuum* L	Siddikee et al. (2011)
Brevibacterium epidermidis RS15 *Micrococcus yunnanensis* RS222	ACC deaminase activity	Reduction in ethylene production by 47–64 %, ACC concentration by 47–55 % and ACO activity by 18–19 % in salt-stressed (9 % NaCl (w/v)) red pepper seedlings *Capsicum annuum* L	Siddikee et al. (2012)
Streptomyces isolate C	IAA, siderophore, and tricalcium phosphate activities	The amount of auxin increased after adding salt and reached to 4.7 g/mL in 17.5 % NaCl (w/v). Solubilization decreased in presence of NaCl (maximum at 92 mg/L). Applying the bacterial inocula increased the concentration of N, P, Fe, and Mn in wheat (cul. *Chamran*) shoots grown in normal and saline soil	Sadeghi et al. (2012)

(continued)

Table 3.11 (continued)

Beneficial actinobacteria	Proposed mechanism(s)	Test crop/explanation	Reference
Brachybacterium saurashtrense	Nitrogen fixation, IAA, siderophores, and ACC deaminase activities	Root-associated bacterium with plant growth-promoting potential	Gontia et al. (2011)
Streptomyces sp. *Micromonospora* sp.	IAA, nitrate, and extracellular protease and chitinase activities	They were able to grow up to pH 9 (below 8 was not tested) and salinity of 5–10 % NaCl (w/v)	Malviya et al. (2014)
Streptomyces sp. strain PGPA39	IAA, ACC deaminase, and phosphate solubilization activities	Increase in biomass and number of lateral roots of *Arabidopsis* seedlings *in vitro*. A significant increase in plant biomass and chlorophyll content, and a reduction in leaf proline content were observed in PGPA39-inoculated tomato plants under 10.5 % NaCl (w/v) stress	Arunachalam Palaniyandi et al. (2014)
Kocuria turfanensis strain 2M4	IAA, phosphate solubilization activity	It produced 38 μg/mL IAA when growth medium was supplemented with 600 μg/mL of L-tryptophan. In presence of fructose, the highest phosphate solubilization was 12 μg/mL. Siderophore production was 53 % units under iron-free minimal MM9 medium and produced 1.8 μmol/mL ammonia in peptone water broth. It promoted growth of groundnut (*Arachis hypogaea* L.) under non-saline and saline soil. There was increase by 18 % in total plant length and 30 % in fresh biomass observed under non-saline control soil, and 17 % increase in total length of the plant and 13 % increase in fresh biomass under saline soil	Goswami et al. (2014)

(continued)

Table 3.11 (continued)

Beneficial actinobacteria	Proposed mechanism(s)	Test crop/explanation	Reference
Nocardiopsis gilva YIM 90087	Biocontrol	p-terphenyl and its derivatives with antifungal activity against *Fusarium avenaceum*, *Fusarium graminearum*, and *Fusarium culmorum*	Tian et al. (2013a)
Actinomycete strain VITDDK2	Biocontrol	Production of active secondary metabolite against *Aspergillus flavus* and *Aspergillus niger*, as well as chitinolytic activity	Deepika and Kannabiran (2010)
Nocardiopsis terrae YIM 90022	Biocontrol	Antifungal activity against *Pyricularia oryzae* (rice blast disease)	Tian et al. (2014)
Saccharopolyspora salina VITSDK4	Biocontrol	The crude extract exhibited significant antagonistic activity against *Aspergillus niger* and *Aspergillus fumigatusi*	Suthindhiran and Kannabiran (2009)
Actinopolyspora sp. AH1	Biocontrol	The crude extract exhibited antagonistic activity against *Aspergillus niger*, *Aspergillus fumigatus*, *Aspergillus flavus*, *Fusarium oxysporum*, *Penicillium* sp., and *Trichoderma* sp.	Kokare et al. (2004)
Streptomyces violascens IN2-10	Biocontrol	The crude extract exhibited significant antagonistic activity against *Microsporum gypseum*, *Microsporucanis*, *Trichophtyton rubrum*, *Aspergillus flavus*, *Aspergillus fumigatus*, and *Aspergillus niger*	Bisht et al. (2013)

3.10 Conclusion

Although salt are present ubiquitously, excess of it can have many detrimental effects on productivity and efficiently of interested process and therefore, either makes the process expensive or not feasible. The study of halophilic and halotolerant actinobacteria has just begun, but the metabolites obtained from these microorganisms have exhibited many potentials which are suitable for vast industrial and environmental applications. As it is predictable, halophilic and halotolerant actinobacteria or the metabolites derived from these organisms can solve these problems. Nowadays, many applications under saline conditions have been proposed for halophilic and halotolerant actinobacteria which include agriculture productivity enhancement in saline soil, bioremediation of effluents from mining operations, oil and natural gas exploitation, tannery, textile, fabric dye, metallic alloys, petrochemical sectors, and likewise from pollutants, as well as degradation and/or conversion of not ecologically problematic resources like feathers, chitin, cellulose, starch and etc. to value-added compounds. Their other fields of applications are within enzyme industry as well as food and drug technologies. The discovery possibility of novel compounds from these microorganisms due to their unleashed niches is quit high. And if studied more systematically, it could help saline-based industries or consumption of raw material obtained from saline environments. As human population is growing fast then many environmental-friendly and efficient methods are required to save world from pollution-producing technologies or improve agriculture in salty soil or saline water and help to solve hunger in world. Also biological active materials from halophilic actinobacteria can save many lives by decreasing mortality rates of cancer or resistant pathogens caused diseases.

References

Ahmed L, Jensen PR, Freel KC, Brown R, Jones AL, Kim B-Y, Goodfellow M (2013) *Salinispora pacifica* sp. nov., an actinomycete from marine sediments. Antonie Van Leeuwenhoek 103(5):1069–1078

Ahmed I, Kudo T, Abbas S, Ehsan M, Iino T, Fujiwara T, Ohkuma M (2014) *Cellulomonas pakistanensis* sp. nov., a novel moderately halotolerant actinobacteria. Int J Syst Evol Microbiol 64(7):2305–2311

Alias-Villegas C, Jurado V, Laiz L, Miller AZ, Saiz-Jimenez C (2013) *Nocardioides albertanoniae* sp. nov., isolated from Roman catacombs. Int J Syst Evol Microbiol 63(4):1280–1284

Almansouri M, Kinet J, Lutts S (1999) Compared effects of sudden and progressive impositions of salt stress in three durum wheat (*Triticum durum* Desf.) Cultivars. J Plant Physiol 154(5):743–752

Al-Mueini R, Al-Dalali M, Al-Amri IS, Patzelt H (2007) Hydrocarbon degradation at high salinity by a novel extremely halophilic actinomycete. Environ Chem 4(1):5–7

Al-Musallam AA, Al-Zarban SS, Fasasi YA, Kroppenstedt RM, Stackebrandt E (2003) *Amycolatopsis keratiniphila* sp. nov., a novel keratinolytic soil actinomycete from Kuwait. Int J Syst Evol Microbiol 53(3):871–874

Al-Tai AM, Ruan J-S (1994) *Nocardiopsis halophila* sp. nov., a new halophilic actinomycete isolated from soil. Int J Syst Bacteriol 44(3):474–478

Altenburger P, Kämpfer P, Schumann P, Vybiral D, Lubitz W, Busse H-J (2002) *Georgenia muralis* gen. nov., sp. nov., a novel actinobacterium isolated from a medieval wall painting. Int J Syst Evol Microbiol 52(3):875–881

Alvarez-Mico X, Jensen PR, Fenical W, Hughes CC (2013) Chlorizidine, a cytotoxic 5 H-Pyrrolo [2, 1-a] isoindol-5-one-containing alkaloid from a marine *Streptomyces* sp. Org Lett 15(5):988–991

Al-Zarban SS, Abbas I, Al-Musallam AA, Steiner U, Stackebrandt E, Kroppenstedt RM (2002a) *Nocardiopsis halotolerans* sp. nov., isolated from salt marsh soil in Kuwait. Int J Syst Evol Microbiol 52(2):525–529

Al-Zarban SS, Al-Musallam AA, Abbas I, Stackebrandt E, Kroppenstedt RM (2002b) Saccharomonospora halophila sp. nov., a novel halophilic actinomycete isolated from marsh soil in Kuwait. Int J Syst Evol Microbiol 52(2):555–558

Amoozegar MA, Khoshnoodi M, Didari M, Hamedi J, Ventosa A, Baldwin SA (2012) Tellurite removal by a tellurium-tolerant halophilic bacterial strain, *Thermoactinomyces* sp. QS-2006. Ann Microbiol 62(3):1031–1037

Anderson AJ, Dawes EA (1990) Occurrence, metabolism, metabolic role, and industrial uses of bacterial polyhydroxyalkanoates. Microbiol Rev 54(4):450–472

Antunes A, Eder W, Fareleira P, Santos H, Huber R (2003) *Salinisphaera shabanensis* gen. nov., sp. nov., a novel, moderately halophilic bacterium from the brine–seawater interface of the Shaban Deep, Red Sea. Extremophiles 7(1):29–34

Aoyagi T, Hatsu M, Imada C, Naganawa H, Okami Y, Takeuchi T (1992) Pyrizinostatin: a new inhibitor of pyroglutamyl peptidase. J Antibiot 45(11):1795

Aoyama T, Kojima F, Imada C, Muraoka Y, Naganawa H, Okami Y, Takexjchi T, Aoyagi T (1995) Pyrostatins A and B, new inhibitors of N-acetyl-β-d-glucosaminidase, produced by *Streptomyces* sp. SA-3501. J Enzyme Inhib Med Chem 8(4):223–232

Arcamone F, Cassinelli G, Fantini G, Grein A, Orezzi P, Pol C, Spalla C (1969) Adriamycin, 14-hydroxydaimomycin, a new antitumor antibiotic from *S. Peucetius* var. *caesius*. Biotechnol Bioeng 11(6):1101–1110

Arora A, Ha C, Park CB (2004) Inhibition of insulin amyloid formation by small stress molecules. FEBS Lett 564(1):121–125

Arul Jose P, Satheeja Santhi V, Jebakumar SR (2011) Phylogenetic-affiliation, antimicrobial potential and PKS gene sequence analysis of moderately halophilic *Streptomyces* sp. inhabiting an Indian saltpan. J Basic Microbiol 51(4):348–356

Arumugam M, Mitra A, Jaisankar P, Dasgupta S, Sen T, Gachhui R, Mukhopadhyay UK, Mukherjee J (2010) Isolation of an unusual metabolite 2-allyloxyphenol from a marine actinobacterium, its biological activities and applications. Appl Microbiol Biotechnol 86(1):109–117

Arunachalam PS, Damodharan K, Yang SH, Suh JW (2014) *Streptomyces* sp. strain PGPA39 alleviates salt stress and promotes growth of micro tomato plants. J Appl Microbiol 117(3):766–773

Ashok B, Saxena S, Musarrat J (1995) Isolation and characterization of four polycyclic aromatic hydrocarbon degrading bacteria from soil near an oil refinery. Lett Appl Microbiol 21(4):246–248

Asker D, Awad TS, Beppu T, Ueda K (2012) Isolation, characterization, and diversity of novel radiotolerant carotenoid-producing bacteria. In: Barredo JL (ed) Microbial carotenoids from bacteria and microalgae. Humana Press. Method Mol Biol 892:21–60

Asolkar RN, Maskey RP, Helmke E, Laatsch H (2002) Chalcomycin B, a new macrolide antibiotic from the marine isolate *Streptomyces* sp. B7064. J Antibiot 55(10):893–898

Asolkar RN, Jensen PR, Kauffman CA, Fenical W (2006) Daryamides AC, weakly cytotoxic polyketides from a marine-derived actinomycete of the genus *Streptomyces* strain CNQ-085. J Nat Prod 69(12):1756–1759

Asolkar RN, Kirkland TN, Jensen PR, Fenical W (2010) Arenimycin, an antibiotic effective against rifampin-and methicillin-resistant *Staphylococcus aureus* from the marine actinomycete *Salinispora arenicola*. J Antibiot 63(1):37–39

Atkinson DE (1977) Cellular energy metabolism and its regulation. Academic, New York

Atkinson NJ, Urwin PE (2012) The interaction of plant biotic and abiotic stresses: from genes to the field. J Exp Bot 63(10):3523–3543

Austin B (1989) Novel pharmaceutical compounds from marine bacteria. J Appl Microbiol 67(5):461–470

Baltz R (2007) Antimicrobials from actinomycetes: back to the future. Am Soc Microbiol 2(3):125–131

Banat IM, Nigam P, Singh D, Marchant R (1996) Microbial decolorization of textile-dyecontaining effluents: a review. Bioresour Technol 58(3):217–227

Banat IM, Makkar RS, Cameotra S (2000) Potential commercial applications of microbial surfactants. Appl Microbiol Biotechnol 53(5):495–508
Barth S, Huhn M, Matthey B, Klimka A, Galinski E, Engert A (2000) Compatible-solute-supported periplasmic expression of functional recombinant proteins under stress conditions. Appl Environ Microbiol 66(4):1572–1579
Becker K, Schumann P, Wüllenweber J, Schulte M, Weil H-P, Stackebrandt E, Peters G, von Eiff C (2002) *Kytococcus schroeteri* sp. nov., a novel Gram-positive actinobacterium isolated from a human clinical source. Int J Syst Evol Microbiol 52(5):1609–1614
Ben-Dov E, Yosef DZB, Pavlov V, Kushmaro A (2009) *Corynebacterium maris* sp. nov., a marine bacterium isolated from the mucus of the coral Fungia granulosa. Int J Syst Evol Microbiol 59(10):2458–2463
Benz M, Schink B, Brune A (1998) Humic acid reduction by *Propionibacterium freudenreichii* and other fermenting bacteria. Appl Environ Microbiol 64(11):4507–4512
Bernard T, Jebbar M, Rassouli Y, Himdi-Kabbab S, Hamelin J, Blanco C (1993) Ectoine accumulation and osmotic regulation in *Brevibacterium linens*. J Gen Microbiol 139(1):129–136
Bhadra B, Raghukumar C, Pindi PK, Shivaji S (2008) *Brevibacterium oceani* sp. nov., isolated from deep-sea sediment of the Chagos Trench, Indian Ocean. Int J Syst Evol Microbiol 58(1):57–60
Bian J, Li Y, Wang J, Song F-H, Liu M, Dai H-Q, Ren B, Gao H, Hu X, Liu Z-H (2009) *Amycolatopsis marina* sp. nov., an actinomycete isolated from an ocean sediment. Int J Syst Evol Microbiol 59(3):477–481
Bisht GS, Bharti A, Kumar V, Gusain O (2013) Isolation, purification and partial characterization of an antifungal agent produced by salt-tolerant alkaliphilic *Streptomyces violascens* IN2-10. Proc Natl Acad Sci India 83(1):109–117
Blasco R, Martínez-Luque M, Madrid MP, Castillo F, Moreno-Vivián C (2001) *Rhodococcus* sp. RB1 grows in the presence of high nitrate and nitrite concentrations and assimilates nitrate in moderately saline environments. Arch Microbiol 175(6):435–440
Borzenkov I, Milekhina E, Gotoeva M, Rozanova E, Belyaev S (2006) The properties of hydrocarbon-oxidizing bacteria isolated from the oil fields of Tatarstan, Western Siberia, and Vietnam. Microbiology 75(1):66–72
Bouïs D, Kusumanto Y, Meijer C, Mulder NH, Hospers GA (2006) A review on pro-and anti-angiogenic factors as targets of clinical intervention. Pharmacol Res 53(2):89–103
Brandi H, Bachofen R, Mayer J, Wintermantel E (1995) Degradation and applications of polyhydroxyalkanoates. Can J Microbiol 41(13):143–153
Brandl H (2008) Microbial leaching of metals. In: Rehm HJ, Reed G (eds) Biotechnology set, 2nd edn. Wiley-VCH Verlag GmbH, Weinheim, pp 191–224
Bruntner C, Binder T, Pathom-aree W, Goodfellow M, Bull AT, Potterat O, Puder C, Hörer S, Schmid A, Bolek W (2005) Frigocyclinone, a novel angucyclinone antibiotic produced by a *Streptomyces griseus* strain from Antarctica. J Antibiot 58(5):346–349
Buchanan GO, Williams PG, Feling RH, Kauffman CA, Jensen PR, Fenical W (2005) Sporolides A and B: structurally unprecedented halogenated macrolides from the marine actinomycete *Salinispora tropica*. Org Lett 7(13):2731–2734
Butler MS (2005) Natural products to drugs: natural product derived compounds in clinical trials. Nat Prod Res 22(2):162–195
Butler MS, Blaskovich MA, Cooper MA (2013) Antibiotics in the clinical pipeline in 2013. J Antibiot 66(10):571–591
Butt MS, Tahir-Nadeem M, Ahmad Z, Sultan MT (2008) Xylanases and their applications in baking industry. Food Technol Biotechnol 46(1):22–31
Cai M, Zhi XY, Tang SK, Zhang YQ, Xu LH, Li WJ (2008) *Streptomonospora halophila* sp. nov., a halophilic actinomycete isolated from a hypersaline soil. Int J Syst Evol Microbiol 58(7):1556–1560
Cai M, Tang SK, Chen YG, Li Y, Zhang YQ, Li WJ (2009) *Streptomonospora amylolytica* sp. nov. and *Streptomonospora flavalba* sp. nov., two novel halophilic actinomycetes isolated from a salt lake. Int J Syst Evol Microbiol 59(10):2471–2475

Cao YR, Jiang Y, Wu J-Y, Xu L-H, Jiang C-L (2009) *Actinopolymorpha alba* sp. nov., isolated from a rhizosphere soil. Int J Syst Evol Microbiol 59(9):2200–2203

Capon RJ, Skene C, Lacey E, Gill JH, Wicker J, Heiland K, Friedel T (2000) Lorneamides A and B: two new aromatic amides from a southern Australian marine actinomycete. J Nat Prod 63(12):1682–1683

Carr GJ, Ferguson SJ (1990) Nitric oxide formed by nitrite reductase of *Paracoccus denitrificans* is sufficiently stable to inhibit cytochrome oxidase activity and is reduced by its reductase under aerobic conditions. BBA Bioenerg 1017(1):57–62

Carro L, Zúñiga P, De la Calle F, Trujillo ME (2012) *Streptomyces pharmamarensis* sp. nov. isolated from a marine sediment. Int J Syst Evol Microbiol 62(5):1165–1170

Castignetti D (1986) Siderophores, the iron nutrition of plants, and nitrate reductase. FEBS Lett 209(2):147–151

Cayley S, Lewis B, Record M (1992) Origins of the osmoprotective properties of betaine and proline in *Escherichia coli* K-12. J Bacteriol 174(5):1586–1595

Cervantes FJ, van der Velde S, Lettinga G, Field JA (2000) Quinones as terminal electron acceptors for anaerobic microbial oxidation of phenolic compounds. Biodegradation 11(5):313–321

Chakraborty S, Khopade A, Kokare C, Mahadik K, Chopade B (2009) Isolation and characterization of novel α-amylase from marine *Streptomyces* sp. D1. J Mol Catal B Enzym 58(1):17–23

Chakraborty S, Khopade A, Biao R, Jian W, Liu XY, Mahadik K, Chopade B, Zhang L, Kokare C (2011) Characterization and stability studies on surfactant, detergent and oxidant stable α-amylase from marine haloalkaliphilic *Saccharopolyspora* sp. A9. J Mol Catal B Enzym 68(1):52–58

Chang X, Liu W, Zhang XH (2011) *Spinactinospora alkalitolerans* gen. nov., sp. nov., an actinomycete isolated from marine sediment. Int J Syst Evol Microbiol 61(12):2805–2810

Chang X, Liu W, Zhang XH (2012) *Salinactinospora qingdaonensis* gen. nov., sp. nov., a halophilic actinomycete isolated from a salt pond. Int J Syst Evol Microbiol 62(4):954–959

Charan RD, Schlingmann G, Janso J, Bernan V, Feng X, Carter GT (2004) Diazepinomicin, a new antimicrobial alkaloid from a marine *Micromonospora* sp. J Nat Prod 67(8):1431–1433

Chen HH, Li WJ, Tang SK, Kroppenstedt RM, Stackebrandt E, Xu LH, Jiang CL (2004) *Corynebacterium halotolerans* sp. nov., isolated from saline soil in the west of China. Int J Syst Evol Microbiol 54(3):779–782

Chen YG, Cui XL, Kroppenstedt RM, Stackebrandt E, Wen ML, Xu LH, Jiang CL (2008) *Nocardiopsis quinghaiensis* sp. nov., isolated from saline soil in China. Int J Syst Evol Microbiol 58(3):699–705

Chen HH, Zhao GZ, Park DJ, Zhang YQ, Xu LH, Lee JC, Kim CJ, Li WJ (2009a) *Micrococcus endophyticus* sp. nov., isolated from surface-sterilized *Aquilaria sinensis* roots. Int J Syst Evol Microbiol 59(5):1070–1075

Chen YG, Wang YX, Zhang YQ, Tang SK, Liu ZX, Xiao HD, Xu LH, Cui XL, Li WJ (2009b) *Nocardiopsis litoralis* sp. nov., a halophilic marine actinomycete isolated from a sea anemone. Int J Syst Evol Microbiol 59(11):2708–2713

Chen YG, Tang SK, Zhang YQ, Liu ZX, Chen QH, He JW, Cui XL, Li WJ (2010a) *Zhihengliuella salsuginis* sp. nov., a moderately halophilic actinobacterium from a subterranean brine. Extremophiles 14(4):397–402

Chen YG, Zhang YQ, Tang SK, Liu ZX, Xu LH, Zhang LX, Li WJ (2010b) *Nocardiopsis terrae* sp. nov., a halophilic actinomycete isolated from saline soil. Anton Leeuwen 98(1):31–38

Cho JY, Kwon HC, Williams PG, Kauffman CA, Jensen PR, Fenical W (2006) Actinofuranones A and B, polyketides from a marine-derived bacterium related to the genus *Streptomyces* (actinomycetales). J Nat Prod 69(3):425–428

Cho JY, Williams PG, Kwon HC, Jensen PR, Fenical W (2007) Lucentamycins A-D, cytotoxic peptides from the marine-derived actinomycete *Nocardiopsis lucentensis*. J Nat Prod 70(8):1321–1328

Choi DH, Kim HM, Noh JH, Cho BC (2007) *Nocardioides marinus* sp. nov. Int J Syst Evol Microbiol 57(4):775–779

Choi EJ, Lee SH, Jung JY, Jeon CO (2013) *Brevibacterium jeotgali* sp. nov., isolated from jeotgal, a traditional Korean fermented seafood. Int J Syst Evol Microbiol 63(9):3430–3436

Chomchoei A, Pathom-aree W, Yokota A, Kanongnuch C, Lumyong S (2011) *Amycolatopsis thailandensis* sp. nov., a poly (l-lactic acid)-degrading actinomycete, isolated from soil. Int J Syst Evol Microbiol 61(4):839–843

Chun J, Bae KS, Moon EY, Jung SO, Lee HK, Kim SJ (2000) *Nocardiopsis kunsanensis* sp. nov., a moderately halophilic actinomycete isolated from a saltern. Int J Syst Evol Microbiol 50(5):1909–1913

Chung KT, Stevens SE (1993) Degradation azo dyes by environmental microorganisms and helminths. Environ Toxicol Chem 12(11):2121–2132

Clydesdale FM (1993) Color as a factor in food choice. Crit Rev Food Sci 33(1):83–101

Collins M, Jones D, Keddie R, Sneath P (1980) Reclassification of *Chromobacterium iodinum* (Davis) in a redefined genus *Brevibacterium* (Breed) as *Brevibacterium iodinum* nom. rev.; comb. nov. J Gen Microbiol 120(1):1–10

Collins M, Farrow J, Goodfellow M, Minnikin D (1983) *Brevibacterium casei* sp. nov. and *Brevibacterium epidermidis* sp. nov. Syst Appl Microbiol 4(3):388–395

Collins MD, Lawson PA, Labrenz M, Tindall BJ, Weiss N, Hirsch P (2002) *Nesterenkonia lacusekhoensis* sp. nov., isolated from hypersaline Ekho Lake, East Antarctica, and emended description of the genus Nesterenkonia. Int J Syst Evol Microbiol 52(4):1145–1150

Cooney J, Marks H, Smith AM (1966) Isolation and identification of canthaxanthin from *Micrococcus roseus*. J Bacteriol 92(2):342–345

Cripps C, Bumpus JA, Aust S (1990) Biodegradation of azo and heterocyclic dyes by *Phanerochaete chrysosporium*. Appl Environ Microbiol 56(4):1114–1118

Cui X-L, Mao P-H, Zeng M, Li WJ, Zhang LP, Xu LH, Jiang CL (2001) *Streptimonospora salina* gen. nov., sp. nov., a new member of the family *Nocardiopsaceae*. Int J Syst Evol Microbiol 51(2):357–363

Cui X, Schumann P, Stackebrandt E, Kroppenstedt RM, Pukall R, Xu L, Rohde M, Jiang C (2004) *Myceligenerans xiligouense* gen. nov., sp. nov., a novel hyphae-forming member of the family *Promicromonosporaceae*. Int J Syst Evol Microbiol 54(4):1287–1293

Dai HQ, Wang J, Xin YH, Pei G, Tang SK, Ren B, Ward A, Ruan JS, Li WJ, Zhang LX (2010) *Verrucosispora sediminis* sp. nov., a cyclodipeptide-producing actinomycete from deep-sea sediment. Int J Syst Evol Microbiol 60(8):1807–1812

Dalvi S, Azetsu S, Patrauchan MA, Aktas DF, Fathepure BZ (2012) Proteogenomic elucidation of the initial steps in the benzene degradation pathway of a novel halophile, *Arhodomonas* sp. strain *Rozel*, isolated from a hypersaline environment. Appl Environ Microbiol 78(20):7309–7316

Dasari VRRK, Muthyala MKK, Nikku MY, Donthireddy SRR (2012) Novel pyridinium compound from marine actinomycete, *Amycolatopsis alba* var. nov. DVR D4 showing antimicrobial and cytotoxic activities *in vitro*. Microbiol Res 167(6):346–351

Dastager SG, Li WJ, Agasar D, Sulochana MB, Tang SK, Tian XP, Zhi XY (2007) *Streptomyces gulbargensis* sp. nov., isolated from soil in Karnataka, India. Anton Leeuwen 91(2):99–104

Deepika L, Kannabiran K (2010) Isolation and characterization of antagonistic actinomycetes from marine soil. J Microb Biochem Technol 2:001–006

del Carmen MCM, Göker M, Pötter G, Rohde M, Spröer C, Schumann P, Gorbushina AA, Klenk HP (2013) *Geodermatophilus africanus* sp. nov., a halotolerant actinomycete isolated from Saharan desert sand. Antonie Van Leeuwenhoek 104(2):207–216

Delgado O, Quillaguamán J, Bakhtiar S, Mattiasson B, Gessesse A, Hatti-Kaul R (2006) *Nesterenkonia aethiopica* sp. nov., an alkaliphilic, moderate halophile isolated from an Ethiopian soda lake. Int J Syst Evol Microbiol 56(6):1229–1232

Demain AL, Sanchez S (2009) Microbial drug discovery: 80 years of progress. J Antibiot 62(1):5–16

Desai JD, Banat IM (1997) Microbial production of surfactants and their commercial potential. Microbiol Mol Biol Rev 61(1):47–64

Dharmaraj S, Ashokkumar B, Dhevendaran K (2009) Food-grade pigments from *Streptomyces* sp. isolated from the marine sponge *Callyspongia diffusa*. Food Res Int 42(4):487–492

Donia M, Hamann MT (2003) Marine natural products and their potential applications as anti-infective agents. Lancet Infect Dis 3(6):338–348

Du ZJ, Jordan EM, Rooney AP, Chen GJ, Austin B (2010) *Corynebacterium marinum* sp. nov. isolated from coastal sediment. Int J Syst Evol Microbiol 60(8):1944–1947
Du HJ, Zhang YQ, Liu HY, Su J, Wei YZ, Ma BP, Guo BL, Yu LY (2013) *Allonocardiopsis opalescens* gen. nov., sp. nov., a new member of the suborder *Streptosporangineae*, from the surface-sterilized fruit of a medicinal plant. Int J Syst Evol Microbiol 63(3): 900–904
Duckworth AW, Grant S, Grant WD, Jones BE, Meijer D (1998) *Dietzia natronolimnaios* sp. nov., a new member of the genus *Dietzia* isolated from an East African soda lake. Extremophiles 2(3):359–366
El-Gendy MM, Shaaban M, Shaaban KA, El-Bondkly AM, Laatsch H (2008) Essramycin: a first triazolopyrimidine antibiotic isolated from nature. J Antibiot 61(3):149–157
Engelhardt K, Degnes KF, Kemmler M, Bredholt H, Fjærvik E, Klinkenberg G, Sletta H, Ellingsen TE, Zotchev SB (2010) Production of a new thiopeptide antibiotic, TP-1161, by a marine *Nocardiopsis* species. Appl Environ Microb 76(15):4969–4976
European Food Safety Authority (2010) EFSA panel on food additives and nutrient sources added to food (ANS), scientific opinion on the re-evaluation of canthaxanthin (E 161 g) as a food additive. EFSA J 8(10):1852–1894
Everest GJ, le Roes-Hill M, Omorogie C, Cheung SK, Cook AE, Goodwin CM, Meyers PR (2013) *Amycolatopsis umgeniensis* sp. nov., isolated from soil from the banks of the Umgeni River in South Africa. Anton Leeuwen 103(3):673–681
Evtushenko LI, Akimov V, Dobritsa SV, Taptykova SD (1989) A new species of actinomycete, *Amycolata alni*. Int J Syst Bacteriol 39(1):72–77
Evtushenko LI, Taran VV, Akimov VN, Kroppenstedt RM, Tiedje JM, Stackebrandt E (2000) *Nocardiopsis tropica* sp. nov., *Nocardiopsis trehalosi* sp. nov., nom. rev. and *Nocardiopsis dassonvillei* subsp. *albirubida* subsp. nov., comb. nov. Int J Syst Evol Microbiol 50(1):73–81
Fang C, Zhang J, Pang H, Li Y, Xin Y, Zhang Y (2011) *Nocardiopsis flavescens* sp. nov., an actinomycete isolated from marine sediment. Int J Syst Evol Microbiol 61(11):2640–2645
Fathepure BZ (2014) Recent studies in microbial degradation of petroleum hydrocarbons in hypersaline environments. Front Microbiol 5:1–16
FDA (2014) Listing of color additives exempt from certification, 73.75 Canthaxanthin, vol 1. 1 Apr 2014
FDA news release (2013) US Food and Drug Administration. 2013-02-04. Retrieved 22 Feb 2014
Feher M, Schmidt JM (2003) Property distributions: Differences between drugs, natural products, and molecules from combinatorial chemistry. J Chem Inf Comput Sci 43(1):218–227
Feling RH, Buchanan GO, Mincer TJ, Kauffman CA, Jensen PR, Fenical W (2003) Salinosporamide A: a highly cytotoxic proteasome inhibitor from a novel microbial source, a marine bacterium of the new genus *Salinospora*. Angew Chem Int Ed Engl 42(3):355–357
Fiedler HP, Bruntner C, Riedlinger J, Bull AT, Knutsen G, Goodfellow M, Jones A, Maldonado L, Pathom-Aree W, Beil W (2008) Proximicin A, B and C, novel aminofuran antibiotic and anticancer compounds isolated from marine strains of the actinomycete *Verrucosispora*. J Antibiot 61(3):158–163
Francis CA, Obraztsova AY, Tebo BM (2000) Dissimilatory metal reduction by the facultative anaerobe *Pantoea agglomerans* SP1. Appl Environ Microbiol 66(2):543–548
Fu P, Wang S, Hong K, Li X, Liu P, Wang Y, Zhu W (2011) Cytotoxic bipyridines from the marine-derived actinomycete *Actinoalloteichus cyanogriseus* WH1-2216-6. J Nat Prod 74(8):1751–1756
Fu P, Yang C, Wang Y, Liu P, Ma Y, Xu L, Su M, Hong K, Zhu W (2012a) Streptocarbazoles A and B, two novel indolocarbazoles from the marine-derived actinomycete strain *Streptomyces* sp. FMA Org Lett 14(9):2422–2425
Fu Y, Li Q, Liu K, Xu Y, Wang Y, Jiao N (2012b) *Oceanitalea nanhaiensis* gen. nov., sp. nov., an actinobacterium isolated from seawater. Int J Syst Evol Microbiol 62(10):2490–2494
Fudou R, Jojima Y, Seto A, Yamada K, Kimura E, Nakamatsu T, Hiraishi A, Yamanaka S (2002) *Corynebacterium efficiens* sp. nov., a glutamic-acid-producing species from soil and vegetables. Int J Syst Evol Microbiol 52(4):1127–1131

Gagnon K, Lenz R, Farris R, Fuller R (1992) Crystallization behavior and its influence on the mechanical properties of a thermoplastic elastomer produced by *Pseudomonas oleovorans*. Macromolecules 25(14):3723–3728
Galinski E (1993) Compatible solutes of halophilic eubacteria: molecular principles, water-solute interaction, stress protection. Experientia 49(6–7):487–496
Galinski EA, Pfeiffer HP, Trüper HG (1985) 1, 4, 5, 6-Tetrahydro-2-methyl-4-pyrimidinecarboxylic acid. Eur J Biochem 149(1):135–139
Gandhimathi R, Kiran GS, Hema T, Selvin J, Raviji TR, Shanmughapriya S (2009) Production and characterization of lipopeptide biosurfactant by a sponge-associated marine actinomycetes *Nocardiopsis alba* MSA10. Bioprocess Biosyst Eng 32(6):825–835
Gao X, Lu Y, Xing Y, Ma Y, Lu J, Bao W, Wang Y, Xi T (2012) A novel anticancer and antifungus phenazine derivative from a marine actinomycete BM-17. Microbiol Res 167(10):616–622
Gavrish EY, Krauzova V, Potekhina N, Karasev S, Plotnikova E, Altyntseva O, Korosteleva L, Evtushenko L (2004) Three new species of *Brevibacteria*, *Brevibacterium antiquum* sp. nov., *Brevibacterium aurantiacum* sp. nov., and *Brevibacterium permense* sp. nov. Microbiol 73(2):176–183
Genilloud O, González I, Salazar O, Martín J, Tormo JR, Vicente F (2011) Current approaches to exploit actinomycetes as a source of novel natural products. Ind J Microbiol Biotechnol 38(3):375–389
Gerber NN, Stahly DP (1975) Prodiginine (prodigiosin-like) pigments from *Streptoverticillium rubrireticuli*, an organism that causes pink staining of polyvinyl chloride. J Appl Microbiol 30(5):807–810
Gesheva V, Stackebrandt E, Vasileva-Tonkova E (2010) Biosurfactant production by halotolerant *Rhodococcus fascians* from Casey station, Wilkes land, Antarctica. Curr Microbiol 61(2):112–117
Ghai R, McMahon KD, Rodriguez-Valera F (2012) Breaking a paradigm: cosmopolitan and abundant freshwater actinobacteria are low GC. Environ Microbiol Rep 4(1):29–35
Ghanem NB, Mabrouk ME, Sabry SA, El-Badan DE (2005) Degradation of polyesters by a novel marine *Nocardiopsis aegyptia* sp. nov.: application of Plackett-Burman experimental design for the improvement of PHB depolymerase activity. J Gen Appl Microbiol 51(3):151–158
Glick BR (1995) The enhancement of plant growth by free-living bacteria. Can J Microbiol 41(2):109–117
Glick BR, Penrose DM, Li J (1998) A model for the lowering of plant ethylene concentrations by plant growth-promoting bacteria. J Theor Biol 190(1):63–68
Gochnauer MB, Leppard GG, Komaratat P, Kates M, Novitsky T, Kushner DJ (1975) Isolation and characterization of *Actinopolyspora halophila*, gen. et sp. nov., an extremely halophilic actinomycete. Can J Microbiol 21(10):1500–1511
Gohel SD, Singh SP (2012) Purification strategies, characteristics and thermodynamic analysis of a highly thermostable alkaline protease from a salt-tolerant alkaliphilic actinomycete, *Nocardiopsis alba* OK-5. J Chromatogr B 889:61–68
Gontia I, Kavita K, Schmid M, Hartmann A, Jha B (2011) *Brachybacterium saurashtrense* sp. nov., a halotolerant root-associated bacterium with plant growth-promoting potential. Int J Syst Evol Microbiol 61(12):2799–2804
Gorajana A, Kurada BV, Peela S, Jangam P, Vinjamuri S, Poluri E, Zeeck A (2005) 1-hydroxy-1-norresistomycin, a new cytotoxic compound from a marine actinomycete, *Streptomyces chibaensis* AUBN1/7. J Antibiot 58(8):526–529
Gorajana A, Vinjamuri S, Kurada BV, Peela S, Jangam P, Poluri E, Zeeck A (2007) Resistoflavine, cytotoxic compound from a marine actinomycete, *Streptomyces chibaensis* AUBN 1/7. Microbiol Res 162(4):322–327
Gordon HT, Bauernfeind JC, Furia TE (1982) Carotenoids as food colorants. Crit Rev Food Sci 18(1):59–97
Goswami D, Pithwa S, Dhandhukia P, Thakker JN (2014) Delineating *Kocuria turfanensis* 2M4 as a credible PGPR: a novel IAA-producing bacteria isolated from saline desert. J Plant Interact 9(1):566–576

Govender L, Naidoo L, Setati ME (2013) *Nesterenkonia suensis* sp. nov., a haloalkaliphilic actinobacterium isolated from a salt pan. Int J Syst Evol Microbiol 63(1):41–46

Graf R, Anzali S, Buenger J, Pfluecker F, Driller H (2008) The multifunctional role of ectoine as a natural cell protectant. Clin Dermatol 26(4):326–333

Groth I, Schumann P, Rainey F, Martin K, Schuetze B, Augsten K (1997a) *Bogoriella caseilytica* gen. nov., sp. nov., a new alkaliphilic actinomycete from a soda lake in Africa. Int J Syst Bacteriol 47(3):788–794

Groth I, Schumann P, Rainey F, Martin K, Schuetze B, Augsten K (1997b) *Demetria terragena* gen. nov., sp. nov., a new genus of actinomycetes isolated from compost soil. Int J Syst Bacteriol 47(4):1129–1133

Groth I, Schumann P, Schütze B, Gonzalez JM, Laiz L, Saiz-Jimenez C, Stackebrandt E (2005) *Isoptericola hypogeus* sp. nov., isolated from the Roman catacomb of Domitilla. Int J Syst Evol Microbiol 55(4):1715–1719

Gu Q, Paściak M, Luo H, Gamian A, Liu Z, Huang Y (2007) *Ruania albidiflava* gen. nov., sp. nov., a novel member of the suborder *Micrococcineae*. Int J Syst Evol Microbiol 57(4):809–814

Guan TW, Tang SK, Wu JY, Zhi XY, Xu LH, Zhang LL, Li WJ (2009) *Haloglycomyces albus* gen. nov., sp. nov., a halophilic, filamentous actinomycete of the family Glycomycetaceae. Int J Syst Evol Microbiol 59(6):1297–1301

Guan TW, Liu Y, Zhao K, Xia ZF, Zhang XP, Zhang LL (2010a) *Actinopolyspora xinjiangensis* sp. nov., a novel extremely halophilic actinomycete isolated from a salt lake in Xinjiang, China. Anton Leeuwen 98(4):447–453

Guan TW, Zhao K, Xiao J, Liu Y, Xia ZF, Zhang XP, Zhang LL (2010b) *Brevibacterium salitolerans* sp. nov., an actinobacterium isolated from salt-lake sediment. Int J Syst Evol Microbiol 60(12):2991–2995

Guan TW, Wu N, Xia ZF, Ruan JS, Zhang XP, Huang Y, Zhang LL (2011a) *Saccharopolyspora lacisalsi* sp. nov., a novel halophilic actinomycete isolated from a salt lake in Xinjiang, China. Extremophiles 15(3):373–378

Guan TW, Xia ZF, Xiao J, Wu N, Chen ZJ, Zhang LL, Zhang XP (2011b) *Glycomyces halotolerans* sp. nov., a novel actinomycete isolated from a hypersaline habitat in Xinjiang, China. Anton Leeuwen 100(1):137–143

Guan TW, Xia ZF, Tang SK, Wu N, Chen ZJ, Huang Y, Ruan JS, Li WJ, Zhang LL (2012) *Amycolatopsis salitolerans* sp. nov., a filamentous actinomycete isolated from a hypersaline habitat. Int J Syst Evol Microbiol 62(1):23–27

Guan TW, Teng Y, Yang LL, Zhang XP, Che ZM (2013a) *Isoptericola salitolerans* sp. nov., a halotolerant filamentous actinobacterium isolated from a salt lake, China. Extremophiles 17(3):471–476

Guan TW, Wei B, Zhang Y, Xia ZF, Che ZM, Chen XG, Zhang LL (2013b) *Actinopolyspora lacussalsi* sp. nov., an extremely halophilic actinomycete isolated from a salt lake. Int J Syst Evol Microbiol 63(8):3009–3013

Guan TW, Wu N, Tang SK, Che ZM, Chen XG (2013c) *Myceligenerans salitolerans* sp. nov., a halotolerant actinomycete isolated from a salt lake in Xinjiang, China. Extremophiles 17(1):147–152

Guan TW, Zhao HP, Che ZM, Zhang XP, Zhang LL (2013d) *Actinopolyspora dayingensis* sp. nov., a novel halophilic actinomycete isolated from a hypersaline lake. Anton Leeuwen 104(5):787–792

Gübitz G, Haltrich D, Latal B, Steiner W (1997) Mode of depolymerisation of hemicellulose by various mannanases and xylanases in relation to their ability to bleach softwood pulp. Appl Microbiol Biotechnol 47(6):658–662

Gupta R, Gigras P, Mohapatra H, Goswami VK, Chauhan B (2003) Microbial α-amylases: a biotechnological perspective. Process Biochem 38(11):1599–1616

Halpern M, Shakéd T, Pukall R, Schumann P (2009) *Leucobacter chironomi* sp. nov., a chromate-resistant bacterium isolated from a chironomid egg mass. Int J Syst Evol Microbiol 59(4):665–670

Hamada M, Tamura T, Ishida Y, Suzuki KI (2009) *Georgenia thermotolerans* sp. nov., an actinobacterium isolated from forest soil. Int J Syst Evol Microbiol 59(8):1875–1879

Hamada M, Tamura T, Yamamura H, Suzuki KI, Hayakawa M (2012) *Lysinimicrobium mangrovi* gen. nov., sp. nov., an actinobacterium isolated from the rhizosphere of a mangrove. Int J Syst Evol Microbiol 62(8):1731–1735

Hamada M, Shibata C, Tamura T, Yamamura H, Hayakawa M, Suzuki KI (2013a) *Janibacter cremeus* sp. nov., an actinobacterium isolated from sea sediment. Int J Syst Evol Microbiol 63(10):3687–3690

Hamada M, Tamura T, Shibata C, Yamamura H, Hayakawa M, Schumann P, Suzuki KI (2013b) *Paraoerskovia sediminicola* sp. nov., an actinobacterium isolated from sea sediment, and emended description of the genus *Paraoerskovia*. Int J Syst Evol Microbiol 63(7):2637–2641

Hamada M, Tamura T, Yamamura H, K-i S, Hayakawa M (2013c) *Demequina flava* sp. nov. and *Demequina sediminicola* sp. nov., isolated from sea sediment. Int J Syst Evol Microbiol 63(1):249–253

Hamedi J, Mohammadipanah F, von Jan M, Pötter G, Schumann P, Spröer C, Klenk HP, Kroppenstedt RM (2010) Nocardiopsis sinuspersici sp. nov., isolated from sandy rhizospheric soil. Int J Syst Evol Microbiol 60(10):2346–2352

Hamedi J, Mohammadipanah F, Pötter G, Spröer C, Schumann P, Göker M, Klenk HP (2011) *Nocardiopsis arvandica* sp. nov., isolated from sandy soil. Int J Syst Evol Microbiol 61(5):1189–1194

Hamedi J, Mohammadipanah F, Ventosa A (2013) Systematic and biotechnological aspects of halophilic and halotolerant actinomycetes. Extremophiles 17(1):1–13

Hameş-Kocabaş EE, Uzel A (2007) Alkaline protease production by an actinomycete MA1-1 isolated from marine sediments. Ann Microbiol 57(1):71–75

Han SK, Nedashkovskaya OI, Mikhailov VV, Kim SB, Bae KS (2003) *Salinibacterium amurskyense* gen. nov., sp. nov., a novel genus of the family *Microbacteriaceae* from the marine environment. Int J Syst Evol Microbiol 53(6):2061–2066

Han X-X, Cui C-B, Gu Q-Q, Zhu W-M, Liu H-B, Gu J-Y, Osada H (2005) ZHD-0501, a novel naturally occurring staurosporine analog from *Actinomadura* sp. 007. Tetrahedron Lett 46(36):6137–6140

Han X, Luo X, Zhang L (2014) *Glycomyces fuscus* sp. nov. and *Glycomyces albus* sp. nov., two novel actinomycetes isolated from a hypersaline habitat in Xinjiang, China. Int J Syst Evol Microbiol. doi:10.1099/ijs. 0.061788-061780

Hardt IH, Jensen PR, Fenical W (2000) Neomarinone, and new cytotoxic marinone derivatives, produced by a marine filamentous bacterium (actinomycetales). Tetrahedron Lett 41(13):2073–2076

Harishchandra RK, Wulff S, Lentzen G, Neuhaus T, Galla H-J (2010) The effect of compatible solute ectoines on the structural organization of lipid monolayer and bilayer membranes. Biophys Chem 150(1):37–46

Hawas UW, Shaaban M, Shaaban KA, Speitling M, Maier A, Kelter G, Fiebig HH, Meiners M, Helmke E, Laatsch H (2009) Mansouramycins A–D, cytotoxic isoquinolinequinones from a marine *Streptomycete*. J Nat Prod 72(12):2120–2124

Hayakawa Y, Shirasaki S, Kawasaki T, Matsuo Y, Adachi K, Shizuri Y (2007a) Structures of new cytotoxic antibiotics, piericidins C7 and C8. J Antibiot 60(3):201–203

Hayakawa Y, Shirasaki S, Shiba S, Kawasaki T, Matsuo Y, Adachi K, Shizuri Y (2007b) Piericidins C7 and C8, new cytotoxic antibiotics produced by a marine *Streptomyces* sp. J Antibiot 60(3):196–200

He H, Ding WD, Bernan VS, Richardson AD, Ireland CM, Greenstein M, Ellestad GA, Carter GT (2001) Lomaiviticins A and B, potent antitumor antibiotics from *Micromonospora lomaivitiensis*. J Am Chem Soc 123(22):5362–5363

He J, Xu Y, Sahu MK, Tian XP, Nie GX, Xie Q, Zhang S, Sivakumar K, Li WJ (2012) *Actinomadura sediminis* sp. nov., a marine actinomycete isolated from mangrove sediment. Int J Syst Evol Microbiol 62(5):1110–1116

Heyrman J, Verbeeren J, Schumann P, Devos J, Swings J, De Vos P (2004) *Brevibacterium picturae* sp. nov., isolated from a damaged mural painting at the Saint-Catherine chapel (Castle Herberstein, Austria). Int J Syst Evol Microbiol 54(5):1537–1541

Hirsch P, Mevs U, Kroppenstedt RM, Schumann P, Stackebrandt E (2004) Cryptoendolithic actinomycetes from Antarctic sandstone rock samples: *Micromonospora endolithica* sp. nov. and two isolates related to *Micromonospora coerulea* Jensen 1932. Syst Appl Microbiol 27(2):166–174

Hohmann C, Schneider K, Bruntner C, Irran E, Nicholson G, Bull AT, Jones AL, Brown R, Stach JE, Goodfellow M (2009) Caboxamycin, a new antibiotic of the benzoxazole family produced by the deep-sea strain *Streptomyces* sp. NTK 937. J Antibiot 62(2):99–104

Holden P, LaMontagne M, Bruce A, Miller W, Lindow S (2002) Assessing the role of *Pseudomonas aeruginosa* surface-active gene expression in hexadecane biodegradation in sand. Appl Environ Microbiol 68(5):2509–2518

Holmes P (1988) Biologically produced (R)-3-hydroxy-alkanoate polymers and copolymers. In: Bassett D (ed) Developments in crystalline polymers. Springer, New York, pp 1–65

Holt J, Krieg N, Sneath P, Staley J, Williams S (1994) Bergey's manual of determinative bacteriology Williamsons and Wilkins. 9th ed. Baltimore: Williams & Wilkins

Hong Y-G, Guo J, Xu Z-C, Xu M-Y, Sun G-P (2007) Humic substances act as electron acceptor and redox mediator for microbial dissimilatory azoreduction by *Shewanella decolorationis* S12. J Microbiol Biotechnol 17(3):428–437

Hozzein WN, Goodfellow M (2008) *Nocardiopsis arabia* sp. nov., a halotolerant actinomycete isolated from a sand-dune soil. Int J Syst Evol Microbiol 58(11):2520–2524

Hozzein WN, Goodfellow M (2011) *Actinopolyspora egyptensis* sp. nov., a new halophilic actinomycete. Afr J Microbiol Res 5(2):100–105

Hozzein WN, Li WJ, Ali MIA, Hammouda O, Mousa AS, Xu LH, Jiang CL (2004) *Nocardiopsis alkaliphila* sp. nov., a novel alkaliphilic actinomycete isolated from desert soil in Egypt. Int J Syst Evol Microbiol 54(1):247–252

Hu Y, Espindola APD, Stewart NA, Wei S, Posner BA, MacMillan JB (2011) Chromomycin SA analogs from a marine-derived *Streptomyces* sp. Bioorg Med Chem 19(17):5183–5189

Huang Y, Paściak M, Liu Z, Xie Q, Gamian A (2004) *Amycolatopsis palatopharyngis* sp. nov., a potentially pathogenic actinomycete isolated from a human clinical source. Int J Syst Evol Microbiol 54(2):359–363

Huang SX, Zhao LX, Tang SK, Jiang CL, Duan Y, Shen B (2009) Erythronolides H and I, new erythromycin congeners from a new halophilic actinomycete *Actinopolyspora* sp. YIM90600. Org Lett 11(6):1353–1356

Huang Z, Sheng XF, Zhao F, He LY, Huang J, Wang Q (2012) *Isoptericola nanjingensis* sp. nov., a mineral-weathering bacterium. Int J Syst Evol Microbiol 62(4):971–976

Hughes CC, MacMillan JB, Gaudêncio SP, Jensen PR, Fenical W (2009) The ammosamides: structures of cell cycle modulators from a marine-derived *Streptomyces* species. Angew Chem Int Ed Engl 48(4):725–727

Hughes CC, Kauffman CA, Jensen PR, William F (2010) Structures, reactivities antibiotic properties of the marinopyrroles A-F. J Org Chem 75(10):3240–3250

Imada C (2005) Enzyme inhibitors and other bioactive compounds from marine actinomycetes. Antonie Van Leeuwenhoek 87(1):59–63

Imada C, Simidu U (1988) Isolation and characterization of an α-amylase inhibitor producing actinomycete from marine environment. Nippon Suisan Gakkaishi 54(10):1839–1845

Inbar L, Lapidot A (1988) The structure and biosynthesis of new tetrahydropyrimidine derivatives in actinomycin D producer *Streptomyces parvulus*. Use of 13C-and 15N-labeled L-glutamate and 13C and 15N NMR spectroscopy. J Biol Chem 263(31):16014–16022

Itoh T, Kinoshita M, Aoki S, Kobayashi M (2003) Komodoquinone A, a novel neuritogenic anthracycline, from marine *Streptomyces* sp. KS3. J Nat Prod 66(10):1373–1377

Ivanova EP, Christen R, Alexeeva YV, Zhukova NV, Gorshkova NM, Lysenko AM, Mikhailov VV, Nicolau DV (2004) *Brevibacterium celere* sp. nov., isolated from degraded thallus of a brown alga. Int J Syst Evol Microbiol 54(6):2107–2111

Jamil M, Deog BL, Kwang YJ, Ashraf M, Sheong CL, Eui SR (2006) Effect of salt (NaCl) stress on germination and early seedling growth of four vegetables species. J Cent Eur Agric 7(2):273–282

Jarerat A, Pranamuda H, Tokiwa Y (2002) Poly (L-lactide)-degrading activity in various actinomycetes. Macromol Biosci 2(9):420–428

Jayakumar R, Prabaharan M, Nair S, Tamura H (2010) Novel chitin and chitosan nanofibers in biomedical applications. Biotechnol Adv 28(1):142–150

Jemal A, Murray T, Ward E, Samuels A, Tiwari RC, Ghafoor A, Feuer EJ, Thun MJ (2005) Cancer statistics, 2005. CA-Cancer J Clin 55(1):10–30

Jendrossek D (2001) Microbial degradation of polyesters. In: Babel W, Steinbüchel A (eds) Biopolyesters. Springer, Berlin Heidelberg, pp 293–325

Jeong SY, Shin HJ, Kim TS, Lee HS, SK P, Kim HM (2006) Streptokordin, a new cytotoxic compound of the methylpyridine class from a marine-derived *Streptomyces* sp. KORDI-3238. J Antibiot 59(4):234–240

Jiang Z-D, Jensen PR, Fenical W (1999) Lobophorins A and B, new antiinflammatory macrolides produced by a tropical marine bacterium. Bioorg Med Chem Lett 9(14):2003–2006

Jones BE, Grant WD, Duckworth A, Schumann P, Weiss N, Stackebrandt E (2005) *Cellulomonas bogoriensis* sp. nov., an alkaliphilic cellulomonad. Int J Syst Evol Microbiol 55(4):1711–1714

Jose PA, Jebakumar SRD (2012) Phylogenetic diversity of actinomycetes cultured from coastal multipond solar saltern in Tuticorin, India. Aquat Biosyst 8:23–32

Jose PA, Jebakumar SRD (2014) Unexplored hypersaline habitats are sources of novel actinomycetes. Front Microbiol 5:242

Jurado V, Miller AZ, Alias-Villegas C, Laiz L, Saiz-Jimenez C (2012) *Rubrobacter bracarensis* sp. nov., a novel member of the genus *Rubrobacter* isolated from a biodeteriorated monument. Syst Appl Microbiol 35(5):306–309

Kageyama A, Takahashi Y, Yasumoto-Hirose M, Kasai H, Shizuri Y, Omura S (2007) *Janibacter corallicola* sp. nov., isolated from coral in Palau. J Gen Appl Microbiol 53(3):185–189

Kämpfer P, Busse H-J, Rainey FA (2002) *Nocardiopsis compostus* sp. nov., from the atmosphere of a composting facility. Int J Syst Evol Microbiol 52(2):621–627

Kämpfer P, Schäfer J, Lodders N, Busse HJ (2010) *Brevibacterium sandarakinum* sp. nov., isolated from a wall of an indoor environment. Int J Syst Evol Microbiol 60(4):909–913

Kango N, Jain SC (2011) Production and properties of microbial inulinases: recent advances. Food Biotechnol 25(3):165–212

Kanoh K, Matsuo Y, Adachi K, Imagawa H, Nishizawa M, Shizuri Y (2005) Mechercharmycins A and B, cytotoxic substances from marine-derived *Thermoactinomyces* sp. YM3-251. J Antibiot 58(4):289–292

Kappler A, Benz M, Schink B, Brune A (2004) Electron shuttling via humic acids in microbial iron (III) reduction in a freshwater sediment. FEMS Microbiol Ecol 47(1):85–92

Kashyap P, Sabu A, Pandey A, Szakacs G, Soccol CR (2002) Extra-cellular l-glutaminase production by *Zygosaccharomyces rouxii* under solid-state fermentation. Process Biochem 38(3):307–312

Kati H, Ince I, Demir I, Demirbag Z (2010) *Brevibacterium pityocampae* sp. nov., isolated from caterpillars of *Thaumetopoea pityocamp* (Lepidoptera, Thaumetopoeidae). Int J Syst Evol Microbiol 60(2):312–316

Khan ST, Harayama S, Tamura T, Ando K, Takagi M, Kazuo SY (2009) *Paraoerskovia marina* gen. nov., sp. nov., an actinobacterium isolated from marine sediment. Int J Syst Evol Microbiol 59(8):2094–2098

Khan ST, Tamura T, Takagi M, Shin-ya K (2010) *Streptomyces tateyamensis* sp. nov., *Streptomyces marinus* sp. nov. and *Streptomyces haliclonae* sp. nov., isolated from the marine sponge *Haliclona* sp. Int J Syst Evol Microbiol 60(12):2775–2779

Khandeparker R, Numan MT (2008) Bifunctional xylanases and their potential use in biotechnology. Ind J Microbiol Biotechnol 35(7):635–644

Khopade A, Biao R, Liu X, Mahadik K, Zhang L, Kokare C (2012a) Production and stability studies of the biosurfactant isolated from marine *Nocardiopsis* sp. B4. Desalination 285:198–204

Khopade A, Ren B, Liu XY, Mahadik K, Zhang L, Kokare C (2012b) Production and characterization of biosurfactant from marine *Streptomyces* species B3. J Colloid Interface Sci 367(1):311–318

Kim SB, Goodfellow M (1999) Reclassification of *Amycolatopsis rugosa* Lechevalier et al. 1986 as *Prauserella rugosa* gen. nov., comb. nov. Int J Syst Bacteriol 49(2):507–512
Kim SB, Nedashkovskaya OI, Mikhailov VV, Han SK, Kim KO, Rhee MS, Bae KS (2004) *Kocuria marina* sp. nov., a novel actinobacterium isolated from marine sediment. Int J Syst Evol Microbiol 54(5):1617–1620
Kim HM, Choi DH, Hwang CY, Cho BC (2008) *Nocardioides salarius* sp. nov., isolated from seawater enriched with zooplankton. Int J Syst Evol Microbiol 58(9):2056–2064
Kim SH, Yang HO, Sohn YC, Kwon HC (2010) *Aeromicrobium halocynthiae* sp. nov., a taurocholic acid-producing bacterium isolated from the marine ascidian *Halocynthia roretzi*. Int J Syst Evol Microbiol 60(12):2793–2798
Kim J, Roh SW, Choi JH, Jung MJ, Nam YD, Kim MS, Park EJ, Shin KS, Bae JW (2011) *Dietzia alimentaria* sp. nov., isolated from a traditional Korean food. Int J Syst Evol Microbiol 61(9):2254–2258
Kim J, Srinivasan S, You T, Bang JJ, Park S, Lee SS (2013) *Brevibacterium ammoniilyticum* sp. nov., an ammonia-degrading bacterium isolated from sludge of a wastewater treatment plant. Int J Syst Evol Microbiol 63(3):1111–1118
Kiran GS, Sabarathnam B, Selvin J (2010a) Biofilm disruption potential of a glycolipid biosurfactant from marine *Brevibacterium casei*. FEMS Immunol Med Microbiol 59(3):432–438
Kiran GS, Thomas TA, Selvin J (2010b) Production of a new glycolipid biosurfactant from marine *Nocardiopsis lucentensis* MSA04 in solid-state cultivation. Colloids Surf B: Biointerfaces 78(1):8–16
Kiran GS, Sabarathnam B, Thajuddin N, Selvin J (2014) Production of glycolipid biosurfactant from sponge-associated marine actinobacterium *Brachybacterium paraconglomeratum* MSA21. J Surfactant Deterg 17(3):531–542
Kitamoto D, Isoda H, Nakahara T (2002) Functions and potential applications of glycolipid biosurfactants—from energy-saving materials to gene delivery carriers. J Biosci Bioeng 94(3):187–201
Kloepper J, Scher F, Laliberte M, Tipping B (1986) Emergence-promoting rhizobacteria: description and implications for agriculture. In: Swinburne TR (ed) Iron, siderophores, and plant diseases, vol 117. Springer, New York, pp 155–164
Kokare C, Mahadik K, Kadam S, Chopade B (2004) Isolation, characterization and antimicrobial activity of marine halophilic *Actinopolyspora* species AH1 from the west coast of India. Curr Sci 86(4):593–597
Kokare C, Kadam S, Mahadik K, Chopade B (2007) Studies on bioemulsifier production from marine *Streptomyces* sp. S1. Indian J Biotechnol 6(1):78–84
Kovács G, Burghardt J, Pradella S, Schumann P, Stackebrandt E, Màrialigeti K (1999) *Kocuria palustris* sp. nov. and *Kocuria rhizophila* sp. nov., isolated from the rhizoplane of the narrow-leaved cattail (*Typha angustifolia*). Int J Syst Bacteriol 49(1):167–173
Krinsky NI (1998) The antioxidant and biological properties of the carotenoidsa. Ann N Y Acad Sci 854(1):443–447
Krishnakumar S, Alexis Arjan R, Ravikumar S (2011) Extracellular production of L-glutaminase by marine alkalophilic *Streptomyces* sp.-SBU1 isolated from Cape Comorin coast. Indian J Mar Sci 40(5):717
Kuhn DA, Starr MP (1960) *Arthrobacter atrocyaneus*, n. sp., and its blue pigment. Arch Microbiol 36(2):175–181
Kumar CG, Takagi H (1999) Microbial alkaline proteases: from a bioindustrial viewpoint. Biotechnol Adv 17(7):561–594
Kumar L, Singh B, Adhikari DK, Mukherjee J, Ghosh D (2012) A thermoalkaliphilic halotolerant esterase from *Rhodococcus* sp. LKE-028 (MTCC 5562): enzyme purification and characterization. Process Biochem 47(6):983–991
Kundu S, Das S, Mondai N, Lyla P, Khan SA (2008) Evaluation of halophilic actinomycete *Actinopolyspora* sp. for osmolyte production. Res J Microbiol 3(1):47–50
Kurahashi M, Fukunaga Y, Sakiyama Y, Harayama S, Yokota A (2010) *Euzebya tangerina* gen. nov., sp. nov., a deeply branching marine actinobacterium isolated from the sea cucumber

Holothuria edulis, and proposal of *Euzebyaceae* fam. nov., *Euzebyales* ord. nov. and *Nitriliruptoridae subclassis* nov. Int J Syst Evol Microbiol 60(10):2314–2319
Kurzbaum E, Zimmels Y, Armon R (2010) Isolation of a halotolerant *Streptomyces* sp. from a constructed wetland that biodegrade phenol and various biopolymers. Actinomycetologica 24(2):31–38
Kushner DJ (1978) Life in high salt and solute concentrations: halophilic bacteria. In: Kushner DJ (ed) Microbial life in extreme environments. Academic, London, pp 317–368
Kuznetsov V, Zaitseva T, Vakulenko L, Filippova S (1992) *Streptomyces albiaxialis* sp. nov., a new petroleum hydrocarbon-degrading species of thermotolerant and halotolerant *Streptomyces*. Microbiology 61(1):62–67
Kwon HC, Kauffman CA, Jensen PR, Fenical W (2006) Marinomycins AD, antitumor-antibiotics of a new structure class from a marine actinomycete of the recently discovered genus *Marinispora*. J Am Chem Soc 128(5):1622–1632
Kyeremeh K, Acquah KS, Sazak A, Houssen W, Tabudravu J, Deng H, Jaspars M (2014) Butremycin, the 3-hydroxyl derivative of ikarugamycin and a protonated aromatic tautomer of 5′-methylthioinosine from a Ghanaian *Micromonospora* sp. K310. Mar Drugs 12(2):999–1012
Lakshmipathy TD, Prasad AA, Kannabiran K (2010) Production of biosurfactant and heavy metal resistance activity of *Streptomyces* sp. VITDDK3-a novel halo tolerant actinomycetes isolated from saltpan soil. Adv Biol Res 4(2):108–115
Lam KS (2006) Discovery of novel metabolites from marine actinomycetes. Curr Opin Microbiol 9(3):245–251
Lechevalier M, Prauser H, Labeda D, Ruan J-S (1986) Two new genera of nocardioform actinomycetes: *Amycolata* gen. nov. and *Amycolatopsis* gen. nov. Int J Syst Bacteriol 36(1):29–37
Lee SD (2006a) *Amycolatopsis jejuensis* sp. nov. and *Amycolatopsis halotolerans* sp. nov., novel actinomycetes isolated from a natural cave. Int J Syst Evol Microbiol 56(3):549–553
Lee SD (2006b) *Brevibacterium samyangense* sp. nov., an actinomycete isolated from a beach sediment. Int J Syst Evol Microbiol 56(8):1889–1892
Lee SD (2008) *Brevibacterium marinum* sp. nov., isolated from seawater. Int J Syst Evol Microbiol 58(2):500–504
Lee DW, Lee SD (2008) *Aeromicrobium ponti* sp. nov., isolated from seawater. Int J Syst Evol Microbiol 58(4):987–991
Lee DW, Lee SD (2010) *Koreibacter algae* gen. nov., sp. nov., isolated from seaweed. Int J Syst Evol Microbiol 60(7):1510–1515
Lee SD, Kim ES, Min KL, Lee WY, Kang SO, Hah YC (2001) *Pseudonocardia kongjuensis* sp. nov., isolated from a gold mine cave. Int J Syst Evol Microbiol 51(4):1505–1510
Lee H-S, Shin HJ, Jang KH, Kim TS, Oh K-B, Shin J (2005) Cyclic peptides of the nocardamine class from a marine-derived bacterium of the genus *Streptomyces*. J Nat Prod 68(4):623–625
Lee DW, Hyun CG, Lee SD (2007) *Nocardioides marinisabuli* sp. nov., a novel actinobacterium isolated from beach sand. Int J Syst Evol Microbiol 57(12):2960–2963
Levine WG (1991) Metabolism of azo dyes: implication for detoxication and activation. Drug Metab Rev 23(3–4):253–309
Li MG, Li WJ, Xu P, Cui XL, Xu LH, Jiang CL (2003a) Nocardiopsis xinjiangensis sp. nov., a halophilic actinomycete isolated from a saline soil sample in China. Int J Syst Evol Microbiol 53(1):317–321
Li WJ, Tang SK, Stackebrandt E, Kroppenstedt RM, Schumann P, Xu LH, Jiang CL (2003b) *Saccharomonospora paurometabolica* sp. nov., a moderately halophilic actinomycete isolated from soil in China. Int J Syst Evol Microbiol 53(5):1591–1594
Li WJ, Xu P, Tang SK, Xu LH, Kroppenstedt RM, Stackebrandt E, Jiang CL (2003c) *Prauserella halophila* sp. nov. and *Prauserella alba* sp. nov., moderately halophilic actinomycetes from saline soil. Int J Syst Evol Microbiol 53(5):1545–1549
Li WJ, Xu P, Zhang LP, Tang SK, Cui XL, Mao PH, Xu LH, Schumann P, Stackebrandt E, Jiang C-L (2003d) *Streptomonospora alba* sp. nov., a novel halophilic actinomycete, and emended description of the genus *Streptomonospora* Cui et al. 2001. Int J Syst Evol Microbiol 53(5):1421–1425

Li WJ, Chen HH, Xu P, Zhang YQ, Schumann P, Tang SK, Xu LH, Jiang CL (2004a) *Yania halotolerans* gen. nov., sp. nov., a novel member of the suborder *Micrococcineae* from saline soil in China. Int J Syst Evol Microbiol 54(2):525–531

Li W-J, Chen H-H, Zhang Y-Q, Schumann P, Stackebrandt E, Xu L-H, Jiang C-L (2004b) Nesterenkonia halotolerans sp. nov. and Nesterenkonia xinjiangensis sp. nov., actinobacteria from saline soils in the west of China. Int J Syst Evol Microbiol 54(3):837–841

Li WJ, Park DJ, Tang SK, Wang D, Lee JC, Xu LH, Kim CJ, Jiang CL (2004c) Nocardiopsis salina sp. nov., a novel halophilic actinomycete isolated from saline soil in China. Int J Syst Evol Microbiol 54(5):1805–1809

Li F, Maskey RP, Qin S, Sattler I, Fiebig HH, Maier A, Zeeck A, Laatsch H (2005a) Chinikomycins A and B: isolation, structure elucidation, and biological activity of novel antibiotics from a marine *Streptomyces* sp. isolate M045#, 1. J Nat Prod 68(3):349–353

Li WJ, Chen HH, Kim CJ, Park DJ, Tang SK, Lee JC, Xu LH, Jiang CL (2005b) *Microbacterium halotolerans* sp. nov., isolated from a saline soil in the west of China. Int J Syst Evol Microbiol 55(1):67–70

Li WJ, Chen HH, Kim CJ, Zhang YQ, Park DJ, Lee JC, Xu LH, Jiang CL (2005c) *Nesterenkonia sandarakina* sp. nov. and *Nesterenkonia lutea* sp. nov., novel actinobacteria, and emended description of the genus *Nesterenkonia*. Int J Syst Evol Microbiol 55(1):463–466

Li WJ, Schumann P, Zhang YQ, Xu P, Chen GZ, Xu LH, Stackebrandt E, Jiang CL (2005d) Proposal of *Yaniaceae* fam. nov. and *Yania flava* sp. nov. and emended description of the genus *Yania*. Int J Syst Evol Microbiol 55(5):1933–1938

Li DH, Zhu TJ, Liu HB, Fang YC, Gu QQ, Zhu WM (2006a) Four butenolides are novel cytotoxic compounds isolated from the marine-derived bacterium, *Streptoverticillium luteoverticillatum* 11014. Arch Pharm Res 29(8):624–626

Li WJ, Kroppenstedt RM, Wang D, Tang S-K, Lee J-C, Park D-J, Kim C-J, Xu L-H, Jiang C-L (2006b) Five novel species of the genus *Nocardiopsis* isolated from hypersaline soils and emended description of *Nocardiopsis salina* Li et al. 2004. Int J Syst Evol Microbiol 56(5):1089–1096

Li J, Zhao GZ, Zhang YQ, Klenk HP, Pukall R, Qin S, Xu LH, Li WJ (2008a) *Dietzia schimae* sp. nov. and *Dietzia cercidiphylli* sp. nov., from surface-sterilized plant tissues. Int J Syst Evol Microbiol 58(11):2549–2554

Li WJ, Zhang Y-Q, Schumann P, Liu HY, Yu LY, Zhang YQ, Stackebrandt E, Xu LH, Jiang CL (2008b) *Nesterenkonia halophila* sp. nov., a moderately halophilic, alkalitolerant actinobacterium isolated from a saline soil. Int J Syst Evol Microbiol 58(6):1359–1363

Li Y, Tang SK, Chen YG, Wu JY, Zhi XY, Zhang YQ, Li WJ (2009) *Prauserella salsuginis* sp. nov., *Prauserella flava* sp. nov., *Prauserella aidingensis* sp. nov. and *Prauserella sediminis* sp. nov., isolated from a salt lake. Int J Syst Evol Microbiol 59(12):2923–2928

Li S, Tian X, Niu S, Zhang W, Chen Y, Zhang H, Yang X, Zhang W, Li W, Zhang S (2011) Pseudonocardians A–C, new diazaanthraquinone derivatives from a deap-sea actinomycete *Pseudonocardia* sp. SCSIO 01299. Mar Drugs 9(8):1428–1439

Li J, Long LJ, Yang LL, Xu Y, Wang FZ, Li QX, Zhang S, Li WJ (2012a) *Janibacter alkaliphilus* sp. nov., isolated from coral *Anthogorgia* sp. Anton Leeuw 102(1):157–162

Li J, Yang J, Zhu WY, He J, Tian XP, Xie Q, Zhang S, Li WJ (2012b) *Nocardiopsis coralliicola* sp. nov., isolated from the gorgonian coral, *Menella praelonga*. Int J Syst Evol Microbiol 62(7):1653–1658

Li Y, Chen F, Dong K, Wang G (2013) *Actinotalea ferrariae* sp. nov., isolated from an iron mine, and emended description of the genus *Actinotalea*. Int J Syst Evol Microbiol 63(9):3398–3403

Liao Z-L, Tang S-K, Guo L, Zhang Y-Q, Tian X-P, Jiang C-L, Xu L-H, Li W-J (2009) *Verrucosispora lutea* sp. nov., isolated from a mangrove sediment sample. Int J Syst Evol Microbiol 59(9):2269–2273

Lin W, Li L, Fu H, Sattler I, Huang X, Grabley S (2005) New cyclopentenone derivatives from an endophytic *Streptomyces* sp. isolated from the mangrove plant *Aegiceras comiculatum*. J Antibiot 58(9):594–598

Lin C, Lu C, Shen Y (2010) Three new 2-pyranone derivatives from mangrove endophytic actinomycete strain *Nocardiopsis* sp. A00203. Rec Nat Prod 4(4):176–179

Liu ZP, Wu JF, Liu ZH, Liu SJ (2006) *Pseudonocardia ammonioxydans* sp. nov., isolated from coastal sediment. Int J Syst Evol Microbiol 56(3):555–558

Liu R, Zhu T, Li D, Gu J, Xia W, Fang Y, Liu H, Zhu W, Gu Q (2007) Two indolocarbazole alkaloids with apoptosis activity from a marine-derived actinomycete Z2039-2. Arch Pharm Res 30(3):270–274

Liu WH, Li MG, Li YQ, Zhao JY, Ding ZG, Yang PW, Cui XL, Wen ML (2008) Cytotoxic metabolites of *Streptimonospora salina*. Chem Nat Compd 44(4):503–505

Liu J, Zhang Z, Liu Z, Zhu H, Dang H, Lu J, Cui Z (2011) Production of cold-adapted amylase by marine bacterium *Wangia* sp. C52: optimization, modeling, and partial characterization. Mar Biotechnol 13(5):837–844

Lovley DR, Coates JD, Blunt-Harris EL, Phillips EJ, Woodward JC (1996) Humic substances as electron acceptors for microbial respiration. Nature 382(6590):445–448

Lovley D, Fraga JL, Blunt-Harris EL, Hayes L, Phillips E, Coates JD (1998) Humic substances as a mediator for microbially catalyzed metal reduction. Acta Hydrochim Hydrobiol 26(3):152–157

Lu J, Ma Y, Liang J, Xing Y, Xi T, Lu Y (2012) Aureolic acids from a marine-derived *Streptomyces* sp. WBF16. Microbiol Res 167(10):590–595

Lu W-D, Li A-X, Guo Q-L (2014) Production of novel alkalitolerant and thermostable inulinase from marine actinomycete *Nocardiopsis* sp. DN-K15 and inulin hydrolysis by the enzyme. Ann Microbiol 64(2):441–449

Luo HY, Miao LH, Fang C, Yang PL, Wang YR, Shi PJ, Yao B, Fan YL (2008) *Nesterenkonia flava* sp. nov., isolated from paper-mill effluent. Int J Syst Evol Microbiol 58(8):1927–1930

Luo HY, Wang YR, Miao LH, Yang PL, Shi PJ, Fang CX, Yao B, Fan YL (2009) *Nesterenkonia alba* sp. nov., an alkaliphilic actinobacterium isolated from the black liquor treatment system of a cotton pulp mill. Int J Syst Evol Microbiol 59(4):863–868

Mabrouk MM, Sabry SA (2001) Degradation of poly (3-hydroxybutyrate) and its copolymer poly (3-hydroxybutyrate-co-3-hydroxyvalerate) by a marine *Streptomyces* sp. SNG9. Microbiol Res 156(4):323–335

Macherla VR, Liu J, Bellows C, Teisan S, Nicholson B, Lam KS, Potts BC (2005) Glaciapyrroles A, B, and C, pyrrolosesquiterpenes from a *Streptomyces* sp. isolated from an Alaskan marine sediment. J Nat Prod 68(5):780–783

Mackay MA, Norton RS, Borowitzka LJ (1984) Organic osmoregulatory solutes in *cyanobacteria*. J Gen Microbiol 130(9):2177–2191

Maheshwari DK (2013) Bacteria in agrobiology: disease management. Springer, Heidelberg

Maheshwari R, Bharadwaj G, Bhat MK (2000) Thermophilic fungi: their physiology and enzymes. Microbiol Mol Biol Rev 64(3):461–488

Maldonado LA, Fenical W, Jensen PR, Kauffman CA, Mincer TJ, Ward AC, Bull AT, Goodfellow M (2005) *Salinispora arenicola* gen. nov., sp. nov. and *Salinispora tropica* sp. nov., obligate marine actinomycetes belonging to the family *Micromonosporaceae*. Int J Syst Evol Microbiol 55(5):1759–1766

Malviya N, Yandigeri MS, Yadav AK, Solanki MK, Arora DK (2014) Isolation and characterization of novel alkali-halophilic actinomycetes from the Chilika brackish water lake, India. Ann Microbiol 64(4):1829–1838

Manam RR, Teisan S, White DJ, Nicholson B, Grodberg J, Neuteboom ST, Lam KS, Mosca DA, Lloyd GK, Potts BC (2005) Lajollamycin, a nitro-tetraene spiro-β-lactone-γ-lactam antibiotic from the marine actinomycete *Streptomyces nodosus*. J Nat Prod 68(2):240–243

Manivasagan P, Sivasankar P, Venkatesan J, Sivakumar K, Kim S-K (2014) Optimization, production and characterization of glycolipid biosurfactant from the marine actinobacterium, *Streptomyces* sp. MAB36. Bioprocess Biosyst Eng 37(5):783–797

Margesin R, Schinner F (2001) Potential of halotolerant and halophilic microorganisms for biotechnology. Extremophiles 5(2):73–83

Martin GD, Tan LT, Jensen PR, Dimayuga RE, Fairchild CR, Raventos-Suarez C, Fenical W (2007) Marmycins A and B, cytotoxic pentacyclic C-glycosides from a marine sediment-derived actinomycete related to the genus *Streptomyces*. J Nat Prod 70(9):1406–1409

Martínez-Luque M, Dobao M, Castillo F (1991) Characterization of the assimilatory and dissimilatory nitrate-reducing systems in *Rhodobacter*: a comparative study. FEMS Microbiol Lett 83(3):329–333

Maskey RP, Helmke E, Fiebig H-H, Laatsch H (2002) Parimycin: isolation and structure elucidation of a novel cytotoxic 2, 3-dihydroquinizarin analogue of gamma-indomycinone from a marine streptomycete isolate. J Antibiot 55(12):1031–1035

Maskey RP, Helmke E, Laatsch H (2003a) Himalomycin A and B: isolation and structure elucidation of new fridamycin type antibiotics from a marine *Streptomyces* isolate. J Antibiot 56(11):942–949

Maskey RP, Kock I, Helmke E, Laatsch H (2003b) Isolation and structure determination of phenazostatin D, a new phenazine from a marine actinomycete isolate *Pseudonocardia* sp. B6273. Z Naturforsch B 58(7):692–694

Maskey RP, Li F, Qin S, Fiebig HH, Laatsch H (2003c) Chandrananimycins A approximately C: production of novel anticancer antibiotics from a marine *Actinomadura* sp. isolate M048 by variation of medium composition and growth conditions. J Antibiot 56(7):622–629

Maskey RP, Helmke E, Kayser O, Fiebig HH, Maier A, Busche A, Laatsch H (2004) Anti-cancer and antibacterial trioxacarcins with high anti-malaria activity from a marine streptomycete and their absolute stereochemistry. J Antibiot 57(12):771–779

Matsumoto A, Nakai K, Morisaki K, Ōmura S, Takahashi Y (2010) *Demequina salsinemoris* sp. nov., isolated on agar media supplemented with ascorbic acid or rutin. Int J Syst Evol Microbiol 60(5):1206–1209

Mayak S, Tirosh T, Glick BR (2004) Plant growth-promoting bacteria confer resistance in tomato plants to salt stress. Plant Physiol Biochem 42(6):565–572

Mayilraj S, Krishnamurthi S, Saha P, Saini H (2006) *Rhodococcus kroppenstedtii* sp. nov., a novel actinobacterium isolated from a cold desert of the Himalayas, India. Int J Syst Evol Microbiol 56(5):979–982

McArthur KA, Mitchell SS, Tsueng G, Rheingold A, White DJ, Grodberg J, Lam KS, Potts BC (2008) Lynamicins A–E, chlorinated bisindole pyrrole antibiotics from a novel marine actinomycete. J Nat Prod 71(10):1732–1737

Meklat A, Bouras N, Zitouni A, Mathieu F, Lebrihi A, Schumann P, Spröer C, Klenk H-P, Sabaou N (2012) *Actinopolyspora algeriensis* sp. nov., a novel halophilic actinomycete isolated from a Saharan soil. Extremophiles 16(5):771–776

Meklat A, Bouras N, Zitouni A, Mathieu F, Lebrihi A, Schumann P, Spröer C, Klenk H-P, Sabaou N (2013a) *Actinopolyspora mzabensis* sp. nov., a halophilic actinomycete isolated from an Algerian Saharan soil. Int J Syst Evol Microbiol 63(10):3787–3792

Meklat A, Bouras N, Zitouni A, Mathieu F, Lebrihi A, Schumann P, Spröer C, Klenk H-P, Sabaou N (2013b) *Actinopolyspora righensis* sp. nov., a novel halophilic actinomycete isolated from Saharan soil in Algeria. Anton Leeuwen 104(3):301–307

Meklat A, Bouras N, Zitouni A, Mathieu F, Lebrihi A, Schumann P, Spröer C, Klenk H-P, Sabaou N (2013c) *Actinopolyspora saharensis* sp. nov., a novel halophilic actinomycete isolated from Saharan soil in Algeria. Anton Leeuwen 103(4):771–776

Meklat A, Bouras N, Riba A, Zitouni A, Mathieu F, Rohde M, Schumann P, Spröer C, Klenk HP, Sabaou N (2014) *Streptomonospora algeriensis* sp. nov., a halophilic actinomycete isolated from soil in Algeria. Anton Leeuw 106(2):287–292

Mendelssohn IA, Sorrell BK, Brix H, Schierup H-H, Lorenzen B, Maltby E (1999) Controls on soil cellulose decomposition along a salinity gradient in a *Phragmites australis* wetland in Denmark. Aquat Bot 64(3):381–398

Mertz FP, Yao RC (1990) *Saccharopolyspora spinosa* sp. nov. isolated from soil collected in a sugar mill rum still. Int J Syst Bacteriol 40(1):34–39

Miller ED, Kauffman CA, Jensen PR, Fenical W (2007) Piperazimycins: cytotoxic hexadepsipeptides from a marine-derived bacterium of the genus *Streptomyces*. J Org Chem 72(2):323–330

Mishra D, Rhee YH (2014) Microbial leaching of metals from solid industrial wastes. J Microbiol 52(1):1–7
Mishra M, Thakur IS (2011) Purification, characterization and mass spectroscopic analysis of thermo-alkalotolerant β-1, 4 endoxylanase from *Bacillus* sp. and its potential for dye decolorization. Int Biodeter Biodegr 65(2):301–308
Mitchell SS, Nicholson B, Teisan S, Lam KS, Potts BC (2004) Aureoverticillactam, a novel 22-atom macrocyclic lactam from the marine actinomycete *Streptomyces aureoverticillatus*. J Nat Prod 67(8):1400–1402
Mohammadipanah F, Hamedi J, Spröer C, Montero-Calasan CC Md, Schumann P, Klenk HP (2014) *Promicromonospora iranensis* sp. nov., an actinobacterium isolated from rhizospheric soil. Int J Syst Evol Microbiol 64(9):3314–3319
Mona EM (2008) Statistical optimization of medium components for chromate reduction by halophilic *Streptomyces* sp. MS-2. Afr J Microbiol Res 2(5):103–109
Morais PV, Francisco R, Branco R, Chung AP, Da Costa MS (2004) *Leucobacter chromiireducens* sp. nov., and *Leucobacter aridicollis* sp. nov., two new species isolated from a chromium contaminated environment. Syst Appl Microbiol 27(6):646–652
Morais PV, Paulo C, Francisco R, Branco R, Paula Chung A, Da Costa MS (2006) *Leucobacter luti* sp. nov., and *Leucobacter alluvii* sp. nov., two new species of the genus *Leucobacter* isolated under chromium stress. Syst Appl Microbiol 29(5):414–421
Moreno-Vivián C, Cabello P, Martínez-Luque M, Blasco R, Castillo F (1999) Prokaryotic nitrate reduction: molecular properties and functional distinction among bacterial nitrate reductases. J Bacteriol 181(21):6573–6584
Mulligan CN, Yong RN, Gibbs BF (2001) Heavy metal removal from sediments by biosurfactants. J Hazard Mater 85(1):111–125
Nadeem SM, Naveed M, Zahir ZA, Asghar HN (2013) Plant–microbe interactions for sustainable agriculture: fundamentals and recent advances. In: Arora NK (ed) Plant microbe symbiosis: fundamentals and advances. Springer, New Delhi, pp 51–103
Nagata S, Wang YB (2001) Accumulation of ectoine in the halotolerant *Brevibacterium* sp. JCM 6894. J Biosci Bioeng 91(3):288–293
Nagata S, Wang Y, Oshima A, Zhang L, Miyake H, Sasaki H, Ishida A (2008) Efficient cyclic system to yield ectoine using *Brevibacterium* sp. JCM 6894 subjected to osmotic downshock. Biotechnol Bioeng 99(4):941–948
Nesterenko O, Nogina T, Kasumova S, Kvasnikov E, Batrakov S (1982) *Rhodococcus luteus* nom. nov. and *Rhodococcus maris* nom. nov. Int J Syst Bacteriol 32(1):1–14
Newman DJ, Cragg GM (2007) Natural products as sources of new drugs over the last 25 years. J Nat Prod 70(3):461–477
Newman DJ, Cragg GM (2012) Natural products as sources of new drugs over the 30 years from 1981 to 2010. J Nat Prod 75(3):311–335
Ningthoujam DS, Kshetri P, Sanasam S, Nimaichand S (2009) Screening, identification of best producers and optimization of extracellular proteases from moderately halophilic alkalithermotolerant indigenous actinomycetes. World J Appl Sci 7(7):907–916
Niu S, Li S, Chen Y, Tian X, Zhang H, Zhang G, Zhang W, Yang X, Zhang S, Ju J (2011) Lobophorins E and F, new spirotetronate antibiotics from a South China Sea-derived *Streptomyces* sp. SCSIO 01127. J Antibiot 64(11):711–716
Njikam E, Schiewer S (2012) Optimization and kinetic modeling of cadmium desorption from citrus peels: a process for biosorbent regeneration. J Hazard Mater 213:242–248
Novakova R, Odnogova Z, Kutas P, Feckova L, Kormanec J (2010) Identification and characterization of an indigoidine-like gene for a blue pigment biosynthesis in *Streptomyces aureofaciens* CCM 3239. Folia Microbiol 55(2):119–125
Nyyssölä A, Leisola M (2001) *Actinopolyspora halophila* has two separate pathways for betaine synthesis. Arch Microbiol 176(4):294–300
Oh DC, Williams PG, Kauffman CA, Jensen PR, Fenical W (2006) Cyanosporasides A and B, chloro-and cyano-cyclopenta [a] indene glycosides from the marine actinomycete *Salinispora pacifica*. Org Lett 8(6):1021–1024

Onraedt A, De Muynck C, Walcarius B, Soetaert W, Vandamme E (2004) Ectoine accumulation in *Brevibacterium epidermis*. Biotechnol Lett 26(19):1481–1485
Onraedt AE, Walcarius BA, Soetaert WK, Vandamme EJ (2005) Optimization of ectoine synthesis through fed-batch fermentation of *Brevibacterium epidermis*. Biotechnol Prog 21(4):1206–1212
Oren A (2002) Diversity of halophilic microorganisms: environments, phylogeny, physiology, and applications. J Ind Microbiol Biotechnol 28(1):56–63
Oren A (2008) Microbial life at high salt concentrations: phylogenetic and metabolic diversity. Saline Syst 4(2):13
Pal S, Maity P (1992) Antineoplastic activities of purified bacterial glutaminase on transplanted tumor systems. Indian J Cancer Chemother 13:73–76
Palleroni N, Reichelt K, Mueller D, Epps R, Tabenkin B, Bull D, Schüep W, Berger J (1978) Production of a novel red pigment, rubrolone, by *Streptomyces echinoruber* sp. nov. I. Taxonomy, fermentation and partial purification. J Antibiot 31(12):1218–1225
Pan E, Jamison M, Yousufuddin M, MacMillan JB (2012) Ammosamide D, an oxidatively ring opened ammosamide analog from a marine-derived *Streptomyces variabilis*. Org Lett 14(9):2390–2393
Pandey A, Nigam P, Soccol C, Soccol V, Singh D, Mohan R (2000) Advances in microbial amylases. Biotechnol Appl Biochem 31:135–152
Park SC, Baik KS, Kim MS, Chun J, Seong CN (2008) *Nocardioides dokdonensis* sp. nov., an actinomycete isolated from sand sediment. Int J Syst Evol Microbiol 58(11):2619–2623
Pasti-Grigsby M, Paszczynski A, Goszczynski S, Crawford D, Crawford R (1992) Influence of aromatic substitution patterns on azo dye degradability by *Streptomyces* spp. and *Phanerochaete chrysosporium*. Appl Environ Microbiol 58(11):3605–3613
Pastor JM, Salvador M, Argandoña M, Bernal V, Reina-Bueno M, Csonka LN, Iborra JL, Vargas C, Nieto JJ, Cánovas M (2010) Ectoines in cell stress protection: uses and biotechnological production. Biotechnol Adv 28(6):782–801
Pathom-Aree W, Nogi Y, Sutcliffe IC, Ward AC, Horikoshi K, Bull AT, Goodfellow M (2006) *Dermacoccus abyssi* sp. nov., a piezotolerant actinomycete isolated from the Mariana Trench. Int J Syst Evol Microbiol 56(6):1233–1237
Patterson JW (1987) Metals speciation separation and recovery, vol 1. CRC Press, Michigan, USA
Peltola JS, Andersson MA, Kämpfer P, Auling G, Kroppenstedt RM, Busse H-J, Salkinoja-Salonen MS, Rainey FA (2001) Isolation of toxigenic *Nocardiopsis* strains from indoor environments and description of two new *Nocardiopsis* species, *N. exhalans* sp. nov. and *N. umidischolae* sp. nov. Appl Environ Microbiol 67(9):4293–4304
Peng F, Liu Z, Wang L, Shao Z (2007) An oil-degrading bacterium: *Rhodococcus erythropolis* strain 3C-9 and its biosurfactants. J Appl Microbiol 102(6):1603–1611
Peng F, Wang Y, Sun F, Liu Z, Lai Q, Shao Z (2008) A novel lipopeptide produced by a Pacific Ocean deep-sea bacterium, *Rhodococcus* sp. TW53. J Appl Microbiol 105(3):698–705
Percival Zhang YH, Himmel ME, Mielenz JR (2006) Outlook for cellulase improvement: screening and selection strategies. Biotechnol Adv 24(5):452–481
Pernetti M, Palma LD (2005) Experimental evaluation of inhibition effects of saline wastewater on activated sludge. Environ Technol 26(6):695–704
Philp JC, Bamforth SM, Singleton I, Atlas RM (2005) Environmental pollution and restoration: a role for bioremediation. In: Atlas RM, Philp JC (eds) Bioremediation: applied microbial solutions for real-world environmental cleanup. ASM Press, Washington, DC, pp 1–48
Phipps RK, Blunt JW, Cole AL, Munro MH (2004) Anthracycline derivatives from a marine-derived New Zealand Streptomycete. Arkivoc 10:94–100
Pimentel-Elardo SM, Tiro LP, Grozdanov L, Hentschel U (2008) *Saccharopolyspora cebuensis* sp. nov., a novel actinomycete isolated from a Philippine sponge (*Porifera*). Int J Syst Evol Microbiol 58(3):628–632
Pindi PK, Manorama R, Begum Z, Shivaji S (2010) *Arthrobacter antarcticus* sp. nov., isolated from an Antarctic marine sediment. Int J Syst Evol Microbiol 60(10):2263–2266

Plotnikova E, Altyntseva O, Kosheleva I, Puntus I, Filonov A, Gavrish EY, Demakov V, Boronin A (2001) Bacterial degraders of polycyclic aromatic hydrocarbons isolated from salt-contaminated soils and bottom sediments in salt mining areas. Microbiology 70(1):51–58

Plotnikova E, Yastrebova O, Anan'ina L, Dorofeeva L, Lysanskaya VY, Demakov V (2011) Halotolerant bacteria of the genus *Arthrobacter* degrading polycyclic aromatic hydrocarbons. Russ J Ecol 42(6):502–509

Prabahar V, Dube S, Reddy G, Shivaji S (2004) *Pseudonocardia antarctica* sp. nov. an actinomycetes from McMurdo dry valleys, Antarctica. Syst Appl Microbiol 27(1):66–71

Prabhu J, Schauwecker F, Grammel N, Keller U, Bernhard M (2004) Functional expression of the ectoine hydroxylase gene (thpD) from *Streptomyces chrysomallus* in *Halomonas elongata*. Appl Environ Microbiol 70(5):3130–3132

Prudhomme J, McDaniel E, Ponts N, Bertani S, Fenical W, Jensen P, Le Roch K (2008) Marine actinomycetes: a new source of compounds against the human malaria parasite. PLoS One 3(6):e2335

Pusecker K, Laatsch H, Helmke E, Weyland H (1997) Dihydrophencomycin methyl ester, a new phenazine derivative from a marine streptomycete. J Antibiot 50(20):479–483

Qin S, Chen HH, Klenk HP, Kim CJ, Xu LH, Li WJ (2010) *Saccharopolyspora gloriosae* sp. nov., an endophytic actinomycete isolated from the stem of *Gloriosa superba* L. Int J Syst Evol Microbiol 60(5):1147–1151

Raafat N, Tharwat E, Radwan E (2011) Improving wheat grain yield and its quality under salinity conditions at a newly reclaimed soil by using different organic sources as soil or foliar applications. J Appl Sci Res 7(1):42–55

Rahman PK, Gakpe E (2008) Production, characterisation and applications of biosurfactants-review. Biotechnol 7(2):360–370

Rainey F, Klatte S, Kroppenstedt R, Stackebrandt E (1995) *Dietzia*, new genus including *Dietzia maris* comb. nov., formerly *Rhodococcus maris*. Int J Syst Bacteriol 45(1):32–36

Raja S, Ganesan S, Sivakumar K, Thangaradjou T (2010) Screening of marine actinobacteria for amylase enzymes inhibitors. Ind J Microbiol 50(2):233–237

Ramesh S, Rajesh M, Mathivanan N (2009) Characterization of a thermostable alkaline protease produced by marine *Streptomyces fungicidicus* MML1614. Bioprocess Biosyst Eng 32(6):791–800

Rao MB, Tanksale AM, Ghatge MS, Deshpande VV (1998) Molecular and biotechnological aspects of microbial proteases. Microbiol Mol Biol Rev 62(3):597–635

Ray L, Suar M, Pattnaik AK, Raina V (2013) *Streptomyces chilikensis* sp. nov., a halophilic streptomycete isolated from brackish water sediment. Int J Syst Evol Microbiol 63(8):2757–2764

Reddy G, Aggarwal R, Matsumoto G, Shivaji S (2000) *Arthrobacter flavus* sp. nov., a psychrophilic bacterium isolated from a pond in McMurdo Dry Valley, Antarctica. Int J Syst Evol Microbiol 50(4):1553–1561

Reddy N, Ramakrishna D, Raja Gopal S (2011) A morphological, physiological and biochemical studies of marine *Streptomyces rochei* (MTCC 10109) showing antagonistic activity against selective human pathogenic microorganisms. Asian J Biol Sci 4(1):1–14

Ren W, Zhang F, Yang X, Tang S, Ming H, Zhou E, Yin Y, Zhang Y, Li W (2013) Purification and properties of a SDS-resistant xylanase from halophilic *Streptomonospora* sp. YIM 90494. Cellulose 20(4):1947–1955

Riegal E, Bissinger H (2003) Industrial fermentation: principles, processes and products. In: Kent JA (ed) Riegel's handbook of industrial chemistry. Kluwer Academic/Plenum, New York, pp 963–1045

Riis V, Kleinsteuber S, Babel W (2003) Influence of high salinities on the degradation of diesel fuel by bacterial consortia. Can J Microbiol 49(11):713–721

Roberts MF (2005) Organic compatible solutes of halotolerant and halophilic microorganisms. Saline Syst 1(5):1–30

Roberts J, Holcenberg JS, Dolowy WC (1970) Antineoplastic activity of highly purified bacterial glutaminases. Nature (London) 227:1136–1137

Roychoudhury A, Haussinger D, Oesterhelt F (2012) Effect of the compatible solute ectoine on the stability of the membrane proteins. Protein Pept Lett 19(8):791–794

Ruan JS, Al-Tai AM, Zhou ZH, Qu LH (1994) Actinopolyspora iraqiensis sp. nov., a new halophilic actinomycete isolated from soil. Int J Syst Bacteriol 44(4):759–763

Runmao H (1987) *Saccharomonospora azurea* sp. nov., a new species from soil. Int J Syst Bacteriol 37(1):60–61

Sabu A, Chandrasekaran M, Pandey A (2000) Biopotential of microbial glutaminases. Chim Oggi 18(11–12):21–25

Sadeghi A, Karimi E, Dahaji PA, Javid MG, Dalvand Y, Askari H (2012) Plant growth promoting activity of an auxin and siderophore producing isolate of *Streptomyces* under saline soil conditions. World J Microbiol Biotechnol 28(4):1503–1509

Sadeghi A, Soltani BM, Jouzani GS, Karimi E, Nekouei MK, Sadeghizadeh M (2014) Taxonomic study of a salt tolerant *Streptomyces* sp. strain C-2012 and the effect of salt and ectoine on *lon* expression level. Microbiol Res 169(2):232–238

Sagar S, Esau L, Hikmawan T, Antunes A, Holtermann K, Stingl U, Bajic VB, Kaur M (2013) Cytotoxic and apoptotic evaluations of marine bacteria isolated from brine-seawater interface of the Red Sea. BMC Complement Altern Med 13(1):29

Saha M, Jaisankar P, Das S, Sarkar KK, Roy S, Besra SE, Vedasiromani JR, Ghosh D, Sana B, Mukherjee J (2006) Production and purification of a bioactive substance inhibiting multiple drug resistant bacteria and human leukemia cells from a salt-tolerant marine actinobacterium sp. isolated from the Bay of Bengal. Biotechnol Lett 28(14):1083–1088

Saha S, Dhanasekaran D, Shanmugapriya S, Latha S (2013) *Nocardiopsis* sp. SD5: a potent feather degrading rare actinobacterium isolated from feather waste in Tamil Nadu, India. J Basic Microbiol 53(7):608–616

Sánchez López JM, Martínez Insua M, Pérez Baz J, Fernández Puentes JL, Cañedo Hernández LM (2003) New cytotoxic indolic metabolites from a marine *Streptomyces*. J Nat Prod 66(6):863–864

Sánchez C, Méndez C, Salas JA (2006) Indolocarbazole natural products: occurrence, biosynthesis, and biological activity. Nat Prod Res 23(6):1007–1045

Sass AM, Sass H, Coolen MJ, Cypionka H, Overmann J (2001) Microbial communities in the chemocline of a hypersaline deep-sea basin (Urania basin, Mediterranean Sea). Appl Environ Microbiol 67(12):5392–5402

Sato S, Iwata F, Mukai T, Yamada S, Takeo J, Abe A, Kawahara H (2009) Indoxamycins A–F. Cytotoxic tricycklic polypropionates from a marine-derived actinomycete. J Org Chem 74(15):5502–5509

Satpute SK, Banat IM, Dhakephalkar PK, Banpurkar AG, Chopade BA (2010) Biosurfactants, bioemulsifiers and exopolysaccharides from marine microorganisms. Biotechnol Adv 28(4):436–450

Saurav K, Kannabiran K (2012) Cytotoxicity and antioxidant activity of 5-(2, 4-dimethylbenzyl) pyrrolidin-2-one extracted from marine *Streptomyces* VITSVK5 spp. Saudi J Biol Sci 19(1):81–86

Schippers A, Bosecker K, Willscher S, Spröer C, Schumann P, Kroppenstedt RM (2002) *Nocardiopsis metallicus* sp. nov., a metal-leaching actinomycete isolated from an alkaline slag dump. Int J Syst Evol Microbiol 52(6):2291–2295

Schleissner C, Pérez M, Losada A, Rodriguez P, Crespo C, Zúniga P, Fernández R, Reyes F, de la Calle F (2011) Antitumor actinopyranones produced by *Streptomyces albus* POR-04-15-053 isolated from a marine sediment. J Nat Prod 74(7):1590–1596

Schumacher RW, Talmage SC, Miller SA, Sarris KE, Davidson BS, Goldberg A (2003) Isolation and structure determination of an antimicrobial ester from a marine sediment-derived bacterium. J Nat Prod 66(9):1291–1293

Schwarzenbach RP, Stierli R, Lanz K, Zeyer J (1990) Quinone and iron porphyrin mediated reduction of nitroaromatic compounds in homogeneous aqueous solution. Environ Sci Technol 24(10):1566–1574

Seghal Kiran G, Anto Thomas T, Selvin J, Sabarathnam B, Lipton A (2010) Optimization and characterization of a new lipopeptide biosurfactant produced by marine *Brevibacterium aureum* MSA13 in solid state culture. Biores Technol 101(7):2389–2396

Selvakumar G, Kim K, Hu S, Sa T (2014) Effect of salinity on plants and the role of arbuscular mycorrhizal fungi and plant growth-promoting rhizobacteria in alleviation of salt stress. In: Ahmad P, Wani MR (eds) Physiological mechanisms and adaptation strategies in plants under changing environment. Springer, New York, pp 115–144

Selvin J, Shanmugha Priya S, Seghal Kiran G, Thangavelu T, Sapna Bai N (2009) Sponge-associated marine bacteria as indicators of heavy metal pollution. Microbiol Res 164(3):352–363

Senthilkuraar S (2005) Mercury resistant halophilic actinomycetes from the salt marsh environment of Vellar estuary, southeast coast of India. SEA 3(8.5):1.2–2.3

Severin J, Wohlfarth A, Galinski EA (1992) The predominant role of recently discovered tetrahydropyrimidines for the osmoadaptation of halophilic eubacteria. J Gen Microbiol 138(8):1629–1638

Sharma S, Pant A (2000) Biodegradation and conversion of alkanes and crude oil by a marine *Rhodococcus*. Biodegradation 11(5):289–294

Sharma S, Pant A (2001) Crude oil degradation by a marine actinomycete *Rhodococcus* sp. Ind J Mar Sci 30(3):146–150

Shi Z, Luo G, Wang G (2012) *Cellulomonas carbonis* sp. nov., isolated from coal mine soil. Int J Syst Evol Microbiol 62(8):2004–2010

Shin HJ, Jeong HS, Lee HS, Park SK, Kim HM, Kwon HJ (2007) Isolation and structure determination of streptochlorin, an antiproliferative agent from a marine-derived *Streptomyces* sp. 04DH110. J Microbiol Biotechnol 17(8):1403–1406

Shin HJ, Kim TS, Lee HS, Park JY, Choi IK, Kwon HJ (2008) Streptopyrrolidine, an angiogenesis inhibitor from a marine-derived *Streptomyces* sp. KORDI-3973. Phytochem 69(12):2363–2366

Siddikee M, Chauhan P, Anandham R, Han G-H, Sa T (2010) Isolation, characterization, and use for plant growth promotion under salt stress, of ACC deaminase-producing halotolerant bacteria derived from coastal soil. J Microbiol Biotechnol 20(11):1577–1584

Siddikee MA, Glick BR, Chauhan PS, Sa T (2011) Enhancement of growth and salt tolerance of red pepper seedlings (*Capsicum annuum* L.) by regulating stress ethylene synthesis with halotolerant bacteria containing 1-aminocyclopropane-1-carboxylic acid deaminase activity. Plant Physiol Biochem 49(4):427–434

Siddikee MA, Chauhan PS, Sa T (2012) Regulation of ethylene biosynthesis under salt stress in red pepper (*Capsicum annuum* L.) by 1-aminocyclopropane-1-carboxylic acid (ACC) deaminase-producing halotolerant bacteria. J Plant Growth Reg 31(2):265–272

Sivaramakrishnan S, Gangadharan D, Nampoothiri KM, Soccol CR, Pandey A (2006) a-Amylases from microbial sources–an overview on recent developments. Food Technol Biotechnol 44(2):173–184

Soria-Mercado IE, Prieto-Davo A, Jensen PR, Fenical W (2005) Antibiotic terpenoid chloro-dihydroquinones from a new marine actinomycete. J Nat Prod 68(6):904–910

Sorokin D, Kolganova T (2014) Bacterial chitin utilization at halophilic conditions. Extremophiles 18(2):243–248

Sorokin DY, van Pelt S, Tourova TP, Evtushenko LI (2009) *Nitriliruptor alkaliphilus* gen. nov., sp. nov., a deep-lineage haloalkaliphilic actinobacterium from soda lakes capable of growth on aliphatic nitriles, and proposal of *Nitriliruptoraceae* fam. nov. and *Nitriliruptorales* ord. nov. Int J Syst Evol Microbiol 59(2):248–253

Stackebrandt E, Koch C, Gvozdiak O, Schumann P (1995) Taxonomic dissection of the genus *Micrococcus*: *Kocuria* gen. nov., *Nesterenkonia* gen. nov., *Kytococcus* gen. nov., *Dermacoccus* gen. nov., and *Micrococcus* Cohn 1872 gen. emend. Int J Syst Bacteriol 45(4):682–692

Starr MP (1958) The blue pigment of *Corynebacterium insidiosum*. Arch Mikrobiol 30(4):325–334

Steinbüchel A (1991) Polyhydroxyalkanoic acids. Byrom D (ed) Biomaterials, Macmillan Publishers Ltd, Basingstoke, pp 123–213

Steinbüchel A, Hein S (2001) Biochemical and molecular basis of microbial synthesis of polyhydroxyalkanoates in microorganisms. In: Babel W, Steinbüchel A (eds) Biopolyesters. vol 71. Springer, Berlin Heidelberg, pp 81–123

Stevenson FJ (1994) Humus chemistry: genesis, composition, reactions. John Wiley & Sons, USA

Strap JL (2011) Actinobacteria–plant interactions: a boon to agriculture. In: Maheshwari DK (ed) Bacteria in agrobiology: plant growth responses. Springer, Berlin, pp 285–307

Stritzke K, Schulz S, Laatsch H, Helmke E, Beil W (2004) Novel caprolactones from a marine streptomycete. J Nat Prod 67(3):395–401

Subramani R, Aalbersberg W (2012) Marine actinomycetes: an ongoing source of novel bioactive metabolites. Microbiol Res 167(10):571–580

Sudhir P, Murthy S (2004) Effects of salt stress on basic processes of photosynthesis. Photosynthetica 42(2):481–486

Sugimoto S, Kato T, Ito M, Sakata N, Tsuchida T, Matsumoto A, Takahashi Y (2011) *Branchiibius hedensis* gen. nov., sp. nov., an actinobacterium isolated from a Japanese codling (*Physiculus japonicus*). Int J Syst Evol Microbiol 61(5):1195–1200

Sun CH, Wang Y, Wang Z, Zhou JQ, Jin WZ, You XF, Gao H, Zhao LX, Si SY, Li X (2007) Chemomicin A, a new angucyclinone antibiotic produced by *Nocardia mediterranei* subsp. *kanglensis* 1747–64. J Antibiot 60(3):211–215

Suthindhiran K, Kannabiran K (2009) Cytotoxic and antimicrobial potential of actinomycete species *Saccharopolyspora salina* VITSDK4 isolated from the Bay of Bengal Coast of India. Am J Infect Dis 5(2):90–98

Suthindhiran K, Jayasri MA, Dipali D, Prasar A (2014) Screening and characterization of protease producing actinomycetes from marine saltern. J Basic Microbiol 54(10):1098–1109

Sutthiwong N, Fouillaud M, Valla A, Caro Y, Dufossé L (2014) Bacteria belonging to the extremely versatile genus *Arthrobacter* as novel source of natural pigments with extended hue range. Food Res Int 65(8):56–162

Syed DG, Tang SK, Cai M, Zhi XY, Agasar D, Lee JC, Kim CJ, Jiang CL, Xu LH, Li WJ (2008) *Saccharomonospora saliphila* sp. nov., a halophilic actinomycete from an Indian soil. Int J Syst Evol Microbiol 58(3):570–573

Syed DG, Agasar D, Pandey A (2009) Production and partial purification of α-amylase from a novel isolate *Streptomyces gulbargensis*. J Ind Microbiol Biotechnol 36(2):189–194

Takeuchi M, Fang C-X, Yokota A (1995) Taxonomic study of the genus *Brachybacterium*: proposal of *Brachybacterium conglomeratum* sp. nov., nom. rev., *Brachybacterium paraconglomeratum* sp. nov., and *Brachybacterium rhamnosum* sp. nov. Int J Syst Bacteriol 45(1):160–168

Tang SK, Tian XP, Zhi XY, Cai M, Wu JY, Yang LL, Xu LH, Li WJ (2008a) *Haloactinospora alba* gen. nov., sp. nov., a halophilic filamentous actinomycete of the family Nocardiopsaceae. Int J Syst Evol Microbiol 58(9):2075–2080

Tang S-K, Wang Y, Schumann P, Stackebrandt E, Lou K, Jiang C-L, Xu L-H, Li W-J (2008b) *Brevibacterium album* sp. nov., a novel actinobacterium isolated from a saline soil in China. Int J Syst Evol Microbiol 58(3):574–577

Tang S-K, Wang Y, Cai M, Zhi X-Y, Lou K, Xu L-H, Jiang C-L, Li W-J (2009a) *Saccharopolyspora halophila* sp. nov., a novel halophilic actinomycete isolated from a saline lake in China. Int J Syst Evol Microbiol 59(3):555–558

Tang SK, Wang Y, Chen Y, Lou K, Cao LL, Xu LH, Li WJ (2009b) *Zhihengliuella alba* sp. nov., and emended description of the genus *Zhihengliuella*. Int J Syst Evol Microbiol 59(8):2025–2032

Tang SK, Wang Y, Lou K, Mao PH, Xu LH, Jiang CL, Kim CJ, Li WJ (2009c) *Kocuria halotolerans* sp. nov., an actinobacterium isolated from a saline soil in China. Int J Syst Evol Microbiol 59(6):1316–1320

Tang SK, Wang Y, Wu JY, Cao LL, Lou K, Xu LH, Jiang CL, Li WJ (2009d) *Saccharopolyspora qijiaojingensis* sp. nov., a halophilic actinomycete isolated from a salt lake. Int J Syst Evol Microbiol 59(9):2166–2170

Tang SK, Wang Y, Guan TW, Lee JC, Kim CJ, Li WJ (2010a) *Amycolatopsis halophila* sp. nov., a halophilic actinomycete isolated from a salt lake. Int J Syst Evol Microbiol 60(5):1073–1078

Tang SK, Wang Y, Lee JC, Lou K, Park DJ, Kim CJ, Li WJ (2010b) *Georgenia halophila* sp. nov., a halophilic actinobacterium isolated from a salt lake. Int J Syst Evol Microbiol 60(6):1317–1421

Tang SK, Wang Y, Zhang H, Lee JC, Lou K, Kim CJ, Li WJ (2010c) *Haloechinothrix alba* gen. nov., sp. nov., a halophilic, filamentous actinomycete of the suborder *Pseudonocardineae*. Int J Syst Evol Microbiol 60(9):2154–2158

Tang S-K, Wu J-Y, Wang Y, Schumann P, Li W-J (2010d) *Yimella lutea* gen. nov., sp. nov., a novel actinobacterium of the family *Dermacoccaceae*. Int J Syst Evol Microbiol 60(3):659–663

Tang S-K, Zhi X-Y, Wang Y, Wu J-Y, Lee J-C, Kim C-J, Lou K, Xu L-H, Li W-J (2010e) *Haloactinobacterium album* gen. nov., sp. nov., a halophilic actinobacterium, and proposal of Ruaniaceae fam. nov. Int J Syst Evol Microbiol 60(9):2113–2119

Tang SK, Wang Y, Klenk HP, Shi R, Lou K, Zhang YJ, Chen C, Ruan JS, Li WJ (2011a) *Actinopolyspora alba* sp. nov. and *Actinopolyspora erythraea* sp. nov., isolated from a salt field, and reclassification of *Actinopolyspora iraqiensis* Ruan et al. 1994 as a heterotypic synonym of *Saccharomonospora halophila*. Int J Syst Evol Microbiol 61(7):1693–1698

Tang SK, Zhi XY, Wang Y, Shi R, Lou K, Xu LH, Li WJ (2011b) *Haloactinopolyspora alba* gen. nov., sp. nov., a halophilic filamentous actinomycete isolated from a salt lake, with proposal of Jiangellaceae fam. nov. and Jiangellineae subord. nov. Int J Syst Evol Microbiol 61(1):194–200

Tatar D, Sazak A, Guven K, Cetin D, Sahin N (2013) *Amycolatopsis cihanbeyliensis* sp. nov., a halotolerant actinomycete isolated from a salt mine. Int J Syst Evol Microbiol 63(10):3739–3743

Tatar D, Guven K, Spröer C, Klenk HP, Sahin N (2014) *Streptomyces iconiensis* sp. nov. and *Streptomyces smyrnaeus* sp. nov., two halotolerant actinomycetes isolated from Tuz (Salt) Lake and Camalti Saltern in Turkey. Int J Syst Evol Microbiol. 64(9):3126–3133

Thumar J, Singh SP (2007) Two-step purification of a highly thermostable alkaline protease from salt-tolerant alkaliphilic *Streptomyces clavuligerus* strain Mit-1. J Chromatogr B 854(1):198–203

Tian XP, Zhi XY, Qiu YQ, Zhang YQ, Tang SK, Xu LH, Zhang S, Li WJ (2009) *Sciscionella marina* gen. nov., sp. nov., a marine actinomycete isolated from a sediment in the northern South China Sea. Int J Syst Evol Microbiol 59(2):222–228

Tian XP, Long LJ, Wang FZ, Xu Y, Li J, Zhang J, Zhang CS, Zhang S, Li WJ (2012a) *Streptomyces nanhaiensis* sp. nov., a marine Streptomycete isolated from a deep-sea sediment. Int J Syst Evol Microbiol 62(4):864–868

Tian XP, Xu Y, Zhang J, Li J, Chen Z, Kim CJ, Li WJ, Zhang CS, Zhang S (2012b) *Streptomyces oceani* sp. nov., a new obligate marine actinomycete isolated from a deep-sea sample of seep authigenic carbonate nodule in South China Sea. Anton Leeuwen 102(2):335–343

Tian SZ, Pu X, Luo G, Zhao LX, Xu LH, Li WJ, Luo Y (2013a) Isolation and characterization of new p-terphenyls with antifungal, antibacterial, and antioxidant activities from halophilic actinomycete *Nocardiopsis gilva* YIM 90087. J Agric Food Chem 61(12):3006–3012

Tian XP, Long LJ, Li SM, Zhang J, Xu Y, He J, Li J, Wang FZ, Li WJ, Zhang CS (2013b) *Pseudonocardia antitumoralis* sp. nov., a deoxynyboquinone-producing actinomycete isolated from a deep-sea sediment. Int J Syst Evol Microbiol 63(3):893–899

Tian S, Yang Y, Liu K, Xiong Z, Xu L, Zhao L (2014) Antimicrobial metabolites from a novel halophilic actinomycete *Nocardiopsis terrae* YIM 90022. Nat Prod Res 28(5):344–346

Tomita K, Nakakita Y, Hoshino Y, Numata K, Kawaguchi H (1987) New genus of the actinomycetales: *Streptoalloteichus hindustanus* gen. nov., nom. rev.; sp. nov., nom. rev. Int J Syst Bacteriol 37(3):211–213

Tonouchi A, Kitamura K, Fujita T (2013) *Brevibacterium yomogidense* sp. nov., isolated from a soil conditioner made from poultry manure. Int J Syst Evol Microbiol 63(2):516–520

Toyoda H, Utsumi R (1991) Method for the prevention of *Fusarium* diseases and microorganisms used for the same. U.S. Patents, 4,988,586

Traiwan J, Park MH, Kim W (2011) *Serinicoccus chungangensis* sp. nov., isolated from tidal flat sediment, and emended description of the genus *Serinicoccus*. Int J Syst Evol Microbiol 61(6):1299–1303

Trujillo ME, Kroppenstedt RM, Schumann P, Martínez-Molina E (2006) *Kribbella lupini* sp. nov., isolated from the roots of *Lupinus angustifolius*. Int J Syst Evol Microbiol 56(2):407–411

Tseng M, Liao HC, Chiang WP, Yuan GF (2011) *Isoptericola chiayiensis* sp. nov., isolated from mangrove soil. Int J Syst Evol Microbiol 61(7):1667–1670

Ue H, Matsuo Y, Kasai H, Yokota A (2011) *Demequina globuliformis* sp. nov., *Demequina oxidasica* sp. nov. and *Demequina aurantiaca* sp. nov., actinobacteria isolated from marine environments, and proposal of *Demequinaceae* fam. nov. Int J Syst Evol Microbiol 61(6):1322–1329

Valdman E, Leite S (2000) Biosorption of Cd, Zn and Cu by *Sargassum* sp. waste biomass. Bioproc Eng 22(2):171–173

Van Der Maarel MJ, Van Der Veen B, Uitdehaag J, Leemhuis H, Dijkhuizen L (2002) Properties and applications of starch-converting enzymes of the α-amylase family. J Biotechnol 94(2):137–155

Van Maanen J, Welle IJ, Hageman G, Dallinga JW, Mertens P, Kleinjans J (1996) Nitrate contamination of drinking water: relationship with HPRT variant frequency in lymphocyte DNA and urinary excretion of N-nitrosamines. Environ Health Perspect 104(5):522

Vasanthabharathi V, Lakshminarayanan R, Jayalakshmi S (2011) Melanin production from marine *Streptomyces*. Afr J Biotechnol 10(54):11224–11234

Venil CK, Lakshmanaperumalsamy P (2009) An insightful overview on microbial pigment, prodigiosin. Electr J Biol 5(3):49–61

Ventosa A, Nieto JJ, Oren A (1998) Biology of moderately halophilic aerobic bacteria. Microbiol Mol Biol Rev 62(2):504–544

Verdine GL (1996) The combinatorial chemistry of nature. Nature 384(6604):11–13

Veyisoglu A, Sazak A, Cetin D, Guven K, Sahin N (2013) *Saccharomonospora amisosensis* sp. nov., isolated from deep marine sediment. Int J Syst Evol Microbiol 63(10):3782–3786

Vilar VJ, Botelho C, Boaventura RA (2007) Copper desorption from *Gelidium* algal biomass. Water Res 41(7):1569–1579

Voisard C, Keel C, Haas D, Dèfago G (1989) Cyanide production by *Pseudomonas fluorescens* helps suppress black root rot of tobacco under gnotobiotic conditions. EMBO J 8(2):351

Volesky B (1990) Biosorption of heavy metals. CRC Press, Boca Raton

Wang JC (1985) DNA topoisomerases. Ann Rev Biochem 54(1):665–697

Wang YM, Zhang ZS, Xu XL, Ruan JS, Wang Y (2001) *Actinopolymorpha singaporensis* gen. nov., sp. nov., a novel actinomycete from the tropical rainforest of Singapore. Int J Syst Evol Microbiol 51(2):467–473

Wang Y, Wu C, Wang X, Zhou S (2009) The role of humic substances in the anaerobic reductive dechlorination of 2, 4-dichlorophenoxyacetic acid by *Comamonas koreensis* strain CY01. J Hazard Mater 164(2):941–947

Wang F, Tian X, Huang C, Li Q, Zhang S (2010a) Marinactinones A–C, new γ-pyrones from marine actinomycete *Marinactinospora thermotolerans* SCSIO 00606. J Antibiot 64(2):189–192

Wang J, Li Y, Bian J, Tang SK, Ren B, Chen M, Li WJ, Zhang LX (2010b) *Prauserella marina* sp. nov., isolated from ocean sediment of the South China Sea. Int J Syst Evol Microbiol 60(4):985–989

Wang SL, Liang TW, Yen YH (2011a) Bioconversion of chitin-containing wastes for the production of enzymes and bioactive materials. Carbohydr Polym 84(2):732–742

Wang Y, Tang SK, Li Z, Lou K, Mao PH, Jin X, Klenk HP, Zhang LX, Li WJ (2011b) *Myceligenerans halotolerans* sp. nov., an actinomycete isolated from a salt lake, and emended description of the genus *Myceligenerans*. Int J Syst Evol Microbiol 61(4):974–978

Warwick S, Bowen T, McVeigh H, Embley TM (1994) A phylogenetic analysis of the family *Pseudonocardiaceae* and the genera *Actinokineospora* and *Saccharothrix* with 16S rRNA sequences and a proposal to combine the genera *Amycolata* and *Pseudonocardia* in an emended genus *Pseudonocardia*. Int J Syst Bacteriol 44(2):293–299

Wauters G, Avesani V, Laffineur K, Charlier J, Janssens M, Van Bosterhaut B, Delmée M (2003) *Brevibacterium lutescens* sp. nov., from human and environmental samples. Int J Syst Evol Microbiol 53(5):1321–1325

Wei RB, Xi T, Li J, Wang P, Li FC, Lin YC, Qin S (2011) Lobophorin C and D, new kijanimicin derivatives from a marine sponge-associated actinomycetal strain AZS17. Mar Drugs 9(3):359–368

Wieser M, Denner EB, Kämpfer P, Schumann P, Tindall B, Steiner U, Vybiral D, Lubitz W, Maszenan A, Patel B (2002) Emended descriptions of the genus *Micrococcus*, *Micrococcus luteus* (Cohn 1872) and *Micrococcus lylae* (Kloos et al. 1974). Int J Syst Evol Microbiol 52(2):629–637

Williams PG, Buchanan GO, Feling RH, Kauffman CA, Jensen PR, Fenical W (2005) New cytotoxic salinosporamides from the marine actinomycete *Salinispora tropica*. J Org Chem 70(16):6196–6203

Williams PG, Asolkar RN, Kondratyuk T, Pezzuto JM, Jensen PR, Fenical W (2007a) Saliniketals A and B, bicyclic polyketides from the marine actinomycete *Salinispora arenicola*. J Nat Prod 70(1):83–88

Williams PG, Miller ED, Asolkar RN, Jensen PR, Fenical W (2007b) Arenicolides A-C, 26-membered ring macrolides from the marine actinomycete *Salinispora arenicola*. J Org Chem 72(14):5025–5034

Willscher S, Bosecker K (2001) Studies on the leaching behaviour of heterotrophic microorganisms isolated from a basic slag dump. Hydrometal 71(1–2):257–264

Woo SG, Cui Y, Kang MS, Jin L, Kim KK, Lee ST, Lee M, Park J (2012) *Georgenia daeguensis* sp. nov., isolated from 4-chlorophenol enrichment culture. Int J Syst Evol Microbiol 62(7):1703–1709

Wu SJ, Fotso S, Li F, Qin S, Kelter G, Fiebig HH, Laatsch H (2006) N-carboxamido-staurosporine and selina-4 (14), 7 (11)-diene-8, 9-diol, new metabolites from a marine *Streptomyces* sp. J Antibiot 59(6):331–337

Wu YH, Wu M, Wang CS, Wang XG, Yang JY, Oren A, Xu XW (2008) *Microbacterium profundi* sp. nov., isolated from deep-sea sediment of polymetallic nodule environments. Int J Syst Evol Microbiol 58(12):2930–2934

Wu CY, Zhuang L, Zhou SG, Li FB, He J (2011) *Corynebacterium humireducens* sp. nov., an alkaliphilic, humic acid-reducing bacterium isolated from a microbial fuel cell. Int J Syst Evol Microbiol 61(4):882–887

Wu CY, Chen N, Li H, Li QF (2014) *Kocuria rosea* HN01, a newly alkaliphilic humus-reducing bacterium isolated from cassava dreg compost. J Soil Sediment 14(2):423–431

Xi L, Zhang L, Ruan J, Huang Y (2012) Description of *Verrucosispora qiuiae* sp. nov., isolated from mangrove swamp sediment, and emended description of the genus *Verrucosispora*. Int J Syst Evol Microbiol 62(7):1564–1569

Xia ZF, Guan TW, Ruan JS, Huang Y, Zhang LL (2013) *Longimycelium tulufanense* gen. nov., sp. nov., a filamentous actinomycete of the family *Pseudonocardiaceae*. Int J Syst Evol Microbiol 63(8):2813–2818

Xiao J, Luo Y, Xie S, Xu J (2011) *Serinicoccus profundi* sp. nov., an actinomycete isolated from deep-sea sediment, and emended description of the genus *Serinicoccus*. Int J Syst Evol Microbiol 61(1):16–19

Xing K, Qin S, Bian GK, Zhang YJ, Zhang WD, Dai CC, Liu CH, Li WJ, Jiang JH (2012) *Pseudonocardia nantongensis* sp. nov., a novel endophytic actinomycete isolated from the coastal halophyte *Tamarix chinensis* Lour. Anton Leeuwen 10(4):659–667

Xing K, Liu W, Zhang YJ, Bian GK, Zhang WD, Tamura T, Lee JS, Qin S, Jiang JH (2013) *Amycolatopsis jiangsuensis* sp. nov., a novel endophytic actinomycete isolated from a coastal plant in Jiangsu, China. Anton Leeuwen 103(2):433–439

Xu Y, He J, Tian X-P, Li J, Yang LL, Xie Q, Tang SK, Chen YG, Zhang S, Li WJ (2012) *Streptomyces glycovorans* sp. nov., *Streptomyces xishensis* sp. nov. and *Streptomyces abyssalis* sp. nov., isolated from marine sediments. Int J Syst Evol Microbiol 62(10):2371–2377

Xu S, Yan L, Zhang X, Wang C, Feng G, Li J (2014) *Nocardiopsis fildesensis* sp. nov., a novel actinomycete isolated from Antarctica soil. Int J Syst Evol Microbiol 64(1):174–179

Xue Y, Sun X, Zhou P, Liu R, Liang F, Ma Y (2003) *Gordonia paraffinivorans* sp. nov., a hydrocarbon-degrading actinomycete isolated from an oil-producing well. Int J Syst Evol Microbiol 53(5):1643–1646

Yamamura H, Lisdiyanti P, Ridwan R, Ratnakomala S, Sarawati R, Lestari Y, Triana E, Kartina G, Widyastuti Y, Ando K (2010a) *Dietzia timorensis* sp. nov., isolated from soil. Int J Syst Evol Microbiol 60(2):451–454

Yamamura H, SY O, Ishida Y, Otoguro M, Tamura T, Hayakawa M (2010b) *Nocardiopsis nikkonensis* sp. nov., isolated from a compost sample. Int J Syst Evol Microbiol 60(12):2967–2971

Yang LL, Tang SK, Zhang YQ, Zhi XY, Wang D, Xu LH, Li WJ (2008) *Thermobifida halotolerans* sp. nov., isolated from a salt mine sample, and emended description of the genus *Thermobifida*. Int J Syst Evol Microbiol 58(8):1821–1825

Yassin A, Galinski E, Wohlfarth A, Jahnke KD, Schaal K, Trüper H (1993) A new actinomycete species, *Nocardiopsis lucentensis* sp. nov. Int J Syst Bacteriol 43(2):266–271

Yassin A, Rainey F, Burghardt J, Gierth D, Ungerechts J, Lux I, Seifert P, Bal C, Schaal K (1997) Description of *Nocardiopsis synnemataformans* sp. nov., elevation of *Nocardiopsis alba* subsp. *prasina* to *Nocardiopsis prasina* comb. nov., and designation of *Nocardiopsis antarctica* and *Nocardiopsis alborubida* as later subjective synonyms of *Nocardiopsis dassonvillei*. Int J Syst Bacteriol 47(4):983–988

Yassin A, Spröer C, Hupfer H, Siering C, Klenk HP (2009) *Nocardiopsis potens* sp. nov., isolated from household waste. Int J Syst Evol Microbiol 59(11):2729–2733

Yewale T, Singhal RS, Vaidya AA (2013) Immobilization of inulinase from *Aspergillus niger* NCIM 945 on chitosan and its application in continuous inulin hydrolysis. Biocatal Agric Biotechnol 2(2):96–101

Yi H, Chun J (2004a) *Nocardioides aestuarii* sp. nov., isolated from tidal flat sediment. Int J Syst Evol Microbiol 54(6):2151–2154

Yi H, Chun J (2004b) *Nocardioides ganghwensis* sp. nov., isolated from tidal flat sediment. Int J Syst Evol Microbiol 54(4):1295–1299

Yi H, Schumann P, Sohn K, Chun J (2004) *Serinicoccus marinus* gen. nov., sp. nov., a novel actinomycete with L-ornithine and L-serine in the peptidoglycan. Int J Syst Evol Microbiol 54(5):1585–1589

Yi H, Schumann P, Chun J (2007) *Demequina aestuarii* gen. nov., sp. nov., a novel actinomycete of the suborder *Micrococcineae*, and reclassification of *Cellulomonas fermentans* Bagnara et al. 1985 as *Actinotalea fermentans* gen. nov., comb. nov. Int J Syst Evol Microbiol 57(1):151–156

Yoon JH, Lee KC, Kang SS, Kho YH, Kang KH, Park YH (2000) *Janibacter terrae* sp. nov., a bacterium isolated from soil around a wastewater treatment plant. Int J Syst Evol Microbiol 50(5):1821–1827

Yoon JH, Lee CH, Oh TK (2005) *Aeromicrobium alkaliterrae* sp. nov., isolated from an alkaline soil, and emended description of the genus *Aeromicrobium*. Int J Syst Evol Microbiol 55(5):2171–2175

Yoon JH, Jung SY, Kim W, Nam SW, Oh TK (2006) *Nesterenkonia jeotgali* sp. nov., isolated from jeotgal, a traditional Korean fermented seafood. Int J Syst Evol Microbiol 56(11):2587–2592

Yoshida M, Matsubara K, Kudo T, Horikoshi K (1991) *Actinopolyspora mortivallis* sp. nov., a moderately halophilic actinomycete. Int J Syst Bacteriol 41(1):15–20

Yu L, Lai Q, Yi Z, Zhang L, Huang Y, Gu L, Tang X (2013) *Microbacterium sediminis* sp. nov., a psychrotolerant, thermotolerant, halotolerant and alkalitolerant actinomycete isolated from deep-sea sediment. Int J Syst Evol Microbiol 63(1):25–30

Yuan LJ, Zhang YQ, Guan Y, Wei YZ, Li QP, Yu LY, Li WJ, Zhang YQ (2008) *Saccharopolyspora antimicrobica* sp. nov., an actinomycete from soil. Int J Syst Evol Microbiol 58(5):1180–1185

Yumoto I, Nakamura A, Iwata H, Kojima K, Kusumoto K, Nodasaka Y, Matsuyama H (2002) *Dietzia psychralcaliphila* sp. nov., a novel, facultatively psychrophilic alkaliphile that grows on hydrocarbons. Int J Syst Evol Microbiol 52(1):85–90

Zammit CM, Mangold S, rao Jonna V, Mutch LA, Watling HR, Dopson M, Watkin EL (2012) Bioleaching in brackish waters—effect of chloride ions on the acidophile population and proteomes of model species. Appl Microbiol Biotechnol 93(1):319–329

Zhang YQ, Schumann P, Li WJ, Chen GZ, Tian XP, Stackebrandt E, Xu LH, Jiang CL (2005) *Isoptericola halotolerans* sp. nov., a novel actinobacterium isolated from saline soil from Qinghai Province, north-west China. Int J Syst Evol Microbiol 55(5):1867–1870

Zhang YQ, Schumann P, Yu LY, Liu HY, Zhang YQ, Xu LH, Stackebrandt E, Jiang CL, Li WJ (2007) *Zhihengliuella halotolerans* gen. nov., sp. nov., a novel member of the family *Micrococcaceae*. Int J Syst Evol Microbiol 57(5):1018–1023

Zhang D-C, Liu H-C, Xin Y-H, Yu Y, Zhou P-J, Zhou Y-G (2008) *Salinibacterium xinjiangense* sp. nov., a psychrophilic bacterium isolated from the China no. 1 glacier. Int J Syst Evol Microbiol 58(12):2739–2742

Zhang F, Hu SN, Chen JJ, Lin LB, Wei YL, Tang SK, Xu LH, Li WJ (2012a) Purification and partial characterisation of a thermostable xylanase from salt-tolerant *Thermobifida halotolerans* YIM 90462 T. Process Biochem 47(2):225–228

Zhang W, Liu Z, Li S, Yang T, Zhang Q, Ma L, Tian X, Zhang H, Huang C, Zhang S (2012b) Spiroindimicins A–D: new bisindole alkaloids from a deep-sea-derived actinomycete. Org Lett 14(13):3364–3367

Zhang YJ, Tang SK, Shi R, Klenk HP, Chen C, Yang LL, Zhou Y, Li WJ (2012c) *Salinisphaera halophila* sp. nov., a moderately halophilic bacterium isolated from brine of a salt well. Int J Syst Evol Microbiol 62(9):2174–2179

Zhang DF, Pan HQ, He J, Zhang XM, Zhang YG, Klenk HP, Hu JC, Li WJ (2013a) Description of *Streptomonospora sediminis* sp. nov. and *Streptomonospora nanhaiensis* sp. nov., and reclassification of *Nocardiopsis arabia* Hozzein & Goodfellow 2008 as *Streptomonospora arabica* comb. nov. and emended description of the genus *Streptomonospora*. Int J Syst Evol Microbiol 63(12):4447–4455

Zhang YG, Wang HF, Liu Q, Hozzein WN, Wadaan MA, Cheng J, Chen YJ, Zhang YM, Li WJ (2013b) *Streptomyces fukangensis* sp. nov., a novel alkaliphilic actinomycete isolated from a saline-alkaline soil. Anton Leeuwen 104(6):1227–1233

Zhang YJ, Zhang WD, Qin S, Bian GK, Xing K, Li YF, Cao CL, Jiang JH (2013c) *Saccharopolyspora dendranthemae* sp. nov. a halotolerant endophytic actinomycete isolated from a coastal salt marsh plant in Jiangsu, China. Anton Leeuwen 103(6):1369–1376

Zhang DF, Wang HF, Xiong ZJ, Tian XP, Liu L, Zhang XM, Jiang Z, Zhang S, Li WJ (2014a) *Mariniluteicoccus flavus* gen. nov., sp. nov., a new member of the family *Propionibacteriaceae*, isolated from a deep-sea sediment. Int J Syst Evol Microbiol 64(3):1051–1056

Zhang YG, Liu Q, Wang HF, Zhang DF, Zhang YM, Park DJ, Kim CJ, Li WJ (2014b) *Haloactinopolyspora alkaliphila* sp. nov., and emended description of the genus *Haloactinopolyspora*. Int J Syst Evol Microbiol 64:1945–1951. doi:10.1099/ijs.0.062646-062640

Zhao GZ, Li J, Qin S, Zhang YQ, Zhu WY, Jiang CL, Xu LH, Li WJ (2009) *Micrococcus yunnanensis* sp. nov., a novel actinobacterium isolated from surface-sterilized *Polyspora axillaris* roots. Int J Syst Evol Microbiol 59(10):2383–2387

Zhao GZ, Li J, Huang HY, Zhu WY, Park DJ, Kim CJ, Xu LH, Li WJ (2011a) *Pseudonocardia kunmingensis* sp. nov., an actinobacterium isolated from surface-sterilized roots of *Artemisia annua* L. Int J Syst Evol Microbiol 61(9):2292–2297

Zhao GZ, Li J, Huang HY, Zhu WY, Xu LH, Li WJ (2011b) *Nonomuraea rhizophila* sp. nov., an actinomycete isolated from rhizosphere soil. Int J Syst Evol Microbiol 61(9):2141–2145

Zhou G, Luo X, Tang Y, Zhang L, Yang Q, Qiu Y, Fang C (2008) *Kocuria flava* sp. nov. and *Kocuria turfanensis* sp. nov., airborne actinobacteria isolated from Xinjiang, China. Int J Syst Evol Microbiol 58(6):1304–1307

Zvyagintseva I, Poglazova M, Gotoeva M, Belyaev S (2001) Effect on the medium salinity on oil degradation by nocardioform bacteria. Microbiol 70(6):652–656

Chapter 4
Biosynthesis of Nanoparticles from Halophiles

Pallavee Srivastava and Meenal Kowshik

Abstract Nanobiotechnology is a multidisciplinary branch of nanotechnology which includes fabrication of nanomaterials using biological approaches. Many bacteria, yeast, fungi, algae and viruses have been used for synthesis of various metallic, metal sulfide, metal oxide and alloy nanoparticles, since the first report on biosynthesis of cadmium sulfide quantum dots by *Candida glabrata* and *Schizosaccharomyces pombe* in 1989. These nanofactories offer a better size control through compartmentalization in the periplasmic space and vesicles, and are usually capped by stabilizing cellular metabolites. Halophiles depending on their salt requirements may be classified as slight, moderate and extreme halophiles. They are found in marine and/or hypersaline environments. These organisms are known to encounter metals in their environment as the econiches they inhabit serve as ecological sinks for metals. Metal based nanoparticle synthesis by halophilic organisms is in its infancy and has only been reported in few organisms. This chapter aims to shed light on the various halophilic organisms and their by-products that have been exploited for nanomaterial synthesis, the mechanisms that may be involved in the nanomaterial fabrication and the possible applications of the fabricated nanoparticles. A special section would be dedicated for the bioavailability of metals to halophiles under varying salinity conditions.

Keywords Nanobiotechnology • Biosynthesis • Nanoparticles • Metal • Haloarchaea • Bio-availability • Quantum-dots • NADH-dependent nitrate reductase

4.1 Introduction to Nanotechnology

Nanotechnology is an interdisciplinary technology seeking to explore the unique advantages of manipulating the structure of materials at a scale of individual atoms, molecules and their organized aggregates. It involves creation and exploitation of materials with structural features in between those of atoms and bulk materials

P. Srivastava • M. Kowshik (✉)
Department of Biological Sciences, BITS Pilani,
K K Birla Goa Campus, Zuarinagar, Goa 403726, India
e-mail: meenal@goa.bits-pilani.ac.in

D.K. Maheshwari, M. Saraf (eds.), *Halophiles*, Sustainable
Development and Biodiversity 6, DOI 10.1007/978-3-319-14595-2_4

(Rao and Cheetham 2001). Nanofabrication is of interest as it results in materials with unique size dependent optical, physico-chemical, electronic, mechanical, magnetic, and biological properties. These novel properties are a result of the large surface area to volume ratio, large surface energy, spatial confinement and reduced imperfections. Nanomaterials can either be synthesized by a top-down approach, where the bulk materials are gradually broken down to nano-dimensions or a bottom-up approach, where the atoms or molecules are assembled into nano-structures. The top-down approach is used by various physical methods of nanomaterial fabrication such as sputter deposition, laser ablation, attrition and pyrolysis. The chemical and biological routes of nanoparticles synthesis utilize the bottom-up approach.

Nanoparticles may be classified as organic (carbon based) and inorganic which include magnetic, metal based and semiconductor nanoparticles. Metallic nanoparticles are characterized by the excitation of the surface plasmons that result in optical properties hardly achievable in other optical materials. Semiconducting nanoparticles and quantum dots exhibit change in electronic properties with decrease in size due to quantum-confinement and increase in band gap. The magnetic moment per atom and magnetic anisotropy of the magnetic nanoparticles are distinct from their bulk counterparts due to finite size and surface effects. Therefore, nanoparticles find applications in fields such as biomedicine, bio-labelling, bio-imaging, drug delivery, photovoltaics, solar cells, photocatalysis, and data storage (Bera et al. 2010; Garcia 2011; Issa et al. 2013).

4.2 Nanobiotechnology

Nanobiotechnology is an interdisciplinary field of science that studies the application of biological material for synthesis of nanomaterials and design of nano-devices and nano-systems. Various biological methods of nano-fabrication include microbial synthesis and biomimetic synthesis using viral particles, S-layers, DNA and proteins as templates. Even though nanoparticles have been synthesized using myriad physical and chemical methods, the processes have various disadvantages. The physical routes of nanofabrication are usually energy (maintenance of high temperature and pressure) and capital (requirement of expensive instrumentation) intensive. Even though the chemical synthesis results in large yield of nanoparticles in short periods of time with a fairly good control over the size and shape, contamination from the toxic precursor chemicals and the generation of hazardous by-products are the major drawbacks (Kowshik et al. 2003). Due to their extremely small size, the nanomaterials possess large surface energies and have the tendency to agglomerate, unless passivated using certain stabilizing agents to maintain their size. The passivating agents often used are toxic and reduce the applicability of the nanomaterials. Accordingly, the need for green synthesis of metallic nanoparticles was recognized and since then biological resources have been explored and exploited for fabrication of nanomaterials using environmentally benign methods.

4.3 Nanoparticles Synthesis by Micro-organisms

Biological synthesis of nanoparticles using micro-organisms is a clean, biocompatible, non-toxic and an eco-friendly method. This method of nanofabrication exploits the highly structured physical and biosynthetic activities of microbial cells. Microbes when exposed to metals, overcome the toxicity using various metal resistance mechanisms. These mechanisms are a result of enzymatic detoxification of the metals and/or extrusion of the metal through the energy-dependent ion efflux membrane transporters present in the cell. Micro-organisms thereby transform toxic metals to their non-toxic nanosized form; or the soluble form to the insoluble nanosized form, during these detoxification processes. Interactions between metals and microbes are well documented and have been exploited for various applications in the fields of biomineralization, bioleaching and bioremediation (Klaus-Joerger et al. 2001). Many magnetotactic bacteria are known to naturally produce nanostructured mineral crystals that have properties similar to chemically synthesized materials (Baumgartner et al. 2013). Such nano-factories exercise strict control over size, shape and composition through compartmentalization in the periplasmic space and vesicles. Additionally, the microbially synthesized nanoparticles are usually capped by stabilizing cellular metabolites produced during the process of synthesis. The rate of synthesis and the size of the nanoparticles can be manipulated to an extent by controlling parameters like pH, temperature, substrate concentration and time of exposure of the substrate (Gericke and Pinches 2006; Kang et al. 2008; Gurunathan et al. 2009). Numerous bacteria, yeast, fungi, algae and viruses have been used for synthesis of various metallic, metal sulphide, metal oxide and alloy nanoparticles. The synthesis may either be an intra- or an extra-cellular process depending upon the nature and the metabolic activity of the organism. The forthcoming sections will describe the various halophilic organisms and their by-products that synthesize nanoparticles, the mechanisms involved and applications of the synthesized nano-materials.

4.3.1 Nanoparticles Synthesis by Halophiles

Halophiles are salt loving organisms that flourish in saline environments and can be classified as slightly, moderately or extremely halophilic depending on their sodium chloride requirements (DasSarma and DasSarma 2012). The saline environments that halophiles inhabit include the marine and estuarine environments, solar salterns, salt lakes, brines and saline soils. Marine environments could be good source of metal tolerant microbes as most of these organisms exist at the bottom of the sea, and contribute towards biogeochemical cycling of inorganic elements. Besides, the marine econiche is continuously exposed to metallic pollution due to volcanic eruptions, natural weathering of the rocks, anthropogenic activities such as mining, combustion of fuels and industrial and urban sewage. Estuaries and solar salterns may also contain high concentrations of metals as they serve as effective traps for river borne metals (Chapman and Wang 2001). Thus, halophiles are continuously exposed to metals and could be exploited for nanoparticle synthesis.

Nanoparticles synthesis by halophiles is in its infancy and has been reported in few organisms like bacteria (*Halomonas salina*, *H. Maura*, *H. eurihaline*, *Pseudomonas* sp., *Idiomarina* sp. PR-58-8), yeasts (*Pichia capsulate*, *Rhodospiridium diobovatum*), fungi (*Penicillium fellutatum*, *Thraustochytrium* sp., *Aspergillus niger*), algae (*Navicula atomus*, Didesmis gallica) and archaea (*Halococcus salifodinae* BK3, *H. salifodinae* BK6). Table 4.1 gives the details of nanoparticles synthesized by halophiles. The exopolysaccharides secreted by the halophiles have also been exploited for synthesis of composite nanoparticles for various drug delivery applications. The exopolysaccharide Mauran, secreted by halophilic bacteria *H. maura* stabilizes ZnS:Mn^{2+} quantum dots (QDs) enhancing the biocompatibility and reducing the cytotoxicity of the QDs. A significantly improved genetic system for displaying foreign proteins on *Halobacterium* gas vesicle for generating gas vesicle nanoparticles has been developed that could be exploited for various biological applications (DasSarma et al. 2013).

4.3.1.1 Halophilic Bacteria and Nanofabrication

Reports on nanoparticles synthesis by halophilic bacteria and their metabolites are mostly confined to metallic nanoparticles. A highly silver tolerant halophilic marine bacterium *Idiomarina* sp. PR 58-8 synthesizes intra-cellular crystalline silver nanoparticles with an average particle size of 26 nm. Non-protein thiols that are known to be expressed in response to metal stress are involved in metal tolerance (Seshadri et al. 2012). Similarly, a novel halophilic strain of *Pseudomonas* sp. 591786 can also synthesize polydisperse intra-cellular silver nanoparticles with sizes ranging from 20 to 100 nm (Muthukannan and Karuppiah 2011). The halophilic proteobacteria *Halomomas salina* can synthesize anisotropic gold nanoparticles under acidic conditions and spherical nanoparticles under alkaline conditions. The nanoparticle synthesis is extracellular and the NADH-dependent nitrate reductase is involved in the silver reduction and nanoparticle synthesis (Shah et al. 2012). Similarly, two halophilic strains of *Bacillus megaterium*, BSB6 and BSB12 isolated from Bhitarkanika mangrove soils, synthesize spherical selenium nanoparticles both intra- and extra-cellularly with an average size of 200 nm (Fig. 4.1a). The mechanism involved for the reduction of selenite to selenium, however remains unexplored (Mishra et al. 2011).

Two strains of halophilic bacteria *H. eurihalina* ATCC 49336 and *H. maura* ATCC 700995 convert graphene oxide (GO) to graphene sheet under both aerobic and anaerobic conditions. Reports on biological synthesis of graphene by microorganisms are scarce. The microbially reduced GO sheet exhibits an increased conductivity as compared to chemically reduced GO and is bio-compatible. Such bio-compatible Graphene sheets may be used for green electronics and biological

Table 4.1 The inorganic nanoparticles biosynthesized by various halophiles

Halophilic	Organisms	Type	Intra-/ extra-cellular	Size (nm)	Shape	Application	Mechanism	Reference
Bacteria	*Idiomarina* sp. PR 58-8	Ag	Intra-	26	–	–	NP-SH	Seshadri et al. (2012)
	Pseudomonas sp. 591786	Ag	Intra-	20–100	Spherical	–	–	Muthukannan and Karuppiah (2011)
	Halomonas salina	Au	Extra-	30–100	Anisotropic; spherical	–	NADH-NR	Shah et al. (2012)
	Bacillus megaterium BSB6 and BSB12	Se	Intra-; extra-	~200	Spherical	–	–	Mishra et al. (2011)
Yeast	*Pichia capsulata*	Ag	Intra-	50–100	–	–	–	Manivannan et al. (2010)
	Rhodosporidium diobovatum	PbS	Intra-	2–5	Spherical	–	NP-SH	Seshadri et al. (2011)
Fungi	*Penicillium fellutatum*	Ag	Extra-	5–25	Mostly spherical	–	70 kDa protein	Kathiresan et al. (2009)
	Aspergillus niger	Ag	Extra-	5–35	Mostly spherical	Anti-bacterial	70 kDa protein	Kathiresan et al. (2010)
Algae	*Sargassum wightii*	Ag	Extra-	8–12	Plannar	–	–	Singaravelu et al. (2007)
	S. wightii	Au	Extra-	30–100	Isotropic spheres	–	NADH-NR	Oza et al. (2012)
	S. longifolium	Ag	Extra-	–	Spherical	Anti-fungal	–	Rajeshkumar et al. (2014)
	Pterocladia capillacae; Jania rubins; Ulva faciata; Colpmenia sinusa	Ag	Extra-	–	–	Anti-bacterial	Polysaccharides	El-Rafie et al. (2013)
	Navicula atomus; Diadesmis gallica	Au	Extra-	9; 22	Spherical	–	–	Schrofel et al. 2011
Archaea	*Halococcus salifodinae* BK3	Ag	Intra-	12	Mostly spherical	Anti-bacterial	NADH-NR	Srivastava et al. (2013)
	H. salifodinae	Ag	Intra-	12	Mostly spherical	Anti-bacterial	NADH-NR	Srivastava et al. (2014)

(–) Not determined, NP-SH: Non-Protein Thiols, NADH-NR: NADH-dependent Nitrate reductase

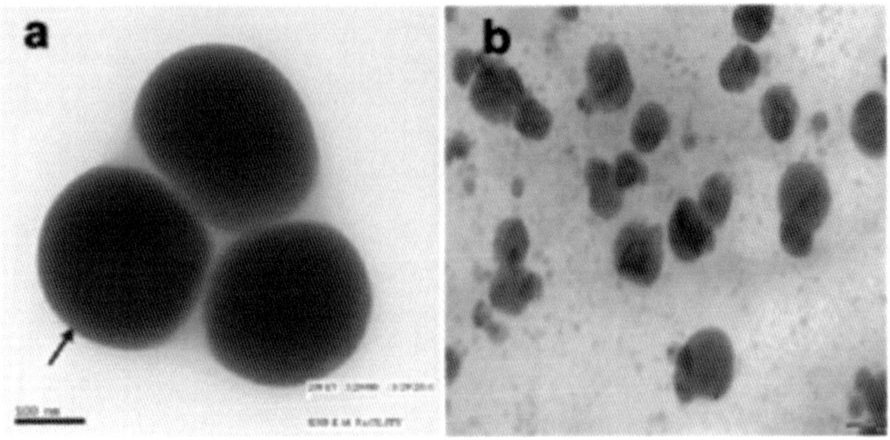

Fig. 4.1 TEM images of the (**a**) Selenium nanoparticles biosynthesized by halophilic bacteria *Bacillus megaterium* BSB12 (Mishra et al. 2011); (**b**) Silver nanoparticles biosynthesized by haloarchaeon *Halococcus salifodinae* BK6

applications such as detection of cancer biomarkers, encapsulation of enzymes and nanoparticles (Myung et al. 2011; Raveendran et al. 2013a).

Bacterial metabolites or products such as polysaccharides/bio-flocculants and enzymes (nitrate reductase) are also being used as reducing agents to synthesize inorganic nanoparticles. In most cases the synthesized nanoparticles are capped by reducing agents. MBSF17, a polysaccharide bio-flocculant produced by a halophilic bacterium *Bacillus subtilis* MSBN17 reduces silver nitrate to synthesize spherical silver nanoparticles in reverse micelles. The electrostatic forces between the amino groups of the polysaccharide MSBF17 and the silver ions in the solution are proposed to be the driving force for the formation and stabilization of the silver nanoparticles. The carboxyl, hydroxyl and methoxyl groups of MSBF17 form a coating on the silver nanoparticles thereby stabilizing them. These nanoparticles exhibit anti-microbial activity against a host of pathogenic organisms (Sathiyanarayanan et al. 2013).

Besides the various inorganic and organic nanoparticles, the exopolysaccharides of the halophilic bacteria have also been utilized for fabrication of polymer hybrid nanomaterials. The highly sulphated anionic exopolysaccharide, Mauran (MR) secreted by the halophilic bacterium *H. maura*, is well characterized and has been successfully used for generation of such hybrid nanomaterials. MR exhibits characteristic viscoelastic, pseudoplastic and thixotropic behaviour and has an ability to withstand harsh conditions, which make it an ideal candidate for material science research. The high sulphate content imparts immunomodulating and anti-cancer properties to MR. Thus, MR can be used for various biomedical applications due to their biological and the physicochemical properties. MR-Chitosan (MR/CH) hybrid nanoparticles (Fig. 4.2) fabricated via the ionic-gelation technique when used for

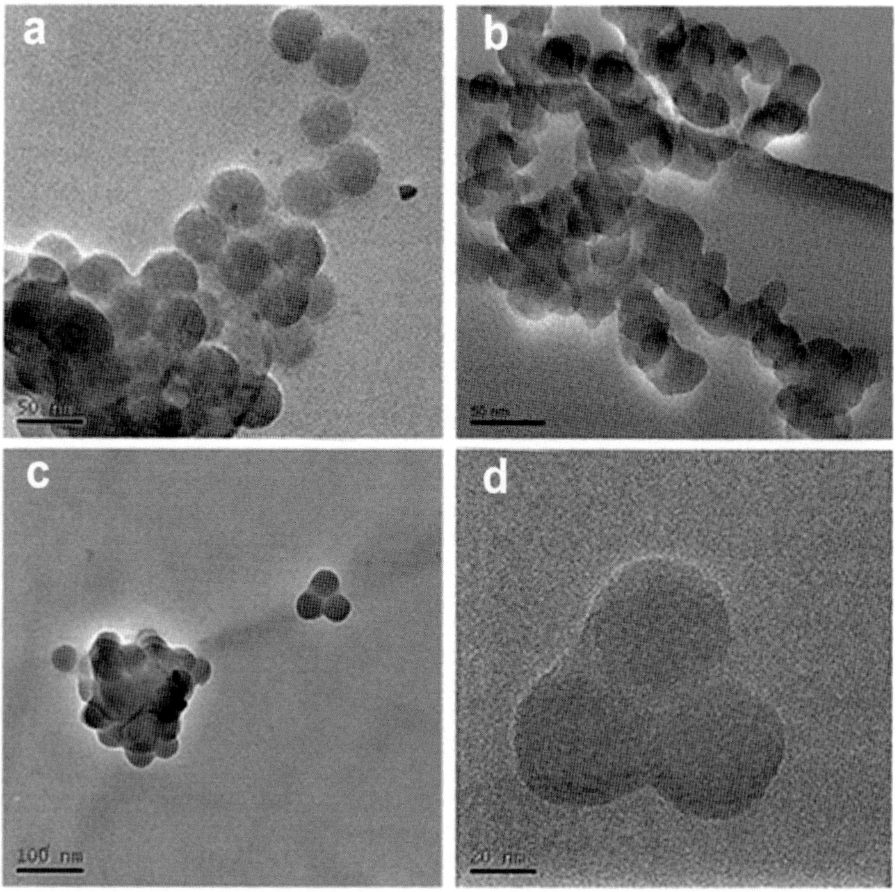

Fig. 4.2 TEM micrographs (**a**–**d**) depicting the morphology of the Mauran-Chitosan nanoparticles fabricated using the exopolysaccharide mauran secreted by the halophilic bacteria *Halomonas maura* (Raveendran et al. 2013b)

encapsulation of drugs exhibits controlled and sustained drug release and bio-compatibility (Raveendran et al. 2013b). Similarly, electrospun MR-Poly Vinyl Alcohol (MR-PVA) nanofibre membranes (Fig. 4.3) boost the cellular adhesion, migration, proliferation and differentiation, properties desirable for tissue engineering applications (Raveendran et al. 2013c). MR may be used in augmenting the biocompatibility of quantum dots that are usually cytotoxic (Fig. 4.4). MR conjugated with $ZnS:Mn^{2+}$ QDs at a concentration of 0.05 mg results in a drastic increase in cell viability as compared to cell viability of bare QDs (Raveendran et al. 2014). Therefore in addition to fabrication of polymer based nanomaterials, MR can also be used in conjugation with inorganic nanoparticles for enhancing their biological applications.

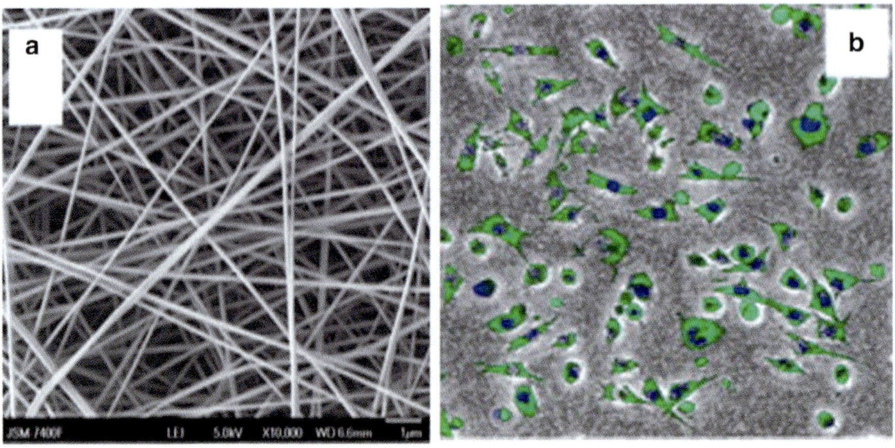

Fig. 4.3 (**a**) SEM micrographs of the MR/PVA nanofibres fabricated using the exopolysaccharide mauran (×10,000) (**b**) Confocal microscopy images of L292 cells attached and proliferating on MR/PVA nanofibres (merged microtracker green and DAPI stained images) (Raveendran et al. 2013c)

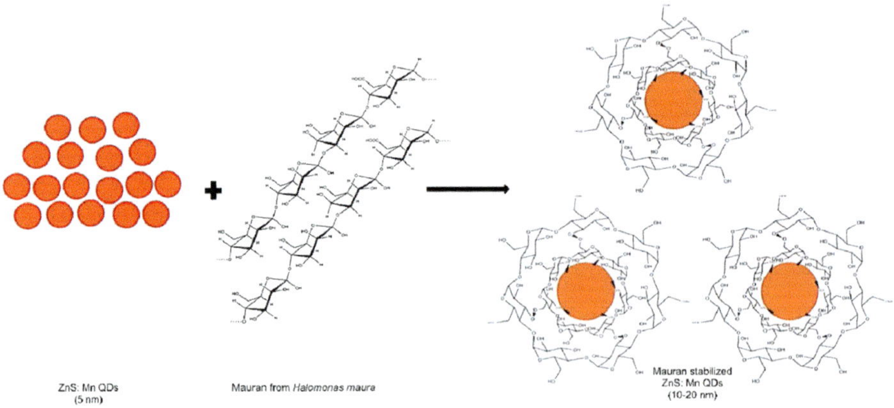

Fig. 4.4 Schematic representation of stabilization of QDs by Mauran, an exopolysaccharide secreted by *H. maura* (Raveendran et al. 2014)

4.3.1.2 Nanoparticles and Halophilic Yeast and Fungi

A few halophilic yeasts and fungi are known to synthesize nanomaterials. *Pichia capsulata*, a mangrove derived halophilic yeast, is capable of synthesizing silver nanoparticles extra-cellularly (Manivannan et al. 2010). *Rhodospiridium diobovatum*, a marine yeast, synthesizes lead sulphide nanoparticles intracellulary with the help of non-protein thiols (Seshadri et al. 2011). Similarly, a 70 kDa protein present in the cell filtrate of the marine fungus *Penicillium fellutatum* exposed to silver nitrate is involved in the extracellular nanoparticles synthesis

(Kathiresan et al. 2009). Silver nanoparticles are also synthesized by the halophilic fungi *Thraustochytrium* sp. and *Aspergillus niger*. Silver nanoparticles synthesized by *Aspergillus niger* exhibit antibacterial activity against pathogenic bacteria which can be further enhanced by passivating them with PVA (Kathiresan et al. 2010).

4.3.1.3 Halophilic Algae in Nanoparticles Synthesis

The reports on nanoparticles synthesis by halophilic algae are few and mostly recent. All studies so far are on extracellular synthesis of inorganic (metallic) nanoparticles. The marine brown algae *Sargassum wightii* synthesizes stable gold (30–100 nm) and silver nanoparticles (8–12 nm) when its extract is exposed to gold chloride and silver nitrate, respectively (Singaravelu et al. 2007; Oza et al. 2012). Similarly, extracts of *S. longifolium* can reduce silver nitrate to spherical silver nanoparticles that exhibit excellent antifungal activity. The extracts contain various active molecules rich in hydroxyl groups or carboxyl groups that are responsible for the reduction of the metallic ion (Rajeshkumar et al. 2014). The water soluble exopolysaccharides extracted from the marine algae *Pterocladia capillacae*, *Jania rubins, Ulva faciata* and *Colpmenia sinusa* reduce silver ions to silver nanoparticles. The exopolysaccharides also serve as the stabilizing agent. These nanoparticles have been used to make antibacterial cotton fabrics (El-Rafie et al. 2013). EPS-gold and silica-gold bio-nanocomposites can be generated using the diatoms *Navicula atomus* and *Diadesmis gallica*. The diatoms when grown in presence of tetrachloroaureate, reduce it to gold nanoparticles that are associated with the diatom frustules and exopolysaccharides (EPS) excreted by the diatom cells. The gold bio-nanocomposites may find applications in the field of catalysis (Schrofel et al. 2011).

4.3.1.4 Nanoparticles and Haloarchaea

Haloarchaea are known to encounter metals in their natural habitat, yet reports on metal tolerance and nanoparticles synthesis by haloarchaea are few. With the exception of two organisms, *H. salifodinae* BK3 and *H. salifodinae* BK6, there are no other reports on metallic nanoparticles synthesis by haloarchaea. The intracellular synthesis of silver nanoparticles by *H. salifodinae* BK3 and BK6 (Fig. 4.1b) involves the enzyme NADH- dependent nitrate reductase that helps in silver ion reduction. These organisms adapt to the metal stress and thus their growth kinetics parameters in presence of silver nitrate are similar to that of organism grown without silver nitrate. These silver nanoparticles exhibit good antibacterial activity against both Gram-positive and Gram-negative bacteria (Srivastava et al. 2013, 2014).

Gas vesicle nanoparticles (GVNPs), that may be engineered for various biotechnological applications are the buoyant gas vesicles widely distributed among bacteria and archaea. These organelles that naturally promote floatation are present in abundance in haloarchaea. These vesicles are plasmid encoded with the genetic cluster *gvp*MLKJIHGFEDACNO involved in gas vesicle formation (DasSarma 1989; DasSarma et al. 1987; Halladay et al. 1993; DasSarma and Arora 1997). The proteins encoded by the gene clusters include the GvpA, J and M of Pfam741 family, involved in gas vesicle membrane formation, and GvpF and L, coiled-coil protein (Pfam 6386) involved in the nucleation process of nanoparticles due to their self-associative properties (Jones et al. 1991; Shukla and DasSarma 2004). Genes corresponding to these proteins have been found in other organisms as well, with the exception of *gvpC* gene, which is found only in haloarchaea and cyanobacteria (van Keulen et al.2005). In the haloarchaeon *Halobacterium* sp. NRC-1, the GvpC protein is hydrophilic and insertion mutations within this gene results in gas vesicles with altered shape and size (Fig. 4.5) (DasSarma et al. 1994). Thus, by genetic manipulation of *gvpC* gene, the gas vesicles may be made to express different proteins or display antigens, thereby increasing their applications in the field of biotechnology. A new *Halobacterium* sp. NRC-1 derived host strain and a series of smaller, more versatile plasmid expression vectors have been constructed. These represent a significantly improved genetic system for expression of GvpC-fusion proteins. For example, active *Gaussia princeps* luciferase enzyme can be fused to GvpC that would result in the expression of the luciferase enzyme on the surface of the GVNPs (DasSarma et al. 2013). Such GVNPs may be used for antigen display and vaccine development.

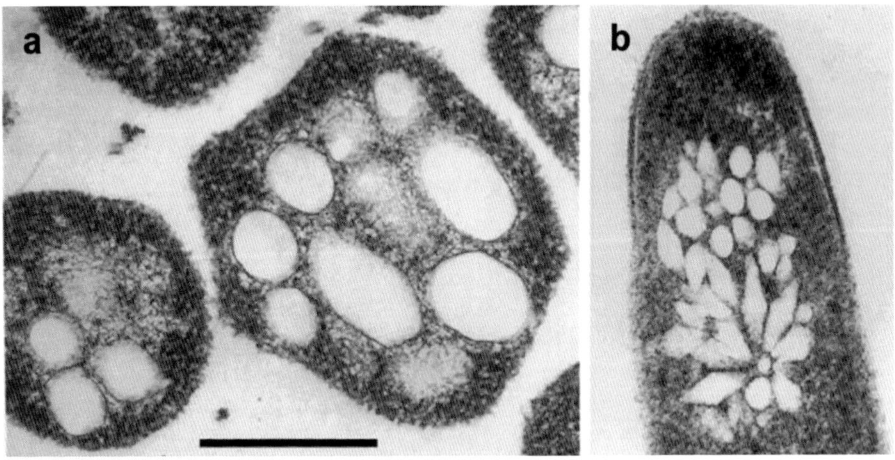

Fig. 4.5 TEM micrographs showing thin sections of (**a**) *Halobacterium* sp. NRC-1 and (**b**) SD109 (pFM104gvpC::κ1) mutant. Gas vesicles can be seen as empty oval or spindle shaped regions. SD109 (pFM104gvpC::κ1) mutant exhibits gas vesicles with altered phenotypes. (Scale bar = 325 nm) (DasSarma et al. 2013)

4.4 Bio-availability of Metals to Halophiles

Metallic nanoparticle synthesis is possible only when the metal is biologically available to halophiles that inhabit saline environments. Bio-availability of metals in such saline environments is determined by the metal species which in turn depends on their solubility. Factors such as alkalinity, pH, hardness (presence of Ca/Mg), natural dissolved organic matter and redox potential of the saline environment determine the solubility of the metal species. Metals that are strongly complexed with a ligand are non-labile and are not available for biological uptake. Labile metal species are the weak complexes formed between a metal and a ligand and are biologically available (Campbell 1995; Markich et al. 2001).

The type of heavy metal and the inorganic/organic ligands present in an econiche determines the nature of the metal complex formed. The ionic radii of a metal plays an important role in this complexation. Metals with small ionic radii such as Zn and Cu easily complex with ligands containing oxygen, such as OH^-, CO_2^{3-}, HCO^{3-} and SO_4^{2-} to form inorganic complexes. Metals with large ionic radii such as Hg, Cd and Ag are easily ionized and preferentially form chloro-complexes. Some of these inorganic complexes dissociate easily into their free ionic forms which bind to the membrane transporters of a micro-organism and move across the cell membrane. Organic-metal complexes cannot be transported across cell membranes and are not available for the micro-organisms for uptake (Nürnberg 1983; Roux et al. 1998; Tipping et al. 1998). Inorganic complexes predominate saline environments. Table 4.2 shows the type of inorganic species formed at different salinities for five major metals.

Table 4.3 summarizes the bio-availability of various metal-chloro complexes in saline environments. Even though metal bio-availability, uptake and toxicity decreases in presence of natural dissolved organic ligands, salinity may alter these

Table 4.2 Effect of salinity on the type of complexes formed by five major metals

Salinity conditions	Silver	Cadmium	Mercury	Zinc	Copper
Estuarine (variable salinity)	$AgCl^0$, $AgHS^0$, $AgCl^{2-}$, $AgCl_3^{2-}$, $AgCl_4^{3-}$	$CdCl_2$, $CdCl^+$	$HgCl^0$, $HgCl^-$, $HgCl_4^{2-}$	Zn^{2+}, $ZnCl_2$, $ZnCO_3$, $Zn(HCO_3)_2$, $Zn(OH)_2$, $ZnSO_4$	$CuCl_2$, CO_3^{2-} and OH^- complexes
Sea water (3.5 % salinity)	$AgCl^0$, $AgHS^0$	$CdCl^+$	$HgCl^-$	Zn^{2+}, $ZnCl_2$, $ZnCO_3$, $Zn(HCO_3)_2$, $Zn(OH)_2$, $ZnSO_4$	CO_3^{2-} and OH^- complexes
Hypersaline (5–35 % salinity)	$AgCl^0$, $AgCl^{2-}$, $AgCl_3^{2-}$, $AgCl_4^3$	$CdCl_2$, $CdCl^+$	$HgCl^0$, $HgCl^-$, $HgCl_4^{2-}$	$ZnCl_2$	$CuCl_2$

Table 4.3 Bio-availability of chloro complexes in varying saline conditions

Type of chloro-complex	Biological availability		
	Hypersaline	Sea water	Estuarine
Insoluble complexes			
Strong (Un-dissociating)	Unavailable	Present in low amounts; unavailable	Unavailable
Weak (dissociating)	Available	Available	Available
Soluble complexes			
Strong	Unavailable	Unavailable/available[b]	Unavailable
Strong lipophilic	Available	Not present	Not present
Weak	Available	Available	Available
Biosorbed complexes	Available	Unavailable	Unavailable
Complexes sorbed on abiotic ligands	Available	Available	Available[a]

[a]Depends upon the salinity of the water body, during ocean water intermixing, chloro-complexes are available

[b]Depends upon the metal involved in complexation

properties influencing the metal bio-availability. For example, in river water Cd may either exist in free cationic form or as CO^{3-} complex, both of which are bioavailable. In oceanic waters (salinity 3.5 %) it exists as bioavailable highly soluble $CdCl_2$, whereas in estuarine waters (varying salinity), it forms a strong $CdCl^+$ complex which is biologically unavailable (Turner 1987; Jensen and Bro-Ramussen 1992; Peakall and Burger 2003). Similarly, Ag and Hg form soluble chloro-complexes in hypersaline and estuarine environments that are bioavailable to the organism. These complexes are lipophilic and can readily move across the cell membrane. Metals like Zn and Cu that exist as $ZnCl^+$ and $CuCl^+$ at higher salinities, co-precipitate due to a decrease in the net negative charge on macromolecular suspended particles and therefore are not available for uptake. Fe, Co, Ni and Mn on the other hand, form bioavailable weak chloro-complexes in saline environments that dissociate easily to their ionic forms (Byrne 2002). Besides the solubility of metal complex, presence of biotic ligands in saline environments may also limit the bioavailability of metals. Biotic ligands are the receptors present on an organism to which metal binds and the toxic effects of metals are manifested. Fish gill surfaces, algal membranes, phytoplanktons etc. have various ion transporters (Na^+ and Ca^{2+}) and metal receptors that act as biotic ligands. Once bound to a biotic ligand, the metal is rendered unavailable for uptake and the binding remains unaffected by salinity. However, if the metal binds to an abiotic ligand (sediments, particulate matter), a change in salinity may result in solubilisation of the metal making it bioavailable. For instance, an increase in salinity results in desorption of the Ag complexed with suspended sediments to form bioavailable Ag-chloro-complexes. However, if Ag is biosorbed, the increase in salinity does not result in desorption of the Ag and metal remains unavailable for uptake by micro-organisms (Sanders and Abbe 1987; Paquin et al. 2002).

The chemical species of the metal greatly influences its toxicity towards micro-organisms. The metal has to be biologically available to a microbe to exert its toxicity and only bio-available metals get detoxified by the organisms to form metallic nanoparticles. Salinity plays a major role in the bioavailability of metals to halophiles. Better understanding in metal speciation at varying salinities is important during the search for novel halophilic organisms for metal nanoparticles biosynthesis.

4.5 Conclusion

Halophiles inhabit saline environments that are exposed to metallic pollution and therefore possess metal resistance mechanisms. These mechanisms are usually involved in metal reduction and nanoparticles synthesis. However, the mechanisms involved in nanoparticles synthesis by halophiles remain unknown. Thus, research in nanoparticle synthesis by halophiles requires further studies to understand the mechanisms involved in the synthesis process. This understanding will enable researchers to design the synthesis process in an intelligent way to maximize the nanoparticle yield. The bioavailability of metals is influenced by the salinity of the growth medium. Therefore, studying the influence of the media components on the metal bioavailability may help in designing better biosynthesis protocols.

References

Baumgartner J, Morin G, Menguy N, Gonzalez TP, Widdrat M, Cosmidis J, Faivr D (2013) Magnetotactic bacteria form magnetite from a phosphate-rich ferric hydroxide via nanometric ferric (oxyhydr)oxide intermediates. Proc Natl Acad Sci U S A 110:14883–14888

Bera D, Qian L, Tseng T-K, Holloway PH (2010) Quantum dots and their multimodal applications: a review. Materials 3:2260–2345

Byrne RH (2002) Inorganic speciation of dissolved elements in seawater: the influence of pH on concentration ratios. Geochem Trans 3:11–16

Campbell PGC (1995) Interactions between trace metals and aquatic organisms: a critique of the free-ion activity model. In: Tessier A, Turner DR (eds) Metal speciation and bioavailability in aquatic systems. Wiley, Chichester, pp 45–102

Chapman PM, Wang F (2001) Assessing sediment contamination in estuaries. Environ Toxicol Chem 20:3–22

DasSarma S (1989) Mechanisms of genetic variability in *Halobacterium halobium*: the purple membrane and gas vesicle mutations. Can J Microbiol 35:65–72

DasSarma S, Arora P (1997) Genetic analysis of gas vesicle gene cluster in haloarchaea. FEMS Microbiol Lett 153:1–10

DasSarma S, DasSarma P (2012) Halophiles. In: eLs. Wiley, Chichester. doi:10.1002/9780470015902.a0000394.pub3

DasSarma S, Damerval T, Jones JG, Tandeau de Marsac N (1987) A plasmid encoded gas vesicle protein gene in a halophilic archaebacterium. Mol Microbiol 1:365–370

DasSarma S, Arora P, Lin F, Molinari E, Yin LR (1994) Wild-type gas vesicle formation requires at least ten genes in the *gvp* gene cluster of *Halobacterium halobium* plasmid pNRC100. J Bacteriol 176:7646–7652

DasSarma S, Karan R, DasSarma P, Barnes S, Ekulona F, Smith B (2013) An improved genetic system for bioengineering buoyant gas vesicle nanoparticles from Haloarchaea. BMC Biotechnol 13:112, http://www.biomedcentral.com/1472-6750/13/112

El-Rafie HM, El-Rafie HM, Zahran MK (2013) Green synthesis of silver nanoparticles using polysaccharides extracted from marine macro algae. Carbohydr Polym 96:403–410

Garcia MA (2011) Surface plasmons in metallic nanoparticles: fundamentals and applications. J Phys D Appl Phys 44:283001

Gericke M, Pinches A (2006) Biological synthesis of metal nanoparticles. Hydrometallurgy 83:132–140

Gurunathan S, Kalishwaralal K, Vaidyanathan R, Venkataraman D, Pandian SR, Muniyandi J, Hariharan N, Eom SH (2009) Biosynthesis, purification and characterization of silver nanoparticles using *Escherichia coli*. Coll Surf B Biointerf 74:328–335

Halladay JT, Jones JG, Lin F, MacDonald AB, DasSarma S (1993) The rightward gas vesicle operon in *Halobacterium* plasmid pNRC100: identification of the *gvpA* and *gvpC* gene products by use of antibody probes and genetic analysis of the region downstream of *gvpC*. J Bacteriol 175:684–692

Issa B, Obaidat IM, Albiss BA, Haik Y (2013) Magnetic nanoparticles: surface effects and properties related to biomedicine applications. Int J Mol Sci 14:21266–21305

Jensen A, Bro-Rasmussen F (1992) Environmental cadmium in Europe. Rev Environ Contam Toxicol 125:101–118

Jones JG, Young DC, DasSarma S (1991) Structure and organization of the gas vesicle gene cluster on the *Halobacterium halobium* plasmid pNRC100. Gene 102:1017–1022

Kang SH, Bozhilov KN, Myung NV, Mulchandani A, Chen W (2008) Microbial synthesis of CdS nanocrystals in genetically engineered *E. coli*. Angew Chem Int Ed 47:5186–5189

Kathiresan K, Manivannan S, Nabeel MA, Dhivya B (2009) Studies on silver nanoparticles synthesized by a marine fungus, *Penicillium fellutanum* isolated from coastal mangrove sediment. Coll Surf B Biointerf 71:133–137

Kathiresan K, Alikunhi NM, Pathmanaban S, Nabikhan A, Kandasamy S (2010) Analysis of antimicrobial silver nanoparticles synthesized by coastal strains of *Escherichia coli* and *Aspergillus niger*. Can J Microbiol 56:1050–1059

Klaus-Joerger T, Joerger R, Olsson E, Granqvist CG (2001) Bacteria as workers in the living factory: metal-accumulating bacteria and their potential for materials science. Trends Biotechnol 19:15–20

Kowshik M, Ashtapure S, Kharazzi S, Vogel W, Urban J, Kulkarni SK, Paknikar KM (2003) Extracellular synthesis of silver nanoparticles by a silver-tolerant yeast strain MKY3. Nanotechnology 14:95–100

Manivannan S, Alikunhi NM, Kandasamy K (2010) In vitro synthesis of silver nanoparticles by marine yeasts from coastal mangrove sediment. Adv Sci Lett 3:1–6

Markich SJ, Brown PL, Batley GE, Apte SC, Stauber JL (2001) Incorporating metal speciation and bioavailability into water quality guidelines for protecting aquatic ecosystems. Aust J Ecotoxicol 7:109–122

Mishra RR, Prajapati S, Das J, Dangar TK, Das N, Thatoi H (2011) Reduction of selenite to red elemental selenium by moderately halotolerant *Bacillus megaterium* strains isolated from Bhitarkanika mangrove soil and characterization of reduced product. Chemosphere 84:1231–1237

Muthukannan R, Karuppiah B (2011) Rapid synthesis and characterization of silver nanoparticles by novel *Pseudomonas* sp. "ram bt-1". J Ecobiotechnol 3:24

Myung S, Solanki A, Kim C, Park J, Kim KS, Lee K-B (2011) Graphene-encapsulated nanoparticle-based biosensor for the selective detection of cancer biomarkers. Adv Mater 23:2221–2225. doi:10.1002/adma.201100014

Nürnberg HW (1983) Investigation on heavy metal speciation in natural waters by voltammetric procedures. Fresen Z Anal Chem 316:557–565

Oza G, Pandey S, Shah R, Sharon M (2012) A mechanistic approach for biological fabrication of crystalline gold nanoparticles using marine algae, *Sargassum wightii*. Eur J Exp Biol 2:505–512

Paquin PR, Zoltay V, Winfield RP, Wu KB, Mathew R, Santore RC, Di Toro DM (2002) Extension of the biotic ligand model of acute toxicity to a physiologically-based model of the survival time of rainbow trout (*Oncorhynchus mykiss*) exposed to silver. Comp Biochem Physiol C Toxicol Pharmacol 133:305–343

Peakall D, Burger J (2003) Methodologies for assessing exposure to metals: speciation, bioavailability of metals, and ecological host factors. Ecotoxicol Environ Saf 56:110–121

Rajeshkumar S, Malarkodi C, Paulkumar K, Vanaja M, Gnanajobitha G, Annadurai G (2014) Algae mediated green fabrication of silver nanoparticles and examination of its antifungal activity against clinical pathogens. Int J Met. doi:10.1155/2014/692643, http://dx.doi.org/

Rao CNR, Cheetham AK (2001) Science and technology of nanomaterials: current status and future prospects. J Mater Chem 11:2887–2894

Raveendran S, Chauhan N, Nakajima Y, Toshiaki H, Kurosu S, Tanizawa Y, Tero R, Yoshida Y, Hanajiri T, Maekawa T, Ajayan PM, Sandhu A, Kumar DS (2013a) Ecofriendly route for the synthesis of highly conductive graphene using extremophiles for green electronics and bioscience. Part Part Syst Charact 30:573–578

Raveendran S, Poulose AC, Yoshida Y, Maekawa T, Kumar DS (2013b) Bacterial exopolysaccharide based nanoparticles for sustained drug delivery, cancer chemotherapy and bioimaging. Carbohydr Polym 91:22–32

Raveendran S, Dhandayuthapani B, Nagaoka Y, Yoshida Y, Maekawa T, Kumar DS (2013c) Biocompatible nanofibers based on extremophilic bacterial polysaccharide, Mauran from *Halomonas maura*. Carbohydr Polym 92:1225–1233

Raveendran S, Girija AR, Balasubramanian S, Ukai T, Yoshida Y, Maekawa T, Kumar DS (2014) Green approach for augmenting biocompatibility to quantum dots by extremophilic polysaccharide conjugation and nontoxic bioimaging. ACS Sustain Chem Eng. doi:10.1021/sc500002g

Roux L, Roux S, Appriou P (1998) Behaviour and speciation of metallic species Cu, Cd, Mn and Fe during estuarine mixing. Mar Poll Bull 36:56–64

Sanders J, Abbe G (1987) The role of suspended sediments and phytoplankton in the partitioning and transport of silver in estuaries. Cont Shelf Res 7:1357–1361

Sathiyanarayanan G, Kiran GS, Selvin J (2013) Synthesis of silver nanoparticles by polysaccharide bioflocculant produced from marine *Bacillus subtilis* MSBN17. Coll Surf B: Biointerf 102:13–20

Schrofel A, Kratosova G, Bohunicka G, Dobrocka E, Vavra I (2011) Biosynthesis of gold nanoparticles using diatoms—silica-gold and EPS-gold bionanocomposite formation. J Nanopart Res 13:3207–3216

Seshadri S, Saranya K, Kowshik M (2011) Green synthesis of lead sulfide nanoparticles by the lead resistant marine yeast, *Rhodosporidium diobovatum*. Biotechnol Prog 7:1464–1469

Seshadri S, Prakash A, Kowshik M (2012) Biosynthesis of silver nanoparticles by marine bacterium, Idiomarina sp. p R58–8. Bull Mater Sci 35:1201–1205

Shah R, Oza G, Pandey S, Sharon M (2012) Biogenic fabrication of gold nanoparticles using *Halomonas salina*. J Microbiol Biotechnol Res 2:485–492

Shukla HD, DasSarma S (2004) Complexity of gas vesicle biogenesis in *Halobacterium* sp. strain NRC-1: identification of five new proteins. J Bacteriol 186:3182–3186

Singaravelu G, Arockiamary JS, Ganesh Kumar V, Govindaraju K (2007) A novel extracellular synthesis of monodisperse gold nanoparticles using marine alga, *Sargassum wightii* Greville. Coll Surf B: Biointerf 57:97–101

Srivastava P, Braganca J, Ramanan SR, Kowshik M (2013) Synthesis of silver nanoparticles using haloarchaeal isolate *Halococcus salifodinae* BK3. Extremophiles 17:821–831

Srivastava P, Braganca J, Ramanan SR, Kowshik M (2014) Green Synthesis of Silver Nanoparticles by Haloarchaeon *Halococcus salifodinae* BK6. Adv Mater Res 938:236–241

Tipping E, Lofts S, Lawlor AJ (1998) Modelling the chemical speciation of trace metals in the surface waters of the Humber system. Sci Total Environ 210:63–77

Turner DR (1987) Speciation and cycling of arsenic, cadmium, lead and mercury in natural waters. In: Hutchinson TC, Meema KM (eds) Lead, mercury, cadmium and arsenic in the environment. Wiley, New York, pp 175–186

van Keulen G, Hopwood DA, Dijkhuizen L, Sawers RG (2005) Gas vesicles in actinomycetes: old buoys in novel habitats. Trends Microbiol 13:350–354

Chapter 5
Halophilic Microorganisms and Their Biomolecules: Approaching into Frame of Bio(Nano) Technologies

Mădălin Enache, Roxana Cojoc, and Masahiro Kamekura

Abstract The increasing human society development claims for products, services and technologies and new and complex interactions between human and biological ecosystems. The capacity of biological ecosystems to regenerate their resources and to answer to these new interactions appears to be generally affected by various factors like climate change, loss of biodiversity etc. Thus, ecosystems that have been considered hostile to normal life forms were investigated extensively in last 20 years, looking for organisms harboring metabolic pathways governed by enzymes functional in extreme conditions like high or low temperature, ionic strength, pH values or combinations of these physico-chemical conditions. Among these hostile ecosystems, the saline environments appear to represent an interesting spring of novel halophilic microorganisms. Several bio-molecules currently produced by these halophilic organisms, i.e. enzymes, halocins (halobacterial proteins with antibiotic activities), exopolysaccharides showed biological activity in harsh conditions. Combination of these bio-molecules with various nanomaterials like thin-layers, nanotubes, nanospheres results in novel compounds possessing both biological properties of biomolecules and physico-chemical characteristics of nanomaterials. The present chapter deals with the main biomolecules produced by both halophilic archaea and bacteria revealing their potential implications in some nanotechnologies. The interaction of moderately halophilic bacteria with nanostructures like titanate nanotubes and silica microtubes are presented and discussed.

Keywords Halophilic microorganisms • Salt • Halophile biomolecules • Bionanotechnologies

---

M. Enache (✉) • R. Cojoc
Department of Microbiology, Institute of Biology Bucharest, Romanian Academy, Splaiul Independentei 296, P.O. Box 56-53, Bucharest, Romania
e-mail: madalin.enache@ibiol.ro; roxana.cojoc@ibiol.ro

M. Kamekura
Halophiles Research Institute, Noda 278-0043, Chiba-ken, Japan
e-mail: masahirokamekura@krc.biglobe.ne.jp

D.K. Maheshwari, M. Saraf (eds.), *Halophiles*, Sustainable Development and Biodiversity 6, DOI 10.1007/978-3-319-14595-2_5

5.1 Introduction

The increasing human society development claims for new products, services and technologies and new and complex interactions between human and biological ecosystems. The capacity of biological ecosystems to regenerate their resources and to answer to these new interactions appears to be generally affected by various factors like climate change, loss of biodiversity (Murariu 2012). Thus, in last 20 years ecosystems considered hostile to normal life forms were investigated extensively; high or low temperature, pH values, high ionic strength, or combinations of these physico-chemical conditions. Between these, the salty environments appear to represent an interesting spring of novel organisms, halophilic microorganisms or simply halophiles, harboring metabolic pathways functional in saline condition.

Saline environments characterized by huge and mostly endemic biodiversity are distributed worldwide as salt lakes, salt mines and salted soils. The salinity of these areas is generally understood by biologists in terms of sodium chloride content. This compound represents over 90 % of the total salt content in many cases and the presence of other compounds in terms of influence to physical or chemical parameters of the saline area could be considered as low. On the other hand, these compounds could have a significant influence on the diversity of halophilic microorganisms if we considered the high concentrations of magnesium in Dead Sea or carbonates in soda lakes such as Magadi, in Kenya, Wadi Natrum in Egypt, Sambhar, India. The predominant microorganisms in the saline environments are represented by both halophilic archaea and bacteria.

Currently, over 182 species distributed in 48 validly described genera in *Halobacteriaceae* cover aerobic halophiles and a few anaerobic and halophilic methanogens are known in *Archaea* domain. The number of halophiles in *Bacteria* domain is higher and they are distributed in many groups (phyla). Representatives of halophilic bacteria are included in Phylum Bacteroidetes, Cyanobacteria, Proteobacteria, Firmicutes and Sulphur-Green bacteria. Furthermore, halophilic Fungi, Plants, Ciliates and Flagellates have also been known in *Eucarya* domain. These organisms cope with harsh environmental conditions by two strategies, namely "salt-in" and "compatible solutes" (Oren 1999a, b, 2002a, b). The salts are accumulated inside of cells to osmolarity equivalent with external environment in first strategy, and synthesis of organic molecules known as compatible solutes occurs or they are accumulated from environments if they are present in the second strategy (Oren 1999a, b). Special adaptations are requested for proteins and enzymes in case of first strategy, and the most well-known fact is the increasing acidic amino-acids residues on their surface (Lanyi 1974; Graziano and Merlino 2014).

The present chapter deals with the main biomolecules produced by both halophilic archaea and bacteria and their potential implications in some nanotechnologies. Our recent work on the interaction of moderately halophilic bacteria with nanostructures like titanate nanotubes and silica microtubes are also presented and discussed.

5.2 Halophilic Enzymes

The life in saline and hypersaline environments has been investigated intensively for many years focusing on diversity of microorganisms and controlling mechanisms involved at high salinity. The enzymes of halophilic microorganisms involved in such mechanisms could be of potential interest for several biotechnologies or agriculture fields. Researches on halophilic enzymes are also attracted to putative impact on the possible life forms in space (Satyanarayana et al. 2005). Halophilic enzymes are biologically active in saline media where generally the structure and function of enzymes are critically affected. In order to keep their functionality at high salt content, the surface of halophilic enzymes is highly negatively charged, and spheres of hydration is clustered to the protein surface to prevent their degradation (Lanyi 1974; Paul et al. 2008). In this context the number of hydrophobic amino-acids is decreasing at the surface and hydrophilic amino-acids residues increase at the protein surface (Bolhuis et al. 2008). In another word, halophilic enzymes have high acidic residues of amino-acids on their surface, the content of lysine is low, high number of salt bridges in their structure and a low hydrophobicity are observed (Mevarech et al. 2000; Fukuchi et al. 2003). The unique properties of halophilic enzymes, requirement of salt for the stability and activity, high resistance to several denaturation methods (Karan et al. 2012) and their ability to perform catalytic activity at low water activity or in organic solvents (non-aqueous media) attracted the interest for research in this area (Tokunaga et al. 2008; Enache and Kamekura 2010; Oren 2010; Karan et al. 2012). The most well investigated haloezymes are hydrolases such as amylases (Amoozegar et al. 2003), lipases and esterases, xylanases, chitinases, proteases, cellulases, nucleases, etc. (Mellado and Ventosa 2003; Oren 2010; Moreno et al. 2013).

5.3 Halocins

Beside enzymes and bacteriorhodopsin, halocin is one of the best characterized proteins produced by halophilic archaea (Shand et al. 1999; O'Connor and Shand 2002). Their presence as protein with antagonic activity against halobacteria was signaled for first time in 1982 by Rodriguez-Valera et al. (1982). The halocins delivered to hypersaline environment by synthesizing cells could be considered as biochemistry way to compete for nutrients. There is an argument, however, that the contribution of halocins in the competition is probably negligible, since no halocin activity was detected in any of brines taken in Israel, U.S.A and Spain (Kis-Papo and Oren 2000). In laboratory, the halocins showed similar behavior as bacteriocins. Similar with BLIAS (Bacteriocin Like Inhibitory Substances produced by Gram positive bacteria) the syntheses of halocins are not induced by intercalating agents and have the effect against huge numbers of halophiles. Several studies were performed on halocin H4 (Cheung et al. 1997), H6 (Meseguer et al. 1995; Torreblanca et al. 1989), Hal R1 (Rdest and Sturm 1987) and Hf1 (Enache et al. 2004). The studies concluded that some halocin (H4) have bactericidal and bacteriolytic effect, lost activity in the absence of sodium chloride and by protease or thermal treatments (80 °C) and have molecular weight over 100 kDa. Halocin Hf1 (Enache et al. 2004) was produced by a haloarchaeal strain isolated from Bride Cave salt lake (in Slanic, Prahova, Romania). This strain has been identified to belong to *Haloferax* genus based on biochemical and molecular taxonomic investigations. High amount of halocin Hf1 was obtained in media with high salinity. The maximum antimicrobial activity has recorded at the start of stationary phase of growth (after 72 h of cultivation) followed by a slightly decreasing intensity in later stages of growth. The purified halocin Hf1 showed maximum activity at 25–30 °C, pH 7 and 3M sodium chloride, features characteristic for proteins of halophilic archaea (Enache et al. 2004). Antagonisms studies performed by Torreblanca et al. (1994) concluded that "production of halocin is a practically universal feature of archaeal halophiles". Similar conclusions were drawn in similar studies (Enache et al. 1999; Birbir and Eryilmaz 2004; Salgaonkar et al. 2012). A fragment of DNA coding for halocin H4 has been isolated for first time from plasmid DNA of *Haloferax mediterranei* (Cheung et al. 1997). They concluded that halocins could act as transporter proteins and their functionality is dependent or regulated function of ionical power of environments.

5.4 Exopolysaccharides Produced by Halophiles

5.4.1 Exopolysaccharides Synthesized by Extremely Halophilic Archaea

Some extremely halophilic microorganisms belonging to Archaea domain, such as *Haloferax mediterranei,* present a particular biotechnological potential due to the ability to produce exopolysaccharides. Generally, the biosynthetic activity is initiated in the early stages of the cell growth, and reaches the maximum value at the beginning of the stationary phase, after which the synthesis activity slows down (Anton et al. 1988). In the case of *Hfx. mediteranei,* the exopolysaccharide synthesis was favored by an aeration rate lower than the optimum for growth, the use of glucose as a sole source of carbon and energy and the cultivation in a continuous-culture system (Anton et al. 1988; Rodriguez-Valera et al. 1991; Hezayen et al. 2000).

The extracellular polymer produced by *Hfx. mediterranei* is an acidic, anionic heteropolysaccharide comprising glucose, mannose, galactose, aminoglucides, uronic acid and a considerable amount of sulphate (6 % of the dry weight of the polymer). It combines the rheological properties (viscosity and high pseudo-plasticity) with a remarkable resistance to extremely high salt concentrations, temperature and pH, suggesting the possibility of wide use in various practical applications (Rodriguez-Valera et al. 1991; Rodriguez-Valera 1994).

5.4.2 *Exopolysaccharides Synthesized by Halotolerant and Moderately Halophilic Bacteria*

The specific physiological characteristics of halotolerant and moderately halophilic bacteria, as well as their ability to produce a variety of bioproducts superior to the ones synthesized by nonhalophilic species, have led to a number of studies on exopolysaccharides production by these microorganisms.

In 1986, Pfiffner et al. isolated from oil wells associated environments some halotolerant, facultative anaerobic strains of *Bacillus* sp. able to produce exopolysaccharides from various substrates (fructose, glucose, mannose, cellobiose, maltose, sucrose, arabinose, galactose, xylose, mannitol, starch), up to the salinity of 12 % NaCl.

Quesada et al. (1993) isolated from samples of hypersaline soil located near Alicante (Spain) 28 moderately halophilic strains of *Volcaniella eurihalina* (later reclassified as *Halomonas eurihalina*) able to produce exopolisaccharides. Of these, strain F2-7 was remarkable for producing a polyanionic exopolysaccharide with high thermal stability, pseudoplasticity and viscosity (especially at low pH values) and resistance to high ionic strength (Quesada et al. 1993; Bejar et al. 1998; Ventosa et al. 1998).

In 1998, the team led by Bejar isolated from hypersaline soil samples collected from Alicante some moderately halophilic bacterial strains belonging to *Halomonas* genus able to produce exopolysaccharides that can serve for biotechnological use. The chemical compositions of exopolysaccharides synthesized by the tested *Halomonas* strains proved to be similar, consisting of three types of neutral sugars (glucose, mannose, rhamnose), small amounts of proteins, hexosamines, uronic acid, acetyl residues and some cations. Moreover, all studied exopolysaccharides showed a high sulfur content (1.3–24.7 %) (Bejar et al. 1998; Margesin and Schinner 2001). Also, all of exopolysaccharides produced by the tested strains presented a high pseudoplasticity, the increase in the shear rate resulting in a decrease of viscosity of the polymer solution (Bejar et al. 1998; Calvo et al. 1998).

Exopolysaccharides containing high amounts of sulfate may be used in the pharmaceutical industry as antivirals, antitumor and anticoagulant substances (Bejar et al. 1998; Ventosa et al. 1998; Margesin and Schinner 2001).

Also, due to the relatively large amount of uronic acid, such polysaccharides may be used in the processes of waste water decontamination (Bejar et al. 1998). Recently, some *Halomonas maura* strains were isolated from hypersaline soil samples collected from a solar saltern in Asilah – Morocco. Those strains produced exopolysaccharides with high viscosity and pseudo-plastic behavior that proved to be more efficient in emulsifying various hydrocarbons (crude oil) than some chemical surfactants. Due to these properties, they can be used as thickeners and emulsifiers in various industrial processes (Bouchotroch et al. 2001; Margesin and Schinner 2001; Llamas et al. 2012).

A moderately halophilic bacteria isolated by our group from rock salt crystal from Slanic, Prahova area showed the capacity to synthesize an exopolysaccharide having high thermal stability, with a melting temperature of 207 °C associated with thermal degradation of the polymer. The polymer production was influenced by composition of culture medium and growth conditions, stirring or static. The polymer was characterized by an emission of fluorescence at 530 nm, and absorption maximum at around 260 nm. Investigations by FT-IR method revealed that amine and sulphate groups were connected to backbone of the sugar polymer (Cojoc et al. 2009).

5.5 Lipids of Archaea and Halophiles: Archaeosome

Membrane lipids of halophilic archaea, as well as other groups of archaea, are distinctively different from those of bacteria in the following major features: the linkage of hydrocarbon to glycerol moiety is ether type, the hydrocarbons are isoprenoid units and chirality of glycerol is sn-1 (sn-glycerol-1-phosphate) (DeRosa et al. 1986; Kamekura 1993; DeRosa 1996; Kates 1997). These unusual qualities attracted a large interest in research for a putative biotechnological exploitation. The major target is represented by their use in archaeozomes (liposome obtained with membrane lipids from archaeal cells).

Archaeosomes constitute a novel family of liposomes that demonstrate higher stabilities to different conditions, in comparison with conventional liposomes. The archaeal lipids show high stability to oxidative stress, low or high temperatures, acidic or alkaline pH, and action of phospholipases and bile salts. The lipid membrane of archaeosomes may have a bilayer form if it is made exclusively from monopolar archaeol (diether) lipids, or may have a monolayer form if made exclusively from bipolar caldarchaeol (tetraether) lipids (Kates 1997), or a combination of monolayers and bilayers if made from caldarchaeol lipids in addition to archaeol lipids or standard bilayer-forming phospholipids. Due to the remarkable molecular and physicochemical properties of archaeal lipids, archaeosomes can be formed using standard procedures at any temperature in the physiological range or lower, making possible the encapsulation of thermally labile compounds (Choquet et al. 1996). Also, they can be prepared and stored in the presence of air/oxygen without any degradation and studies in vitro and in vivo indicated that archaeosomes are safe and do not elicit toxicity in mice. Therefore, the biocompatibility and the superior stability properties of archaeosomes in different conditions offer advantages over conventional liposomes in the manufacture and the applications as drug delivery systems (Sprott 1992; Gambacorta et al. 1995; Patel et al. 2000; Gonzales-Peredes et al. 2011).

5.6 Interaction of Several Nanomaterials with Halophilic Bacteria

Our recent studies (Neagu et al. 2014) revealed the reaction of moderately halophilic bacteria *Virgibacillus halodenitrificans* at the presence of various nanomaterials like titatane nanotubes (Fig. 5.1) or silica microtubes in their culture medium.

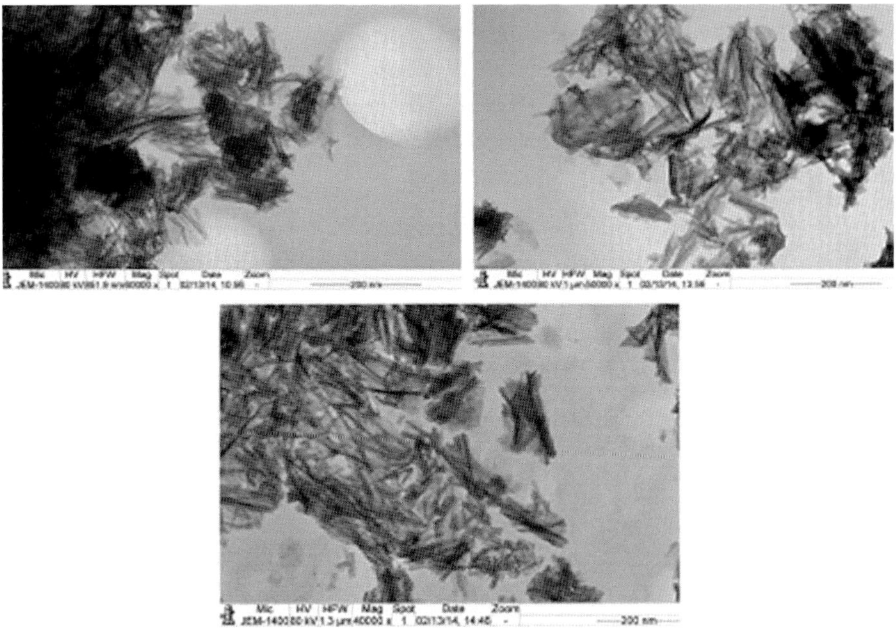

Fig. 5.1 TEM image of titanate nanotubes used in investigations towards *Virgibacillus halodenitrificans*

These materials were subject of several physical or chemicals treatment after synthesis reaction like thermally treatment. The obtained results revealed antimicrobial action of titanate nanostructures against moderately halophilic strains of *Virgibacillus halodenitrificans* and *Bacillus subtillis*. The same investigations showed that post reaction thermal treatment conducted to increasing of antimicrobial properties and authors consider that mechanisms of action are related to bacteriocin-like substances inhibitory properties (Merciu et al. 2009).

In another study, Cojoc et al. (2013) demonstrated the ability of *Virgibacillus halodenitrificans* to grow for short time in culture media with nine milligrams of silica microtubes (Fig. 5.2). The primary reaction of the microbial cells was considered to be the exceeding synthesis of exopolysaccharides. The growth stopped after 24 h. The photoactivity and structure of silica microtubes and their antimicrobial activity is considered to be factors which combined contribution to the first answer of microbial cells to the microtubes. Supplementary data related to total dehydrogenase activity determined with tryphenyl tetrazolium chloride as substrate (Casida et al. 1964) supported activity with low intensity of microtubes towards *Virgibacillus halodenitrificans* and their absence towards *Bacillus subtilis*. The behavior differences between tested strains are considered to be due to either exceeding exopolysaccharides synthesis or photoactivity of silica microtubes (Cojoc et al. 2013).

Another investigation (Enache et al. 2014) showed that commercially available calcined silica microtubes had a slight inhibitory effect towards *Bacillus subtilis* but stimulated the growth of *Virgibacillus halodenitrificans*. If these microtubes are

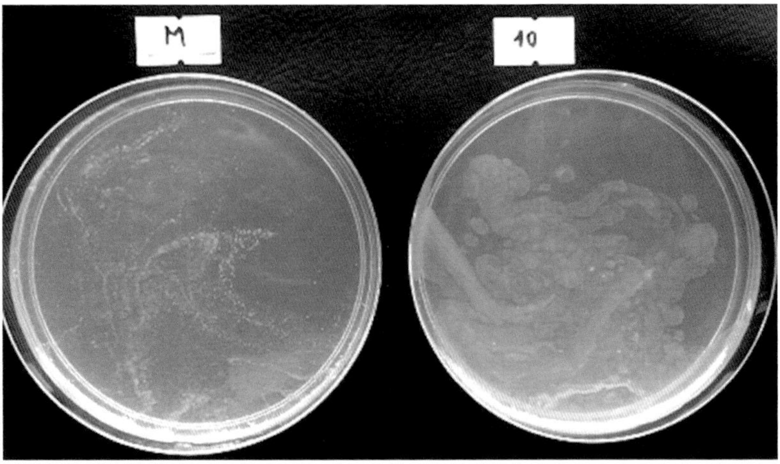

Fig. 5.2 The effect of microtubes against *Virgibacillus halodenitrificans* (*right*) in comparison with control Petri dish (Picture from reference Cojoc et al. 2013 – Reproduced with permission)

thermally treated 1 h at 400 °C in our laboratory, they stimulated the growth of *Bacillus subtilis* but diminished the growth of *Virgibacillus halodenitrificans*. If we used silica nanostructures having spherical shape or those doped with platinum, they showed a positive effect on the growth of tested halophilic bacteria. Similarly to previously studies (Cojoc et al. 2013), the different reactions of investigated halobacterial cells in the presence of calcined or doped silica microstructures is attributed to the structure and composition of nanomaterials and conclusion is supported also by viability of cells and total dehydrogenase activity (Enache et al. 2014). When the calcined silica microtubes, silica nanospheres and platinum doped silica microtubes were thermally treated (400 °C, 1 h), they stimulated the growth of *Virgibacillus halodenitrificans* (Enache et al. 2014). The strain *V. halodenitrificans* has been isolated from hypersaline habitat from Slanic, Prahoca county, Romania, namely Neogene dated subterranean salt rock. The strain is sensitive to chloramphenicol and erithromicin but is able to grow in the presence of neomycin, penicillin, anysomicin and bile salt. This strain is able to grow until to 2M sodium chloride in culture medium. The partial sequence of 16S rRNA showed 99 % similarity with *V. halodenitrificans* AY543168 in BLAST analysis.

The investigation related to immobilization of a halophilic protease isolated from culture medium of a moderately halophilic strain obtained from subterranean rock salt crystal from Slanic, Prahova, Romania on silica and titanate nanostructures revealed that immobilization capacity is three times higher on the titanate nanotubes compared with silica microtubes (Neagu et al. 2014). The differences of composition, morphology and structure of titanate nanotubes and silica microtubes influenced the immobilization capacity. For this study was used the halophilic protease in various steps of purification by treatment with organic solvent, acetone. The biological activity of enzymes after immobilization was not affected by purification step but is higher after immobilization on silica microtubes, as expressed in units per mg of immobilizing material, in spite of low adsorption on this material by electrostatic bridge (Neagu et al. 2014).

5.7 Conclusion

The investigations about interactions between nanostructures and various biological molecules like proteins, vitamins, and lipids of microorganisms attracted a huge interest for the last few decades with the main focus to understand their benefits for human life (Kaur et al. 2013). Nanostructures available nowadays in market are composed of commonly used materials including metal oxides (titanium, zinc and iron). These structures are defined by chemical stability, several electrical or magnetic properties, and also some catalytic properties (Katz et al. 2003; Willner and Katz 2004). These properties suggested the use of nanostructures in a various applications including food component, cosmetic products, antimicrobial agents, biosensor materials, electronic components (Willner and Katz 2004).

Today it is well known that hypersaline and saline environments are the natural or anthropic habitat of a huge number of halophilic microorganisms including bacteria and archaea extremophiles. These organisms represent a valuable resource of enzymes (extremozymes) with stability in harsh conditions of pH or/and ionic strength. Thus, their investigations as biocatalysts in the presence of novel nanomaterials are attractive. Several bio-molecules produced by these halophilic organisms, i.e. enzymes, halocins (halobacterial proteins with antibiotic activities), exopolysaccharides etc. show biological activity in harsh conditions. Combining of these bio-molecules with various nanomaterials like thin-layers, nanotubes, nanospheres results in novel compounds harboring both biological properties of biomolecules and physico-chemical characteristics of nanomaterials. The data summarized in this work argued that some extremophiles, either bacteria or archaea should be regarded as important subjects for novel activities and applications in correlation with new materials and nanomaterials.

References

Amoozegar MA, Malekzadeh F, Malik KA (2003) Production of amylase by newly isolated moderate halophile, *Halobacillus* sp. strain MA-2. J Microbiol Methods 52:353–359

Anton J, Meseguer I, Rodriguez-Valera F (1988) Production of an extracellular polysaccharide by *Haloferax mediterranei*. Appl Environ Microbiol 54:2381–2386

Béjar V, Llamas I, Calvo C, Quesada E (1998) Characterization of exopolysaccharides produced by 19 halophilic strains of the species *Halomonas eurihalina*. J Biotechnol 61:135–141

Birbir M, Eryilmaz S (2004) Prevention of halobacterial damage on hide caused by lipolytic halophilic archaea with halocins. J Soc Leather Technol Chem 88:99–104

Bolhuis A, Kwan D, Thomas JR (2008) In: Siddiqui KS, Thomas T (eds) Protein adaptation in extremophiles. Nova Science Publishers, New York, pp 71–104

Bouchotroch S, Quesada E, del Moral A, Llamas I, Bejar V (2001) *Halomonas maura* sp. nov., a novel moderately halophilic, exopolysaccharide-producing bacterium. Int J Syst Evol Microbiol 51:1625–1632

Calvo C, Martinez-Checa F, Mota A, Bejar V, Quesada E (1998) Effect of cations, pH and sulfate content on the viscosity and emulsifying activity of the *Halomonas eurihalina* exopolysaccharide. J Ind Microbiol Biotechnol 20:205–209

Casida JE Jr, Klein DA, Santoro T (1964) Soil dehydrogenase activity. Soil Sci 98:371–376

Cheung J, Danna JK, O'Connor ME, Price BL, Shand FR (1997) Isolation, sequence and expression of the gene encoding halocin H4, a bacteriocin from the halophilic archeon *Haloferax mediterranei* R4. J Bacteriol 179:548–551

Choquet C, Patel G, Sprott G (1996) Heat sterilization of archaeal liposomes. Can J Microbiol 42:183–186

Cojoc R, Merciu S, Oancea P, Pincu E, Dumitru L, Enache M (2009) Highly thermostable exopolysaccharide produced by the moderately halophilic bacterium isolated from a man-made young salt lake in Romania. Pol J Microbiol 58:289–294

Cojoc R, Anastasescu C, Zaharescu M, Enache M, Neagu S (2013) The growth of moderately halophilic bacteria in the presence of silica microtubes. Drobeta – Ser Stiintele Naturii XXIII:40–51

DeRosa M (1996) Archaeal lipids: structural features and supramolecular organization. Thin Solid Film 284–285:13–17

DeRosa M, Gambacorta A, Gliozzi A (1986) Structure, biosynthesis and physico-chemicals properties of archaebacterial lipids. Microbiol Rev 50:70–80

Enache M, Kamekura M (2010) The halophilic enzyme and their economical values. Rom J Biochem 47:47–59

Enache M, Dumitru L, Faghi AM (1999) Occurrence of halocins in mixed archaebacteria culture. Proc Inst Biol (Ed Rom Acad) II:151–154

Enache M, Faghi AM, Dumitru L, Teodosiu G, Zarnea G (2004) Halocin Hf1 a bacteriocin produced by *Haloferax* sp. GR1. Proc Rom Acad B 6:27–32

Enache M, Neagu S, Anastasescu C, Cojoc R, Zaharescu M (2014) The effects of silica nanostructures on halotolerant microorganisms isolated from rock salt crystal. In: Dascălu D, Cârjă G, Ciurea ML, Catrinel Ion A, (eds) New applications of nanomaterial, Series in "micro and nanoengineering", Ed Acad Rom, Bucharest, pp 51–59

Fukuchi S, Yoshimune K, Wakayama M, Moriguchi M, Nishikawa K (2003) Unique amino acid composition of proteins in halophilic bacteria. J Mol Biol 327:347–357

Gambacorta A, Gliozzi A, Rosa M (1995) Archaeal lipids and their biotechnological applications. World J Microbiol Biotechnol 11:115–131

Graziano G, Merlino A (2014) Molecular bases of protein halotolerance. Biochim Biophys Acta 1844:850–858

Hezayen FF, Rehm BHA, Eberhardt R, Steinbuchel A (2000) Polymer production by two newly isolated extremely halophilic archaea: application of a novel corrosion-resistant bioreactor. Appl Microbiol Biotechnol 54:319–325

Kamekura M (1993) Lipids of halophilic bacteria. In: Biology of halophilic bacteria, pp 135–161

Karan R, Kumar S, Sinha R, Khare SK (2012) Halophilic microorganisms as sources of novel enzymes. In: Satyanarayana T, Johri BN (eds) Microorganisms in sustainable agriculture and biotechnology. Springer, Dordrecht, pp 555–579

Kates M (1997) Diether and tetraether phospholipids and glycolipids as molecular markers for archaebacteria (Archaea). In: Molecular markers in environmental geochemistry. Eganhouse RP (ed), ACS symposium series 671: 35–48

Katz E, Shipway AN, Willner I (2003) In: Schmid G (ed) Nanoparticles – from theory to applications. Wiley-VCH, Weinheim, pp 368–421

Kaur H, Kaur S, Singh M (2013) Biosynthesis of silver nanoparticles by natural precursor from clove and their antimicrobial activity. Biologia 68:1048–1053

Kis-Papo T, Oren A (2000) Halocins: are they involved in the competition between halobacteria in saltern ponds? Extremophiles 4:35–41

Lanyi JK (1974) Salt-dependent properties of proteins from extremely halophilic bacteria. Bacteriol Rev 38:272–290

Llamas I, Amjres H, Mata JA, Quesada E, Béjar V (2012) The potential biotechnological applications of the exopolysaccharide produced by the halophilic bacterium *Halomonas almeriensis*. Molecules 17:7103–7120

Margesin R, Schinner F (2001) Potential of halotolerant and halophilic microorganisms for biotechnology. Extremophiles 5:73–83

Mellado E, Ventosa A (2003) Biotechnological potential of moderately and extremely halophilic microorganisms. In: Barredo JL (ed) Microorganisms for health care, food and enzyme production. Research Signpost, Trivandrum, pp 233–256

Merciu S, Vacaroiu C, Filimon R, Popescu G, Preda S, Anastasescu C, Zaharescu M, Enache M (2009) Nanotubes biologically active in media with high salt concentrations. Biotechnol Biotechnol Eq 23:827–831, special issue

Meseguer I, Torreblanca M, Konishi T (1995) Specific inhibition of the halobacterial Na^+/H^+ antiporter by halocin H6. J Biol Chem 270:6450–6455

Mevarech M, Frolow F, Gloss LM (2000) Halophilic enzymes: protein with a grain of salt. Biophys Chem 86:155–164

Moreno ML, Perez D, Garcia MT, Mellado E (2013) Halophilic bacteria as a source of novel hydrolytic enzymes. Life 3:38–51

Murariu D (2012) Origin and evolution of actual biodiversity. Oltenia-Stud Comunicări Stiinţele Naturii 28:227–242

Neagu S, Preda S, Anastasescu C, Zaharescu M, Enache M, Cojoc R (2014) The functionalization of silica and titanate nanostructures with halotolerant proteases. Rom J Chem 2, 97–103

O'Connor EM, Shand RF (2002) Halocins and sulfolobicins: the emerging story of archaeal protein and peptide antibiotics. J Ind Microbiol Biotechnol 28:23–31

Oren A (1999a) Bioenergetic aspects of halophilism. Microbiol Mol Biol Rev 63:334–348

Oren A (1999b) The halophilic *Archaea* – evolutionary relationships and adaptation to life at high salt concentrations. In: Wasser SP (ed) Evolutionary theory and processes: modern perspectives, papers in honour of eviatar nevo. Kluwer Academic, Dordrecht, pp 345–361

Oren A (2002a) Diversity of halophilic microorganisms: environments, phylogeny, physiology, and applications. J Ind Microbiol Biotechnol 28:56–63

Oren A (2002b) Halophilic microorganisms and their environments. In: Seckbach J (ed) Cellular origin and life in extreme habitats, vol 5. Kluwer Academic, Dordrecht, p 575

Oren A (2010) Industrial and environmental applications of halophilic microorganisms. Environ Technol 31:825–834

Paredes A, Clares-Naveros B, Ruiz-Martinez MA, Durban-Fornieles JJ, Ramos-Cormenzana A, Monteoliva-Sanchez M (2011) Delivery systems for natural antioxidant compounds: archaeosomes and archaeosomal hydrogels characterization and release study. Int J Pharm 421:321–331

Patel GB, Agnew BJ, Deschatelets L, Fleming LP, Sprott GD (2000) In vitro assessment of archaeosome stability for developing oral delivery systems. Int J Pharm 194:39–49

Paul S, Bag SK, Das S, Harvill ET, Dutta C (2008) Molecular signature of hypersaline adaptation: insights from genome and proteome composition of halophilic prokaryotes. Genome Biol 9:R70

Pfiffner SM, McInerney MJ, Jenneman GE, Knapp RM (1986) Isolation of halotolerant, thermotolerant, facultative polymer-producing bacteria and characterization of the exopolymer. Appl Environ Microbiol 51:1224–1229

Quesada E, Béjar V, Calvo C (1993) Exopolysaccharide production by *Volcaniella eurihalina*. Experientia 49:1037–1041

Rdest V, Sturm M (1987) Bacteriocins from halobacteria, protein purification: micro to macro. Alan R. Liss, New York, pp 271–278

Rodriguez-Valera F (1994) Biotechnological potential of halobacteria. Biochem Soc Symp 58:135–147

Rodriguez-Valera F, Juez G, Kushner DJ (1982) Halocins: salt-dependent bacteriocins produced by extremely halophilic rods. Can J Microbiol 28:151–154

Rodriguez-Valera F, Garcia Lillo JA, Anton J, Meseguer I (1991) Biopolymer production by *Haloferax mediterranei*. In: Rodriguez-Valera F (ed) General and applied aspects of halophilic microorganisms. Plenum Press, New York, pp 373–380

Salgaonkar BB, Mani K, Nair A, Gangadharan S, Braganca MJ (2012) Interspecific interactions among member of family halobacteriaceae from natural solar saltern. Probiotics Antimicrob Proteins 4:98–107

Satyanarayana T, Raghukumar C, Shivaji S (2005) Extremophilic microbes: diversity and perspectives. Curr Sci 89:78–90

Shand FR, Price LB, O'Connor ME (1999) Halocins: protein antibiotics from hypersaline environments. In: Oren A (ed) Microbiology and biogeochemistry of hypersaline environments, Series: extreme and unusual environments. CRC, Boca Raton, pp 295–306, serie ed R.H. Vreeland

Sprott GD (1992) Structures of archaebacterial membrane lipids. J Bioenerg Biomembr 24:555–566

Tokunaga H, Arakwa T, Tokunaga M (2008) Engineering of halophilic enzymes: two acidic amino acid residues at the carboxy-terminal region confer halophilic characteristics to *Halomonas* and *Pseudomonas* nucleoside diphosphate kinases. Protein Sci 17:1603–1610

Torreblanca M, Meseguer I, Rodriguez-Valera F (1989) Effects of halocin H_6 on the morphology of sensitive cells. Biochem Cell Biol 68:396–399

Torreblanca M, Meseguer I, Ventosa A (1994) Production of halocin is a practically universal feature of archaea halophilic rods. Lett Appl Microbiol 19:201–205

Ventosa A, Nieto JJ, Oren A (1998) Biology of moderately halophilic aerobic bacteria. Microbiol Mol Biol Rev 62:504–544

Willner I, Katz E (2004) Integrated nanoparticle-biomolecule hybrid systems: synthesis, properties and applications. Angew Chem Int Ed 43:6042–6108

Chapter 6
Environmental Diversity and Biological Survivability of Halophilic Bacteria

Narayanan Rajendran

Abstract Halophiles are among the best model microbes to study cellular adaptation. *Haloferax*, *Haloarcula*, *Haloquadratum*, *Halobacteriales* and many other halophiles living in the Great Salt Lake can survive extremely high salt concentrations. Many of them cannot survive if moved from hypersaline to non-saline habitats, and die from immediate cell lysis. The unique biological survivability of these bacteria in the saline condition is dictated by the phylogenetic dexterity at the given environmental constraints. Microbial adaptation has played a major role when bacteria in all forms branched out into different environmental niches, arising from the hypersaline conditions of the primordial sea. Their pervasive nature, created by ever-changing environmental conditions, was acquired by microbes from millions of years of making. The adaptation of enzymes during the course of their evolutionary development and some metabolic differences helped them expand and achieve environmental diversity. Metabolic processes, like osmoregulation in halotolerant cells, dictate the regulation of the bacterial cell membrane. For example, glycerol metabolism has been linked to osmoregulation in some halophilic microbes. Some mechanisms behind osmoregulation in halophiles are extremely energetic since they live in hostile environments. Pathways of such regulations, including de novo synthesis of solutes anabolically and/or extraction of solutes environmentally, help them to produce secondary metabolites like poly-β-hydroxyalkanoates. How halophiles maintain high metabolic similarity to other non-saline bacteria while showing different survivability under heavy salt stress remains an important question. This review attempts to outline progress in our understanding of their environmental diversity and biological survivability.

Keywords Halophiles • High salinity • Biological survivability • Secondary metabolites

N. Rajendran (✉)
MASC, Kentucky State University, Frankfort, KY 40601, USA
e-mail: narayanan.rajendran@kysu.edu

D.K. Maheshwari, M. Saraf (eds.), *Halophiles*, Sustainable Development and Biodiversity 6, DOI 10.1007/978-3-319-14595-2_6

6.1 Introduction

In recent years, modern tools and techniques have provided insight into the selection and optimization of extremophiles for commercial uses (Tango et al. 2002). The potential application of halophiles in bioconversion processes has emerged as a promising field for future biofuel conversion technology (Oren 2010). The breakdown of biomass materials to gain energy, and the cleanup of contaminated salty lands and aqueous sites are particularly important potential uses as reported (Rodriguez-Valera 1991). The exploration of their biodiversity in extreme salty environments has opened a new era for biotechnologists to use various techniques including genomics and proteomics. For example, in the last 10 years genomic sequencing and proteomic analysis have vastly expanded since the genomics studies of *Halobacterium* NRC-1 was first published in the year 2000 (Ng et al. 2000). Since then, the complete genomic sequences of many halophiles and halotolerant bacteria were published, including *Haloarcula marismortui* (Baliga et al. 2004), *Natronomonas pharaonis* (Falb et al. 2005), *Haloquadratum walsbyi* (Bolhuis et al. 2006), *Halomicrobium mukohataei* (Tindall et al. 2008), *Halorhabdus utahensis* (Bakke et al. 2009), *Halogeometricum borinquense* (Malfatti et al. 2009), *Haloterrigena turkmenica* (Saunders et al. 2009), and *Haloferax volcanii* (Hartman et al. 2010).

Halophiles are salt-loving microbes and multiply quickly in saline environments. They are predominantly concentrated in bacterial group while some fungi (such as *Wallemia ichthyophaga*) and others algae (*Dunaliella salina*). Among the prokaryotes, the domain Archaea has several extremophilic genera of halophiles, like *Haloferax*, *Haloarcula*, *Haloquadratum*, *Halobacteriales* etc. The bacterial domain also has halophiles like *Salinibacter*, but they are less common when compared to archaea. Halophilic bacteria come in many sizes and shapes and live in variable salt stress conditions. For example, the bacterium *Thiomargarita namibiensis*, found in the ocean sediments of the continental shelf of Namibia, is the largest bacteria ever discovered. The second largest bacterium, *Epulopiscium fishelsoni*, was found in the Red Sea within the marine surgeonfish (Angert et al. 1993) (Fig. 6.1a). This

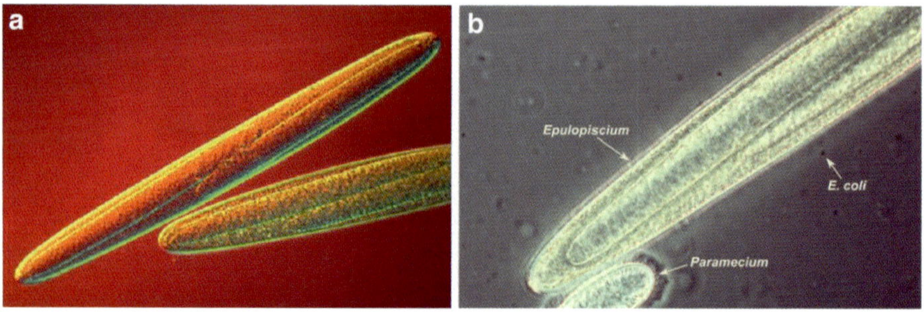

Fig. 6.1 (**a**, **b**) The largest bacterium *Epulopiscium fishelsoni* found in the Red Sea within the marine surgeonfish was isolated by D. A. Angert, and reported in 1993 (**a**) This bacterium is a million times larger than *E. coli*, as shown in the picture* as a dot, and one can see *E. fishelsoni* with the naked eye (**b**) (Picture permission was obtained from Dr. D. Angert, Cornell University, Ithaca, New York)

bacterium is a million times larger than *E. coli*, and one can see it with the naked eye (Fig. 6.1b).

The family Halobacteriaceae, which contains 15 genera, is a member of the domain Archaea and comprises the majority of the prokaryotic extremophiles (Oren 2002a). Due to the presence of halophiles like *Salinibacter*, *Haloquadratum*, *Halobacteriales*, and *Dunaliella salina*, many lake and saltern habitats display a pink or reddish color. Hence, most known halophiles such as *Halobacterium*, *Haloferax*, *Halococcus*, *Halorubrum*, *Halogeometricum*, *Haloterrigena* and *Haloarcula* have become popular models for habitat adaptation studies in the archaeal domain. This review will focus on the exploitation of halophilic bacteria and its environmental adaptability and biological survivability in salt stress conditions.

6.2 Habitat and Environmental Adaptation

Microbial adaptation has played a major role in the existence of Halophiles, especially when bacteria in all forms and shapes branched out into different environmental niches (Rajendran et al. 2008), when they were evolving from the hypersaline primordial sea (Dundas 1998). High salty salterns or lakes represent an exceptional ecosystem in which several bacteria, especially halotolerants, can survive under the high salt pressure. Halophiles can thrive not only in those small salt water bodies such as salterns and lakes but also in large oceans and benthic floors of the sea. Halophilic bacteria are some of the best examples of environmental adaptation. With a heavy salt presence in the surrounding area, these microbes have developed metabolic conditions favorable to their survival. Some may be found within fluid inclusions in salt crystals (Norton and Grant 1988) and others produce compatible solutes to adapt high salt-stress conditions. The requirement for sodium chloride plays a major role in the metabolism of extremely halophilic bacteria, but they also require magnesium. As the growth of halophiles is predominantly dictated by the concentration of salt in the water, most moderate and slight halophilic bacteria do not require magnesium (Grant et al. 2001).

Halophilic bacteria are more abundant than any other groups in high salt conditions such as the Dead Sea (Bodaker et al. 2010), saline lakes in Inner Mongolia (Pagaling et al. 2009), African soda lakes, and deep-sea brines (van der Wielen et al. 2005) etc. The water salinity (Fig. 6.2) for fresh water is measured as less than 0.05 %, brackish water is 0.05–3 %, saline water is 3–5 % and Brine is more than 5 % (Anati 1999). Halophilic bacteria multiply better at the temperature of 28–37 °C with a pH range of 7.0–8.0 at 5–20 % sodium chloride concentration. Based on their salinity habitat and extent of their halotolerance, they were grouped into extremely (15–32 % w/v), moderately (3–15 % w/v), and slightly (1–3 % w/v) halophilic (Oren 2013). The slight halophiles generally prefer 0.3–0.8 M NaCl (Sodium chloride 1.8–4.7 % – seawater is 0.6 M or 3.5 %), moderate halophiles 0.8–3.4 M (4.7–20 %), and extreme halophiles 3.4–5.1 M (20–30 %). Some of the regular

Salty environment and examples		Salinity in % and ppt	
	**Fresh Water:** Lakes, River, Streams, Ponds	Less than 0.5 %	Less than 0.5 ppt
	**Brackish Water:** Estuaries, Mangrove lakes, Brackish swamps	0.5 – 3 %	0.5 - 30 ppt
	**Saline Water:** Ocean, Sea, Salt lakes	3 – 5 %	30 - 50 ppt
	**Brine Water:** Benthic sites, Brine pools	More than 5 %	More than 50 ppt

Fig. 6.2 Four different saline aqueous environments such as Freshwater, Brackish water, Saline water and Brine water were pictured with examples. Their salinity contents were expressed in percentage as well as in the traditional parts per thousand units

halophiles are halotolerant, meaning they do not require high NaCl condition, but can grow under saline conditions (Bowers et al. 2009).

Halophiles can be aerobic or anaerobic (Ollivier et al. 1994). Their cellular uniqueness, adaptation and survivability are dictated by the phylogenetic dexterity at given environmental constraints. For example, the osmotic stress in the surrounding space makes the saline water enter into the cytoplasm through the membrane and non-halophiles may not survive under this stress. However, halophiles and halotolerant microbes maintain turgor pressure at least equal to the ambient pressure (Mustkhimov et al. 2010). In that case, anaerobic halophilic bacteria *Haloanaerobiales*, acetogenic anaerobes such as *Halobacterioides*, *Sporohalobacter*, and *Acetohalobium* can all thrive well in those high salt water bodies. Those that maintain aerobic life are also able to tolerate low salt pressure like Halobacteriaceae, and depend on other forms of nutrients and become phototrophic, fermentative, sulfate-reducing, homoacetogenic, methanogens etc. (Oren 2002a). Some show heterotrophic, phototrophic, photosynthetic, methenogenic or litotrophic nature. Analysis of microbial diversity in lakes and salterns using metagenomic studies (Bodaker et al. 2010) are useful in order to show biological adaptability and nutritional habits.

There are many hypersaline environments found around the globe and different types of halophilic bacteria survive in different salt stresses as presented to them by the environment (Gunde-Cimerman et al. 2005). Within each ecosystem, well-adapted microfloras flourish and in some cases possess unique resident microbiota within themselves. Some live conveniently on the surfaces of large marine ecosystems and some in the far more complex saline systems including the various trophic levels of ocean, deep sea, mangrove saline sites, arid, islands and coastal area, underground salt mines and saline caves. Some reservoirs of natural salt sources such as the Great Salt Lake (Utah), Owens Lake (CA), halite cores from Saline Valley (CA), Thane Papke (Storrs, CT), deep-brines of the Red Sea (Antunes et al. 2003), Xinjiang salt lake, Lake Assal in Djibouti (French Somaliland), Chinese salt mine, Goa salterns in India, salterns of Israel, the Salar de Atacama, Chile, Turkish salt mine, Iranian salt lakes hypersaline lake in Argentina, La Malá saltern near Granada, salterns of Spain, Western Australian hypersaline lake (Lake O'Grady North), Quidam Basin Quaternary sediments, Gabara in the Wadi Natrun, Egypt, brine wells in southwestern China, The South China Sea, a Korean salt flat, Venere Lake, Pantelleria Island Italy, the Yellow Sea, Mongolian salt and soda lakes and south Siberian hypersaline lakes, Mexican soda environments, salted hides), benthic floor of the Mediterranean Sea, the Gulf of Mexico, and soda lakes of the Kulunda Steppe (Sorokin et al. 2010) are ecologically significant as they are considered rich in halophiles.

6.3 Cellular Adaptation and Biological Survivability

Marine microorganisms and halophiles in particular play a major role in the atmospheric oxygen stability. Halophiles perform a recycling function similar to other diverse non-halophiles (Kastritis et al. 2007) in spite of their salt-stress cellular adaptations. The higher rate of oxygen production comes from a significant marine bacterium, *Cyanobacterium*, which predominantly occupies the surface of global saltwater. This individual species, *C. prochlorococcus* is very diverse due to its adaptability and shows 96 variations of strains sampled from southeast of Bermuda in the Atlantic Ocean. Such distinct variable markers within a single species indicates that this species would have originated millions of years ago and diverted into ecologically distinct groups because of the genomic diversity with hypervariable genes (Kashtan 2014). Others, like *Haloferax*, *Haloarcula*, *Haloquadratum*, *Halobacteriales* and many more can survive extremely high salt concentrations and also produce a high rate of oxygen.

Many halophiles cannot show optimal growth if moved from hypersaline to less-saline habitats (Mustkhimov et al. 2010). During the course of evolutionary development in several millions of years, the enzymes and other biomolecules of halotolerant microbes have been modified to function efficiently at high intracellular salt concentrations. Hence they may die from immediate lysis of cell if the salt conditions are changed drastically. For example, in *Haloarcula marismortui*, the behavior of membrane proteins is highly active in high-salt environments in support of ribosome and molecular adaptations of malate dehydrogenase and other enzymes (Madern et al. 2000). The acidic surface of the macromolecule allows protein-salt interactions that avoid water or salt enrichment at the surface of the protein and preserve its solubility (Ebel and Zaccai 2004). In a situation where the concentration of salinity increases, the number of diversified groups of halotolerant bacteria living in the vicinity of such hypersalinity conditions decreases. Since high salinity represents an extreme condition, relatively few microbes can survive in those conditions (Gunde-Cimmerman et al. 2005). Even if the rate of salinity is inconsistent in the aqueous environment, they can often survive due to their defensive mechanisms in order to prevent a desiccation type of dehydration.

The osmosis process is the most common method of preventing water loss among halophiles in a hypersaline stress condition (Mustkhimove et al. 2010). In this way, organic compounds known as osmoprotectants or compatible solutes such as amino acids, sugars, ectoines, betaines etc. are accumulated in the cytoplasm to prevent the desiccation. Some groups of halophiles such as *Salinibacter ruber* of the *Halanerobiales* group adopt another mechanism, using an influx of potassium ions into cytoplasm. Both mechanisms increase the internal osmo-regularity condition of the microbial cell. Such adaptation of cells helps most halophiles to sustain their viability in their native high salt conditions. If they were removed and left in a freshwater environment, their cells often burst and may not survive due to the change of osmotic pressure. The osmoregulatory mechanism with the help of osmoprotectant or compatible solutes is the most adapted cellular mechanism from an evolutionary point of view. Compared to the mechanism of K^+ ion influx into the cytoplasm, the compatible solute mechanism helps halophiles gain more ATPs for other uses. More structural protein molecules are needed in the mechanism of influx of K^+ ions into the cytoplasm. Since the compatible solutes often act as high salt stress protectants, most halophiles follow this method of osmoregulation (Santos et al. 2002).

6.4 Pathways of Regulation and Evolutionary Adaptability

The ubiquitous nature of halophiles in the salty environment was acquired by microbes from millions of years of making. It occurred due to the ever-changing environmental conditions due to the climate variations and is dictated by the regulation of the bacterial cell membrane like osmoregulation in hypersaline natural habitats. Some mechanisms behind osmoregulation in halophiles are extremely energetic (Saum and Müller 2008), and adaptation of enzymes during the course of their evolutionary development, made some differences in environmental diversity (McGenity et al. 2000), and helped halophilic cells to develop a metabolic variation. In order to adapt to stress and damage to proteins, halophiles possess more acidic residues especially glutamate (Kuntz 1971). Glutamate has a water binding nature higher than any other amino acid (Saenger 1987). Such evolutionary development helps halophiles to adapt to the salt stress by binding more compactly with water molecules (Lanyi 1974; Bolhuis et al. 2008). Beyond this, extreme halophiles have increased the number of acidic glutamic and aspartic acids (Kennedy et al. 2001), and charged amino acids (Fukuchi et al. 2003) on the surface of active proteins (Tadeo et al. 2009). Since glutamate residues have a superior water binding capacity over all other amino acids, they are generally found in excess on the surface of halophilic proteins.

The catabolic ability of a microbe depends on the environment (Hough and Danson 1999) where it tries to survive, by overcoming the chemical rejections or autointoxication. For example, *Thermus aquaticus* is a bacterium that lives in hot springs and hydrothermal vents, and Taq polymerase was identified (Chien et al. 1976) as an enzyme able to withstand the protein-denaturing conditions (high temperature) required during PCR (Saiki 1988). In salty conditions extremophiles develop evolutionary variations to have specific hyperenzymatic reactions against given substrates such as highly concentrated sodium chloride. Such ability leads extremophiles to competitively survive better than any other group of microbes in a saltwater body, such as the ocean. Since one third of the biosphere is covered with salty water, it is obvious that halophiles are a dominant group (Gomes and Steiner 2004). Protein hydration needs free water molecules, but in the ocean, water molecules are sequestered in hydrated ionic structures (Danson and Hough 1997). The evolutionary adaptation of halophilic cells to high concentrated salt condition depends on the ability of the catalytic protein dehydrogenase (Britton et al. 2006) or synthetase which they harbor. Such enzymes provide a cover to halophiles or an ability to adapt the influx of salt to regulate the osmotic pressure (DasSarma et al. 2010). The three dimensional structural features of the family of halophilic proteins (Frolow et al. 1996), and their functional stabilities against salt stress supports their survivability (Dym et al. 1995)

6.5 Exploitation of Halophilic Bacteria

6.5.1 Secondary Metabolic Products

Moderate halophiles accumulate high cytoplasmic concentrations of organic compounds to cope with the osmotic stress and to maintain positive turgor pressure (Graf et al. 2008). The natural ability of halophiles to produce and accumulate high concentrations of these low-molecular-weight compounds makes moderate halophiles useful for the biotechnological production of these osmolytes (Santos and da Costa 2002) Some these compatible solutes, especially glycine, betaine and ectoines have gained considerable attention in the recent past. They are good stabilizers of enzymes, nucleic acids, membranes and whole cells (Louis et al. 1994). They are used as stress protectants against high salinity, thermal denaturation, desiccation, and freezing. The industrial uses of these compounds in enzyme technology are used in biosensors and PCR technologies as well as in pharmaceuticals and cosmetics (Vendosa 1995). Salt labile enzymes such as lactate dehydrogenase and phosphofructokinase can be protected by using several compatible solutes such as betaine, trehalose, glycerol, glycine proline, ectoines and hydroxyectoine (Louis et al. 1994). Ectoine is as an active ingredient for many epidermal applications. This medically exploitable product commercially manufactured by a German company BitopAG from *Halomonas elongata* is one the most successful bacterial resources for Ectoine (Graf et al. 2008). Currently ectoine becomes part of compounding industries in medicinal product preparations (Graf et al. 2008). The derivatives of ectoine are used in Microarray and PCR technology as biomolecular protectants (Schnoor et al. 2004). Many of them were used as building blocks for some pharmaceuticals. Novel antibiotics were isolated from Halophilic Actinomycetes. *Nocardiopsis*, *Saccharomonospora*, and *Streptomonospora* were also tested in the context of cytotoxic effect against a range of cancer cell lines.

Some commercially significant products are obtained from halophiles such as β-carotene from *Dunaliella*, bacteriorhodopsin from *Halobacterium*, and ectoine from *Halomonas* (Oren 2010). For example, *Halomonas elongata* is used to produce ectoine, the active ingredient of many cosmetics and skin care products. Ectoine and hydroxyectoine biosynthesis is widely found in halophilic and halotolerant microorganisms, and can only be produced biologically. The production of ectoine for industrial purposes by using *H. halodenitrificans* is known since 1996. The halophilic cells were grown in an anaerobic fed-batch fermentation process, in a synthetic medium at high concentration with glycerol as the carbon source. Once the cell density reached a certain threshold, ectoine was extracted by "bacterial milking", a process previously developed for *Holomonas elongata*, in which an osmotic down-shock applied from 10 to 2 % NaCl, results in excretion of about 80 % of the intracellular ectoine to the surrounding medium (Van-Thuoc et al. 2010). Subsequent exposure of the cells to a hyperosmotic shock from 2 to 10 % NaCl restored the original level of ectoine in a period of 10 h. Thus, a yield of 2 g ectoine per liter of medium per day was obtained. Ectoine and hydroxyectoine can also be produced by the Gram-positive moderate halophiles like *Marinococcus* strain M52. A maximum yield of 35.3 g/l cell dry weight was achieved as reported (Frings et al. 1995). The use of a dialysis reactor, in which cells were grown in an inner chamber fed with fish peptone and glucose, resulted in a dramatic increase in yield up to 132 g of cell dry weight with about 20 % hydroxyectoine obtained (Krahe et al. 1996). Another organism used for the production of ectoine is the actinomycetes *Nocardiopsis lucentensis* A5-1 (Yassin et al. 1993), isolated from a saline soil near Alicante, Spain (Lippert and Galinski 1992).

Halophiles, while surviving under high salt stress, can also produce secondary metabolites like poly-β-hydroxyalkanoates or bioplastics, halophilic enzymes, and biofuel. Some halophilic bacteria such as *Halomonas boliviensis* and members of genus *Haloferax* produce poly-β-hydroxyalkanoates commonly known as bioplastic (Quillaguaman et al. 2010). Many useful enzymes obtained commercially from halophilic microbes such as *Salinibacter* are applied in the detergent and the textile industries. Some enzymes such as proteases, cellulases, lipases, amylases, and mannanases are produced from various halophiles (Min et al. 1993) including but not limited to the group of Haloarchaea. A novel halophilic dehydrogenase was created using site directed mutagenesis from *Halobacterium salinarum* glutamate dehydrogenase (GDH). Due to their high salt and heat tolerance ability, some of them such as *Haloferax volcanii* (Large et al. 2007), are used as industrial biocatalysts and production of alcohol from ketones. Some halophiles like *Alcanivorax dieselolei*, *Marinobacter* and *Halomonas* sp., are able to degrade various polyaromatic hydrocarbons (Fig. 6.2). Another process in which halophiles may contribute in the future is the production of biofuel. Some halophilic bacteria have shown that they can break down cellulosic biomass via fermentation. The yield of ethanol and hydrogen could be a good source of biofuel, which opens up a new line of commercial uses for halophiles.

In recent years the poly-β-hydroxyalkanoate-based bioplastics has stimulated higher expectations for biotechnologists as ingredients for many house-hold products. Such biodegradable polymers are produced from *Halomonas boliviensis* (Quillaguaman et al. 2010). A successful attempt was made by mixing the biopolymer from *H. boliviensis* and the compatible solute ectoine from *H. elongata* with hydroxyectoine to gain a synthetic high-value product (Van-Thuoc 2010). A commercial halophilic strain of *Haloferax* was used to produce Tirel, a bioplastic product by Metabolix. The genus *Haloferax* (*H. volcanii* and *H. mediterranei*) was tested (Han et al. 2009) for its polyhydroxyalkanoate (PHA), and *Haloarcula hispanica* was also compared (Lu et al. 2008) for its poly (3-hydroxybutyrate-co-3-hydroxyvalerate) synthase.

6.5.2 Halophilic Remediation

Biological remediation is well known for its ability to take up and concentrate contaminants in the biological tissues without destroying the environment. Microbial remediation in particular has been developed from a conceptual methodology to a viable technology for contaminant clean-up (Rajendran et al. 2013). The potential applications of halophiles as remediators of toxic chemicals and biological wastes were studied well in the past (Ventosa and Nieto 1995). The halophiles are relatively less exploited and applied in the commercial sectors, compared to other extremophiles with the notable exceptions of β-carotene, bacteriorhodopsin, and ectoine (Oren 2010). Some of those commercially successful products from halophiles are employed as surfactants in olive oil industries to recover hypersaline waste brines, and some were used in fur-curing leather industries. The halophilic bacteria can be used to remove phosphate from saline environments and it is known to be a cost-effective method (Ramos-Coremzama et al. 1991).

Many toxicants persist hundreds of years in large and small bodies of water such as seas and lakes not only because of natural disasters but also due to man-made wastes. These wastes can arise from smelter and mining sites, manufactured gas plants, ammunition factories, run-over contaminants from fertilized farmland, sludge from agriculture, and pollutants from industrial and municipal waste-dumping grounds. These wastes are now reaching the oceans (Rajendran et al. 2008).

Many chemical contaminants, especially hydrocarbon-based petroleum contaminants, are very problematic for the environment (Schwab et al. 1999). Remediation of such contaminants from the land and aqueous sites, using non halophiles, is a very difficult and slow process. Only 25 % of petroleum-contaminated land sites were being bioremediated using *Pseudomonas* (Holden et al. 2002). Beyond this, using those microbes in high salt environments is difficult due the effect of salt on the cell membrane and for other metabolic reasons. The application of halophiles as remediator in petroleum and other contaminants land and aqueous environment will have an impact on cleaning up our chemical toxic sites and will become increasingly important in the cleanup of aquifers and large plumes.

In halophilic remediation process, toxic heavy metals and organic pollutants are the major targets for a potential remediation (Roberts 1987). Many chemicals like polycyclic aromatic hydrocarbon (PAH) cause many health problems (Pradhan et al. 1998). Enhancement of PAH dissipation in soil and aqueous environment is often suggested to be a result of human impact. For example, PAHs is one of the dominant contaminants found in the superfund sites and resources conservation and recovery act (RCRA) sites (Kevein et al. 2006). One of the methods that can be effectively applied for cleaning-up many toxicants like PAHs is using microbes. Other major toxic chemicals and its species are Polychlorinated biphenyls, Chlorinated benzoic acid, Mercury (Hg), Hexachlorobiphenyl, 2,4,6-trinitrotoluene (TNT), Naphthalene, Pyrene, Chloroacetamide herbicides, Benzo(a)pyrene, 3,3′4,4-tetra chloroazobenzene (TCAB), Arsenic (As), metals such as Nickel (Ni), Zinc (Zn), Copper (Cu), Rubidium (Rb), Cesium (Cs), Manganese (Mn), Iron (Fe), Selenium (Se), Chromium (Cr), Cadmium (Cd), Lead (Pb), or radioactive isotopes such as Uranium, Cesium-137, Strontium or Cobalt (Rajendran et al. 2013).

Microbial species uptake, metabolize and accumulate toxic chemicals and metals. Environmental cleaning of such chemical contaminants by bacteria is varied from bacterial species to species, and based on the chemical contaminant in question. Some species act as powerful remediators of the primary chemical contaminants but others remediate only the chemical species of the primary contaminants. For example, uptake of Selenium, Trichloroethylene (TCE) Tetrachloroethylene (PCE) and its speciation such as Trichloroethanol, and Trichloroacetic acid as well as Nitroaromatic compounds such as Aminodinitrotoluene (ADNT), Diaminonitrotoluene (DANT), Hexahydro-1,3,5-trinitro-1,3,5-triazine (RDX) and Octahydro-1,3,5,7-tetranitro-1,3,5,7,-tetraazocine (HMX), Brominated compounds such as Ethylene dibromide, 2,4,5-Trichlorophenoxyacitic acid (2,4,5-T), Carbon Tetrachloride and Tetrachloroethlene, and Dibromochloropropane as well as nonhalogenated compounds such as methyl-t-butyl ether (MTBE) are significantly present in many contaminated sites and aquifers (Rajendran et al. 2013).

Many bacteria can be directly used to remediate toxic wastes including PCBs (Roberts 1987). As the result, the accumulated biomass, similar to that of "commercial ore" can be recycled. It can also be decomposed into manure or disposed-off into a traditional landfill (Rajendran et al. 2008). The association of halophilic microbes and toxic contaminants (Table 6.1) therefore offer a viable means of accomplishing the remediation of contaminated aquifer and other sites

Table 6.1 Halophilic remediators mentioned here are associated with remediation of benzene, toluene, phenolics and various other chemical toxicant groups such as aliphatic hydrocarbon (Example: Octane, Pristane), polycyclic aromatic hydrocarbons (Example: Naphthalene, Anthracene), as given

Toxicants*	Remediator-halophiles	
Acenaphthene	*Actinopolyspora* sp. DPD1	*Halomonas* sp. strain C2SS100
Anthracene	*Alcanivorax* sp. Otet3	*Halomonas* sp. strain IMPC
Benzene	*Arhodomonas* sp. strain Rozel	*Halorubrum ezzemoufense*,
Benzoate	*Arhodomonas* sp strain Seminole	*Marinobacter aquaeolei*
Biphenyl	*Atcanivorax* sp. HA03	*Marinobacter falvimaris Marinobacter hydrocarbonoclasticus*
Cinnamic acid	*Candida tropical*	*Marinobacter lipolyticus Marinobacter nanhaiticus*
Eicosane	*Chromohalobacrer* sp. strain HS-2	*Marinobacter sedimentalis*
Ethylbenzene	*Haloarcula hispanica*	*Marinobacter vinifirmus*,
Fluorene	*Haloarcula vallismortis*	*Modicisalibacter tunisiensis*
Heneicosane	*Haloarcula* sp. strain 01	*Planococcus* sp. strain ZD22
Heptadecane	*Halobacterium piscisalsi*,	*Pseudomonas* sp. C-450R
4-Hydroxybenzoate	*Halobacterium salinarium*,	*Salinicoccus roseus*
Gentisate	*Haloferax* sp.01227	*Thelassobacillus devorans*
Naphthalene	*Halomonas alimentaria*	
Octadecane	*Halomonas campisalis*	
Pentacosane	*Halomonas elongata*	
Phenanthrene	*Halomonas eurihalina*	
Phenol	*Halomonas glaciei*	
Phenyl propionic acid	*Halomonas halodurans*	
Phthalate	*Halomonas halophil*	
Phytane	*Halomonas organivorans*	
Pristane	*Halomonas salina*	
Pyrene	*Halomonas venusta*	
Salicylate		
Tetracosane		
Toluene		
Xylene		

The references for respective halophilic remediator and its target hydrocarbon contaminant were given adequately by Fathepure (2014)

like deep hypersaline anoxic basins (van der Wielen 2005). Such bioremediation processes have many advantages over physical and chemical means of remediation at the open surface of ocean and other large bodies of water such as lakes. They are environmentally safe (Philip et al. 2005), biologically feasible, and economically cheaper to remove contaminants from the crude oil wastes and oil spillage or a variety of other contaminants such as pesticides, solvents. At high risk sites of contaminants and benthic zones, halophilic remediation can be used as supportive method to remove contaminants (Ramos-Cormenzana 1991). Because of the ability of specific halophiles which can reach specific depths and clean-up the last remains of contaminants trapped in the benthic floor halophilic remediation could become a truly viable option.

6.6 Future Application of Halophiles

Problems of chemical contamination and its salvage will continue because of the continuous use of the metals, chemicals and its derivatives in our everyday life in one way or other. The tremendous use of chemicals today will have a profound effect on the environment tomorrow. Due to the synergistic chemical effect on public health (Barbosa et al. 1998), which poses a significant threat to our economy, application of halophilic microbes or its products will dictate the alternative approaches to solve such emerging risks. The other future application of halophiles is the use of halo-tolerant bacteria in breakdown of biomass materials to gain the energy in the form of ethanol as biofuel (Tango and Islam 2002). The current progress towards achieving an outcome-based approach as studied by many marine microbiologists to replace fresh-water algae expected to reduce high consumption of fresh water (as in the case of fresh-water algae) and to speed up the entire biofuel conversion process (Fathepure 2014). Diverting the established algal conversion technologies and accessory approaches towards halophilic bacterial biofuel conversion will not only reduce energy consumption for quick recovery of products but also offer a cost-effective way for making commercial products. The commercial success of ectoine (Bestvater et al. 2008) achieved through the biotechnology and genetic engineering synthesis has opened an array of products from halophiles to be commercialized using fermentation technology. Cloning of ectoine genes made it possible so that 1 day salt tolerance genes could be cloned in stable foods such wheat to make them grow in salty soils or use sea water to grow rice in paddy fields (Min-Yu et al. 1993).

6.7 Conclusion

Halophilic bacteria and their abilities to survive in the extreme salt condition have stretched our understanding not only of how metabolic processes played crucial roles in survivability but also in the intermediate processes involved in the evolutionary adaptability. The current proteomic and genomic studies on the stability, structural and functional analysis aided by modern molecular and genetic engineering approaches paved a way for scientists to develop a family of enzymes and products from Halophiles for commercial purposes. However, how halophiles maintain high metabolic similarities to other non-hypersaline bacteria, but show different survivability under heavy salt stress, remains an important question to be further studied.

References

Anati DA (1999) The salinity of hypersaline brines: concepts and misconceptions. Int J Salt Lake Res 8:55–70

Angert ER, Clements KD, Pace NR (1993) The largest bacterium. Nature 362(6417):239–241

Antunes A, Eder W, Fareleira P, Santos H, Huber R (2003) Salinisphaera shabanensis gen. nov., sp. nov., a novel, moderately halophilic bacterium from the brine-seawater interface of the Shaban Deep, Red Sea. Extremophiles 7:29–34

Bakke P, Carney N, Deloache W, Gearing M, Ingvorsen K, Lotz M, McNair J, Penumetcha P, Simpson S, Voss L, Win M, Heyer LJ, Campbell AM (2009) Evaluation of three automated genome annotations for *Halorhabdus utahensis*. PLoS One 4:e6291

Baliga NS, Bonneau R, Facciotti MT, Pan M, Glusman G, Deutsch EW, Shannon P, Chiu Y, Weng RS, Gan RR, Hung P, Date SV, Marcotte E, Hood L, Ng WV (2004) Genome sequence of *Haloarcula marismortui*: a halophilic archaeon from the Dead Sea. Genome Res 14:2221–2234

Barbosa AC, Silva SR, Dorea JG (1998) Concentration of mercury in hair of indigenous mothers and infants from the Amazon basin. Arch Environ Contam Toxicol 34.100–105

Bestvater T, Louis P, Galinski EA (2008) Heterologous ectoine production in *Escherichia coli*: by-passing the metabolic bottle-neck. Saline Syst 4:12

Bodaker I, Sharon I, Suzuki MT, Reingersch R, Shmoish M, Andreishcheva E, Sogin ML, Rosenberg M, Belkin S, Oren A, Béjà O (2010) The dying Dead Sea: comparative community genomics in an increasingly extreme environment. ISME J 4:399–407

Bolhuis H, Palm P, Wende A, Farb M, Rampp M, Rodriguez-Valera F, Pfeiffer F, Oesterhelt D (2006) The genome of the square archae on *Haloquadratum walsbyi*: life at the limits of water activity. BMC Genomics 7:169

Bolhuis A, Kwan D, Thomas TR (2008) Halophilic adaptations of proteins. In: Siddiqui KS, Thomas T (eds) Protein adaptation in extremophiles. Nova Science, New York, pp 71–104

Bowers KJ, Mesbah NM, Wiegel J (2009) Biodiversity of poly-extremophilic bacteria: does combining the extremes of high salt, alkaline pH and elevated temperature approach a physico-chemical boundary for life? Saline Syst 5:9

Britton KL, Baker PJ, Fisher M, Ruzheinikov S, Gilmour DJ, Bonete MJ, Ferrer J, Pire C, Esclapez J, Rice DW (2006) Analysis of protein solvent interactions in glucose dehydrogenase from the extreme halophile *Haloferax mediterranei*. Proc Natl Acad Sci U S A 2006(103): 4846–4851

Chien A, Edgar DB, Trela JM (1976) Deoxyribonucleic acid polymerase from the extreme thermophile Thermus aquaticus. J Bacteriol 127(3):1550–1557

Danson MJ, Hough DW (1997) The structural basis of protein halophilicity. Comp Biochem Physiol A Physiol 117:307–312

DasSarma P, Coker JA, Huse V, DasSarma S (2010) Halophiles. Biotechnology. In: Flickinger MC (ed) Encyclopedia of industrial biotechnology, bioprocess, bioseparation, and cell technology. Wiley, Hoboken, pp 2769–2777

Dundas I (1998) Was the environment for primordial life hypersaline? Extremophiles 2:375–377

Dym O, Mevarech M, Sussman JL (1995) Structural features that stabilize halophilic malate dehydrogenase from an archaebacterium. Science 267:1344–1346

Ebel C, Zaccai G (2004) Crowding in extremophiles: linkage between solvation and weak protein-protein interactions, stability and dynamics provides insight into molecular adaptation. J Mol Recognit 17:382–389

Falb M, Pfeiffer F, Palm P, Rodewald K, Hickmann V, Tittor J, Oesterhelt D (2005) Living with two extremes: conclusions from the genome sequence of *Natronomonas pharaonis*. Genome Res 15:1336–1343

Fathepure ZB (2014) Recent studies in microbial degradation of petroleum hydrocarbons in hypersaline environment. Front Microbiol 5:173

Frings E, Sauer T, Galinski EA (1995) Production of hydroxyectoine: high cell-density cultivation and osmotic down shock of *Marinococcus* strain M52. J Biotechnol 43:53–61
Frolow F, Harel M, Sussman JL, Mevarech M, Shoham M (1996) Insights into protein adaptation to a saturated salt environment from the crystal structure of a halophilic 2Fe-2S ferredoxin. Nat Struct Biol 3:452–458
Fukuchi S, Yoshimune K, Wakayama M, Moriguchi M, Nishikawa K (2003) Unique amino acid composition of proteins in halophilic bacteria. J Mol Biol 327:347–357
Gomes J, Steiner W (2004) The biocatalytic potential of extremophiles and extremozymes. Food Technol Biotechnol 42:223–235
Graf R, Anzali S, Bünger J, Pflücker F, Driller H (2008) The multifunctional role of ectoine as a natural cell protectant. Clin Dermatol 26:326–333
Gunde-Cimerman N, Oren A, Plemenitas A (eds) (2005) Adaptation to life at high salt concentrations in archaea, bacteria, and eukarya. Springer, Dordrecht
Han J, Lu Q, Zhou L, Liu H, Xiang H (2009) Identification of the polyhydroxyalkanoate (PHA)-specific acetoacetyl coenzyme A reductase among multiple FabG paralogs in *Haloarcula hispanica* and reconstruction of the PHA biosynthetic pathway in *Haloferax volcanii*. Appl Environ Microbiol 75:6168–6175
Hartman AL, Norais C, Badger JH, Delmas S, Haldenby S, Madupu R, Robinson J, Khouri H, Ren Q, Lowe TM, Maupin-Furlow J, Pohlschröder M, Daniels C, Pfeiffer F, Allers T, Eisen JA (2010) The complete genome sequence of *Haloferax volcanii* DS2, a model archaeon. PLoS One 5:e9605
He Z, Zhou A, Baidoo E, He Q, Joachimiak MP, Benke P, Phan R, Mukhopadhyay A, Hemme CL, Huang K, Alm EJ, Fields MW, Wall J, Stahl D, Hazen TC, Keasling JD, Arkin AP, Zhou J (2010) Global transcriptional, physiological, and metabolite analyses of the responses of *Desulfovibrio vulgaris* Hildenborough to salt adaptation. Appl Environ Microbiol 76:1574–1586
Holden PA, LaMontagne MG, Bruce AK, Miller WG, Lindow SE (2002) Assessing the role of *Pseudomonas aeruginosa* surface-active gene expression in hexadecane biodegradation in sand. Appl Environ Microbiol 68:2509–2518
Hough DW, Danson MJ (1999) Extremozymes. Curr Opin Chem Biol 1:39–46
Kashtan N (2014) Single-cell genomics reveals hundreds of coexisting subpopulations in wild *Prochlorococcus*. Science 344:416–420
Kastritis PL, Papandreou NC, Hamodrakas SJ (2007) Haloadaptation: insights from comparative modeling studies of halophilic archaeal DHFRs. Int J Biol Macromol 41(4):447–453
Kennedy SP, Ng WV, Salzberg SL, Hood L, DasSarma S (2001) Understanding the adaptation of *Halobacterium* species NRC-1 to its extreme environment through computational analysis of its genome sequence. Genome Res 11:1641–1650
Krahe M, Antranikian G, Markl H (1996) Fermentation of extremophilic microorganisms. FEMS Microbiol Rev 18:271–285
Kuntz ID (1971) Hydration of macromolecules IV: polypeptide conformation in frozen solutions. J Am Chem Soc 93:516–518
Lanyi JK (1974) Salt dependent properties of proteins from extremely halophilic bacteria. Bacteriol Rev 38:272–290
Large A, Stamme C, Lange C, Duan Z, Allers T, Soppa J, Lund PA (2007) Characterization of a tightly controlled promoter of the halophilic *Haloferax volcanii* and its use in the analysis of the essential *cct1* gene. Mol Microbiol 66:1092–1106
Lippert K, Galinski EA (1992) Enzyme stabilization by ectoine-type compatible solutes: protection against heating, freezing and drying. Appl Microbiol Biotechnol 37:61–65
Louis P, Trüper HG, Galinski EA (1994) Survival of *Escherichia coli* during drying and storage in the presence of compatible solutes. Appl Microbiol Biotechnol 41:648–688
Lu Q, Han J, Zhou L, Zhou J, Xiang H (2008) Genetic and biochemical characterization of the poly(3-hydroxybutyrate-co-3-hydroxyvalerate) synthase in *Haloferax mediterranei*. J Bacteriol 190:4173–4180
Madern D, Ebel C, Zaccai G (2000) Halophilic adaptation of enzymes. Extremophiles 4:91–98

Malfatti S, Tindall BJ, Schneider S, Faehnrich R, Lapidus A, LaButti K, Copeland A, Glavina Del Rio T, Nolan M, Chen F, Lucas S, Tice H, Cheng J-F, Bruce D, Goodwin L, Pitluck S, Anderson IJ, Pati A, Ivanova N, Mavromatis K, Chen A, Palaniappan K, D'Haeseleer P, Goeker M, Bristow J, Eisen JA, Markowitz V, Hugenholtz P, Kyrpides NC, Klenk H-P, Chain P (2009) Complete genome sequence of *Halogeometricum borinquense* type strain (PR3T). Stand Genomic Sci 1:150–158

McGenity TJ, Gemmell RT, Grant WD, Stan-Lotter H (2000) Origins of halophilic microorganisms in ancient salt deposits. Environ Microbiol 2:243–250

Min-Yu L, Ono H, Takano M (1993) Gene cloning of ectoine synthase from *Halomonas* sp. Annu Rep Int Cent Coop Res Biotechnol Jpn 16:193–200

Mueller EK, Shann JR (2006) PAH dissipation in spiked soil: impact of bioavailability, microbial activity, and trees. Chemosphere 64:1006–1014

Mustakhimov I, Reshetnikov I, Khmelenina AS, Trotsenko VN, Yu A (2010) Regulatory aspects of Ectoine biosynthesis in Halophilic bacteria. Microbiology 79:583–592

Ng WV, Kennedy SP, Mahairas GG, Berquist B, Pan M, Shukla HD, Lasky SR, Baliga NS, Thorsson V, Sbrogna J, Swartzell S, Weir D, Hall J, Dahl TA, Welti R, Goo YA, Leithausen B, Keller K, Cruz R, Danson MJ, Hough DW, Maddocks DG, Jablonski PE, Krebs MP, Angevine BM, Dale H, Isenbarger TA, Peck RF, Pohlschröder M, Spudich JL, Jung K-II, Alam M, Freitas T, Hou S, Daniels CJ, Dennis PP, Omer AD, Ebhardt E, Lowe TM, Liang P, Riley M, Hood L, DasSarma S (2000) Genome sequence of *Halobacterium* species NRC-1. Proc Natl Acad Sci U S A 97:12176–12181

Norton CF, Grant WD (1988) Survival of halobacteria within fluid inclusions in salt crystals. J Gen Microbiol 134:1365–1373

Ollivier B, Caumette P, Garcia J-L, Mah R (1994) Anaerobic bacteria from hypersaline environments. Microbiol Rev 58(1):27–38

Oren A (2002a) Diversity of halophilic microorganisms: environments, phylogeny, physiology and applications. J Ind Microbiol Biotechnol 28:56–63

Oren A (2002b) Molecular ecology of extremely halophilic Archaea and Bacteria. FEMS Microbiol Ecol 39:1–7

Oren A (2010) Industrial and environmental applications of halophilic microorganisms. Environ Technol 31:825–834

Oren A (2013) Life at high salt concentrations. In: Rosenberg E, Delong EF, Lory S, Stackebrandt E, Thompson F (eds) The prokaryotes – prokaryotic communities and ecophysiology. Springer, Berlin, pp 421–440

Pagaling E, Wang HZ, Venables M, Wallace A, Grant WD, Cowan DA, Jones BE, Ma Y, Ventosa A, Heaphy S (2009) Microbial biogeography of six salt lakes in Inner Mongolia, China, and in Argentina. Appl Environ Microbiol 75:5750–5760

Philip JC, Bamforth SM, Singleton I, Atlas RM (2005) Environmental pollution and restoration: a role for bioremediation. In: Atlas RM, Philip J (eds) Bioremediation: applied microbial solutions for real-world environmental cleanup. ASM Press, Washington, DC, pp 1–48

Quillaguaman J, Guzman H, Van-Thuoc D, Hatti-Kaul R (2010) Synthesis and production of polyhydroxyalkanoates by halophiles: current potential and future prospects. Appl Microbiol Biotechnol 85:1687–1696

Rajendran N (2013) The state of the rhizoinhabitants in bridging the gap between plant productivity and persuasiveness during remediation. In: Maheshwari DK (ed) Bacteria in agronomy: crop productivity. Springer, Berlin, pp 463–496

Rajendran N, Saraf M, Tank N (2008) Rhizomediation in soil for sustainable agriculture. In: Maheshwari DK, Dubey RC (eds) Potential microorganisms for sustainable agriculture: a techno-commercial perspective, 19th edn. I. K. International Publishing House, New Delhi, pp 364–393

Ramos-Cormenzana A (1991) Halophilic organisms and the environment. In: Rodriguez-Valera F (ed) General and applied aspects of halophilic microorganisms. Plenum Press, New York, pp 15–24

Roberts L (1987) Discovering microbes with a taste for PCBs. Science 237:975–977

Rodriguez-Valera F (ed) (1991) General and applied aspects of halophilic microorganisms. Plenum Press, New York, pp 40–44

Saenger W (1987) Structure and dynamics of water surrounding biomolecules. Annu Rev Biophys Biophys Chem 16:93–114

Saiki RK (1988) Primer-directed enzymatic amplification of DNA with thermostable DNA polymerase. Science 239(4839):487–491

Santos H, da Costa MS (2002) Compatible solutes of organisms that live in hot saline environments. Environ Microbiol 4:501–509

Saum SH, Müller V (2008) Regulation of osmoadaptation in the moderate halophile *Halobacillus halophilus*: chloride, glutamate and switching osmolyte strategies. Saline Syst 4:4–6

Saunders E, Tindall BJ, Fahnrich R, Lapidus A, Copeland A, Glavina Del Rio T, Lucas S, Chen F, Tice H, Cheng J-F, Han C, Detter JC, Bruce D, Goodwin L, Chain P, Pitluck S, Pati A, Ivanova N, Mavromatis K, Chen A, Palaniappan K, Land M, Land M, Hauser l, Chang Y-J, Jeffries CD, Brettin T, Rohde M, Goker M, Bristow J, Eisen JA, Markowitz V, Hugenholtz P, Klenk H-P, Kyrpides NC (2009) Complete genome sequence of *Haloterrigena turkmenica* type strain ($4k^T$). Stand Genomic Sci 2:107–116

Schnoor M, Voß P, Cullen P, Böking T, Galla HJ, Galinski EA, Lorkowski S (2004) Characterization of the synthetic compatible solute homoectoine as a potent PCR enhancer. Biochem Biophys Res Commun 322:867–872

Schwab AP, Su J, Wetzel S, Pekarek S, Banks MK (1999) Extraction of petroleum hydrocarbons from soil by mechanical shaking. Environ Sci Technol 33:1940–1945

Sorokin DY, Rusanov II, Pimenov NV, Tourova TP, Abbas B, Muyzer G (2010) Sulfidogenesis at extremely haloalkaline conditions in soda lakes of Kulunda Steppe (Altai, Russia). FEMS Microbiol Ecol 73:278–290

Tadeo X, Lopez-Mendez B, Trigueros T, Lain A, Castano D, Millet O (2009) Structural basis for the amino acid composition of proteins from halophilic archaea. PLoS Biol 7

Tango MSA, Islam MR (2002) Potential for extremophiles for biotechnological and petroleum applications. Energy Sources 24:543–559

Tindall BJ, Schneider S, Lapidus A, Copeland A, Glavina del Rio T, Nolan M, Lucas S, Chen F, Tice H, Cheng J-F, Saunders E, Bruce D, Goodwin L, Pitluck S, Mikhailova N, Pati A, Mavromatis K, Ivanova N, Chen A, Palaniappan K, Chain P, Land M, Hauser L, Chang Y-J, Jeffries CD, Brettin T, Han C, Rohde M, Goeker M, Bristow J, Eisen JA, Markowitz V, Hugenholtz P, Klenk H-P, Kyrpides NC, Detter JC (2009) Complete genome sequence of *Halomicrobium mukohataei* type strain (arg-2^T). Stand Genomic Sci 1:270–277

van der Wielen PW, Bolhuis H, Borin S, Daffonchio D, Corselli C, Giuliano L, D'Auria G, de Lange GJ, Huebner A, Varnavas SP, Thomson J, Tamburini C, Marty D, McGenity TJ, Timmis KN, BioDeep Scientific Party (2005) The enigma of prokaryotic life in deep hypersaline anoxic basins. Nature 307:121–123

Van-Thuoc D, Guzman H, Quillaguaman J, Hatti-Kaul R (2010) High productivity of ectoines by *Halomonas boliviensis* using a combined two-step fed-batch culture and milking process. J Biotechnol 147:46–51

Ventosa A, Nieto JJ (1995) Biotechnological applications and potentialities of halophilic microorganisms. World J Microbiol Biotechnol 11:85–94

Yassin AF, Galinski EA, Wohlfarth A, Jahnke K-D, Schaal KP, Trüper HG (1993) A new actinomycete species, *Nocardiopsis lucentensis* sp. nov. Int J Syst Bacteriol 43:266–271

Chapter 7
Investigating the Quorum Sensing System in Halophilic Bacteria

Tommonaro Giuseppina, Abbamondi Gennaro Roberto, Toksoy Oner Ebru, and Nicolaus Barbara

Abstract Halophiles are organisms that inhabit hypersaline environments and require hypersaline conditions to grow (from 2–5 % NaCl up to above 20–30 % NaCl). Halophiles are mostly prokaryotes, which are able to thrive in such harsh conditions because of their capability to balance the osmotic pressure of the environment through different mechanisms. Bacteria mainly produce high concentrations of compatible solutes, small water-soluble organic molecules, which do not interfere with cell metabolism (amino acids and their derivatives, sugar and polyols). Archaea and some halophilic anaerobic bacteria accumulate high concentration of salts intracellularly (KCl). Halophilic bacteria produce biomolecules with great potential for biotechnology: exopolysaccharides, compatible solutes (which can be used as stabilizers of biomolecules) and stable enzymes (with potential use as biocatalysts). Quorum sensing (QS) could influence the production of these biomolecules, thus a better understanding of halophilic bacterial communication mechanisms can help to improve the yields of these biotechnological processes. QS contributes to environmental adaptation via different models, by inducing the expression of genes that are required for the well-being of the organisms in a given milieu. To date, different kinds of autoinducers were detected in halophilic microorganisms (diketopiperazines, N-Acyl homoserine lactones, autoinducer-2), but the role of QS in this group of extremophiles needs to be elucidated yet. This chapter presents the recent advances in our understanding of bacterial intercommunication systems in halophilic bacteria.

Keywords Quorum sensing • Halophiles • AHLs • Autoinducers • Quorum quenching • Extremophiles

---

T. Giuseppina (✉) • A.G. Roberto • N. Barbara
National Research Council of Italy, Institute of Biomolecular Chemistry, Pozzuoli, Naples, Italy
e-mail: gtommonaro@icb.cnr.it; gennaroroberto.abbamondi@icb.cnr.it; bnicolaus@icb.cnr.it

T.O. Ebru
Department of Bioengineering, Marmara University, Istanbul, Turkey
e-mail: ebru.toksoy@marmara.edu.tr

D.K. Maheshwari, M. Saraf (eds.), *Halophiles*, Sustainable Development and Biodiversity 6, DOI 10.1007/978-3-319-14595-2_7

7.1 Introduction

Quorum sensing (QS) is the term adopted to describe the mechanism by which microbial communities coordinate gene expression at high cell density. It was observed for the first time by Nealson et al. in 1970. They investigated the bioluminescence production in the *Photobacterium fischeri* strain MAV, then reclassified as *Aliivibrio fischeri* (Urbanczyk et al. 2007), a marine bacterium associated with the Hawaiian bobtail squid *Euprymna scolopes*. They observed that the appearance of luminescence started during the exponential phase of growth, while it was repressed or inactive in freshly inoculated cultures. On the base of the experimental results, they concluded that the key step of the control of bioluminescence production in *A. fischeri* was the transcription of luciferase gene (or operon). The authors could not determine if it was a negative or positive control, but they described the phenomenon by which the synthesis of luciferase was activated at high cell density as "autoinduction" (Nealson et al. 1970). Further studies focused on the understanding of this communication system, starting from the chemical characterization of the autoinducer responsible for the activation of bacterial luciferase synthesis. The cell-free medium of *A. fischeri* cultures was extracted with ethyl acetate, then the extract was purified by means of chromatographic techniques and chemically characterized by high-resolution ^{1}H nuclear magnetic resonance spectroscopy, infrared spectroscopy, and high-resolution mass spectrometry data. The structure was determined as N-(3-oxohexanoyl) homoserine lactone (3-oxo-C6-HSL) (Eberhard et al. 1981). The term "Quorum Sensing" (QS) was adopted for the first time by Fuqua et al. in 1994 to describe that kind of cell-to-cell communication mechanism. Bacteria synthesize small diffusible compounds (termed "autoinducer") and release them in the surrounding environment where they accumulate. When the concentration of these substances reaches a critical threshold, the bacterial communities "sense" the reached "quorum" and co-ordinate gene expression and subsequent phenotypic outcomes (Fuqua et al. 1994). Since the discovery of QS, different QS mechanisms were described within the three domains of life: Bacteria, Archaea and Eukaryota. Focusing on microbial communication, it is possible to generalize that in Gram-negative bacteria, the signalling system is mainly based on N-acyl homoserine lactones (AHL), while in Gram-positive bacteria, QS is mainly mediated by peptides (Figs. 7.1 and 7.2) The AHL-dependent QS-system, also known as autoinducer-1 (AI-1), is the most known and studied. It is based on two crucial genes: *luxI* and

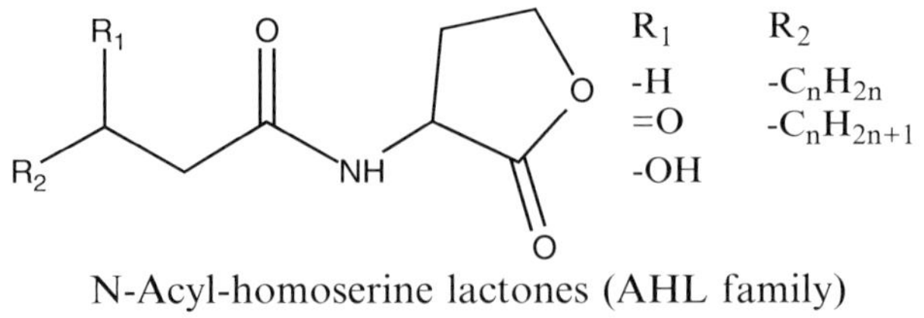

Fig. 7.1 General chemical structure of Acyl Homoserine Lactones (AHLs)

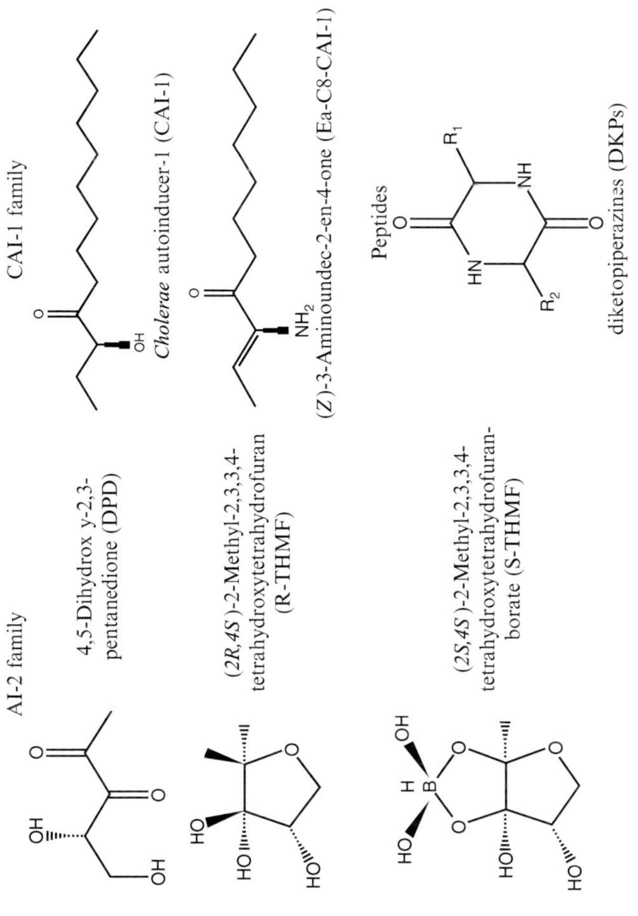

Fig. 7.2 Examples of some QS molecules different from AHLs

luxR. The transcription of *luxI* and its homologues leads to the synthesis of the autoinducers (acylated homoserine lactones); LuxR and its homologues are the signal receptors, they are responsible for the detection of AHLs and they modulate the expression of the target genes as transcription factors. The QS-dependent target genes often incorporate those required for the synthesis of the autoinducers, thus this system provide an auto-regulatory mechanism for amplifying signal molecule production (Williams et al. 2007). AHLs have the same central ring structure (homoserine lactone), but they differ in the side-chain. The acyl chain may vary in length and possess oxo or hydroxyl groups, and this variety allows AHLs to be species specific. Gram-negative bacteria can also employ 2-alkyl-4-quinolones (AQs), long-chain fatty acids, fatty acid methyl esters and autoinducer-2 (AI-2) (Atkinson and Williams 2009). AI-2 QS system is also utilized by some Gram-positive bacteria and it is dependent on furanosyl borate diesters. Contrary to AHLs, whose structure can considerably vary, in AI-2 QS-systems, the chemical products of the signal synthase (LuxS) are the same in all AI-2-producing bacteria, but products detected by the signal receptor (LuxP) can differ, e.g. S-2-methyl-2,3,3,4-tetrahydroxytetrahydrofuranborate (S-THMF-borate) or R-2-methyl-2,3,3,4-tetrahydroxytetrahydrofuran (R-THMF) (Fig. 7.2). Unravelling the role of AI-2 in bacterial Kingdom is complicated, because of the dual role that it plays as signal molecule and in metabolic pathway. The *luxS* gene, in fact, is part of an operon that encodes enzymes involved in the activated methyl cycle (AMC) (Pereira et al. 2013). AMC is an important metabolic pathway for the recycling of S-adenosylmethionine (SAM), and it is interesting that it is also a common substrate for the synthesis of different autoinducers, not only AI-2 but also AHLs and CAI (cholerae AI-1) (Tang and Zhang 2014). Physiological activities related to QS systems are various and include biofilm development, virulence expression, plasmid conjugation, motility, growth inhibition, etc., that are associated with a variety of vital functions which sustain life (Williams 2007). The expression of these QS-dependent phenotypic outcomes contributes to environmental adaptation in different models (such as bacteria found in the rhizosphere) (Joint et al. 2007). It is possible to speculate that autoinducers may control the expression of genes that are required for the well-being of organisms in extreme conditions, in particular, QS could be involved in the strategy evolved by Bacteria and Archaea to thrive in hypersaline environments. The diversity of life in extreme environments is surprising. The term "extreme" refers to habitat in which physical or chemical parameters are prohibitive for the majority of the organisms, such as temperature, pH, salt concentration and hydrostatic pressure. Within this group, halophiles are distinguished by their adaptation to hypersaline conditions, that are environments in which salt concentrations are in excess of sea water (from 2–5 % NaCl up to above 20–30 % NaCl). This kind of microorganisms actually require high salt concentrations for growth. A clear distinction must be made between thalassohaline and athalassohaline environments. In thalassohaline habitat, the hypersaline milieu is generated by the evaporation of sea water, thus sodium chloride is found to be the most abundant salt and the proportion between the salt composition is comparable to the one of sea water. There are many similarities between the communities of moderately halophilic bacteria that inhabit thalassohaline hypersaline environments and the ones present in seawater. Different marine aerobic heterotrophic bacteria could in fact grow at a NaCl concentration of up to 20 % (in some cases also at 30 % NaCl). On the contrary, in athalassohaline environments sea water salt composition is not kept (DasSarma and Priya 2012). Dead Sea is an example of athalassohaline in which a higher concentration of divalent cations (Mg^{2+} and Ca^{2+}) compared to monovalent cations (Na^{+} and K^{+}) was found. A nonmarine proportion can be also generated by the precipitation of NaCl, and the consequent concentration of potassium and magnesium salts. Ventosa et al. (1998) divided the environments in which moderately halophilic bacteria thrive in five categories: salt lakes and brines; saline soils; cold saline habitats; alkaline saline habitats; salted fish, meat, and other foods; unusual habitats (such as desert plants and desert animals). Prokaryotes are the most abundant constituents of halophiles. Every living cell is subject to the availability of liquid water as solvent for metabolic reactions, thus halophiles evolved different strategies to adapt to such harsh conditions. Halophilic bacteria can thrive in hypersaline habitat because of their capability to balance the osmotic pressure of the environment through different mechanisms. Bacteria mainly produce high concentrations of compatible solutes, small water-soluble organic molecules which do not interfere with cell metabolism (amino acids and their derivatives, sugar and polyols). Archaea and some halophilic anaerobic bacteria accumulate high concentration of salts intracellularly (KCl) (Saum and Müller 2008). Only one group of bacteria adopted this "salt-in" strategy, the order Halanaerobiales, that are fermentative and homoacetonic anaerobes. Otherwise, the main strategy adopted by bacteria is to accumulate organic solutes, keeping their cytoplasm at least isoosmotic with the hypersaline milie. The most common organic solutes synthesized or only accumulated from the medium by halophiles are ectoine and glycine betaine, with the cyclic amino acid derivatives ectoine and hydroxyectoine found to be the most abundant in the domain Bacteria (Oren 2008).

7.2 Quorum Sensing in Halophilic Bacteria

Halophilic bacteria are salt-loving microorganisms, able to thrive in habitats characterized by high salt concentrations. This condition is often correlated with high pH, some halophiles can in fact survive at a pH level of >9 (alkaliphiles). There are also examples of halophilic bacteria that are able to thrive at high temperatures (thermophiles) (Montgomery et al. 2013). It is known that QS plays a role in the processes involved in bacterial adaptation and growth in complex environmental niches. As an example, the surfaces and the internal spaces of marine sponges are a unique niche for the isolation of diverse bacteria and fungi. Abbamondi et al. (2014) described the detection of signal molecules belonging to the class of diketopiperazines (cyclic dipeptides) in different bacteria isolated from marine sponges. *Vibrio* sp. and *Pseudoalteromonas* sp., associated with the same sponge (*Dysidea avara*), showed contrasting responses respectively as activator/inhibitor of cell-to-cell communication mechanism in specific QS-biosensors. This finding led them to speculate that QS can play a role in the mutual control on the growth of different microorganisms in the same host (Abbamondi et al. 2014). *Acidithiobacillus ferrooxidans* is an extremely acidophilic bacterium, capable to resist at high concentrations of Cu^{2+} that can cause serious damage to cells. The effect of (5Z)-4-bromo-5-(bromomethylene)-2(5H)-furanone (FUR) on the Cu^{2+} resistance of *A. ferrooxidans* was investigated and results showed a dose dependent inhibition of Cu^{2+} resistance by FUR, starting from a concentration of 0,01 μg ml^{-1}. The role of QS in heavy metal resistance needs further investigation, but the collected data suggest that QS signal system might be implicated in Cu^{2+} resistance mechanism (Wenbin et al. 2011). The human body can also be considered as an "extreme environment", because of the wide range of acidic, oxic and anoxic conditions. *Helicobacter pylori* is a Gram-negative pathogen of the human stomach that causes gastritis, peptic ulcer disease, gastric adenocarcinoma, and low-grade gastric lymphoma. In this model, the regulation of flagella gene transcription, and consequently normal levels of motility essential for the infection, is regulated by AI-2 signaling (Rader et al. 2007). Therefore, it is possible to speculate that the autoinducers may control the expression of genes that are required for the well-being of the organisms in extreme conditions, in particular QS could be involved in the strategy evolved by halophilic bacteria to thrive in hypersaline environments. Five bacterial strains (genera *Marinobacter* and *Halomonas*) and one Archaea strain (*Haloterrigena*) were isolated from an hypersaline cyanobacterial mat, desert wadi in south eastern Oman. The strains were tested for their ability to synthesize bioactive compounds, potentially relevant for biotechnological use as antifouling agents. Different studies confirm that compounds with Quorum Quenching (strategies employed by various bacteria and fungi to degrade acylated homoserine lactones and other lactone-containing compounds) activity can be used as antifouling agents. Ethyl acetate extracts of total hypersaline mats inhibited QS in *Agrobacterium tumefaciens* NTL4 (pZLR4) and *S. enterica* S235. Four DKPs were isolated from *Marinobacter* sp. (SK-3): cyclo(L-Pro-L-Phe), cyclo(L-Pro-L-Leu), cyclo(L-Pro-L-isoLeu) and cyclo(L-Pro-D-Phe). Cyclo(L-Pro-L-Leu) and cyclo(L-Pro-L-isoLeu) inhibited QS violacein production in CV-017; cyclo(L-Pro-L-Phe), cyclo(L-Pro-L-Leu) and cyclo(L-Pro-L-isoLeu) inhibited bioluminescense production in *E. coli* pSB401 reporter (Abed et al. 2013). AHL-QS system is widespread within Halomonadaceae. Forty-three strains of that family were analysed for their ability to activate QS in *A. tumefaciens* NTL4 (pZLR4) that is sensitive to AHLs with medium-to-long acyl chains. All the analysed strains activated NTL4 bioreporter. Further analysis by PCR and DNA sequencing approaches demonstrated that most of the studied species contained a LuxI homolog. TLC-overlay test showed that the most predominant AHL molecule was C_6-HLS. The chemical characterization of C_6-HLS was confirmed in *Halomonas anticariensis* $FP35^T$ by means chromatography/mass spectrometry (GM/MS) and electrospray ionization tandem mass spectrometry (ESI MS/MS) (Tahrioui et al. 2013b). AHL production by moderately halophilic bacteria was described for the first time by Llamas et al. in 2005. In that study, a growth-phase dependent N-acyl homoserine lactones production was observed in the *Halomonas* species (*H. maura*, *H. eurihalina*, *H. ventosae* and *H. anticariensis*).

The analogous growth-phase dependent exopolysaccharide production led the research group to hypothesize that EPS synthesis could be regulated by AHL-based QS. Four AHLs were isolated from *H. anticariens* FP35^T, and identified by means GM/MS and ESI MS/MS: N-butanoyl homoserine lactone (C_4- HLS), N-exanoyl homoserine lactone (C_6- HLS), N-octanoyl homoserine lactone (C_8- HLS) and N-dodecanoyl homoserine lactone (C_{12}- HLS). Actually, an increase of EPS production was observed in the culture of the analysed strains supplemented with exogenous AHLs (Llamas et al. 2005). A deeply investigation of AHL-QS system in *H. anticariens* FP35^T was performed by the same group in 2011. They described a system composed of *lux*R/*lux*I homologues: *han*R (the putative transcriptional regulator gene) and *han*I (the AHL synthase gene) (Tahrioui et al. 2011). The draft genome sequence of *H. anticariens* FP35^T was then deposited at DDBJ/EMBL/GenBank as a starting point for further studies to investigate the role of QS-regulatory system in the adaptation to hypersaline environments (Tahrioui et al. 2013a). Production of autoinducer-2 (AI-2) was detected in the moderately halophilic bac terium *Halobacillus halophilus. H. halophilus* carries a homologue of *Lux*S that is upregulated by the chloride concentration, but its production is also regulated by salinity and growth phase. *Lux*S is involved in the activated methyl cycle, and one of the products of the *Lux*S-catalyzed reaction leads to AI-2 production. In this study, the authors suggested a potential role of QS in salt precipitation (chloride regulon) (Sewald et al. 2007). *Halomonas pacifica* and *Marinobacter hydrocarbonoclasticus* can cause serious environmental problems through their vital roles in microfouling processes. These two halophilic marine microorganisms produce significant amounts of AI-2 activity at early exponential to stationary phases. Liquiat et al. investigated the effect of QS inhibitors (QSI) on biofilm formation. Penicillic acid and patulin inhibited AI-2 based QS in *H. pacifica*, but additional research is needed to elucidate the implication of QS in biofilm formation in marine isolates (Liaqat et al. 2014). *Vibrio vulnificus* molecular communication system was deeply investigated over the past 10 years. This halophilic bacterium is a marine pathogen that causes human diseases such as fatal septicaemia and necrotizing wound infections. A screening of signal molecules production by *V. vulnificus* was at first performed by Kim et al. in 2003. They observed AI-2 production and identified the gene responsible for its synthesis (*lux*S_{Vv}). The maximal AI-2 activity was detected during mid-exponential to early stationary growth phase that is the same growth rage in which the production of haemolysin was detected. After the early stationary phase, a metalloprotease begins to be produced. These observations lead the authors to speculate a possible role of QS in the co-ordination of haemolysin/protease expression. A *lux*S_{Vv} mutation that resulted in the loss of AI-2 synthesis, caused a decrease in protease level and an increase in haemolysin synthesis. On the basis of these results they suggested that the virulence expression in *V. Vulnificus* was regulated by an AI-2 based QS system (Kim et al. 2003). Iron availability is a crucial factor in the pathogenicity of many bacteria and it is frequently involved in the expression of virulence-associated properties. Pathogens can counteract iron-limited conditions by the synthesis of high iron affinity compounds such as siderophores. These compounds transport iron across the cell envelope and release it once inside the cell, in this way iron becomes available for utilization by the cell. Siderophores production is one of the most important virulence factors in an infection. *V. vulnificus* produces vulnibactin (a phenolate-type siderophore), and its biosynthesis is regulated by QS. Genes *vvs*A and *vvs*B are co-transcribed and are involved in the production of vulnibactin. Wen et al. (2012) identified smcR as the main regulatory protein for QS signalling in *V. vulnificus*, repressing vvsAB transcription. They suggested that this mechanism could have a role in the management of bacterial metabolic economy. It is clear that at high cell density, bacterial growth is slowed, thus iron (and consequently siderophores) are not required (Wen et al. 2012). *V. vulnificus* virulence is mainly due to the two protein toxins that it secretes: a metalloprotease (gene *vvp*E) and a cytolysin (gene *vvp*A). The synthesis of these virulence factors is regulated by QS through small diffusible signal molecules of the autoinducer 2 (AI-2) type. In this model, *lux*O encodes a central response regulator of the QS circuit. Elgaml et al. (2014) described that the disruption of *lux*O leads to an increase of metalloprotease level and a decrease of hemolysin production. Temperature was also identified as an important virulence factor. The authors described that the optimal temperature for protease production was 26 °C, while for the cytolysin was 37 °C. In that way *V. vulnificus* senses the temperature in the small intestine and produces enough amount of hemolysin, that accelerates the bacterial invasion to the blood-stream (Elgaml et al. 2014). The role of QS in *Vibrionaceae* family was also investigated by Israil et al. (2008). Nine halophilic bacteria from water sources were subject of this study: two *V. alginolyticus* no. 229 and 1560, three *V. fischeri* no. 10, 754 and 898, two *V. anguillarum* no. 798 and 1545 and two *V. parahaemolyticus* no. 1442 and 1671. They evaluated the effects of exogenous QS soluble mediators on bacterial growth. Sterile cell free cultures were prepared for each analysed strain at the logarithmic phase (filter sterilized by 0.22 μm membrane filtration). Ninety-six plastic well microplates were prepared by adding an aliquot of the homologous bacterial filtrate (and consequently autoinducers accumulated in it) in the respective bacterial culture. The viable cell counts were performed in order to establish the bacterial growth curves. Results demonstrated a possible role of autoinducers in the modulation of the multiplication rate and growth curve of the selected strains (Israil et al. 2009). Wang et al. (2013) detected multiple QS systems in the halophilic bacterium *Vibrio fluvialis*. It can be detected in aquatic environments, and it is classified as a human enteric pathogen, causing diarrhoea, dehydration, abdominal pain, fever and vomiting. The authors identified three kinds of molecules involved in QS systems: 3-oxo-C10-HLS, CAI-1 and AI-2 signal molecules. *V. fluvialis* synthesizes a protease and a hemolysis as virulence factors. In this study, production of the protease is regulated by a CAI-1/AI-2 QS-system, while the transcription of the gene for the expression of the hemolysin is activated by AHL-QS system and repressed by CAI-1/AI-2 QS system. This finding showed the existence of a multiple QS network that co-ordinate virulence factors production, and consequently pathogenesis, in *V. fluvialis* (Wang et al. 2013). The more representative findings are summarized in Tables 7.1 and 7.2.

Table 7.1 Halophilic microorganisms producing QS molecules and their corresponding activity

Bacterial strain	Signal molecules detected	Activity	Reference
Halomonadaceae (43 strains)	C_6-HLS (most predominant AHL)	Activation of QS in *Agrobacterium tumefaciens* NTL4 (pZLR4)	Tahrioui et al. (2013b)
Halomonas anticariensis FP35^T	C_4-HLS, C_6-HLS, C_8-HLS, C_{12}-HLS	Potential role in EPS production	Llamas et al. (2005)
			Tahrioui et al. (2011, 2013a, b)
Halobacillus halophilus	Autoinducer-2 (AI-2)	Potential role in salt precipitation (chloride regulon)	Sewald et al. (2007)
Vibrio vulnificus	Autoinducer-2 (AI 2)	Biosynthesis of vulnibactin	Kim et al. (2003)
		Metalloprotease and cytolysin expression	Wen et al. (2012)
			Elgaml et al. (2014)
V. alginolyticus no. 229 and 1560, *V. fischeri* no. 10, 754 and 898, *V. anguillarum* no. 798 and 1545 and *V. parahaemolyticus* no. 1442 and 1671	–	Possible role of exogenous autoinducers in the modulation of the multiplication rate and growth curve of the selected strains	Israil et al. (2009)
Vibrio fluvialis	3-oxo-C10-HLS, CAI-1 and AI-2	Protease production regulated by a CAI-1/AI-2 QS-system, hemolysin expression activated by AHL- system and repressed by CAI-1/AI-2 QS system	Wang et al. (2013)
Natronococcus occultus (archaeon)	Autoinducer molecules, most likely belonging to the AHL class	Correlation between signal molecules production and the production/activation of an extracellular protease	Paggi et al. (2003)
Haloterrigena hispanica (archaeon)	cyclo-(D-Pro-L-Tyr), cyclo-(L-Pro-L-Tyr), cyclo-(L-Pro-L-Val), cyclo-(L-Pro-L-Phe) and cyclo-(L-Pro-L-isoLeu)	Supernatant extract: activation QS in pSB401 and NTL4 (pCF218; pCF372)	Tommonaro et al. (2012)
		Cyclo-(L-Pro-L-Val): activation QS in NTL4 (pCF218; pCF372) and *Gfp* production in DM27	

7.2.1 Quorum Quenching

In recent years, several chemicals and enzymes have been identified that interfere with the key metabolites of QS mechanism. These Quorum Quenching (QQ) molecules that prevent cell-cell communications are considered as a promising way to

Table 7.2 Halophilic microorganisms producing QQ molecules and their corresponding activity

Bacterial strain	Signal molecules detected	Activity	Reference
Five bacterial strains (genera *Marinobacter* and *Halomonas*) and one archaea strain (*Haloterrigena*)	–	Inhibition of QS in *Agrobacterium tumefaciens* NTL4 (pZLR4) and *S. enterica* S235	Abed et al. (2013)
Marinobacter sp. (SK-3)	cyclo(L-Pro-L-Phe), cyclo (L-Pro-L-Leu), cyclo(L-Pro-L-isoLeu) and cyclo(L-Pro-D-Phe)	Cyclo(L-Pro-L-Leu) and cyclo(L-Pro-L-isoLeu): inhibition of QS violacein production in CV-017	Abed et al. (2013)
		cyclo(L-Pro-L-Phe), cyclo (L-Pro-L-Leu) and cyclo(L-Pro-L-isoLeu): inhibition of bioluminescense production in *E. coli* pSB401 reporter	
Halomonas pacifica and *Marinobacter hydrocarbonoclasticus*	Autoinducer-2 (AI-2)	Penicillic acid and patulin inhibited AI-2 based QS in *H. pacifica*	Liaqat et al. (2014)
Halobacillus salinus	N-(2′-phenylethyl)-isobutyramide, 3-methyl-N-(2′-phenylethyl)-butyramide	Bioluminescence inhibition against *Vibrio harveyi*	Teasdale et al. (2009)
		Isolated compound: inhibition violacein biosynthesis by CV026 and Gfp production by JB525	
Five bacterial strains (genera *Bacillus* or *Halobacillus*)	–	Bioluminescence inhibition against *Vibrio harveyi*	Teasdale et al. (2011)
		Inhibition violacein biosynthesis by *Chromobacterium violaceum*	
Haloterrigena saccharevitan (archaeon)	Quorum quenching compounds	Inhibition of violacein production CV017 (supernatant and water extract of cell pellet)	Abed et al. (2013)
Vibrio fluvialis	3-oxo-C10-HLS, CAI-1 and AI-2	Protease production regulated by a CAI-1/AI-2 QS-system, hemolysin expression activated by AHL- system and repressed by CAI-1/AI-2 QS system	Wang et al. (2013)

control bacterial diseases (Hong et al. 2012; Tang and Zhang 2014). Some of these compounds are analogous to AHLs or HSLs and work as competitive antagonists such as the furanones produced by the macroalgae *Delisea pulchra* (Hentzer et al. 2002) as well as their synthetic derivatives (Martinelli et al. 2004). Others act as inhibitors of enzymes that catalyze key steps in QS mechanism. The antimicrobial Triclosan and antiparasitic agent Closantel are two well known examples. There are also enzymes that function as QQ molecules. Acyl-homoserine lactonase (EC 3.1.1.81) enzyme that hydrolyzes the lactone bond of AHLs was first isolated from *Bacillus* sp. and was shown to quench QS-dependent bacterial infection (Dong et al. 2001). Besides this AHL lactonase, another enzyme, AHL acylases (EC 3.5.2.5) was found to hydrolyze the amide linkage between the acyl chain and the homoserine moiety of AHL molecules (Leadbetter and Greenberg 2000). Other examples include mammalian lactonases, named paraoxonases (PONs), isolated from human airway epithelia (Billecke et al. 2000) and enzymes that do not hydrolyze AHLs but modify them chemically such as the AHL oxidoreductases (Uroz et al. 2005). Several QQ molecules have been isolated from halophiles (Teasdale et al. 2011). Whereas the tumonoic acids isolated from the marine cyanobacterium *Blennothrix cantharidosmum* were found to inhibit the bioluminescence of *V. harveyi* BB120 without affecting bacterial growth (Clark et al. 2008), another marine cyanobacterium *Lyngbya majuscula* was found to produce compounds that inhibited the QS of *P. aeruginosa* (Dobretsov et al. 2010). In another study, inhibition of both bioluminescence in *V. harveyi* BB120 and lipopolysaccharide-stimulated nitric oxide production in the murine macrophage cell line RAW264.7 suggested a dual-inhibitory effect for the honaucins A–C isolated from the marine cyanobacterium *Leptolyngbya crossbyana* (Choi et al. 2012). QQ molecules from the halophilic bacteria *Halobacillus salinus* (Teasdale et al. 2009), *Bacillus cereus* and *Marinobacter* sp. SK-3 (Abed et al. 2013; Teasdale et al. 2011) were also reported. Besides the furanones of red alga *D. pulchra*, a mixture of floridoside, betonicine and isethionic acid produced by the marine red alga *Ahnfeltiopsis flabelliformes* (Liu et al. 2008) as well as several other marine microalgae (Natrah et al. 2011) were attributed with potential QQ activity. The more representative findings are summarized in Table 7.2.

7.3 Halophilic Archaea

The halophilic Archaea are microorganisms able to grow from around 8 % (1.5 M) to about 36 % (5 M) sodium chloride. They are Gram-negative, cocci or rods and they have at least one plasmid. Their optimal temperature of growth is around 37 °C, even if they are able to grow in a range of 15 °C to approximatively 45–50 °C, and many of their enzymes conserve a good activity in this range (Grant et al. 2001). Some of halophilic Archaea use complex amino acids as their carbon and nitrogen sources, while other members, like the genera *Haloquadratum* and *Halosimplex*, use pyruvate and NH_4^+ as carbon and nitrogen sources, respectively (Burns et al. 2007). Because haloarchaea live in high salt environment, they produce proteins constituted more in acidic amino acids than basic amino acids. These proteins require high salt concentrations for the optimum in activity (Mevarech et al. 2000) and they have gained the attention of several researchers for their biotechnological application in diverse fields (agriculture, wastewater treatment, medical field and bioplastic). The interest has been directed mainly towards specific enzymes (glycosyl hydrolases, proteases, lipases and esterases) and products (biopolymers and surfactants) produced by haloarchaea with a possible application in industrial processes (Margesin and Schinner 2001). Sodium chloride concentration ranging from 1.5 to 4–5 M is necessary for growth of haloarchaea. This high concentration of salt has represented a problem for the industrial fermentation and biotransformation system. For this reason, the development of novel bioreactor and fermentation systems have been necessary to recover bioproducts from haloarchaea (Hezayen et al. 2000). Many of the above mentioned enzymes are extracellular, then it is easier their isolation and purification. Moreover, because many haloarchaea are able to grow at low salinity, it is possible to decrease the sodium chloride concentration manipulating the growth medium, obtaining a good production of enzymes (Purdy et al. 2004; Di Donato et al. 2011). Because the enzymes produced by haloarchaea are able to function in high-reduced water conditions, they can function in different hydrophobic solvents (Ventosa and Oren 1996; Usami et al. 2005), then this aspect represent an advantage for the applications of haloarchaea in industry processes.

7.3.1 *Quorum Sensing in Halophilic Archaea*

Despite the large number of studies on QS, very few papers have addressed QS communication in Archaea. The first study about this field date at 2003. Paggi et al. (2003) reported the detection of autoinducer molecules, most likely belonging to the AHL class in the haloalkaliphilic archaeon *Natronococcus occultus*. They also demonstrated correlation between these molecules and the production/activation of an extracellular protease produced by cells of *N. occultus* (Paggi et al. 2003). Then, AHLs-based quorum sensing system was detected in a methanogenic archaeon; which in turn suggested the existence of a possible universal mechanism of communication among prokaryotes (Zhang et al. 2012). An archaeal strain was isolated from an hypersaline cyanobacterial mat, desert wadi in south eastern Oman. 16S rRNA-based phylogenetic reconstruction showed a 99.6 % sequence similarity with the archeon *Haloterrigena saccharevitans*. Quorum Quenching bioassays were performed with the aim to select strains with potential antifouling ability; inhibition of QS can in fact disrupt biofilm formation in marine environment. Results demonstrated that the analysed *Haloterrigena* sp. was able to synthesize bioactive compounds with Quorum Quenching activity. In particular, supernatant and water extract of cell pellet exhibited some QS-inhibitory proprieties against *C. violeaceum* CV017 (inhibition of violacein production) (Abed et al. 2013). The only report about the detection of diketopiperazines (DKPs) from a halophilic archaeon was published by Tommonaro et al. (2013). They investigated QS mechanism in *Haloterrigena hispanica*, isolated from Fuente de Piedra salt lake, Spain. Its optimal growth parameters were 3.4 M NaCl concentration, 0.2 M Mg^{2+} concentration, pH 7.0 and 50 °C, thus it was classified as an extremely halophilic archaeon. *H. hispanica* supernatant activated QS in AHL bioreporter pSB401 (activated by mid chain AHLs) and NTL4 (pCF218; pCF372). Five DKPs were isolate by means chromatographic techniques and chemically characterized according to NMR/EIMS data (correspondent to data present in literature): cyclo-(D-Pro-L-Tyr), cyclo-(L-Pro-L-Tyr), cyclo-(L-Pro-L-Val), cyclo-(L-Pro-L-Phe) and cyclo-(L-Pro-L-isoLeu). Among these five DKPs, only cyclo-(L-Pro-L-Val) showed QS-inducing properties in TLC-overlay test, activating NTL4 (pCF218; pCF372) QS system. This compound also induced green fluorescent protein (Gfp) production in *Vibrio anguillarum* DM27 strain. Despite the role of DKPs in the metabolic pathway of *H. hispanica* is not clear, the authors speculate that they may be involved in the mechanisms evolved by Archaea to thrive in such harsh conditions (hypersaline habitat) (Tommonaro et al. 2012). The more representative findings are summarized in Tables 7.1 and 7.2.

7.4 Biotechnological Applications Quorum Sensing

QS is known to regulate important microbial pathways that include antibiotic synthesis, pollutant biodegradation, and bioenergy production, which are highly relevant to human health. Biotechnological applications of QS cover various fields including medicine, agriculture, marine biology and biotechnology.

7.4.1 Medical Applications

Inappropriate and excessive use of antibiotics as well as the selective pressures imposed on the pathogens resulted in the constant emergence of antibiotic resistant strains. This in turn led to the emergence of new antimicrobial strategies that target the virulence factors rather than the vital mechanisms of pathogens (Clatworthy et al. 2007). In pathogens like *Burkholderia* species, *P. aeruginosa*, *Erwinia carotovora* and *Vibrio* species, biofilm formation is regulated by QS and since QS is not a vital mechanism for the growth of these pathogens, inhibiting QS would only prevent virulence factors rather than killing the bacteria (Tang and Zhang 2014). This makes QS an ideal target for antivirulence therapy due to the delayed emergence of resistance that would result from imposing less selective pressure to the pathogens. QS-based synthetic multicellular systems are also interesting applications where novel functions non-existent in nature can be created by integrating basic functions into complex functional circuits via genetic engineering (Endy 2005; Basu et al.

2005). In one study, a cell population control mechanism is constructed by coupling QS with protein production. The cell density is followed by means of QS signals and after a certain population threshold density, the expression of a recombinant protein is induced. Such a population-based gene expression system can be used for the self-regulated production and subsequent harvesting of exogenous cytotoxic proteins (You et al. 2004). In a similar system, *E. coli* cells were engineered to attack cancer cells in a density-dependent fashion and only when tumor – indicative conditions are present (Anderson et al. 2006). This methodology could be applied to develop bacteria that can sense the microenvironment of a tumor and attack the cancerous cells by excreting a cytotoxic agent or drug. In another system, bacteria are engineered to sense the pathogenic QS signals and then to secrete a toxin and hence kill the pathogen (Saeidi et al. 2011).

7.4.2 Control of Biofouling

Biofouling is the attachment and subsequent growth of a biological community on surfaces. In marine biotechnology, biofouling is the undesirable accumulation of microorganisms, plants, and animals on artificial surfaces immersed in sea water. This biological settlement in ships results in the deterioration of the coating, unnatural transport of species and economic burdens associated with the additional fuel required for the high frictional resistance (Yebra et al. 2004). Attachment of marine invertebrates and plants is believed to be primarily caused by the bacterial biofilm that is formed immediately on the submerged surfaces (Rice et al. 1999). To control biofouling, a conventional strategy is the use of anti-fouling paints that contain biocides such as tributyl tin (TBT), copper, zinc pyrithione, Irgarol 1051, diuron, Sea-Nine 211, chlorothalonil or dichlofluanid however these chemicals are known to have ecotoxic effects and therefore there is a constant search for more ecofriendly additives (Garg et al. 2014). A viable alternative is the QS-directed control of biofilm formation. In literature, there are numerous QQ producer organisms isolated from marine environments and marine organisms are known to respond to bacterial QS signals (Skindersoe et al. 2008). Use of QQ molecules for the interruption of QS by inhibiting the signal generation, degrading the generated signals or by suppression of QS receptors, could make the submerged surfaces incompatible for biofilm formation and hence prevent the attachment and colonization of invertebrate larvae and algal spores and associated biofouling (Garg et al. 2014). Membrane bioreactors (MBR) are bioreactors with membrane filtration units for biomass retention. They have become an effective solution to transform wastewaters into high quality effluents suitable for discharge. During the operation, attachment of bacteria and microbial products to the membrane surface lead to biofilm formation which in turn leads to the clogging of the membrane surface and loss in permeability. This phenomenon is called as ‘membrane fouling’ and it is one of the major limitations of the MBR process (Lin et al. 2014). Some strategies for prevention include cleaning or exchange of membranes and incorporation of antimicrobial agents to the membrane. However, recently, use of QQ molecules or bacteria excreting QQ molecules gain increasing popularity (Lade et al. 2014). These QQ systems can be used to prevent the biofilm formation without affecting bacterial growth and therefore have a high potential towards solving the biofouling problem. In another study, AHL acylase enzymes were immobilized onto magnetic nanoparticles and then crosslinked to get a network of acylase activity in magnetically separable mesoporous silica which showed highly effective antifouling activity against biofilm maturation of *P. aeruginosa* PAO1 on the membrane surface (Lee et al. 2014).

7.4.3 Agricultural Applications

One application area of QS involves their use in hindering the bacterial infection development in plants. From the two approaches applied, in the first approach, genes encoding AHL synthase (Toth et al. 2004) or AHL-degrading (Dong et al. 2001) enzymes are expressed in plant tissues whereas in the second approach, AHL-degrading bacteria are employed for plant protection (Jafra et al. 2006). Expression of the AHL-synthase genes resulted in contradictory observations (Czajkowski and Jafra 2009), however, expressing AiiA lactonase from *Bacillus thuringiensis* in potato and tobacco plants improved their resistance against infection by *Pectobacterium carotovorum* subsp. *carotovorum* (Dong et al. 2001). Nevertheless, since both strategies relied on transgenic plants, such an application can only have a limited use when the restrictions are considered. Another example is the combined use of AHL-degrading enzymes and QS inhibitors to disrupt the intercellular communication and hence to control the associated pathogenesis (Park et al. 2005).

7.5 Conclusion

The halophilic microorganisms represent a rich source of novel products (enzymes, surfactants, biopolymers) for biotechnological applications. The production of these products, however, is still challenging, because the yield and the difficulty of their purification and characterization on industrial scale. Moreover, new culturing methods (novel bioreactors, new cultural media) are necessary to avoid the corrosion due to high salt concentration at which halophilic bacteria grow. The Quorum Sensing mechanism play a great role in controlling a several microbial cell activities, such as enzymes production, biofilm formation and virulence. Many studies are directed towards to understanding QS and the development of strategies to manipulate this mechanism with the aim to direct some microbial activities. At now, there are still few reports about the study of QS in halophilic microorganisms, and more needs to be investigated in this topic to improve the understanding of link between chemical communication and bioproducts production by halophiles. The challenge is still to be met.

Acknowledgment The authors thank the "First executive programme of the agreement between the government of the republic of Italy and the government of the republic of Turkey on scientific and technical cooperation", entitled: "Extremophiles for next-generation biofuels" (MAE-TUBITAK No. M00192-111Y295).

References

Abbamondi GR, De Rosa S, Iodice C, Tommonaro G (2014) Cyclic dipeptides produced by marine sponge-associated bacteria as quorum sensing signals. Nat Prod Commun 9(2):229–232

Abed RM, Dobretsov S, Al-Fori M, Gunasekera SP, Sudesh K, Paul VJ (2013) Quorum-sensing inhibitory compounds from extremophilic microorganisms isolated from a hypersaline cyanobacterial mat. J Ind Microbiol Biotechnol 40:759–772

Anderson JC, Clarke EJ, Arkin AP, Voigt CA (2006) Environmentally controlled invasion of cancer cells by engineered bacteria. J Mol Biol 355:619–627

Atkinson S, Williams P (2009) Quorum sensing and social networking in the microbial world. J R Soc Interface 6:959–978

Basu S, Gerchman Y, Collins CH, Arnold FH, Weiss R (2005) A synthetic multicellular system for programmed pattern formation. Nature 434:1130–1134

Billecke S, Draganov D, Counsell R, Stetson P, Watson C, Hsu C, La Du BN (2000) Human serum paraoxonase (PON1) isozymes Q and R hydrolyze lactones and cyclic carbonate esters. Drug Metab Dispos 28:1335–1342

Burns DG, Janssen PH, Itoh T, Kamekura M, Li Z, Jensen G, Rodriguez-Valera FE, Bolhuis Hand Dyall-Smith ML (2007) *Haloquadratum walsbyi* gen. nov., sp. nov., the square haloarchaeon of Walsby, isolated from saltern crystallizers in Australia and Spain. IJSEM 57:387–392

Choi H, Mascuch SJ, Villa FA, Byrum T, Teasdale ME, Smith JE, Preskitt LB, Rowley DC, Gerwick L, Gerwick WH (2012) Honaucins A–C, potent inhibitors of inflammation and bacterial quorum sensing: synthetic derivatives and structure-activity relationships. Chem Biol 19:589–598

Clark BR, Engene N, Teasdale ME, Rowley DC, Matainaho T, Valeriote FA, Gerwick WH (2008) Natural products chemistry and taxonomy of the marine cyanobacterium *Blennothrix cantharidosmum*. J Nat Prod 71:1530–1537

Clatworthy AE, Pierson E, Hung DT (2007) Targeting virulence: a new paradigm for antimicrobial therapy. Nat Chem Biol 3:541–548

Czajkowski Rand Jafra S (2009) Quenching of acyl-homoserine lactone-dependent quorum sensing by enzymatic disruption of signal molecules. Acta Biochim Pol 56(1):1–16

DasSarma S, DasSarma P (2012) Halophiles. In: eLS. JohnWiley&Sons, Ltd:Chichester

Di Donato P, Fiorentino G, Anzelmo G, Tommonaro G, Nicolaus B, Poli A (2011) Re-use of vegetable wastes as cheap substrates for extremophile biomass production. Waste Biomass Valoriz 2:103–111

Dobretsov S, Teplitski M, Alagely A, Gunasekera SP, Paul VJ (2010) Malyngolide from the cyanobacterium *Lyngbya majuscule* interferes with quorum sensing circuitry. Environ Microbiol Rep 2:739–744

Dong YH, Wang LH, Xu JL, Zhang HB, Zhang XF et al (2001) Quenching quorum-sensing-dependent bacterial infection by an N-acyl homoserine lactonase. Nature 411:813–817

Eberhard A, Burlingame AL, Eberhard C, Kenyon GL, Nealson KH, Oppenheimer NJ (1981) Structural identification of autoinducer of *Photobacterium fischeri* luciferase. Biochemistry 20(9):2444–2449

Elgaml A, Higaki K, Miyoshi S (2014) Effects of temperature, growth phase and luxO-disruption on regulation systems of toxin production in *Vibrio vulnificus* strain L-180, a human clinical isolate. World J Microbiol Biotechnol 30(2):681–691

Endy D (2005) Foundations for engineering biology. Nature 438:449–453

Fuqua WC, Winans SC, Greenberg EP (1994) Quorum sensing in bacteria—the LuxR–LuxI family of cell density responsive transcriptional regulators. J Bacteriol 176(2):269–275

Garg N, Manchanda G, Aditya K (2014) Bacterial quorum sensing: circuits and applications. Antonie Van Leeuwenhoek 105(2):289–305

Grant WD, Kamekura M, McGenity TJ, Ventosa A (2001) Class III. Halobacteria class. nov. In: Boone DR, Castenholz RW, Garrity GM (eds) Bergey's manual of systematic bacteriology, vol 1. Springer, New York, pp 294–334

Hentzer M, Riedel K, Rasmussen TB, Heydorn A, Andersen JB, Parsek MR, Rice SA, Eberl L, Molin S, Hoiby N, Kjelleberg S, Givskov M (2002) Inhibition of quorum sensing in *Pseudomonas aeruginosa* biofilm bacteria by a halogenated furanone compound. Microbiology 148:87–102

Hezayen FF, Rehm BHA, Eberhardt R, Steinbüchel A (2000) Polymer production by two newly isolated extremely halophilic Archaea: application of a novel corrosion-resistant bioreactor. Appl Microbiol Biotechnol 54:319–325

Hong KW, Koh CL, Sam CK, Yin WF, Chan KG (2012) Quorum quenching revisited-from signal decays to signalling confusion. Sensors (Basel) 12:4661–4696

Israil AM, Chifiriuc MC, Delcaru C, Iordache C, Pelinescu D, Sasarman E (2009) Evaluation and monitoring of quorum sensing soluble mediators implicated in the regulation of bacterial growth in *Vibrio* strains. Rom Biotechnol Lett 14(2):4211–4224

Jafra S, Przysowa J, Czajkowski R, Michta A, Garbeva P, van der Wolf JM (2006) Detection and characterization of N-acyl homoserine lactone-degrading bacteria from the potato rhizosphere. Can J Microbiol 52:1006–1015

Joint I, Downie JA, Williams P (2007) Bacterial conversations: talking, listening and eavesdropping. An introduction. Philos Trans R Soc Lond B Biol Sci 362(1483):1115–1117

Kim SY, Lee SE, Kim YR, Kim CM, Ryu PY, Choy HE, Chung SS, Rhee JH (2003) Regulation of *Vibrio vulnificus* virulence by the LuxS quorum-sensing system. Mol Microbiol 48(6):1647–1664

Lade H, Diby P, Kweon JH (2014) Quorum quenching mediated approaches for control of membrane biofouling. Int J Biol Sci 10(5):550–565

Leadbetter JR, Greenberg EP (2000) Metabolism of acyl-homoserine lactone quorum-sensing signals by *Variovoraxparadoxus*. J Bacteriol 182:6921–6926

Lee B, Yeon K, Shim J, Kim S, Lee C, Lee J, Kim J (2014) Effective antifouling using quorum-quenching acylase stabilized in magnetically-separable mesoporous silica. Biomacromolecules 15(4):1153–1159

Liaqat I, Bachmann RT, Edyvean RG (2014) Type 2 quorum sensing monitoring, inhibition and biofilm formation in marine microorganisms. Curr Microbiol 68(3):342–351

Lin H, Meijia Z, Fangyuan W, Fangang M, Bao-Qiang L, Huachang H, Jianrong C, Weijue G (2014) A critical review of extracellular polymeric substances (EPSs) in membrane bioreactors: characteristics, roles in membrane fouling and control strategies. J Membr Sci 460:110–125

Liu HB, Koh KP, Kim JS, Seo Y, Park S (2008) The effects of betonicine, floridoside, and isethionic acid from the red alga *Ahnfeltiopsis flabelliformis* on quorum-sensing activity. Biotechnol Bioprocess Eng 13:458–463

Llamas I, Quesada E, Martínez-Cánovas MJ, Gronquist M, Eberhard A, González JE (2005) Quorum sensing in halophilic bacteria: detection of N-acyl-homoserine lactones in the exopolysaccharide-producing species of *Halomonas*. Extremophiles 9(4):333–341

Margesin R, Schinner F (2001) Potential of halotolerant and halophilic microorganisms for biotechnology. Extremophiles 5:73–83

Martinelli D, Grossmann G, Sequin U, Brandl H, Bachofen R (2004) Effects of natural and chemically synthesized furanones on quorum sensing in *Chromobacterium violaceum*. BMC Microbiol 4:25. doi:10.1186/1471-2180-4-25

Mevarech M, Frolow F, Gloss LM (2000) Halophilic enzymes: proteins with a grain of salt. Biophys Chem 86:155–164

Montgomery K, Charlesworth JC, LeBard R, Visscher PT, Burns BP (2013) Quorum sensing in extreme environments. Life 3:131–148

Natrah FMI, Kenmegne MM, Wiyoto W, Sorgeloos P, Bossier P, Defoirdt T (2011) Effects of micro-algae commonly used in aquaculture on acyl-homoserine lactone quorum sensing. Aquaculture 317:53–57

Nealson KH, Platt T, Woodland Hastings J (1970) Cellular control of the synthesis and activity of the bacterial luminescent system. J Bacteriol 104(1):313–322

Oren A (2008) Microbial life at high salt concentrations: phylogenetic and metabolic diversity. Saline Syst 4:2

Paggi RA, Martone CB, Fuqua C, De Castro RE (2003) Detection of quorum sensing signals in the haloalkaliphilic archaeon *Natronococcus occultus*. FEMS Microbiol Lett 221(1):49–52

Park SY, Kang HO, Jang HS, Lee JK, Koo BT, Yum DY (2005) Identification of extracellular N-acylhomoserine lactone acylase from a *Streptomyces* sp. and its application to quorum quenching. Appl Environ Microbiol 71:2632–2641

Pereira CS, Thompson JA, Xavier KB (2013) AI-2-mediated signalling in bacteria. FEMS Microbiol 37(2):156–181

Purdy KJ, Cresswell-Maynard TD, Nedwell DB, McGenity TJ, Grant WD, Timmis KN, Embley TM (2004) Isolation of haloarchaea that grow at low salinities. Environ Microbiol 6:591–595

Rader BA, Campagna SR, Semmelhack MF, Bassler BL, Guillemin K (2007) The quorum-sensing molecule autoinducer 2 regulates motility and flagellar morphogenesis in Helicobacter pylori. J Bacteriol 189(17):6109–6117

Rice SA, Givskov M, Steinberg P, Kjelleberg S (1999) Bacterial signals and antagonists: the interaction between bacteria and higher organisms. J Mol Microbiol Biotechnol 1:23–31

Saeidi N, Wong CK, Lo TM, Nguyen HX, Ling H, Leong SSJ, Poh CL, Chang MW (2011) Engineering microbes to sense and eradicate *Pseudomonas aeruginosa*, a human pathogen. Mol Syst Biol 7:521

Saum SH, Müller V (2008) Regulation of osmoadaptation in the moderate halophile *Halobacillus halophilus*: chloride, glutamate and switching osmolyte strategies. Saline Syst 4:4

Sewald X, Saum SH, Palm P, Pfeiffer F, Oesterhelt D, Müller V (2007) Autoinducer-2-producing protein LuxS, a novel salt- and chloride-induced protein in the moderately halophilic bacterium *Halobacillus halophilus*. Appl Environ Microbiol 73(2):371–379

Skindersoe ME, Ettinger-Epstein P, Rasmussen TB, Bjarnsholt T, de Nys R, Givskov M (2008) Quorum sensing antagonism from marine organisms. Mar Biotechnol 10:56–63

Tahrioui A, Quesada E, Llamas I (2011) The hanR/hanI quorum-sensing system of *Halomonas anticariensis*, a moderately halophilic bacterium. Microbiology 157(12):3378–3387

Tahrioui A, Quesada E, Llamas I (2013a) Draft genome sequence of the moderately halophilic gammaproteobacterium Halomonas anticariensis FP35T. Genome Announc 1(4):e00497–13. doi:10.1128/genomeA.00497-13

Tahrioui A, Schwab M, Quesada E, Llamas I (2013b) Quorum sensing in some representative species of halomonadaceae. Life 3:260–275

Tang K, Zhang XH (2014) Quorum quenching agents: resources for antivirulence therapy. Mar Drugs 12(6):3245–3282

Teasdale ME, Liu JY, Wallace J, Akhlaghi F, Rowley DC (2009) Secondary metabolites produced by the marine bacterium *Halobacillus salinus* that inhibit quorum sensing-controlled phenotypes in Gram-negative bacteria. Appl Environ Microbiol 75:567–572

Teasdale ME, Donovan KA, Forschner-Dancause SR, Rowley DC (2011) Gram-positive marine bacteria as a potential resource for the discovery of quorum sensing inhibitors. Mar Biotechnol 13:722–732

Tommonaro G, Abbamondi GR, Iodice C, Tait K, De Rosa S (2012) Diketopiperazines produced by the halophilic archaeon, *Haloterrigena hispanica*, activate AHL bioreporters. Microb Ecol 63(3):490–495

Toth IK, Newton JA, Hyman LJ, Lees AK, Daykin M, Ortori C, Williams P, Fray RG (2004) Potato plants genetically modified to produce N-acylhomoserine lactones increase susceptibility to soft rot erwiniae. Mol Plant Microbe Interact 17:880–887

Urbanczyk H, Ast JC, Higgins MJ, Carson J, Dunlap PV (2007) Reclassification of *Vibrio fischeri*, *Vibrio logei*, *Vibrio salmonicida* and *Vibrio wodanis* as *Aliivibrio fischeri* gen. nov., comb. nov., *Aliivibrio logei* comb. nov., *Aliivibrio salmonicida* comb. nov. and *Aliivibrio wodanis* comb. nov. Int J Syst Evol Microbiol 57(12):2823–2829

Uroz S, Chhabra SR, Camara M, Williams P, Oger P, Dessaux Y (2005) *N*-Acylhomoserine lactone quorum-sensing molecules are modified and degraded by *Rhodococcuserythropolis* W2 by both amidolytic and novel oxidoreductase activities. Microbiology 151:3313–3322

Usami R, Fukushima T, Mizuki T, Yoshida Y, Inoue A, Horikoshi K (2005) Organic solvent tolerance of halophilic archaea, *Haloarcula* strains: eVects of NaCl concentration on the tolerance and polar lipid composition. J Biosci Bioeng 99:169–174

Ventosa A, Oren A (1996) *Halobacterium salinarum* nom. corrig., a name to replace *Halobacterium salinarium* (Elazari-Volcani) and to include *Halobacterium halobium* and *Halobacterium cutirubrum*. Int J Syst Bacteriol 46:347

Ventosa A, Nieto JJ, Oren A (1998) Biology of moderately halophilic aerobic bacteria. Microbiol Mol Biol Rev 62(2):504–544

Wang Y, Wang H, Liang W, Hay AJ, Zhong Z, Kan B, Zhu J (2013) Quorum sensing regulatory cascades control *Vibrio fluvialis* pathogenesis. J Bacteriol 195(16):3583–3589

Wen Y, Kim IH, Son JS, Lee BH, Kim KS (2012) Iron and quorum sensing coordinately regulate the expression of vulnibactin biosynthesis in *Vibrio vulnificus*. J Biol Chem 287(32):26727–26739

Wenbin N, Dejuan Z, Feifan L, Lei Y, Peng C, Xiaoxuan Y, Hongyu L (2011) Quorum-sensing system in *Acidithiobacillus ferrooxidans* involved in its resistance to Cu^{2+}. Lett Appl Microbiol 53(1):84–91

Williams P (2007) Quorum sensing, communication and cross-kingdom signalling in the bacterial world. Microbiology 153(12):3923–3938

Williams P, Winzer K, Chan WC, Cámara M (2007) Look who's talking: communication and quorum sensing in the bacterial world. Philos Trans R Soc Lond B Biol Sci 362(1483):1119–1134

Yebra DM, Kiil S, Dam-Johansen K (2004) Antifouling technology- past, present and future steps towards efficient and environmentally friendly antifouling coatings. Prog Org Coat 50:75–104

You L, Cox RS III, Weiss R, Arnold FH (2004) Programmed population control by cell–cell communication and regulated killing. Nature 428:868–871

Zhang G, Zhang F, Ding G, Li J, Guo X, Zhu J, Zhou L, Cai S, Liu X, Luo Y, Zhang G, Shi W, Dong X (2012) Acyl homoserine lactone-based quorum sensing in a methanogenic archaeon. ISME J 6(7):1336–1344

Chapter 8
Site Specific Bioinoculants for Sustainable Agriculture in Coastal Saline Soil

Sudipta Tripathi, Shilajit Barua, and Kalyan Chakrabarti

Abstract Salinity is a natural stressor, extensively limiting land use for agriculture in coastal regions throughout the world. Beside plant growth, salinity also has an impact on microbial activity and diversity. Saline habitats have a unique microbial population, adapted to such environment. Recent developments in soil microbiology have improved the potential of saline land use. This involves use of salt tolerant or halophilic bioinoculants to improve nutrient mobilization in saline soil thereby favoring plant growth. Microbial community structure in terrestrial saline habitats varied from halophilic archaea, Gram-negative, Gram-positive bacteria to photosynthetic cyanobacteria. Many of them have been artificially augmented and applied directly to saline soils to improve nutrient status of such soil. One of the most interesting findings in this field of research is spatial and temporal variation in microbial community structure, which was related to variation in salinity of the microenvironment. Modern methods of studying environmental microbiology involves culture independent molecular techniques, biomarker based and in situ activity analyses, which assisted to have a better and detailed look on microbiological process in coastal saline environments. This review aims to recollect findings related to microbiology of coastal saline soils, with special emphasis to studies related to development of bioinoculants compatible for agriculture in coastal saline environments.

Keywords Saline soil • Soil quality • Eco-friendly agriculture • N_2 fixation • PGPR • Bioinoculant

S. Tripathi (✉) • K. Chakrabarti
Institute of Agric Science, University of Calcutta, 51/2 Hazra Road, Kolkata 700019, India
e-mail: sudipta_t@yahoo.com

S. Barua
Department of Microbiol, Vijaygarh Jyotish Ray College, Kolkata, India

D.K. Maheshwari, M. Saraf (eds.), *Halophiles*, Sustainable Development and Biodiversity 6, DOI 10.1007/978-3-319-14595-2_8

8.1 Introduction

In order to support the ever increasing population in the world, a remarkable rise in the productivity of Agric systems had been observed in latter part of the twentieth century.

Today, the global production of food is 145 % greater than that of 1960 (Pretty 2008). Though the present level of Agric output suffices feeding of global population, political, economic and social challenges prevent feeding of population living in crippling poverty (Hazell and Wood 2008). Technological advances enabled agriculture to intensify its production practices, which relied on the application of huge amounts of inorganic fertilizer that effectively meets the demand of crops throughout the growing season and removes the requirement for a fertility enhancing cycle in crop rotations (Cummings and Orr 2011). Significant benefits with respect to Agric food productivity and global food security have been achieved by chemical fertilization. However, such practices had compromised the sustainability of this technology to ensure that food production keeps pace with the increasing population. It has been proved that extensive chemical fertilization leads to the degradation of Agric soils resulting in loss of organic matter, compaction, increase in salinity, leaching of inorganic nitrate, along with associated costs such as fuel requirements and the loss of water resources (Kibblewhite et al. 2008). Chemical fertilization not only had a detrimental effect on soil health but also on the beneficial soil microbial communities and the plants cultivated on these soils.

In order to overcome this ecologically hazardous situation, Agric management systems embracing the principles of sustainability are now being practiced. However, prompt and urgent implementation of such technologies is essential. This would be an ecologically healthy alternative to current conventional Agric management systems, that relies on application of fertilizers and pesticides with expansion of Agric land and indefinite usage of machine (Kitzes et al. 2008). A high demand for organic produce by the present-day health conscious society and sporadic attempts, being made by farmers all over the world to detoxify the land by switching over to organic farming dispensing chemical fertilizers, pesticides, fungicides and herbicides (Chavada 2010a) are significant steps towards sustainable agriculture.

Improvement of the beneficial associations between microorganisms and plants, particularly in the rhizosphere, is an area of research of global interest (Kloepper 1994). Microbes present in soil are the principal mediators of elemental geochemical cycling, thereby affecting plant nutrient status of soil. Alterations in the biological or chemical environment of the rhizosphere are expected to influence plant health as many rhizosphere microorganisms mobilize nutrients, produce phytohormones and suppress pathogens (Tripathi et al. 2002). They affect plant growth in different ways. For example, *Azotobacter* can affect plant growth directly, either by fixing nitrogen or by producing plant growth promoting hormones or indirectly by change in the microflora of the rhizosphere (Sinegani and Sharifi 2004).

Bio-fertilizer are bio-organic substance that have the capability to improve soil fertility significantly by fixing atmospheric nitrogen both in isolation with or without plant root. These can solubilize insoluble phosphate and produce plant growth hormone (auxin, indolacitic acid) (Chavada 2010b).

The term ‘Biofertilizer’ in India specifies fertilizers to meet the nutritional requirements of a crop through microbiological means, while other countries use the term “microbial inoculants”. These are usually carrier based microbial preparations containing beneficial microorganisms in a viable state intended for seed or soil application. They enhance plant growth through nutrient uptake and/or growth hormone production. Important and popular microbial inoculants in our country are those that supplement nitrogen, phosphorus and plant growth promoting rhizobacteria (PGPR) (Brahmaprakash and Sahu 2012).

A major drawback of biofertilizer application is the unpredictability of its potentiality in improving crop productivity (Yadav et al. 1998). The biological nature of biofertilizers and their susceptibility to abiotic factors is responsible for their highly inconsistent performance. Many microorganisms have a short shelf-life, and there is a lack of suitable carrier material for restoration and longevity in actual field conditions. In general, the response of crops to biofertilizer inoculation is highly variable, location-specific and management dependent. Hence, in order to harness potential benefit of bio fertilizer in commercial agriculture, the consistency of their performance needs to be improved (Acharya and Sharma 2007). A bacterial strain found to be ideal at one location, might be ineffective at another due to competition from native soil microbes, poor aeration, high temperature, soil moisture, acidity, salinity, alkalinity and presence of other toxic elements, etc. (Johnston and Syers 1998). Therefore in order to develop efficient biofertilizer, it is essential to perform location-specific research to identify and isolate new biofertilizers, study their characteristics and developmental needs such as growth requirements and multiplication techniques, and study the influence of other factors on their large-scale use under field conditions (Banta and Mendoza 1984).

Substantial Agric lands in the tropics and subtropics are affected by salinity. It is estimated that 20 % of the irrigated land in the world is presently affected by salinity (Yeo 1998). Salinity represents a very serious problem since salinization suppresses plant growth, particularly in arid and semiarid areas (Parida and Das 2005). The loss of farmable land due to salinization is directly in conflict with the needs of the world population, which is projected to increase by 1.5 billion over the next 20 years and the challenge of maintaining the world food supplies (Joseph and Jini 2010).

Microorganisms play a significant role in nutrient transformation in soil. They are the source and sink of plant nutrients in soil and instrumental in organic matter dynamics. Although salinity regulates soil microbial parameters (El-Shinnawi and Frankenberger 1988; Tripathi et al. 2006) a large population of microorganisms, mostly bacteria, survives in saline soils. Bacterial populations of saline soils are halophilic/salt tolerant representatives of bacteria inhabiting arable non-problem soils. Their ability to adapt to fluctuating osmolality in the environment is of fundamental importance for their survival (Rodriguez-Valera 1988).

Saline habitats are nitrogen poor (Sprent and Sprent 1990). N-input in the form of chemical N fertilizers though preferable, caused considerable land degradation and detrimentally affected microbial process in such environment. Therefore, organic N-input is very important in this environment. An increasing supply of N through biological nitrogen fixation may increase crop production in such environments (Zahran 1997). Hence, most of the biofertilizers (bioinoculants) developed for saline soil agriculture are nitrogen fixing biofertilizers (Chavada et al. 2010a, b). Butale et al. (2010) reported experimental usage of haloalkaliphilic non-symbiotic diazotrophs isolated from Lonar lake, India in improving crop productivity in saline soils. *Azolla* as a biofertilizer offer improved soil quality and enhance rice productivity in these saline soils (Szuster et al. 2010).

Numerous literatures exist concerning the ecology of free-living diazotrophs and nitrogenase activity in normal arable soils. However, reports in this respect from saline soils, in general (El-Sinnawi and Frankenberger 1988; Wollenweber and Zechmeister-Boltenstern 1989; Saleena et al. 2002): and coastal saline soils of West Bengal (Sengupta and Chaudhuri 1991; Sarkar et al. 1995) in particular, are scarce. Understanding factors affecting diazotrophic populations and nitrogen fixation in soil is important for beneficial utilization of such microbes for efficient crop production.

Plant growth-promoting rhizobacteria (PGPR) are now-a-days used for reducing chemical inputs in agriculture. Application of PGPR for improving crop productivity under salinity stress has also been reported. They includes diazotrophic bacteria, *Pseudomonas* sp, *Bacillus* sp and arbuscullar mycorrhizal fungi (Singh et al. 2011).

Enumeration of dinitrogen fixing populations represent a fraction of total microbial population in soil (Sekiguchi 2006) and fails to throw light on diversity of diazotrophic population in soil. For better resolution of diazotrophic community structure in soil, Widmer et al. (1999) adopted polymerase chain reaction- restriction fragment length polymorphism (PCR-RFLP) of *nifH* gene. Poly et al. (2001) studied similarities and differences in the structures of the *nifH* gene pools of six different soils and five soil fractions by PCR-RFLP technique. Molecular approach to study the diversity of diazotrophs is primarily based on PCR amplification of *nifH* gene. Currently the *nifH* database constitutes one of the largest gene database (Zehr et al. 2003): with little information from saline soils (Chelius and Lepo 1999).

Saline soils might have their own microbial communities adapted to saline environments (Zahran 1997). For efficient use of the bacterial diazotrophic communities in coastal agriculture, it seems prudent to use site-specific N-fixing microbial inoculants which are naturally adapted to a particular ecosystem. Such efficient strains can be utilized to improve the capability of soil to mobilize and supply nutrients to crops naturally.

8.2 World Salinization with Emphasis on India

Land degradation is a matter of grave concern as far as increasing demand of global food production is concerned. It is estimated that about 15 % of the total land area of the world has been degraded by soil erosion, physical and chemical degradation, including soil salinization (Wild 2003).

Salinization of soil occurs due to accumulation of water-soluble salts in the soil column (the upper part of a soil profile, including the A and B horizons) or regolith (the layer or mantle of fragmental and unconsolidated rock material, whether residual or transported) to a level that affects Agric productivity (Rengasamy 2006).

As per Richards (1954), a soil is considered saline if the electrical conductivity of its saturation extract (EC_e) is above 4 dSm^{-1} at 25 °C. However, the deleterious effects of soil salinity depend on several factors including plant type, soil-water regime and climatic condition (Maas 1984).

Plant productivity in saline soils is considerably reduced due to improper nutrition of plants in addition to the osmotic and drought stresses (Munn 1993). As per Munns 2002, osmotic effect which reduces the ability of the plant to take up water and ion, majorly affects plant growth. Induction of programmed cell death (PCD) in plant cells due to salinity stress has also been reported (Palavan-Unsal et al. 2005).

According to FAO, the total global area of salt-affected soils including saline and sodic soils was 831 million hectares (Martinez-Beltran and Manzur 2005); extending over all the continents including Africa, Asia, Australasia, and the Americas. The most recent report on global distribution of salinity affected soil is represented in Table 8.1 (Squires and Glenn 2009).

In India, out of an estimated area of 187.7 million hectares of total degraded land, 8.1 million hectares are salt affected and out of which 3.1 million hectares of land are in the coastal regions (Tripathi et al. 2007). Among all the states of India, West Bengal has the largest area (0.82 million hectares) of salt affected soils, covering the coastal districts of North and South 24-Parganas, Haora and East Midnapore in the delta region of the river Ganges (Bandyopadhyay et al. 2001). Coastal West Bengal could be visualized into two parts: one is the mangrove forest area, which is popularly known as the Sunderbans, and the other is the forest cleared Agric land and dwelling places.

The Sunderbans delta consists of conglomeration of islands, about 56 in number, separated by interconnected tidal rivers, creeks and canals linked to innumerable rivers that ultimately flow into the Bay of Bengal. The islands have mudflats and

Table 8.1 World distribution of salt-affected areas

Continent	Saline (Mha)	Sodic (Mha)	Total (Mha)
Africa	122.9	86.7	209.6
South Asia	82.2	1.8	84.0
North and Central Asia	91.4	120.1	211.4
Southeast Asia	20.0	–	20.0
South America	69.4	59.8	129.2
North America	6.2	9.6	15.8
Mexico/Central America	2.0	–	2.0
Australasia	17.6	340.0	357.6
Global total	411.7	617.9	1029.5

Mha million hectares of land

uplands. Only the mudflats are subjected to periodic inundation by tidal seawater, while the uplands remain clear of inundations.

The Sunderbans falls a little south to the Tropic of Cancer, between the latitudes and longitude 21°31′N-22 °30′N and 88°10′E-89°51′E. Since the region is situated almost near Tropic of Cancer, it undergoes defined seasonal changes. The seasons are distinctly categorized as summer (April–May) with higher temperature occasionally accompanied by rains and thunder storms; monsoon (July–October) with an average rainfall of 1,650 mm, and winter (November–February):characterized by cold weather and negligible rainfall. The average humidity is about 82 % and is more or less uniform throughout the year.

The dominant vegetations in the Sunderbans forest area are mangroves of various kinds, and they grow densely on the uplands.

8.3 Physico-chemical Properties of Coastal Saline Soils

All soil types with diverse morphological, physical, chemical and biological properties may be affected by salinity (Rengasamy 2006). Earlier, in a FAO soil bulletin, Abrol et al. (1988) tabulated selected physicochemical properties of coastal and acidic saline soils (Table 8.2)

Soil characteristics of coastal region vary widely from place to place depending on their physiographic locations, climatic conditions, soil forming materials, the characteristics of the ground water at shallow depth etc. Almost all the coastal soils have saline ground water table at shallow depth. Those were mostly heavy textured (Bandyopadhyay et al. 1998).

From pedologic point of view, coastal soils are usually younger soils and are mostly classified as Entisols and Inceptisols (Bandyopadhyay et al. 1998). The soils are grey to greyish black in colour, generally silty clay in texture (10 % sand, 51.2 % silt and 38.8 % clay) and have organic carbon content of 0.98 % (Yadav et al. 1981). Sandy and alkaline soils are found on the islands near the Bay and many other degraded places on the surface. Sand occurs only along the sea face where the strength of the tides and the roughness of the water is more, which prevents fine particles from settling.

The soils are generally saline in nature. Salinity of the soils in this region is variable during different seasons. It reaches maximum in summer, decreases during monsoon (Yadav et al. 1981) and regains to rise during winter.

ECe (electrical conductivity of saturation water extract of soil) of coastal soils mostly vary from 0.5 dSm^{-1} in monsoon to 50 dSm^{-1} or more in summer. In coastal saline soils of West Bengal the soluble salts are mainly comprised of chlorides and sulphates of Na, Mg, Ca and K in the decreasing order of preponderance. Bicarbonate is present in traces in a few soils while carbonate is mostly absent in all the soils (Bandyopadhyay et al. 1988, 2003). Most of the soils are slightly acidic to neutral or slightly alkaline in reaction (Yadav et al. 1983).

Table 8.2 Characteristics of typical saline soils

	Mechanical composition (%)							Composition of the saturation extract					
Depth (cm)	Organic mattet (%)	Clay <2 μ	Silt (2–50 μ)	Sand (50 μ–2 mm)	pH_s	CEC me/ 100 g	ECe dSm^{-1}	Na^+	Ca^{++}	Mg^{++}	Cl^-	SO_4^{--}	SAR
Coastal saline soil. Canning, West Bengal, India (Bhargava, personal communication)													
0–12	0.4	M	26	40	7.2	15	27	156	29	120	213	16	18
12–28	0.3	45	18	17	7.5	18	7	38	11	17	S7	8	10
28–80	0.3	46	36	16	7.5	20	7	43	17	19	73	9	10
80–105	0.3	40	44	17	6.7	16	8	46	16	16	75		12
Acid saline soil, Calicut, Kerala, India (Bhargava, personal communication)													
0–9	4.8	50	21	29	4.4	19		320	30	93	354	102	41
9–20	3.6	40	31	29	4.8	20	13	100	8	30	106	36	23
20–36	2.4	47	33	20	4.8	15	12	82	10	25	86	53	17
36–70	2.6	65	24	11	4.5	26	8	76	6	20	58	65	21

pH_s pH of soil saturated paste, *CEC* cation exchange capacity, *SAR* sodium adsorption ratio

Highly acidic soils with acid sulphate characteristics at surface/subsurface soil horizons were reported in places of Kerala, West Bengal, Orissa and Andaman and Nicobar group of islands (Bandyopadhyay and Bandyopadhyay 1984; Bandyopadhyay and Maji 1995).

Physical and chemical properties of soils of the successional stages of mangrove plant community of the Ganges river estuary, India was reported by Sengupta and Chaudhuri (1991). The pH, EC_e, organic carbon and total nitrogen were in the ranges of 7.2–8.4, 3.0–16.0 (m.mhos cm^{-1}):0.46–1.42 (%) and 0.09–1.17 (%) respectively.

Sarkar et al. (1995) studied the physico-chemical characteristics and chemical properties of mangrove soils from five different locations of Sunderbans viz. Pathankhali Undisturbed, Datter Bhidya Junction, Jhinge Khali, Masjid Bari and Puinjali. The authors reported that the pH of the soils varied from 5.1 in Datter Bhiya Junction to 7.8 in the other soils studied. The EC (dSm^{-1}) varied from 2.1 in Pathankhali to 26.0 in Jhinge Khali. Organic carbon of soil (%) varied from 0.27 in Pathankhali to 1.90 in Datter Bhidya Junction.

Maji et al. (1998) carried out a reconnaissance soil survey in the Sagar Island of the Sunderbans delta region of West Bengal. The surface horizons of the studied soils had pH 6.1–6.5, ECe 7.0–8.2 dSm^{-1} and CEC 18.5–21 cmol (p^+) kg^{-1}. Organic carbon content of soils varied from 6.9 to 7.8 g kg^{-1}.

In a voluminous monograph, Naskar and Mandal (1999) recorded physico-chemical parameters of soil samples from different major mangrove flora in West Bengal. Soils under members of the *Rhizophoraceae* family recorded: pH 8.22–8.36; EC 1.35–3.50 (mmhos cm^{-1}); organic carbon 0.31–0.74 (%); sand 45–74 (%); silt 10–37 (%); clay 14–26 (%). Same values under members of *Sonneratiaceae* and *Avicenniaceae* family were: pH 8.03–8.43 and 8.36–8.50; EC 0.60–4.35 and 1.50–2.50 (mmhos cm^{-1}); organic carbon 0.56–0.79 and 00.41–0.49 (%); sand 30–48 and 65–72 (%); silt 25–42 and 4–14 (%); clay 10–45 and 21–24 (%) respectively.

Tripathi et al. (2007) studied the seasonal variation of selected physico-chemical properties of coastal arable soils of Canning Town, West Bengal, India. The authors reported statistically insignificant variation in average soil pH values between the seasons. Soils collected from different sites exhibited a wide range of pH between 4.8 and 7.8. The average ECe of the soils varied widely between the seasons, with the lowest average value recorded in the monsoon season (2.7 dSm^{-1}) followed by an abrupt increase in the winter season (8.3 dSm^{-1}) which peaked (13.8 dSm^{-1}) in the summer season. Soil samples from different sites showed a wide range (2.2–16.3 dSm^{-1}) of variation in average ECe. The average organic carbon (OC) content of soils in winter (10.1 g kg^{-1}) was significantly higher than that of summer (9.2 g kg^{-1}) and monsoon (9.1 g kg^{-1}) seasons. The soils collected from different sites showed large difference in average OC content within the same season, varying from the lowest value 5.2 g kg^{-1} to the highest value of 14.1 g kg^{-1}. The average TN content of the soils ranged from 0.68 to 1.4 g kg^{-1} during different seasons.

Ramanathan et al. (2008) reported that the pH of Jharkhali and Pakhiralay of Sunderbans region were 7.4 and 7.3 respectively while the EC values of the same soils were 16,250 and 16,000 mS cm^{-1} respectively.

Recently, significant seasonal variation in physico-chemical properties in saline tracts of coastal West Bengal (Barua et al. 2011) and Barhanpur, Maharashtra (Mali et al. 2012) were being reported. Both these reports concluded that soil salinity reaches maximum in summer (as evident by high ECe and dominance of Na^+ and Cl^- resulted in high SAR value) and least during monsoon.

8.4 Microbiology of Saline Soils

Saline soils may have their own microbial communities, which have adapted to such environments. Among them, the bacteria is the most important and dominant inhabitant. Bacteria in saline habitats falls in two categories: Archaebacteria and Eubacteria. The former are extremely halophilic microorganisms which grow optionally at salt saturation (up to 30 % NaCl): while the later included non-saline halotolerant bacteria which have become adapted to these extreme environments (Zahran 1997).

A significantly higher number of bacteria, recovered in medium with high salinity, were found to colonize the hypersaline soils of Spain (Quesada et al. 1982; Del Moral et al. 1987) and Egypt (Zahran 1992).

The strict halophiles only develop in media with higher concentrations of Na^+, whereas the facultative halophiles usually develop in media containing low concentrations of Na^+ (Giambiagi and Lodeiro 1989). The total count of bacteria is usually negatively correlated with the total soluble salts of saline soils (Ragab 1993) but positively correlated with organic carbon contents (Ragab 1993).

The Gram-negative bacteria appear to be much more frequent in saline environments (Quesada et al. 1982; Del Moral et al. 1987). Members of the Gram-negative genera *Vibrio*, *Pseudomonas*, *Acinetobacter* and *Alteromonas* were isolated from saline habitats (Del Moral et al. 1988). The root-nodule bacteria are another group of Gram-negative bacteria which were reported to colonize the saline soils of Greece (Douka et al. 1978).

Bacteria in the saline soils are not confined to definite group of phylogenetically related microbes but represent a group which has evolved in many different groups of organisms. The wide spectrum includes the following representatives: *Halomonas*, *Pseudomonas*, *Vibrio* and *Actinopolyspora* and a whole range of Gram-positive rods and cocci, e.g. *Bacillus*, *Micrococcus* and *Salinicoccus* (Trúper et al. 1991). The gram-positive bacteria are also found in saline habitats, and members of the genera *Bacillus* and *Micrococcus* are dominant among other Gram-positive bacteria in saline soils. Microbial aspects of salt affected soils in India have received very little attention. Bajpai and Gupta (1979) reported sulphur metabolism of chemolithotrophic bacteria and other bacteria from saline and alkaline soils in Uttar Pradesh.

8.5 Free Living Diazotrophs and Dinitrogen Fixation in Coastal Saline Soil

Nitrogen fixation is considered as one of the significant biological processes in soil. It is influenced drastically by environmental factors such as temperature, pH, oxygen, and mineral nutrients (Buresh et al. 1980).

Factors, such as salinity of soil may affect soil fertility by inhibiting the activity of microorganisms mediating the nitrogen turnover. In saline ecosystems, the conditions created by the density of the soil and the water regime are limiting for nitrogen fixation (Rice and Paul 1971).

Iswaran and Sen (1958) made an effort to observe the salt tolerance of *Azotobacter* isolated from soils of different parts of India. Sodium chloride was found to affect nitrogen fixation adversely. The concentration of sodium chloride, which inhibited fixation of nitrogen completely, varied for different strains. The rate of decrease in nitrogen fixation with increase in the concentration of sodium chloride was less for *Azotobacter* isolated from soils of higher salinity than for the organisms from less saline ones. Concentration of sodium chloride, which inhibited fixation of nitrogen, was higher for *Azotobacter* isolated from soils containing more soluble salts than for the organisms isolated from the soils with less soluble salts.

There is immense volume of literature with respect to the ecology of free-living nitrogen fixers. Not many reports are available on impact of salinity on soil microorganisms, in general, and dinitrogen fixation and asymbiotic diazotrophs, in particular (Mishustin and Shillnikova 1971).

The low oxygen tension in saline soils may favor the process of nitrogen fixation, but the diffusion of gases may be impaired at higher density and water regime in saline soil, an effect that might reduce nitrogen fixation (Rice and Paul 1971).

Earlier studies, as reviewed by Mishustin and Shillnikova (1971) revealed that the population of *Azotobacter* decreased sharply with the increase in salinity and the inhibition was related to the composition of the salts. Subsequent studies (Ibrahim 1974; Cervantes and Olivares 1976; Mahmoud et al. 1978) also supported earlier contentions.

Tilak and Krishnamurti (1981) found that *Azospirillum* survived in saline and alkali soils and seed inoculation with this organism in these soils increased yields of a salt tolerant crop like barley in India.

The occurrence and characterization of N_2-fixing *Azospirilla* in some Egyptian soils has been investigated. According to the physiological properties studied, all isolates were classified as members of *Azospirillum brasilense* (Nadia et al. 1984).

Zafar et al. (1987) indicated the presence of *Klebsiella pneumonae* and *Beijerinkia* sp. in the rhizosphere of salt tolerant grass *Leptochha fusca*, and these could tolerate 3 % NaCl in the medium.

Jena et al. (1988) noted that nitrogen fixation in *Azospirillum* sp. from saline soils decreased with an increase in salinity level at Cuttack, India. El-Shinnawi and Frankenberger (1988) studied the nitrogenase activity and population dynamics of isolated-free living aerobic bacterial diazotrophs enriched from soils upon salt

(KCl, NaCl, $CaCl_2$ and $MgCl_2$) addition after 3–24 days of incubation. In all the cases, the applied salts inhibited the population of diazotrophs. However, survival was evident even at higher salt levels. The intensity of inhibition was higher in the salinized culture medium than in soils. Nitrogenase activity, as determined by acetylene reduction method, followed a similar pattern as that of the cell densities of diazotrophs in the salinized soils. The extent of inhibition with regard to cell population density and nitrogenase activities was as follows: (1) among the Cl^- salts $CaCl_2$>KCl>NaCl>$MgCl_2$; (2) co-ion treatments, CO_3^{2-}>Cl^->SO_4^{2-} with the degree of inhibition varying with accompanying cations, Mg^{++} being the least inhibitory; and (3) combination of salt mixtures (homocationic) Cl^- plus $CO_3^=$ > $SO4^=$ and (heteroionic) Na_2CO_3 plus $CaSO_4$ being the most inhibitory, and $MgCO_3$ plus K_2SO_4 being the least.

Xanthobacter lavas and *Alcaligem paradoxus* from strongly saline Takyar soil in USSR showed nitrogenase activity when they were grown in media containing 0.5–2.0 % NaCl. The predominant growth of these bacteria in the complex of diazotrophs from strongly saline Takyar soils and utilization of various substrates has been noted (Kravchenko and Kaliminskaya 1988).

A.brasilense and *A. lipoferum* were isolated from the roots of wheat grown in sandy saline soils in Iraq (Al Maadhidi 1989). Higher rates of nitrogen fixation in saline soils, compared to non-saline soils and Agric soils, were reported (Wollenweber and Zechmeister-Boltenstern 1989). Saline habitats are nitrogen poor (Sprent and Sprent 1990). Various groups of diazotrophic bacteria grow and fix atmospheric nitrogen, but their population and nitrogen fixing ability are affected by numerous environmental factors of which salt stress is important (Rai 1991).

In a study, Sengupta and Chaudhuri (1991) collected soils from representative locations of four distinct successional stages (formative mangrove swamp, developed mangrove swamp, declining ridge mangrove and protected Agric land) from 10 km^2 of the riverine delta. The authors observed that there were 100–300 times as many N_2-fixing aerobic and microaerophilic bacteria in root association compared to plant free sediments. Fourteen isolates, presumed to be N_2-fixing bacteria belonging to different O_2 response groups, apparently different in their morphology, were obtained from 20 different plant species of the mangrove ecosystem. Nine distinct strains of N_2-fixing bacterial species were present in root association of mangrove and associate plants of the ecosystem. There were three different isolates of *Azotobacter*, one of *Klebsiella*, two of *Clostridium* and one of *Azospirillum* and other two isolates could not be identified.

Sarkar et al. (1995) reported the population and some characters of the free-living aerobic heterotrophic N_2-fixing microorganisms in soils of five different islands of the Sunderbans. The islands were Pathankahli (undisturbed) Datterbhidya junction, Jhinge Khali, Masjidbari and Puinjali. Out of the five locations, N_2 fixing population was found only in three. The populations ranged between 2×10^6 and 2×10^3 microorganisms g^{-1} soil. Results further showed that the isolates were mostly rods. All the isolated microorganisms were gram negative and motile. Some of them showed the presence of capsule. Turbidimetric studies of the isolates in the N free media showed that maximum growth occurred after 48 h of incubation. Hassouna et al. (1995) observed negative correlation between the diazotrophic bacteria and salinity.

Several strains of *Bacillus*, *Halomonas* and *Azospirillum. Herbaspirillum* have been reported. The persistence of these bacteria in saline soil emphasizes that the search for new heterotrophic nitrogen-fixing (Gram-negative) halotolerant bacteria colonizing saline soils is worth further investigation (Zahran et al. 1995). Ravikumar and Vittal (1996) reported that N_2-fixing *Azotobacter* is abundant in the mangrove soils.

The diazotrophic and salt tolerant *Azotobacter* was isolated from non-saline habitats, which showed either decreasing nitrogen fixation rates with increasing salinity (>10 % NaCl) or optimum fixation at low salinity (5–10 % NaCl) (Zahran 1997).

Biological nitrogen fixation systems are one of the most effective ways of increasing the N status and sustaining crop production. However, salinity and alkalinity adversely affect not only the physico-chemical properties of soil but also the soil microbial activities. Nitrifying bacteria and *Azotobacter* are very sensitive to salinity. Soil improvement on leaching after subsurface drainage installation in an alluvial saline soil resulted in increased number of *Azotobacter* (Rao 1994; Rao and Barrueco 1998).

In Egypt 156 representative soil samples, particularly from Nile valley, showed the presence of *Azotobacter croococcum* and *Azotobacter vinelandii*. Development of these species in saline soil was improved by amending the soil with 1 % maize straw (Gangawane 2007).

Ramanathan et al. (2008) studied microbial loads of the mangrove soils of the Sunderbans including the N_2-fixing bacterial population. The study included two forest areas-viz. Jharkhali and Pakhiraly and forest cleared soil at Canning. Free living N_2-fixers in Jharkhali and Pakhiralay were 1.6×10^4–1.44×10^4 colony forming units respectively. The populations were lower compared to that of Canning (13.67×10^6 cfu).

Recent scientific works in this regard reports variation in the count of free living diazotrophs and dinitrogen fixation in saline soils of Sunderbans mangroves, West Bengal, India. Response to variation in salinity was studied and the diazotrophic members were concluded to be halotolerant rather than halophillic. Pure culture of diaztrophs were prepared and characterized. Molecular characterization revealed them to be members of diverse group of non-conventional diazotrophic bacteria like *Agrobacterium*, *Pseudomonas*, *Bacillus*, *Vibrio*, etc. (Barua et al. 2008). Barua et al. (2011) reported seasonal variation in the count of free living diazotrophs and dinitrogen fixation in saline Agric soils of coastal West Bengal, India. Detrimental effect of increased salinity in Summer on diazotrophic population and their activity were reported. Pure cultures of diazotrophs obtained were efficiently exhibited nitrogenase (ARA) activity under 1 % salinity stress. Molecular characterization revelaved all of them to be members of genus *Bacillus* (Fig. 8.1).

In a related work, *Azotobacter* sp. with variation in pH of saline soils in Bangladesh were reported. Pure cultures of *Azotobacter* sp. from different salt affected regions of Khulna and Satkhira in Bangladesh were prepared and characterized. The salinity tolerance and N_2 fixation of these isolates were reported and selected isolates were concluded to be prospective members to be used as biofertilizers for agriculture in saline soils (Akhter et al. 2012).

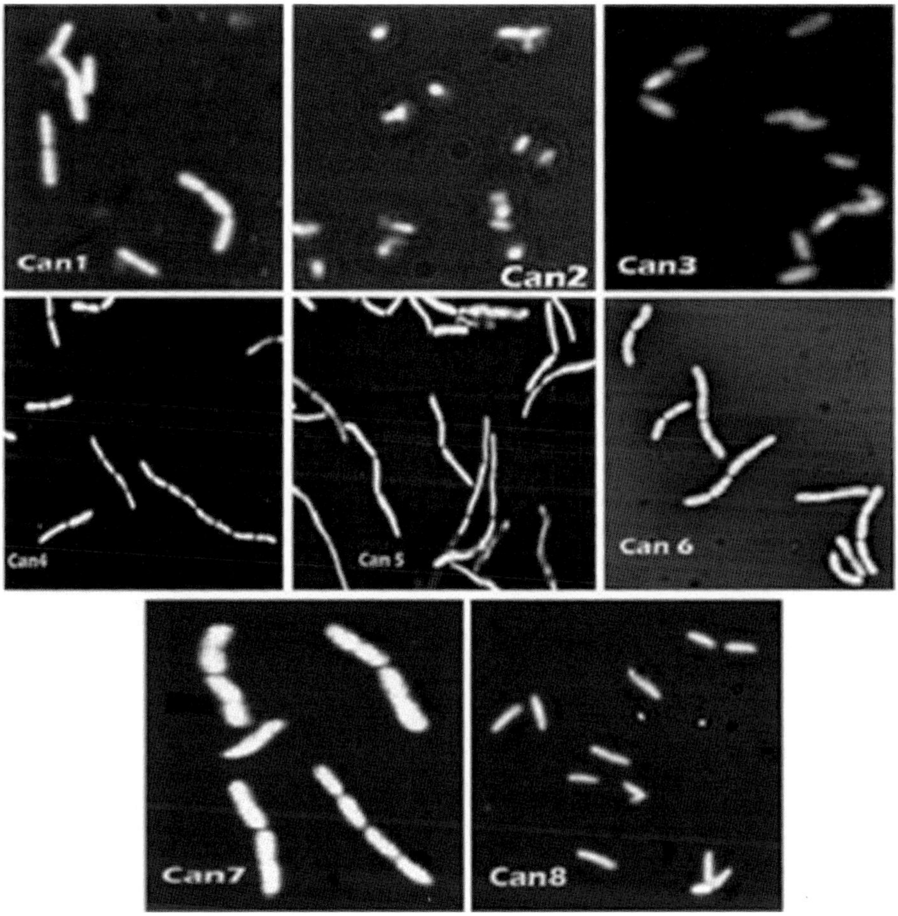

Fig. 8.1 Light microscopic appearance of the isolates by negative staining (Barua et al. 2011)

A strain of *Azospirillum brasilense* (N-30) was isolated from saline soil and its growth pattern, stress tolerance were studied and mass production was optimized. Efficacy of the isolate as potential biofertilizer was established in pot experimentation (Bapurao 2012).

8.6 Acetylene Reduction Activity of Free Living Diazotrophs Under Salinity Stress

Acetylene reduction assay (ARA) is a simple, rapid, sensitive and cost effective method of measuring biological nitrogen fixation. This technique commonly involves the incubation of soil, bacterial culture, whole plants or plant parts in a closed vessel containing about 10 % acetylene for a time period of approximately 0.5–2 h. The reduction of N_2 by the nitrogenase enzyme is inhibited and instead the enzyme reduces acetylene to ethylene. These gases are measured by gas chromatography (Ladha et al. 1992).

Acetylene reduction activity (ARA) of several non-symbiotic diazotrophic strains isolated from saline soils of Egypt ranged from 15 to 85 nmoles of C_2H_2 2 ml^{-1}12 h^{-1} in non-saline media and 10–60 nmoles of C_2H_2 2 ml^{-1}12 h^{-1} at 5 % NaCl in media (Zahran et al. 1995).

Out of 20 diazotrophic strains isolated from the roots of *L. sindicus*, an endemic grass, which grows naturally in sandy soils of the Thar Desert in Barmer and Jaisalmer Districts of Western Rajasthan, India, C3 and C3r1 strains showed considerable acetylene reduction, 18 and 11.2 nmol C_2H_4 min^{-1} mg $protein^{-1}$ respectively (Chowdhury et al. 2007).

Effect of salinity on ARA activity of diazotrophs isolated from saline tracts was much less than on isolates obtained from non-saline soils. Diverse group of diazotrophic bacteria as reported by Barua et al. (2008, 2011) reported salt tolerant diazotrophs exhibited considerable acetylene reduction both under salinity stress (100.6–1.72 nmole of C_2H_4) and stress free conditions (174.47–4.51 nmole of C_2H_4). In both the reports, a stain of *Azotobacter vinelandii* isolated from non-saline environments were used as control which exhibited 96 % decrease in ARA under 1 % salinity stress.

Islam et al. (2010) studied the effect of four different selective N_2-free media on ARA of diazotrophic microorganisms isolated from Korean paddy soil. The ARA varied from as low as 1.8 to as high as 2,845 nmol C_2H_4 min^{-1} mg $protein^{-1}$ in different media. Based on the data, authors showed significant interactions between the isolates and the media with respect to ARA.

In a recent study, effect of salinity on ARA activity of microbial mats from intertidal regions were studied. Analysis were carried out at three different stations at different time points during the incubation period of 72 h and for 5 salinities ranging from 0 to 165 PSU. At all stations, highest activity were observed at half the natural salinity (16.5 PSU) or natural salinity (33 PSU) and decreased at two- and fivefold the natural salinity (66 PSU and 165 PSU, respectively) (Severin et al. 2012).

8.7 Plant Growth Promoting Rhizobacteria and Salinity Stress

Bacteria that exert beneficial effects on plant development known as plant growth promoting rhizobacteria (PGPR) have been reported widely. One of the basic requirements for the effectiveness of PGPR is their ability to colonize hosts rhizosphere, rhizoplane, or the root interior (Glick et al. 2007). The PGPR were applied to various crops to enhance growth, seed emergence and crop yield, and a few such applications were commercialized (Dey et al. 2004; Herman et al. 2008; Minorsky 2008). Benefits derivable from plant – PGPR interactions to includes improvements in seed germination rate, root development, shoot and root weights, yield, leaf area, chlorophyll content, hydraulic activity, protein content, and nutrient uptake – including phosphorus and nitrogen (Adesemoye and Kloepper 2009).

Diazotrophic bacteria are also PGPR, because of their competitive advantage in C-rich and N-poor environments (Kennedy et al. 2004). In addition to fixing nitrogen, diazotrophic bacteria were reported to secrete growth promoting hormones and solubilize insoluble phosphate (Verma et al. 2001; Loganathan and Nair 2004) thereby promoting plant growth. By virtue of such attributes, pre-treatment of seeds with a suspension of Azotobacter isolated from mangrove soils, was shown to improve seed germination and plant growth under laboratory conditions (Ravikumar et al. 2004).

Diverse soil microorganisms including algae (Finnie and Van Staden 1985) fungi (Stein et al. 1990) and bacteria (Muller et al. 1989) are capable of producing physiologically active quantities of auxins, which may exert pronounced effects on plant growth and establishment. Indole acetic acid (IAA) is one of the most physiologically active auxins. It is a common product of L-tryptophan metabolism by several microorganisms including PGPR (Frankenberger and Brunner 1983; Lynch 1985).

Microorganisms inhabiting rhizospheres of various plants are likely to synthesize and release auxin as secondary metabolites because of the rich supplies of substrates exuded from the roots compared with non rhizospheric soils (Strzelczyk and Pokojska-Burdziej 1984).

The quantity of IAA produced by the bacterial isolates varied from strain to strain. Production of IAA by 18 strains of *Acetobacter diazotrophicus*, analyzed quantitatively by the colorimetric Salkowsky assay, was as high as 19–65 μg $ml^{-1.}$

Many bacteria isolated from the rhizosphere have the capacity to synthesize IAA in vitro in the presence or absence of physiological precursors, mainly tryptophan (Caron et al. 1995; Davies 1995).

Microbial isolates from the rhizosphere of different crops appear to have a greater potential to synthesize and release IAA as secondary metabolites because of the relatively rich supply of substrates (Muller et al. 1989; Caron et al. 1995).

Production of IAA by microbial isolates varies greatly among different species and strains and depends on the availability of substrate(s). Two new N_2-fixing species of Gluconacetobacter, *Gluconacetobacter johannae* and *Gulconacetobacter azotocaptans*, isolated from the rhizosphere of coffee plants in Mexico, were found to produce IAA in the range of 5.30–24.51 μM (Fuentes-Ramirez et al. 2001). The range of IAA production in several phosphate solubilizing bacterial isolates with tryptophan was 57–288 μg ml^{-1} culture media (Shahab et al. 2009). However, the above reports in this section are mostly concerning with isolates from normal soils.

Phosphorus exists in soil in various organic and insoluble inorganic forms. Goldstein (1986) reported that many soil microorganisms are able to solubilize insoluble phosphates through the production of organic acids. Notable phosphate solubilizing bacterial genera are *Pseudomonas*, *Bacillus*, *Rhizobium*, *Burkholderia*, *Achromobacter*, *Agrobacterium*, *Microccocus*, *Aereobacter*, *Flavobacterium* and *Erwinia* (Rodríguez and Fraga 1999).

The discovery of mutual relationship between plants and phosphate solubilizing bacteria (PSB):in which bacteria provide soluble phosphate and plants supply root borne carbon compounds (mainly sugars):encouraged the development of new technologies, such as the use of PSB for biofertilization to improve crop yield (Goldstein 1995; Pérez et al. 2007).

Diazotrophic microorganisms showing phosphate solubilizing activity are rarely reported. Verma et al. (2001) identified two endophytic diazotrophic phosphate solubilizers that of *Pantoea agglomerans* (BOX PCR types I and III) by their ability to show clear zones in Pikovskaya medium amended with tricalcium phosphate.

Loganthan and Nair (2004) reported two diazotrophic strains *Swaminathania salitolerans* PA51 and PA12, isolated from mangrove-associated wild rice (*Porteresia coarctata* Tateoka) of three different sites along coastal Tamil Nadu, India. These two strains also exhibited tricalcium phosphate solubilizing activity in Pikovskaya medium.

Soil contains a wide range of phosphate containing organic substrates, which can be a source of P for plant growth. To make this form of P available for plant nutrition, it must be hydrolyzed to inorganic P.

Mineralization of most organic phosphorous compounds is carried out by means of phosphatase enzymes (Rodríguez and Fraga 1999). Burns (1983) studied the activity of various phosphatases in the rhizosphere of maize, barley, and wheat. Phosphatase activity was considerable in the inner rhizosphere at acidic and neutral soil pH.

Significant acid phosphatase activity was observed in the rhizosphere of slash pine in two forested Spodo soils (Fox and Comerford 1992). The major source of phosphatase activity in soil is considered to be of microbial origin (Garcia et al. 1992; Xu and Johnson 1995). Soil bacteria expressing a significant level of acid phosphatases include strains from the genus *Pseudomonas* (Gugi et al. 1991) *Rhizobium* (Abd-Alla 1994) *Enterobacter*, *Serratia*, *Citrobacter*, *Proteus* and *Klebsiella* (Thaller et al. 1995) as well as *Bacillus* (Skrary and Cameron 1998). Several reports on halotolerant phosphate solubilizing bacteria are available. Halotolerant bacteria, capable of growing in media containing 10 % NaCl and reported to exhibit P solubilization on Pikovskaya media were isolated from Yellow Sea near the city of Incheon in the Republic of Korea (Siddikee et al. 2010). A salt tolerant *Bacillus cereus* isolated from the rhizosphere of a facultative halophyte, exhibited phosphate solubilization in addition to several other PGP activities (Chakraborty et al. 2011).

PGPR assisted improvement of plant productivity in salinity affected soils has also been reported. Beneficial microbes can also enhance plants tolerance to salinity stress (Egamberdieva 2008). It has been demonstrated that inoculations with AM fungi improves plant growth under salt stress (Cho et al. 2006). The three PGPR isolates *Pseudomonas alcaligenes* PsA15, *Bacillus polymyxa* BcP26 and *Mycobacterium phlei* MbP18 were able to tolerate high temperatures and salt concentrations and thus confer on them potential competitive advantage to survive in arid and saline soils such as calcisol (Egamberdiyeva 2007).

Barua et al. (2012) screened four potent diazotrophic strains from several isolates obtained from saline tracts of coastal West Bengal. Those isolates exhibited PGP activities like IAA production and phosphate solubilization under salinity stress. The isolates were found capable of solubilizing both organic and inorganic phos-

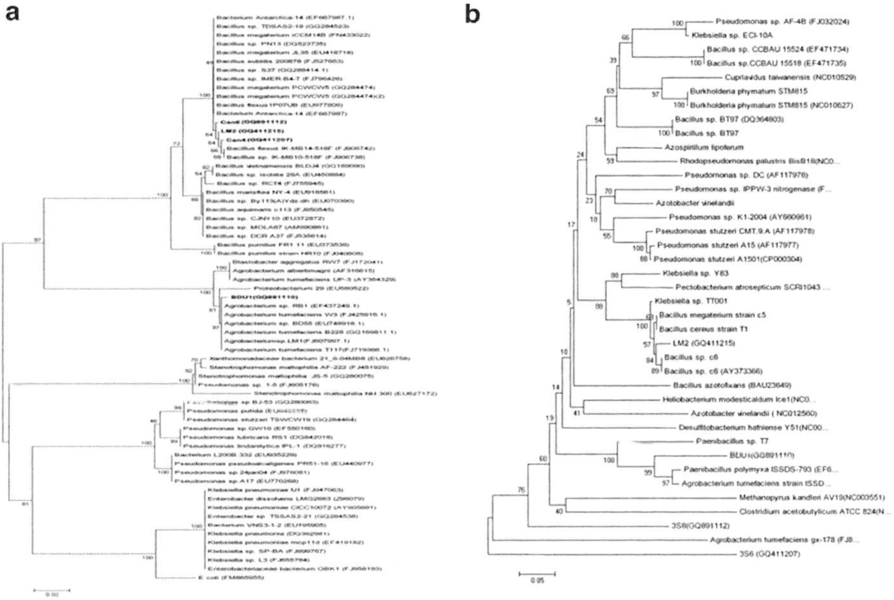

Fig. 8.2 (**a**) Phylogenetic affirmation of the isolates based on 16S rDNA sequence analysis; (**b**) Phylogenetic affirmation of the isolates based on *nif*H sequence (Barua et al. 2012)

phates under salinity stress. Molecular characterization of the strains (SUND_BDU1, SUND_LM2, Can4 and Can6) revealed that they belonged to *Agrobacterium* sp. and 3 *Bacillus* sp. respectively. Phylogenetic affirmation based on 16S rDNA and *nif*H sequencing were also reported (Fig. 8.2a, b).

A salt tolerant strain of *Planococcus rifietoensis* (SAL-15):isolated from alkali-saline soils of Faisalabad, Pakistan, is able to survive high salt concentrations (65 g/l NaCl) and pH (>9). It has been reported to exhibit a mixture of PGP activity, thereby improving plant growth under high salinity stress (Rajput et al. 2013).

Strains of *Pseudomonas* sp. and *Serratia* sp., having multiple PGP activities were examined for their effect on rice germination and seedlings under salt stress. Growth curve and IAA production of the strain was unaffected by salinity stress of 4–16 dS/m. However, probably for excessive IAA production, salt-tolerant *Pseudomonas sp*. ZnCd2003 showed a reduction in germination, root length, shoot length, fresh weight, and dry weight of *Oryza sativa* L. cv. RD6 (Nakbanpote et al. 2014).

8.8 Biofertilizers for Crop Improvement Under Salinity Stress

Several reports exist on the role of bio-inoculants in the improvement of crop production in normal arable soils. Improvement in productivity of field crops like sugarcane (Mirza et al. 2001) onion (Kashyap et al. 2005) Chinese cabbage (Yim et al.

2009) rice (Govindarajan et al. 2008; Jha et al. 2009) and wheat (Zorita and Canigia 2009) due to inoculation by free living diazotrophic bacteria was documented in normal soils. Progress with respect to development of bio-inoculants for problematic soils, in general and saline soils in particular, is not very encouraging. The impact of such bio-inoculants on crop productivity has not been adequately studied. Limited works carried out in this respect.

Azotobacter vinelandii inoculation increased the root growth in *Rhizophora mucronata* by enhancing average root biomass by 98.27 % and root length by 98.57 %. The leaf area was higher by 277.86 % in *R. mucronata* treated with *A. beijerinckii* and by 72.18 % in *R. apiculata* treated with *A. chroococcum* with respect to that of control. The shoot biomass was higher by 29.06 % in *R. mucronata* treated with *A vinelandii* and by 29.9 % in *R. apiculata* inoculated with *A. chroococcum* (Ravikumar et al. 2004).

Three species of *Azotobacter*, viz. *A. chroococcum*, *A.berijerinkii* and *A. vinelandii* was isolated from mangrove rhizosphere sediments. All the three species could tolerate high saline concentrations (up to 35 gl^{-1} and 30 gl^{-1}). These species of A*zotobacter* enhanced the germination and growth of rice and black gram seedlings even at high saline conditions by fixing atmospheric nitrogen and producing phytohormones. Among them, *A. chroococcum* was highly recommended in comparison to other bacterial species. Compared with the existing bio-inoculants, their morphological and biochemical characteristics were similar, except the saline induced effects on growth and physiology (Ravikumar 2008).

Chavada et al. (2010) had carried out plant growth promotion study of three diazotrophic haloalkalotolerant strain with respect to the germination of wheat seeds. Wheat seeds treated with bio-inoculant resulted in rapid germination, compared to un-inoculated control.

Effect of inoculation of halotolerant nonsymbiotic nitrogen fixers from Lonar lake soils, Maharashtra, India, on the growth of *Triticum aestivum* L. was carried out under pot trial in saline soils (Butale et al. 2010). Shoot length of plants, inoculated with bacteria exhibited an increased value of 5.22–12.5 cm with respect to control 4.25 cm, while root length of plants under the same treatment showed an increased value of 4.14–6.0 cm over control 2.5 cm. Dry mass of the plants also increased from (0.008 to 0.010 g as against control 0.003 g).

García de Salamone et al. (2010) reported that the inoculation of a mixture of *Azospirillum* strains increased N accumulation by the plant, representing 16.8 and 52.6 % of the total N demanded by this crop production. These data suggest that a hormonal effect exerted by *Azospirillum*, improved the efficiency of N absorption leading to superior yields of biomass. Inoculation improved plant growth at the grain filling stage by 28 and 50 % in relation to the control; however, no significant increase in grain production could be observed.

With respect to improvement in crop productivity under biotic/abiotic stress, role of PGPR is of paramount importance. The diazotrophic bacterial isolated characterized by Barua et al. (2012) were also examined for their efficacy for improving productivity of rice and ladys' finger in salinity affected Agric fields of coastal West Bengal, India. Though the highest yield attributes were obtained in fields amended with recommended dose of NPK fertilizers, inoculation by one of the isolates (Can6) along with recommended dose of FYM, exhibited considerable increase in productivity when compared with uninoculated control. Mixed inoculum prepared from salt tolerant *Azotobacter* sp. and *Azospirilum* sp. (nitrogen fixing bacteria) isolated from coastal mangrove in Pulau seribu, Indonesia, were tested for their potentiality as Biofertilizer. Gree-house based experimentation revealed that plants inoculated with the bacterial preparation can survive and show better growth to the level of salinity 12.43 dSm^{-1}, inferring that the adverse effect of salinity could be overcome by treatment with the bacterial preparation (Suliasih 2013). In a recent report, 75 % of recommended dose of NPK plus bio-fertilizer like Azolla and/or halotolerant bacteria increased barley productivity in saline soil thereby decreasing 25 % of recommended dose of N, P and K and hence environmental pollution (Sherif and Mohamed 2014).

8.9 Conclusion

Bio-inoculants are low cost organic inputs with socio-ecological benefits for sustainable crop production. Its efficiency largely depends upon its component micro flora, which are extensively location specific. Hence, development of location specific bio-inoculants and popularizing its use among the farmers is a major challenge. Soil salinity is a natural stress that detrimentally affects plant productivity throughout the world. Saline soils have their own microbial community adapted to such environment. Identification and manipulation of PGPR strains in general and diazotrophs in particular from such environment is one of the eco-friendly biotechnologies to overcome the stress. However, the lack of awareness regarding improved protocols of bio-fertilizer applications to the field is one of the few reasons why many useful PGPRs are still beyond the knowledge of ecologists and agriculturists. Hence, the challenge to develop and popularize an inoculant formulation for saline environment with long shelf life and high efficacy is to be addressed.

References

Abd-Alla MH (1994) Phosphatases and the utilization of organic phosphorus by Rhizobium leguminosarum biovar viceae. Lett Appl Microbiol 18:294–296

Abrol IP, Yadav JSP, Massoud FI (1988) Salt-affected soils and their management (No. 39). Food & Agriculture Organization of the United Nations, p 14

Acharya CL, Sharma AR (2007) Management options for increasing nitrogen use efficiency. Agric nitrogen use and its environmental implications. I.K. International Publishing House, India, p 213

Adesemoye AO, Kloepper JW (2009) Plant–microbes interactions in enhanced fertilizer-use efficiency. Appl Microbiol Biotechnol 85(1):1–12

Akhter MS, Hossain SJ, Amir-Hossain SK, Datta RK (2012) Isolation and characterization of salinity tolerant Azotobacter sp. Greener J Biol Sci 2:42–051

Al-Maadhidi JF (1989) Isolation and characterization of Azospirillum spp. from Iraqi wheat cultivars. Kuwait J Sci Eng 16(22):343–348
Bajpai PD, Gupta BR (1979) Effect of salinity and alkalinity of soil on some important microbial activities. Indian Soc Soil Sci 27:197–198
Bandyopadhyay BK, Bandyopadhyay AK (1984) Effect of application of farmyard manure on soil properties and yield of crops in coastal saline soil. J Indian Soc Coast Agric Res 2(3):63–73
Bandyopadhyay BK, Maji B (1995) Nature of acid soil of Sundarban delta and suitability of classifying them as acid sulphate or potential acid sulphate soils. J Indian Soc Soil Sci 43:103–107
Bandyopadhyay AK, Bhargava GP, Rao KVGK, Sen HS, Sinha TS, Bandyopadhyay BK, Biswas CR, Bal AR, Dutt SK, Mandal RC (1988) Coastal saline soils of India and their management, Bull No 13. Central Soil Salinity Res Institute, RRS, Canning Town
Bandyopadhyay BK, Sahu GC, Maji B (1998) Status, nature and composition of organic matter in coastal areas. Bulletin no. 19. Ind Soc Soil Sci 58–67
Bandyopadhyay BK, Sen HS, Maji B, Yadav JSP (2001) Saline and alkali soils and their management, vol 1, Monograph series. Indian Society of Coastal Agric Res. Central Soil Salinity Res Institute, Canning Town, pp 1–66
Bandyopadhyay BK, Maji B, Sen HS, Tyagi NK (2003) Coastal soils of West Bengal – their nature, distribution and characteristics, Bulletin no. 1/2003. Central Soil Salinity Res Institute, Regional Res Station, Canning Town, pp 1–62
Banta S, Mendoza CV (1984) Organic matter and rice. Int Rice Res Inst 616. http://pdf.usaid.gov/pdf_docs/PNAAR182.pdf
Bapurao ND (2012) Potential halotolerant Azospirillum brasilense N-30 as a biofertilizer for saline soils. Acta Biol Indic 1(2):198–206
Barua S, Tripathi S, Ghosh A, Chakraborty A, Ghosh S, Charabarti K (2008) Studies on non symbiotic diazotrophic bacterial populations of the saline soils of Sunderbans. J Pure Appl Microbiol 2(2):355–364
Barua S, Tripathi S, Chakraborty A, Ghosh S, Chakrabarti K (2011) Studies on non-symbiotic diazotrophic bacterial populations of coastal arable saline soils of India. Indian J Microbiol 51(3):369–376
Barua S, Tripathi S, Chakraborty A, Ghosh S, Chakrabarti K (2012) Characterization and crop production efficiency of diazotrophic bacterial isolates from coastal saline soils. Microbiol Res 167(2):95–102
Brahmaprakash GP, Sahu PK (2012) Biofertilizers for sustainability. J Indian Inst Sci 92(1):37–62
Buresh RJ, Casselman ME, Patrick WH Jr (1980) Nitrogen fixation in flooded soil systems: a review. Adv Agron 33:149–192
Burns RG (1983) Extracellular enzyme-substrate interactions in soil. In: Slater JH, Whittenbury R, Wimpenny JWT (eds) Microbes in their natural environment. Cambridge University Press, Cambridge, pp 249–298
Butale SV, Raut AA, Sawant TB, Basu P, Chakraborty HJ, Ganguli S, Krishnappa M (2010) Application of moderately haloalkaliphilic nonsymbiotic diazotrophs of Lonar lake to saline soils. Int J Microbiol Res 2(2):01–04
Caron M, Patten CL, Ghosh S (1995) Effects of plant growth promoting rhizobacteria Pseudomonas putida GR-122 on the physiology of canolla roots. Plant Growth Reg. Soc. Am. 22nd proceeding, Ed. Green DW, July 18–20
Cervantes LA, Olivares J (1976) Microbiological studies of saline soil in El-Salitre. Colomb Rev Latinoam 18:73–77
Chakraborty AP, Dey P, Chakraborty B, Chakraborty U, Roy S (2011) Plant growth promotion and amelioration of salinity stress in crop plants by a salt-tolerant bacterium. Recent Res Sci Technol 3(11):61–70

Chavada NB, Dave A, Patel B (2010a) Study on diazotrophic and IAA producing bacteria isolated from desert soil submit an application for biofertilizer. Int J Appl Biol Pharm Technol 1(3):1067–1071

Chavada NB, Patel R, Vanpuria S, Raval BP, Thakkar PV (2010b) A study on isolated diazotrophic (non-symbiotics) bacteria from saline desert soil as a biofertilizer. Int J Pharm Sci Res 1(1):52–54

Chelius MK, Lepo JE (1999) Restriction fragment length polymorphism analysis of PCR-amplified nifH sequences from wetland plant rhizosphere communities. Environ Technol 20(8):883–889

Cho K, Toler H, Lee J, Ownley B, Stutz JC, Moore JL, Augé RM (2006) Mycorrhizal symbiosis and response of sorghum plants to combined drought and salinity stresses. J Plant Physiol 163(5):517–528

Chowdhury SP, Schmid M, Hartmann A, Tripathi AK (2007) Identification of diazotrophs in the culturable bacterial community associated with roots of Lasiurus sindicus, a perennial grass of Thar Desert, India. Microb Ecol 54(1):82–90

Cummings SP, Orr C (2011) The role of plant growth promoting rhizobacteria in sustainable and low-input graminaceous crop production. In: Plant growth and health promoting bacteria. Ed. Dinesh. K. Maheshwari. Springer, Berlin, p 297–315

Davies PJ (1995) Plant hormones physiology, biochemistry and molecular biol, 2nd edn. Kluwer Academic Publishers, Dordrecht, p 112

Del Moral A, Quesada E, Ramos-Cormenzana A (1987) Distribution and types of bacteria isolated from an inland saltern. Ann Inst Pasteur/Microbiol 138(1):59–66

Del Moral A, Prado B, Quesada E, Garcia T, Ferrer R, Ramos-Cor- menzana A (1988) Numerical taxonomy of moderately halophilic Gram-negative rods from an inland saltern. J Gen Microbiol 134(3):733–741

Dey R, Pal KK, Bhatt DM, Chauhan SM (2004) Growth promotion and yield enhancement of peanut (Arachis hypogaea L) by application of plant growth-promoting rhizobacteria. Microbiol Res 159(4):371–394

Douka CE, Apostalakis CG, Skarloy VD (1978) Studies of Rhizobium meliloti isolated from salt-affected soils. Ann Appl Biol 88(3):457–460

Egamberdieva D (2008) Plant growth promoting properties of rhizobacteria isolated from wheat and pea grown in loamy sand soil. Turk J Biol 32(1):9–15

Egamberdiyeva D (2007) The effect of plant growth promoting bacteria on growth and nutrient uptake of maize in two different soils. Appl Soil Ecol 36(2):184–189

El-Shinnawi MM, Frankenberger WT Jr (1988) Salt inhibition of free living diazotroph population density and nitrogenase activity in soil. Soil Sci 146(3):176–184

Finnie JF, Van Staden J (1985) Effect of seed weed concentrate and Appl hormones on in vitro cultured tomato roots. J Plant Physiol 120:215–222

Fox TR, Comerford NB (1992) Rhizosphere phosphatase activity and phosphatase hydrolysable organic phosphorus in two forested spodosols. Soil Biol Biochem 24:579–583

Frankenberger WT Jr, Brunner W (1983) Methods of detection of auxin-indole acetic acid in soil by high performance liquid chromatography. Soil Sci Soc Am J 47(2):237–241

Fuentes-Ramírez LE, Bustillos-Cristales R, Tapia-Hernandez A, Jiménez-Salgado T, Wang ET, Martínez-Romero E, Caballero-Mellado J (2001) Novel nitrogen-fixing acetic acid bacteria, Gluconacetobacter johannae sp. nov. and Gluconacetobacter azotocaptans sp. nov., associated with coffee plants. Int J Syst Evol Microbiol 51(4):1305–1314

Gangawane LV (2007) Role of nitrogen fixing microorganism in bioremediation of salt affected soils. In: Gangawane LV, Khilare VC (eds) Sustainable environmental management. Daya Publishing House, New Delhi, pp 89–95

Garcia C, Fernandez T, Costa F, Cerranti B, Masciandaro G (1992) Kinetics of phosphatase activity in organic wastes. Soil Biol Biochem 25:361–365

García de Salamone IE, Di Salvo LP, Escobar Ortega JS, Boa Sorte MP, Urquiaga S, Dos Santos Teixeira KR (2010) Field response of rice paddy crop to inoculation with Azospirillum: physi-

ology of rhizosphere bacterial communities and the genetic diversity of endophytic bacteria in different parts of the plants. Plant Soil 336:351–362
Giambiagi N, Lodeiro A (1989) Response of the bacterial microflora of a sodic saline soil to different concentrations of nitrogen and sodium in agar culture. Soil Biol Biochem 21:177–178
Glick BR, Todorovic B, Czarny J, Cheng Z, Duan J, McConkey B (2007) Promotion of plant growth by bacterial ACC deaminase. Crit Rev Plant Sci 26(5–6):227–242
Goldstein AH (1986) Bacterial solubilization of mineral phosphates: historical perspective and future prospects. Am J Alternat Agric 1(02):51–57
Goldstein AH (1995) Recent progress in understanding the molecular genetics and biochemistry of calcium phosphate solubilization by Gram negative bacteria. Biol Agric Hortic 12(2):185–193
Govindarajan M, Balandreau J, Kwon S-W, Weon H-Y, Lakshminarasimhan C (2008) Effects of the inoculation of Burkholderia vietnamensis and related endophytic diazotrophic bacteria on grain yield of rice. Microb Ecol 55(1):21–37
Gugi B, Orange N, Hellio F, Burini JF, Guillou C, Leriche F, Guespin-Michel JF (1991) Effect of growth temperature on several exported enzyme activities in the psychrotrophic bacterium Pseudomonas fluorescens. J Bacteriol 173:3814–3820
Hassouna MS, Madkour MA, Helmi SHE, Yacout SI (1995) Salt tolerance of some free-living nitrogen fixers. Alex J Agric Res 40:389–413
Hazell P, Wood S (2008) Drivers of change in global agriculture. Philos Trans R Soc B: Biol Sci 363(1491):495–515
Herman MAB, Nault BA, Smart CD (2008) Effects of plant growth-promoting rhizobacteria on bell pepper production and green peach aphid infestations in New York. Crop Prot 27:996–1002
Ibrahim AN (1974) Non-symbiotic nitrogen fixation by Azotobacter as influenced by soil salinity. Acta Agron Acad Sci Hung 23:113–119
Islam R, Trivedi P, Madhaiyan M, Seshadri S, Lee G, Yang J, Kim Y, Kim M, Han G, Chauhan PS, Sa T (2010) Isolation, enumeration, and characterization of diazotrophic bacteria paddy soil sample under long-term fertilizer management experiment. Biol Fertil Soil 46(3):261–269
Iswaran V, Sen A (1958) Effect of salinity on nitrogen fixation by Azotobacter Sp. in some Indian soils. J Indian Soc Soil Sci 6:109–113
Jena PK, Adhya TK, Rajaramamohan Rao V (1988) Nitrogen fixation in Azospirillum species isolated from saline paddy soils. Microbios 54:220–221
Jha B, Thakur MC, Gontia I, Albrecht V, Stoffels M, Schmid M, Hartmann A (2009) Isolation, partial identification and application of diazotrophic rhizobacteria from traditional Indian rice cultivars. Eur J Soil Biol 45(1):62–72
Johnston AE, Syers JK (1998) Nutrient management for sustainable crop production in Asia. CAB International, USA, p 284
Joseph B, Jini D (2010) Salinity induced programmed cell death in plants: challenges and opportunities for salt-tolerant plants. J Plant Sci 5:376–390
Kashyap S, Sharma S, Vasudevan P (2005) Effect of native bioinoculants on vegetable crops grown on alkaline soil under field conditions. J New Seeds 7(3):75–90
Kennedy IR, Choudhury ATMA, Kecskes ML (2004) Free-living bacterial diazotrophs in crop-farming systems: can their potential for plant growth promotion be better exploited? Soil Biol Biochem 36(8):1229–1244
Kibblewhite MG, Ritz K, Swift MJ (2008) Soil health in Agric systems. Philos Trans R Soc B: Biol Sci 363(1492):685–701
Kitzes J, Wackernagel M, Loh J, Peller A, Goldfinger S, Cheng D, Tea K (2008) Shrink and share: humanity's present and future ecological footprint. Philos Trans R Soc B: Biol Sci 363(1491):467–475
Kloepper JW (1994) In: Okon Y (ed) Azospirillum/plant associations. CRC Press, Boca Raton, pp 137–166

Kravchenko JK, Kaliminskaya TA (1988) Nitrogen-fixing bacteria strongly saline Takyr soil (in Russian). Microbiologia 52:279–283
Ladha JK, George T, Bohlool BB (Eds) (1992) Biological nitrogen fixation for sustainable agriculture: extended versions of papers presented in the symposium, role of biological nitrogen fixation in sustainable agriculture at the 13th Congress of Soil Science, Kyoto, Japan, 1990 (Vol. 49). Int Rice Res Inst p 140–141
Loganathan P, Nair S (2004) Swaminathania salitolerans gen. nov., sp. nov., a salt-tolerant nitrogen-fixing and phosphate-solubilizing bacterium from wild rice (Porteresia coarctata Tateoka). Int J Syst Evol Microbiol 54(4):1185–1190
Lynch JM (1985) Origin, nature and biological activity of aliphatic substances and growth hormones found in soil. In: Vaughan D, Malcom RE, Nijhoff M (eds) Soil organic matter and biological activity. Dr. W. Junk Publishers, Dordrecht, pp 151–174
Maas EV (1984) Salt tolerance of plants. Handb Plant Sci Agric 2:57–75
Mahmoud SAZ, El-Sawy M, Ishac YZ, El-Safty MM (1978) The effects of salinity and alkalinity on the distribution and capacity of N2-fixation by Azotobacter in Egyptian soils. Ecol Bull NFR 26:99–109
Maji B, Bandyopadhyay BK, Sarkar D, Chatterji S (1998) Morphological and chemical characterization of soils of Sagar Island of the Sundarbans, West Bengal. J Indian Soc Soil Sci 46(1):99–103
Mali BS, Thengal SS, Pate PL (2012) Physico-chemical characteristics of salt affected soil from Barhanpur, MS, India. Ann Biol Res 3(8):4091–4092
Martinez-Beltran J, Manzur CL (2005) Overview of salinity problems in the world and FAO strategies to address the problem. In: Proceedings of the international salinity forum, Riverside, California. USDA-ARS Salinity Lab Riverside, pp 311–313
Minorsky PV (2008) On the inside. Plant Physiol 146:323–324
Mirza MS, Ahmad W, Latif F, Haurat J, Bally R, Normand P, Malik KA (2001) Isolation, partial characterization, and the effect of plant growth-promoting bacteria (PGPB) on micro-propagated sugarcane in vitro. Plant Soil 237:47–54
Mishustin EN, Shillnikova VK (1971) Structure and development of Azotobacter. In: Biological fixation of atmospheric nitrogen. The Macmillan Press, London, pp 213–214. http://books.google.co.in/books/about/Biological_fixation_of_atmospheric_nitro.html?id=W4UMAQAAIAAJ&redir_esc=y
Müller M, Deigele C, Ziegler H (1989) Hormonal interactions in the rhizospheres of maize (Zea mays, L.) and their effect on plant development. Z Pflanzenernähr Bodenkd 152(2):247–254
Munn R (1993) Physiological processes limiting plant growth in saline soils: some dogmas and hypotheses. Plant Cell Environ 16(1):15–24
Munns R (2002) Comparative physiology of salt and water stress. Plant Cell Environ 25(2):239–250
Nadia FE, Fayaz H, Makboul HE, El-Shanwy R (1984) Occurrence and characterization of nitrogen-fixing Azospirilla in some soils of Egypt. Z Pflanzenernähr Bodenkd 147(2):210–217
Nakbanpote W, Panitlurtumpai N, Sangdee A, Sakulpone N, Sirisom P, Pimthong A (2014) Salt-tolerant and plant growth-promoting bacteria isolated from Zn/Cd contaminated soil: identification and effect on rice under saline conditions. J Plant Interact 9(1):379–387
Naskar KR, Mandal RN (1999) Ecology and biodiversity of Indian mangroves. Daya Publishing House, New Delhi, pp 191–192
Palavan-Unsal N, Buyuktuncer ED, Tufekci MA (2005) Programmed cell death in plants. J Cell Mol Biol 4:9–23
Parida AK, Das AB (2005) Salt tolerance and salinity effects on plants: a review. Ecotoxicol Environ Saf 60(3):324–349
Perez E, Miguel S, Marıa MB, Yarzabal LA (2007) Isolation and characterization of mineral phosphate-solubilizing bacteria naturally colonizing a limonitic crust in the south-eastern Venezuelan region. Soil Biol Biochem 39(11):2905–2914

Poly F, Ranjard L, Nazaret S, Gourbière F, Monrozier LJ (2001) Comparison of nifH gene pools in soils and soil microenvironments with contrasting properties. Appl Environ Microbiol 67(5):2255–2262

Pretty J (2008) Agric sustainability: concepts, principles and evidence. Philos Trans R Soc B: Biol Sci 363(1491):447–465

Quesada E, Ventosa A, Rodriguez-Valera F, Ramos-Cormenzana A (1982) Types and properties of some bacteria isolated from hypersaline soils. J Appl Bacteriol 53(2):155–161

Ragab M (1993) Distribution pattern of soil microbial population in salt-affected soils. In: Lieth H, Al-Masoom AA (eds) Towards the rational use of high salinity tolerant plants, vol 1, deliberations about high salinity tolerant plants and ecosystems. Kluwer Academic, Dordrecht, pp 467–472

Rai R (1991) Strain-specific salt tolerance and chemotaxis of Azospirillum brasilense and their associative N_2-fixation with finger millet in saline calcareous soil. Plant Soil 137:55–59

Rajput L, Imran A, Mubeen F, Hafeez FY (2013) Salt-tolerant pgpr strain planococcus rifietoensis promotes the growth and yield of wheat (Triticum aestivum L.) cultivated in saline soil. Pak J Botany 45(6):1955–1962

Ramanathan AL, Singh G, Majumdar J, Samal AC, Chauhan R, Ranjan RK, Rajkumar K, Santra SC (2008) A study of microbial diversity and its interaction with nutrients in the sediments of Sundarban mangroves. Indian J Mar Sci 37:159–165

Rao NSS (1994) Soil microorganisms and plant growth. UI Press, Jakarta

Rao NSS, Barrueco CR (1998) Variability in Casuarina–Frankia symbiosis. In: Rao NSS, Dommergues YR (eds) Microbial interactions in agriculture and forestry. Oxford & IBH, New Delhi, pp 97–113

Ravikumar S (2008) Prospects of marine biofertilizers for saline soil crop cultivation. NVIS CENTRE Newsl 6(2), June. http://www.envismadrasuniv.org/nl200082%20articles%20saline.html

Ravikumar DR, Vittal BPR (1996) Fungal diversity on decomposing biomass of mangrove plant Rhizophora in Pichavaram Estuary, east coast of India. Indian J Mar Sci 25(2):142–144

Ravikumar S, Kathiresan K, Ignatiammal STM, Selvam MB, Shanthy S (2004) Nitrogen-fixing azotobacters from mangrove habitat and their utility as marine biofertilizers. J Exp Mar Biol Ecol 312(1):5–17

Rengasamy P (2006) World salinization with emphasis on Australia. J Exp Bot 57(5):1017–1023

Rice WD, Paul EA (1971) An acetylene reduction assay for measuring nitrogen fixation in waterlogged soil. Can J Microbiol 17(8):1049–1056

Richards LA (1954) Diagnosis and improvement of saline and alkali soils. Soil Sci 78(2):154

Rodriguez H, Fraga R (1999) Phosphate solubilizing bacteria and their role in plant growth promotion. Biotechnol Adv 17(4):319–339

Rodriguez-Valera F (1988) Characteristics and microbial ecology of hypersaline environments. In: Rodriguez-Valera F (ed) Halophilic bacteria, vol 1. CRC Press, Boca Raton, pp 3–30

Saleena LM, Rangarajan S, Nair S (2002) Diversity of Azospirillum strains isolated from rice plants grown in saline and nonsaline sites of coastal agric ecosystem. Microb Ecol 44:271–277

Sarkar MA, Chakrabarti K, Paul AK, Chattopadhyay NC (1995) Studies on some microbiological aspects of different mangrove soils of the Sunderbans. Indian Agric 39(3):159–164

Sekiguchi Y (2006) Yet-to-be cultural microorganisms relevant to methane fermentation processes. Microbes Environ 21(1):1–15

Sengupta A, Chaudhuri S (1991) Ecology of heterotrophic dinitrogen fixation in the rhizosphere of mangrove plant community at the Ganges river estuary in India. Oecologia 87:560–564

Severin I, Confurius-Guns V, Stal LJ (2012) Effect of salinity on nitrogenase activity and composition of the active diazotrophic community in intertidal microbial mats. Arch Microbiol 194(6):483–491

Shahab S, Ahmed N, Khan NS (2009) Indoleacetic acid production and enhanced plant growth promotion by indigenous PSBs. Afr J Agric Res 4(11):1312–1316

Sherif AEA, Mohamed FM (2014) Response of barley grown in saline soil to bio-fertilizer as a partial substitutive of mineral fertilizer. Glob J Sci Res 2(5):144–153

Siddikee MA, Chauhan PS, Anandham R, Han GH, Sa T (2010) Isolation, characterization, and use for plant growth promotion under salt stress, of ACC deaminase-producing halotolerant bacteria derived from coastal soil. J Microbiol Biotechnol 20(11):1577–1584

Sinegani AS, Sharifi Z (2004) Land use effect on the occurrence and distribution of Azotobacter in Hamadan soils, Iran. In: Proceedings of the fourth international Iran & Russia conference, pp 614–618

Singh JS, Pandey VC, Singh DP (2011) Efficient soil microorganisms: a new dimension for sustainable agriculture and environmental development. Agr Ecosyst Environ 140(3):339–353

Skrary FA, Cameron DC (1998) Purification and characterization of a Bacillus licheniformis phosphatase specific for d-alphaglycerphosphate. Arch Biochem Biophys 349:27–35

Sprent JI, Sprent P (1990) Nitrogen-fixing organisms: pure and Appl aspects. Chapman and Hall, London

Squires VR, Glenn EP (2009) Salination, desertification, and soil erosion. In: Squires VR (ed) The role of food, agriculture, forestry and fisheries in human nutrition, vol 3. UNESCO Publishing Eolss Publishers, Oxford, UK, p 104

Stein A, Fortin JA, Vallée G (1990) Enhanced rooting of Picea mariana cuttings by ectomycor rhizal fungi. Can J Bot 68(3):468–470

Strzelczyk E, Pokojska-Burdziej A (1984) Production of auxins and gibberellin like substances by mycorrhizal fungi, bacteria and actinomycetes isolated from soil and mycorhizosphere of pine (Pinus silvestris L.). Plant Soil 81:185–194

Suliasih S (2013) The effect of salt tolerant nitrogen fixing bacteria on the growth of paddy rice (Oryza sativa. L). J Biol Res 19(1):11–14

Szuster B, Hoanh CT, Kam SP, Ismail AM, Noble AD, Borger M (2010) Policy, planning and management at the land-water interface. In: Tropical deltas and coastal zones, ed. C.T. Hoanh et al. CABI Publishing, USA, p 8

Thaller MC, Berlutti F, Schippa S, Iori P, Passariello C, Rossolini GM (1995) Heterogeneous patterns of acid phosphatases containing low-molecular-mass polypeptides in members of the family Enterobacteriaceae. Int J Syst Bacteriol 45(2):255–261

Tilak, KVBR, Krishna Murti, GSR (1981) Nitrogen fixation by Azospirillum spp. in certain saline and saline-alkali soils of India. Zentralblatt für Bakteriologie, Parasitenkunde, Infektionskrankheiten und Hygiene. Zweite Naturwissenschaftliche Abteilung: Mikrobiologie der Landwirtschaft, der Technologie und des Umweltschutzes. 136(8):641–643

Tripathi AK, Verma SC, Ron EZ (2002) Molecular characterization of a salt-tolerant bacterial community in the rice rhizosphere. Res Microbiol 153(9):579–584

Tripathi S, Kumari S, Chakraborty A, Gupta A, Chakrabarti K, Bandyapadhyay BK (2006) Microbial biomass and its activities in salt affected coastal soils. Biol Fertil Soil 42:273–277

Tripathi S, Chakraborty A, Chakrabarti K, Bandyopadhyay BK (2007) Enzyme activities and microbial biomass in coastal soils of India. Soil Biol Biochem 39(11):2840–2848

Trüper HG, Severin J, Wohlfarth A, Muller E, Galinski EA (1991) Halophily, taxonomy, phylogeny and nomenclature. In: Rodriguez VF (ed) General and applied aspects of halophilic microorganisms. Plenum Press, New York, pp 3–7

Verma SC, Ladha JK, Tripathi AK (2001) Evaluation of plant growth promoting and colonization ability of entophytic diazotrophs from deep water rice. J Biotechnol 91:127–141

Widmer F, Shaffer BT, Porteous LA, Seidler RJ (1999) Analysis of nifH gene pool complexity in soil and litter at a Douglas fir forest site in the Oregon Cascade Mountain Range. Appl Environ Microbiol 65:374–380

Wild A (2003) Soils, land and food: managing the land during the twenty-first century. Cambridge University Press, Cambridge

Wollenweber B, Zechmeister-Boltenstern S (1989) Nitrogen fixation and nitrogen assimilation in a temperate saline ecosystem. Bot Acta 102(1):96–105

Xu JG, Johnson RL (1995) Root growth, microbial activity and phosphatase activity in oil-contaminated, remediated and uncontaminated soils planted to barley and field pea. Plant Soil 173:3–10

Yadav JSP, Bandopadhyay AK, Rao KVGK, Sinha TS, Biswas CR, Bandopadhyay BK, Dutta SK (1981) Management of coastal saline soils of Sundarbans. CSSRI-ICAR Bull 7:1–32

Yadav JSP, Bandyopadhyay AK, Bandyopadhyay BK (1983) Extent of coastal saline soils of India. J Indian Soc Coast Agric Res 1:1–6

Yadav RL, Prasad K, Gangwar KS, Dwivedi BS (1998) Cropping systems and resource-use efficiency. Indian J Agric Sci 68(8):548–558

Yeo A (1998) Predicting the interaction between the effects of salinity and climate change on crop plants. Sci Hortic 78(1):159–174

Yim WJ, Poonguzhali S, Madhaiyan M, Palaniappan P, Siddikee MA, Sa T (2009) Characterization of plant-growth promoting diazotrophic bacteria isolated from field grown Chinese cabbage under different fertilization conditions. J Microbiol 47(2):147–155

Zafar Y, Malik KA, Niemann EG (1987) Studies on N_2 fixing bacteria associated with salt tolerant grass, Leptochloa fusca (L.) Kunth. MIRCEN J Appl Microbiol Biotechnol 3(1):45–55

Zahran HH (1992) Characterization of root-nodule bacteria indigenous in the salt-affected soils of Egypt by lipopolysaccharide, protein and plasmid profiles. J Basic Microbiol 32(4):279–287

Zahran HH (1997) Diversity, adaptation and activity of the bacterial flora in saline environments. Biol Fertil Soil 25(3):211–223

Zahran HH, Ahmed MS, Afkar EA (1995) Isolation and characterization of nitrogen-fixing moderate halophilic bacteria from saline soils of Egypt. J Basic Microbiol 35(4):269–275

Zehr JP, Jenkins BD, Short SM, Steward GF (2003) Nitrogenase gene diversity and microbial community structure: a cross- system comparison. Environ Microbiol 5(7):539–554

Zorita MD, Canigia MVF (2009) Field performance of a liquid formulation of Azospirillum brasilienses on dryland wheat productivity. Eur J Soil Biol 45:3–11

Chapter 9
Application of Halotolerant Bacteria to Restore Plant Growth Under Salt Stress

Elhafid Nabti, Michael Schmid, and Anton Hartmann

Abstract High salinity abolishes several stages of plant life ranging from the seed germination step to maturity. Many processes are inhibited, such as phytohormone synthesis and regulation, normal root and shoot development, nutrient uptake, photosynthesis, and DNA replication. Plant growth promoting bacteria (PGPB) are naturally colonizing plants and occur in the rhizosphere or non rhizosphere soil and benefit plant growth by numerous processes. The importance of halotolerant PGPB resides in their ability to adapt to increased salinity by efficient osmoregulatory mechanism to be able to continue regular cell functions. Thus, halotolerant PGPB are able to provide plants with their activities to challenge osmotic stress by supporting them in the restoration of essential activities, e.g., in their hormonal balance. Halotolerant PGPB stimulate plant growth under high salinity by using similar mechanisms like halosensitive PGPB, such as synthesis of indole acetic acid (IAA), gibberellins (GA), cytokinins (CK), abscisic acid (ABA), solubilization of insoluble phosphate, synthesis and excretion of siderophores, and production of ACC-deaminase to reduce high growth inhibitory levels of ethylene occurring in plants at salt stress conditions. Furthermore, some halotolerant PGPB are even able to colonize plants endophytically, produce various antimicrobial metabolites against pathogenic fungi and bacteria, support plant health by improving systemic resistance and contribute to soil fertility and remediation.

Keywords Bacteria • Plant growth • Restoration • Halotolerance • Saline stress

E. Nabti (✉)
Laboratory of Renewable Energies, Group of Biomass and Environment, University of Bejaia, FSNV, Targa Ouzemmour, 06000 Bejaïa, Algeria
e-mail: elhnabti1977@yahoo.fr

M. Schmid • A. Hartmann
Research Unit Microbe-Plant Interactions, Helmholtz Zentrum Muenchen, German Research Center for Environmental Health (GmbH), Ingolstaedter Landstrasse 1, 85764 Neuherberg, Germany
e-mail: anton.hartmann@helmholtz-muenchen.de

D.K. Maheshwari, M. Saraf (eds.), *Halophiles*, Sustainable Development and Biodiversity 6, DOI 10.1007/978-3-319-14595-2_9

9.1 Introduction

Soil salinity constitutes a major obstacle for agriculture in arid and semi-arid regions. Twenty to thirty percent of the world's cultivated area and almost half of the world's irrigated soils are severely affected by the lack of water in prolonged drought conditions, and by increased salinity (Bartels and Sunkar 2005). Furthermore, increases in aridity and salinity seem unavoidable in many countries. Hence, the development and sustainability of profitable agricultural systems are seriously threatened (Cordovilla et al. 1994; Nabti et al. 2007, 2010). Furthermore, all global change scenarios predict a substantial raise of the sea level in the next decades due to global warming. In consequence, salinization of large agricultural land areas at fertile coastal areas, which are also highly populated, is to be expected.

It is well known, that the performance of agriculture under less favorable conditions, such as semiarid or saline conditions, presents already major challenges in many countries. In addition to a proper supply of nutrients, dominating constraints for plant growth are the lack of water and the increasing salinity of soils. Sustainable and cost-effective plant growth becomes even more important due to the need of energy efficient plant growth for biomass and bio-energy production, especially in soils of lower quality, in addition to food production (Nabti et al. 2007).

The natural occurrence of halotolerant bacteria in saline soils opens up a possible important role of these microorganisms also in the interaction with plants under these stress conditions. However, the diversity of these bacteria and the ecology of their interactions with plants are still under investigation and not fully understood. General strategies existing for growth and survival of prokaryotes in environments with elevated osmolarity are well documented. The compatibility of high salt concentrations within the cell cytoplasm requires extensive structural and enzymatic modifications, which is restricted mainly to members of the Halobacteraceae. All other prokaryotes have evolved the accumulation of a specific group of molecules of low molecular mass, termed compatible solutes, as general mechanism to cope with environments of elevated osmolarity. High intracellular concentrations of these osmolytes balance out high external osmolarity and are perfectly compatible or even protect cellular processes (Sleator and Hill 2001). To overcome the salt problem, improvement of management practices by using suitable salt-tolerant plant cultivars is under development (Vinocur and Altman 2005). However, the inoculation with osmotolerant bacteria or rhizobacteria could be a more ready utilizable and sustainable solution to mitigate the detrimental salt-effects on plants by the modulation or amelioration the salt stress (Dodd and Pérez-Alfocea 2012). Some osmotolerant, plant growth promoting bacterial strains (e.g., *Pseudomonas, Rhizobium, Azospirillum, Bacillus and Planococcus*) are being developed on these lines (Creus et al. 1998; Mayak et al. 2004; Bartels and Sunkar 2005; Egamberdiyeva 2005, 2011; Rabie et al. 2005; Sapsirisopa et al. 2009; Rajput et al. 2013).

Recent studies about the utilization of halophilic and halotolerant plant growth promoting bacteria in mitigation of the deleterious effects of salt stress have been performed. There are numerous and diverse, process of mitigation of salt stress in wheat seedlings by halotolerant bacteria (Nabti et al. 2012; Ramadoss et al. 2013), application of osmotolerant PGPR to ameliorate sodium chloride stress on tomato plants (Tank and Saraf 2010; Shen et al. 2012), restoration of growth of salt-affected maize and soybean plants by application with osmotolerant rhizobacteria (Aly et al. 2003; Naz et al. 2009), the improvement of nutrient elements in sunflower under high salinity (Shirmardi et al. 2010), and the increase of halotolerant growth of barley by the inoculation with novel halotolerant rhizobacteria (Cardinale et al. 2014).

9.2 Effect of Salinity

Salinity is considered as a major threat in ecosystem, because this phenomenon increases progressively and quickly, then disturbs and influences the biotic and abiotic soil components.

9.2.1 Saline Stress and Its Effects on Soil Properties

Soil salinity constitutes a natural situation in ecosystems in arid and semi-arid regions. The term salt-affected soil is reserved for soils that are saline or sodic. Saline soils have an excess of soluble salts in the solution within the micro- and macropore space structure. Sodic soils have a surplus of sodium associated with the negatively charged clay particles (Bianco and Defez 2011).

Accumulation of salt in the soil is due to inadequate leaching and drainage of irrigation water. Most of salts present in irrigation water are chlorides, sulfates, carbonates, and bicarbonates of calcium, magnesium, sodium and potassium. Salinity affects the soil structure and adversely affects plant growth and crop yields (Frenkel et al. 1978; Ghadiri et al. 2004). Three major types of salinity based on soil and groundwater processes are known: groundwater associated salinity, transient salinity and irrigation salinity (Rengasamy 2006). Salinity is regarded as the most severe environmental stress, which affects both the soil system and living organisms. Soil structure is defined as the arrangement of the solid particles and aggregates and of the micro-and macropore space between them. Pores are the main soil physical features because most soil processes that have immediate consequences for soil biological activity or soil conservation occur either within the pore space or on the surfaces of the particles that form the pores (Kay 1990; Rengasamy and Olsson 1991; Allotey et al. 2008). It is well known that high salinity leads to negative effects on soil structure: elevated sodium concentrations cause soil dispersion and clay platelets to swell and aggregate. Thus, the forces involved in the binding of clay particles are disrupted under the influence of sodium ions. Soil dispersion causes clay particles to plug soil pores. Therefore, soil permeability for water and air is reduced and surface crusting does occur (Abu-Sharar et al. 1987). It is documented that the presence of water in soil leads to the swelling of soil particles with high smectite clay content and to the hydratation of some minerals as a result of the reduction of the cross-sectional area of soil pores. This process is occurring under high sodium or low salt concentrations and causes dispersion and movement of fine particles within the pores. The particles will be caught in smaller pores and therefore, water and air will be blocked within the soil structure (Roiston et al. 1984).

9.2.2 *Effect of Salinity on Plant Growth*

Saline soil becomes a problem when salt is highly accumulated in the root zone. As a result, plant growth is affected negatively, because plant roots are inhibited to withdraw water from the surrounding soil and thus the water supply is reduced (Rengasamy 2010a). Plant growth is also directly and efficiently inhibited by high salinity of the water. This is due to the high osmotic potential of the soil solution and nutrient uptake deficiencies caused by the abundance of Na^+ and Cl^- in the soil solution (Marschner 1995); Sairam and Tyagi 2004). The presence of these ions in high concentrations decreases the activity of other essential elements in soil and can lead to a reduction in the accessibility and uptake of essential mineral nutrients by plants (Bianco and Defez 2011).

Also, high salinity causes an inhibition and deterioration of several stages of plant life such as seed germination, biosynthesis of phytohormones and other plant growth stimulating factors, photosynthesis by increasing photorespiration and alerting the normal homoeostasis of cells which generates an increased generation of reactive oxygen species. Other affected processes are the maturation of cell structures and plant morphology, root and stem growth, ion and organic solute transport, nutrient uptake as well as general enzymatic activities (Xiong and Zhu 2002). Osmotic stress reduces stem height, root length and dry weight. Under high environmental NaCl concentrations, sodium uptake into roots raises as well, while N, P, K and Mg-uptake decrease and the intracellular ionic equilibrium is disturbed (Gunes et al. 1996; Abdel-Ghaffar et al. 1998). Plants become less capable to take up sufficient water for growth in the presence of high level of dissolved salts in the root zone (Rengasamy 2010b). Under these conditions, plants expend more energy on osmotic adjustment by accumulating compatible solutes (Robinson and Jones 1986; Kempf and Bremer 1998).

Numerous studies showed the negative effects of high salinity of plant growth from the seed germination to maturity (François et al. 1986). The effects of salt stress on wheat and barley were often and profoundly studied by different authors (Sarin and Narayanan 1968; Qureshi et al. 1980; Rachidai et al. 1994; Sharma et al. 2005; Mahmoodzadeh et al. 2013). Other plants were also investigated under stress conditions : pepper (Gunes et al. 1996), sorghum (Gill et al. 2002), alfalfa (Fougere et al. 1991), beans (Cordovilla et al. 1994), maize (Azevedo et al. 2004), lentil (Golezani and Yengabad 2012), soybean (Ahmed and Sandhu 1988), sunflower (Naz and Bano 2013), tomato (Tank and Saraf 2010; Kahlaoui et al. 2011; Jan et al. 2014) cucumber etc. (Ibekwe et al. 2010).

9.3 Strategy of Osmoregulation in Halotolerant Bacteria

When bacterial cells are exposed to high osmolarity or salt stress, the water activity decreases in their cytoplasm (Epstein 1986) and most of their proteins and other macromolecules as well as essential functions are impaired (Bakker et al. 1987). Also, abrupt plasmolysis blocks different physiological processes, such as nutrient uptake or inhibition of DNA-replication and macromolecule biosynthesis (Kogut and Russell 1987; Bartels and Sunkar 2005).

Cellular adaptation to osmotic stress is a fundamental biological process protecting organisms against the lethal effects of dehydratation. Osmoregulation is of great significance in agriculture, since water is the major limiting factor in crop productivity (Le Rudulier et al. 1984). Two strategies of osmoadaptation are adopted in bacteria: at first, maintenance of osmotic equilibrium by keeping cytoplasmic KCl concentrations similar to the surrounding environment and in parallel some physiological modifications to protect vital metabolic functions are key mechanisms; secondly the accumulation of compatible solutes (organic osmolytes) in the cytoplasm (Galinski and Trüper 1982; Galinski and Trüper 1994; Galinski 1995) so as to balance the high osmotic potential in the environment in a physiological compatible manner.

The adaptation of rhizospheric bacteria to high salinity was reviewed in detail by Miller and Wood (1996). Osmoregulation is a general process in soil microorganisms, because the presence of variable salt concentrations in soil would lead to the disruption of numerous processes. For example, a disturbance in the interaction of plants with bacteria is caused by alterations of proteins involved in the initial attachment steps (adsorption and anchoring) of bacteria to plant roots in symbiotic interactions as well as by the inhibition of nodulation and nitrogen fixation activities. Furthermore, a disturbance of the molecular structure of exopolysaccharide (EPS) and lipopolysaccharide (LPS) of bacterial cells surfaces cuts off the molecular signal exchange between bacteria and plant host, bacterial chemotaxis and finally root colonization of surface-associated or endophytic PGPB (Jofré et al. 1998).

Osmoregulation in the nitrogen fixing rhizobacteria of the genus *Azospirillum* is well documented (Hartmann 1988; Riou and Le Rudulier 1990; Hartmann et al. 1991, 1992; Riou et al. 1991; Tripathi et al. 1998). Bacteria of the genus *Azospirillum* accumulate compatible solutes such as gulatamate, proline, glycine betaine, trehalose in order to adapt to the fluctuations in soil salinity. The understanding of the osmoadaptation mechanisms in *Azospirillum* spp. can contribute towards the long-term goal of improving plant-microbe interactions in salinity affected crop productivity (Tilak et al. 2005). Other substances like poly-β-hydroxybutyrate (PHB) may also be involved in the resistance of *Azospirillum brasilense* towards the osmotic stress and root colonization (Kadouri et al. 2003). PHB is an energy and carbon storage material accumulated in response to the limitation of an essential nutrient. The effect of different salt concentrations on growth and PHB accumulation of four different *Sinorhizobium* strains was examined by Arora et al. (2006) suggested that minimum PHB content was accumulated at low or zero salinity, maximum was observed by the salt-tolerant strains at higher salt concentrations which, define key role of PHB in cell protection in saline conditions.

Certain osmotolerant strains of the genus *Rhizobium* showed a different utilization of osmoprotective substances such as dimethylsulfoniopropionate, dimethylsulfonioacetate, glycine betaine, proline, ectoine and choline (Le Rudulier et al. 1983; Bernard et al. 1986; Le Rudulier and Bernard 1986; Talibart et al. 1994; Pichereau et al. 1998; Boncompagni et al. 1999; Zahran 1999; Bouhmouch et al. 2001). Mechanisms of osmotolerance with different pathways are present in the genus *Sinorhizobium* (Shamseldin et al. 2006; Abolhasani et al. 2010). *Sinorhizobium meliloti* takes up osmolytes like trehalose, maltose, cellobiose, gentibiose, turanose and palatinose as cytosolic osmolytes under salt-stress in an early growth phase. These compounds are catabolized during later growth phases and thus contribute to an increase in the cytosolic concentrations of two endogenously-synthesized osmolytes: glutamate and N-acetyl glutaminylglutamine amide (Gouffi and Blanco 2000). Many other osmotolerant PGPB, like *Pseudomonas* spp., *Bacillus* spp., and e. g. *Streptomyces* spp., were isolated and examined for their capacities to challenge osmotic stress in detail (Halverson et al. 2000; Upadhyay et al. 2012).

9.4 Plant Growths Under Salinity Stress Conditions

Under saline or sodic stress, both halophytes and halotolerant rhizospheric and plant-associated bacteria are especially adapted to the particular stress conditions. The functional interactions between the two partners can reduce the effect of the salt pressure (Biro et al. 2002). Previous studies reported that plant growth promotion and amelioration of salinity stress in crop plants by salt-tolerant bacteria could involve different mechanisms such as antioxidant enzymes, phosphate solubilization, siderophores and secretion of various phytohormones (Alizadeh and Parsaeimehr 2011; Chakraborty et al. 2011). Furthermore, the application of halotolerant PGPB reduced the negative effects of saline stress by increasing the leaf's relative water content and enhancing photosynthetic pigment production in both stress and normal conditions (Saghafi et al. 2013). Additionally, the majority of halophilic or halotolerant PGPB are characterized by their capability to induce antioxidant enzymes involved in the tolerance of plants towards severe salt stress (Del Rio et al. 2003; Kohler et al. 2009).

Rabie and Almadini (2005) showed the positive effect of the dual inoculation of the arbuscular mycorrhizal fungus *Glomus clarum* and the diazotroph *Azospirillum brasilense* on salt-tolerance in cowpea plants. They revealed that different vegetative plant growth parameters, protein content, nitrogenase, and phosphatase activities as well as nutrient elements (N, P, K, Ca, Mg) are significantly improved in the presence of *A. brasilense*. Ghorai et al (2013) showed the potential effect of some high salt tolerant PGPB on groundnut seedling under saline conditions. The experiment revealed significant improvement of shoot height, root length and biomass of groundnut seedling after 7 days under salinity stress. The application of halotolerant

PGPB is not only restricted to legumes and cereals, but is also performed on trees such as banana and mangrove forest growing under hard environmental conditions caused by salinity (Maziah et al. 2009; Das et al. 2011). Furthermore, Shukla et al. (2012) demonstrated an increase of NaCl-stress inhibited growth of peanut (*Arachis hypogaea*) after inoculation with the diazotrophic rhizosphere bacterium *Brachybacterium saurastrense* (Gontia et al. 2011) and other halotolerant isolates from the halophyte *Salicornia* (Jha et al. 2012).

The use of Rhizosphere bacteria possessing the traits of PGP under saline stress is becoming prevalent worldwide to achieve sustainable agriculture along with soil reclamation through phytoremediation as well as bioremediation (Tank and Saraf 2010). Additionally, many studies were carried out to highlight the possibility to restore cereal growth under saline conditions by using osmotolerant bacteria. A restoration of salt effected growth of (*Triticum durum*) was sown after inoculation of seeds with the osmotolerant PGPR *Azospirillum brasilense NH* isolated from salt-affected soil was reported by Nabti et al. (2010). The addition of extracts of marine algae *Ulva* increased the protecting effect on salt-stressed plants due to the content of osmolytes in the algae. Chickpea plants irrigated with saline water were inoculated with *Azospirillum brasilense* strain Cd, and a noticeable growth-restoration of nodulation, root and shoot and crop yield occurred under saline conditions (Hamaoui et al. 2001). Similarly, Rojas-Tapias et al (2012) reported significant effects of inoculation with two strains of *Azotobacter* sp. on the amelioration of growth of (*Zea mays*) under high NaCl concentration. The experiment revealed a significant restoration of plant biomass (length and weight), exclusion of Na^+ and K^+, improvement of polyphenol and chlorophyll contents and maintenance of nitrogen fixation and phosphate solubilization activities under saline stress conditions. Concurrent with accumulation of proline, which is regarded as potent osmolyte and indicator of osmotic stress, was greatly diminished.

Other studies revealed the necessity of application of osmotolerant strains of *Rhizobium* to improve common bean production in the Mediterranean area (Bouhmouch et al. 2001). In fact certain *Rhizobium* sp. are able to respond to the challenge to soil salinity by a natural selection imposed by certain legumes, which play an important role in the remediation of this kind of soils (Yadav and Agarwal 1961). Some salt tolerant *pseudomonas* (*P. fluorescens, P. aeruginosa and P. stutzeri*) (<6 % NaCl) of PGPB were selected which showed a significant effect on tomato plant growth under sodium chloride stress due to phosphate solubilization, siderophores production; ethylene reduction and IAA production.

Most recently, research was conducted on the utilization of marine bacteria as salt tolerant PGPB to mitigate the effect of stress on inoculated plants (Kim et al. 2014). Schnell and colleagues isolated a diversity of novel halotolerant bacteria that mitigate the adverse plant growth under salt-stress conditions (Suarez et al. 2014a, b; Cardinale et al. 2014). Upon inoculation of these bacteria to *Hordeum vulgare* cv. Propino was, a remarkable restoration of salt-affected growth was achieved (Schnell, personal communication).

9.4.1 ACC-Deaminase and Halotolerant PGPB

Ethylene is a volatile phytohormone and is known to play an important role in plant growth regulation at very low concentrations such as development of different vegetative plant parts, nodulation, or rooting of cuttings (Davis 2004). It is also involved in the transduction of a signal for the recognition of salt stress (Selvakumar et al. 2012). Plants growing in contaminated soils are often subjected to the combined stress of nutritional deficiency and chemical toxicity resulting. This results in the production of stress ethylene, which leads to growth inhibition of plants and decreases in plant biomass (Maheshwari 2011).

One of the strategies for increasing the efficacy of phytoremediation is based on the use of contaminant–tolerant plant species in combination with PGPB (Gerhardt et al. 2006). These soil microorganisms producing 1-aminocyclopropane-1-carboxylate deaminase (ACC) are capable to promote plant growth by cleaving plant-produced ACC, a precursor in the biosynthesis of ethylene, which decreases the level of ethylene in plants and thereby relieves the blockage in plant growth (Honma and Shimomura 1978; Glick 2005; Aamir et al. 2013). Ethylene production is regulated by different parameters such as light, temperature and IAA. When plant growth is subjected to an environmental stress, ethylene is produced at high concentrations as response to harsh conditions (Glick 2005). In addition, the production of ethylene is indispensable to ensure normal plant growth and development. However, high concentrations of ethylene induced by stressful conditions proved harmful, because it provokes defoliation and inhibits seed germination and other cellular processes involved in normal growth and development. The involvement of ACC-deaminase positive PGPB degrades ACC in the ethylene biosynthesis pathway in the root system. However, the ACC-deaminase level should be at least 100–1,000-fold greater than the ACC-oxidase level. This is likely to be the case, provided that the expression of ACC- oxidase is not induced (Glick et al. 1998). This ACC-deaminase enzymatic activity is regarded as a very important mechanism (Glick and Holgan 2003; Mayak et al. 2004), since decreasing elevated ethylene levels in roots during salinity stress leads to improved plant development (Fig. 9.1) as observed by Blaha et al. (2006). Bacteria containing ACC deaminase have been used successfully in many cases, both in laboratory experiments and in the field, to ameliorate some of the stress experienced by plants used in phytoremediation (Gerhardt et al. 2006).

Many rhizobacteria are described as ACC-deaminase-positive such as (*Azospirillum*, *Pseudomonas*, *Rhizobium* and *Bradyrhizobium*) . They hydrolyze the ethylene precursor ACC as nutrient source for C and N by converting it to ammonia and α-ketoglutarate (Honma and Shimomura 1978; Glick et al. 1998; Saleem et al. 2007). Various halotolerant bacteria isolated from the rhizosphere of halophytic plants in the vicinity of the Yellow Sea harbour ACC-deaminase activity in saline medium (0,85 M NaCl) (Siddikee et al. 2010).

Kausar and Shahzad (2006) determined the positive effect of ACC deaminase in *P. putida* and *P. fluorescens* which reused to restore growth of maize under saline conditions. Bacteria containing ACC-deaminase with a potential effect on growth of

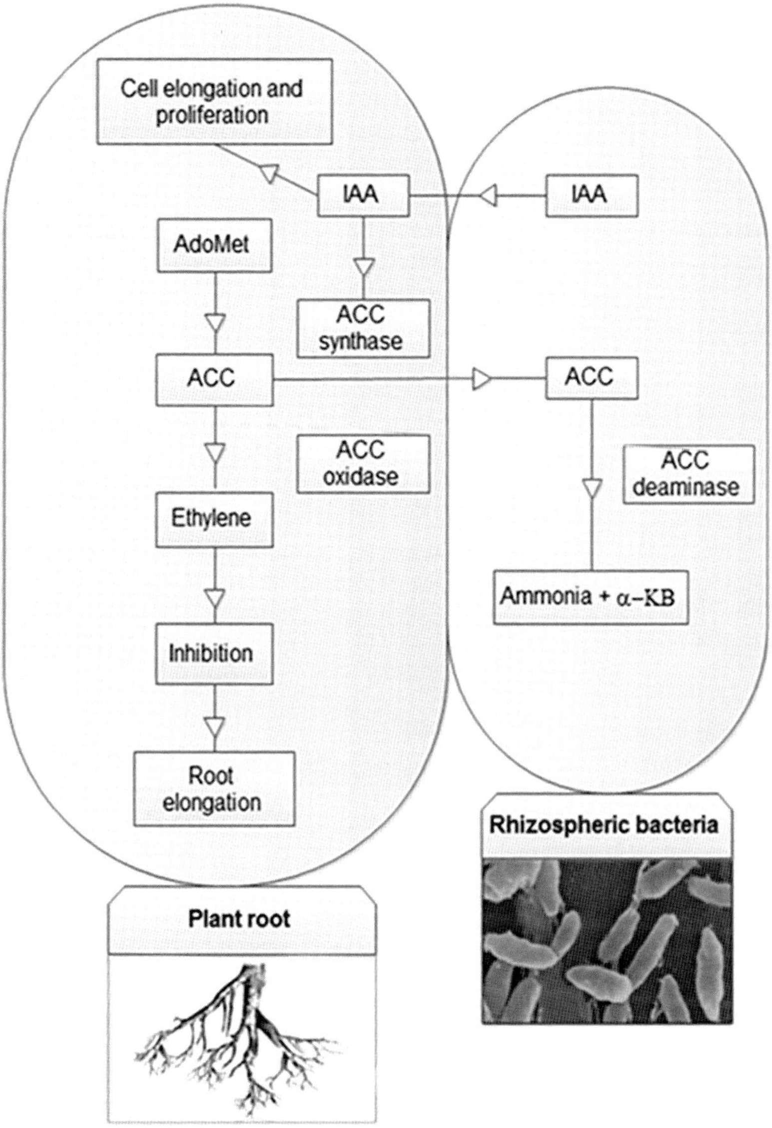

Fig. 9.1 Diagram of ACC-deaminase in bacteria following the model of Glick et al. (1998). The ACC-deaminase in bacteria reduce the ethylene level (inhibition of root elongation) in plants by conversion of ACC to ammonia and α-ketobutyrate

maize under saline conditions were also isolated and characterized by Hussain et al. (2013). Montero-Calasanz et al (2013) confirmed, that the inoculation of *Pantoea* sp. AG9, *Chryseobacterium* sp. AG13, *Chryseobacterium* sp. CT348, *Pseudomonas* sp. CT364 and *Azospirillum brasilense* Cd containing ACC-deaminase significantly induced the rooting in semi-hardwood cuttings of Arbequina, Hojiblanca and Picual cultivars of olive (*Olea europea* L.). Similar results were reported by Husen et al (2011), who were able to restore and enhance soybean growth under acidic and low fertility conditions in field soil by using ACC-deaminase producing *Pseudomonas* sp. strains. Nadeem and coworkers reported, that several halotolerant rhizosphere strains containing ACC-deaminase increased plant height, root length, total biomass and grain, N-, P-, and K-contents, chlorophyll- and carotenoid-concentrations and yields in maize under salt stress (Nadeem et al. 2006a, b, 2007). The authors showed the potential effect of some rhizospheric bacteria on ground-nut seedlings using ACC-deaminase containing *P. fluorescens* strain TDK1. Under high salt conditions, salt-affected plant growth was restored using the wild type *P. fluorescens*, while seedlings inoculated with a ACC-deaminase negative mutant or non-inoculated seedlings were not rescued (Saravanakumar and Samiyappan 2007).

PGPB containing ACC-deaminase in combination with rhizobia improve the growth and nodulation in plants by suppressing the endogenous level of ethylene (Zafar-ul-Hey et al. 2013). Ameliorative effects of some bacteria containing ACC-deaminase were clearly shown after inoculation of squash plants under salt stress (Yildirim et al. 2006). However, some osmotolerant bacteria increasing plant growth under salt stress were shown to lack the ACC deaminase gene, for e.g. in the slightly halophilic rhizospheric bacterium *Azospirillum brasilense* NH which, nevertheless, is able to restore salt-affected growth of wheat plants (Nabti et al. 2010).

9.4.2 Phytohormone Production of PGPB

Plant hormones or phytohormones are a group of naturally occurring, organic substances which influence physiological processes of plants at low concentrations. These hormones affect differentiation and development of plant growth through the regulation of diverse processes. These hormones are also involved in signaling systems of stress (Davis 2004). Plant hormones are divided into different groups such as: indole acetic acid (IAA), gibberelins (GA), cytokinins (CK), abscisic acid (ABA), ethylene and the cofactor pyrroquinoline quinone (PQQ) (Davis 2004; Jha et al. 2013).

It is well documented that the application of salt tolerant PGPB supports plant growth and renders them more tolerant to salt stress by improving their antioxidant status and physiological response by the intervention with several growth promoting substances like IAA, GA_3 and ABA mediated by PGPB (Perrig et al. 2007). More recently, spermidine was described as new protector molecule against stress (Alavi et al. 2013). In their study regarding salt effects on soybean seedlings, Asim et al. (2013) showed a remarkable decrease in IAA content in leaves of stressed plants. They found major changes in endogenous level of phytohormones to occur between 48 and 96 h after inoculation with PGPB. IAA causes morphological changes of roots and the root system leading to improved shoot growth and yields when present at optimal concentrations. Naz and co-workers (2009) isolated and characterized halotolerant bacteria by a high production of proline and the plant hormones IAA, GA3, trans-zeatin riboside and ABA. Inoculation of soybean plants by these halotolerant strains showed a significant improvement in shoot and root length and dry weight under salt stress (20 dS/m NaCl). Thus, these bacteria appear promising as potential bio-fertilizers to that of saline soils.

Nabti et al. (2007, 2010) revealed the restoration of salt-affected growth of durum wheat by the osmotolerant PGPB *A. brasilense* NH, isolated from salt-affected Algerian soil close to the Mediterranean coast. This strain was able to produce high amount of IAA in saline medium (200 mM NaCl), while IAA-production in the salt-sensitive reference strain *A. brasilense* Sp7 was almost completely inhibited at this NaCl-concentration. It is well demonstrated that IAA play a considerable role in plant resistance to salt stress (Ali and Abbas 2003; Kaya et al. 2010, 2013; Khalid et al. 2013; Kang et al. 2014). Many strains of halophilic bacteria belonging to the genus *Halomonas* sp. isolated from the rhizosphere of *Salicornia* plants were characterized and tested for PGP-features under high salt concentration, where a noticeable quantity of IAA was observed (Mapelli et al. 2013). Ul-Hassan and Bano (2014) reported in their study that the level of IAA and ABA increased in leaves of wheat growing in saline soil after a dual inoculation with isolates of *Pseudomonas* sp. and *Bacillus cereus*. Furthermore, Jha and Subramanian (2013) showed clearly the direct and potential effect of some osmotolerant bacteria on germination of paddy seeds under saline conditions. This is probably due to the intervention and supplementation of IAA by the inoculated bacteria which plays an important role in the germination of seeds. Additionally, Raza and Faisal (2013) studying the growth promotion of maize by desiccation tolerant *Micrococcus luteus* chp 37 showed an involvement of IAA and HCN in the restoration and protection of growth of maize under harsh conditions. Inoculation with halotolerant PGPB improved plant growth under saline conditions by the augmentation of auxin (IAA) which caused a reduction of the uptake of toxic ions by plants (Zhang et al. 2008; Chakraborty et al. 2011). Recently, the impact of plant hormones like IAA, ABA and GA_3 on amelioration of salinity stress in crop plant growth as well as the direct implication of phytohormones was investigated (Jha et al. 2013) in restoration and growth enhancement of common bean (*Phaseolus vulgaris*) under high salinity conditions after inoculation with two halotolerant strains of *P. extremorientalis* and *P. chlororaphis* (Egamberdieva 2011). In their study on the utilization of osmotolerant PGPB so as to mitigate deleterious effects of salinity and osmotic stress in *Cucumis sativus*, Kang et al (2014) revealed that stressed plants showed an upregulation of stress-responsive abscisic acid which is not occurring when PGPB were inoculated, whereas salicylic acid and GA were highly produced in PGPB-inoculated plants.

9.4.3 Minerals Uptake, Nitrogen Fixation, Siderophores Production, Phosphate Solubilization and Root Colonization by Halotolerant PGPB

Mineral disturbances under salinity stress reduce plant growth by affecting the availability, transport, and partitioning of nutrients. Salinity can differentially affect the mineral nutrition of plants and cause nutrient deficiencies or imbalances, due to the competition of Na^+ and Cl^- with nutrients such as K^+, Ca^{2+} and NO_3^- (Jouyban 2012). Nevertheless, halotolerant bacteria can help to overcome the turbulence of nutrient uptake by restoration of phosphorus availability, and enrichment of rhizosphere with NO_3^- N (Ul-Hassan and Bano 2014). These bacteria also influence the uptake of nutrient elements (N, P, and K, Ca, Mg, Na Cl, Fe, Zn, Cu Mn) (Shirmardi et al. 2010; Yildirim et al. 2011; Jha and Subramanian 2013). The effects of co-inoculation with *P. fluorescens* and *R. meliloti* strains on nodulation and mineral uptake in alfalfa (*Medicago sativa*) clearly showed restoration of growth and only moderated negative effects of salt and significantly increased plant growth by elevation of P and N and diminution of Na^+ *in planta* (Younesi et al. 2013a). Similar effects were reported by Han and Lee (2005) during their study on the effect of soil salinity on the antioxidant status, photosynthesis, mineral uptake and growth of lettuce. In this experiment, the P, K and N concentrations were increased in lettuce plants; in contrast, Na^+ was decreased. Some observations revealed that the capacity of bacteria to mitigate salt stress is due to the production of polysaccharides binding Na^+ in the root zone (Ashraf et al. 2006; Awad et al. 2012). Sarathambal and Ilamurugu (2013) showed the detailed PGP-features of some salt-tolerant diazotrophic PGPR, which were used to restore growth of paddy under saline conditions. Nutrient uptake (N, P and K) were increased, and in parallel, Na^+ level in plant tissues were reduced. In addition to these characters, these rhizobacteria showed the potential to fix nitrogen and solubilize phosphate under high salinity conditions (Sarathambal and Ilamurugu 2013). Osmotolerant rhizobacteria are known to enhance nutrient uptake in plants, nitrogen fixation, phosphate solubilization and siderophore biosynthesis in saline soils (Bashan et al. 2000; Ibekwe et al. 2010; Chakraborty et al. 2011; Shookietwattana and Maneewan 2012; Aamir et al. 2013; Jha and Subramanian 2013; Younesi et al. 2013b). In addition, the inoculation with osmotolerant PGPB to alleviate the deleterious effects of salt stress conditions on growth and yield of strawberry plants under salinity conditions was examined by Karlidag et al. (2013). Improvement of vegetative growth of plants, chlorophyll content, nutrient element, and yield of strawberry plants under saline stress was highly pronounced. Electrolyte leakage of plants was found reduced and the contents of nutrient elements in leaves and roots were considerably increased – except of sodium (Na^+) and chlorine (Cl^-). Nitrogen fixation, potassium, phosphorus, calcium, magnesium, sulfur, manganese, copper, and iron concentrations were increased besides of N, P, K, Ca, Mg, S, Mn, Cu, and Fe in leaves and roots. The neutralizing effect on salinity and drought stress in plants was explained by ameliorative effects of PGPB on increased water potential and decreased electrolytic leakage in plants. In parallel, Na^+ concentration was decreased, while potassium and phosphate achieved a high level in comparison to the non-inoculated plants (Kang et al. 2014). Some species of halotolerant *Rhizobium* are capable to nodulate under high NaCl concentration. This phenomenon is mainly due to the induction of the *nod* genes in the absence of flavonoid inducers (Guasch-Vidal et al. 2013). A co-inoculation of alfalfa with *Rhizobium* sp. and *P. fluorescens* under saline conditions resulted in a synergetic effect, which was more significant than upon inoculation with each strain separately. This is due to the combination between the nodulating rhizobacteria (*Rhizobium*) and the phosphate-solubilizing bacteria (*Pseudomonas fluorescens*) (Younesi et al. 2013a). Similar results were obtained in the co-inoculation of bean plants with both of the bacterial strains of *Rhizobium* and *Pseudomonas* (Ahmed et al. 2013). Pandey and Maheshwari (2007a, b) studied combination of rhizobacteria for growth promotional activity in *Cajanus cajan* using two species consortia and multispecies consortia and observed exceptional increase in seedling growth in mixed-species, co-inoculated consortium.

Efficient root colonization of plants by PGPB is a central parameter for the enhancement of plant growth in presence or absence of salinity. Bacterial root colonization is fundamentally influenced by the presence of specific bacterial traits required for attachment and subsequent establishment. However, additional abiotic and biotic factors play an important role in root colonization too. When a microbe colonizes a root, the process must be in accordance with an array of external parameters including water content, temperature, pH, soil types (texture, organic matter, micro-aggregate stability, presence of nutrients such as N, P, K, and Fe), composition of root exudates, and the presence of other microorganisms (Ahmad et al. 2011; Maheshwari 2012).

Some endophytic bacterial members of the plant microbiome have multiple functions as they have potentials for biological control, nitrogen fixation and plant growth promotion. This extraordinary feature was found for example in the Gram-negative nitrogen–fixing bacteria of the genus *Azospirillum* living in close association with plant roots (Broek et al. 1998). Within the species *A. brasilense*, an endophytic colonization is known for the strain Sp245, which colonize the root cortex area of wheat plant roots (Rothballer et al. 2003; Schloter and Hartmann 1998). Endophytic colonization of plants by microorganisms is only possible, when the plant is not recognizing the bacterial cells as enemy and does not mobilize defense reactions against it. On the other hand, the bacterium has abilities to defend itself, enter and thrive within a plant successfully, exerting its beneficial effects with less competition with the abundant rhizosphere microbiota (Nabti et al. 2010). Puentea et al. (1999) showed a potential root colonization of mangrove seedlings by strains of *A. brasilense and A. halopraeferens*, irrigated with seawater. A difference of the colonization pattern was observed between the two strains, where *A. brasilense* cells were anchored to the root surfaces and to themselves by a network of fibrillar material. However, *A. halopraeferens* yielded mainly single cells embedded in a thick sheath. The author suggested the feasibility of using terrestrial PGPB for the inoculation of marine plants. An analogous experiment was performed with slightly halophilic *A. brasilense* NH showing an efficient root colonization of durum wheat at 200 mM NaCl (Fig. 9.2). However, the same strain did not show this capacity in the absence of NaCl (Nabti et al. 2010).

In recent years, many studies are concerned with isolation of salt tolerant endophytic bacteria and their application in saline stress alleviation (Chookietwattana and Maneewan 2012; Damodaran et al. 2013; Kannan et al. 2014; Cardinale et al. 2014).

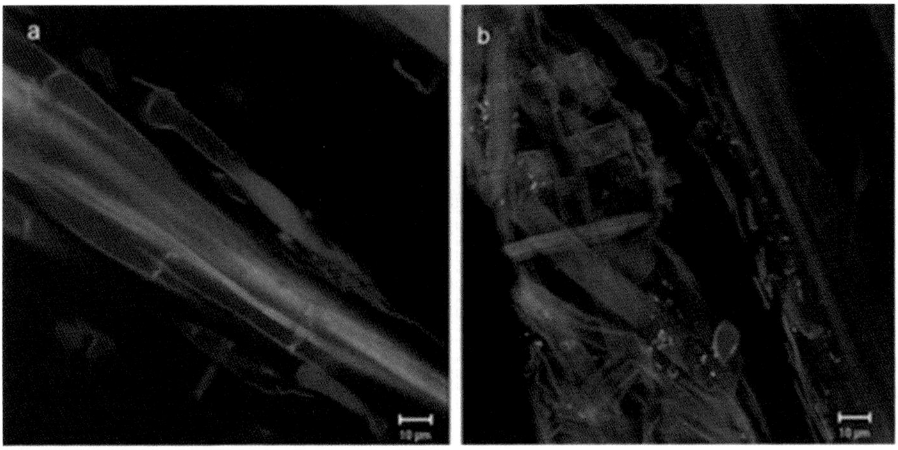

Fig. 9.2 Confocal laser scanning microscopic images of wheat roots inoculated with *A. brasilense* NH in axenic conditions after 4 weeks of growth under saline (200 mM NaCl) and non saline conditions: The roots were fixed in 4 % PFA, FISH-analysis was performed using the probes EUB-338-I, II, III, labelled with FLUOS and the specific probe Abras-1420, labelled with Cy3. *A. brasilense* NH cells are stained in yellow (combination of both fluorescent signals [Cy3=*red*, FLUOS=*green*]). Orthogonal optical sections of a three dimensional confocal image created from a z-stack of xy-scans. (**a**) Shows the root surface colonization by *A. brasilense* NH at 200 mM NaCl. (**b**) No surface root colonization by *A. brasilense* NH in absence of NaCl

9.4.4 Biocontrol and Soil Remediation

In addition to deleterious effects of salinity stress on plant growth, phytopathogenic microbes often successfully attack salt-stressed plants probably due to a weakened defense system or lesions in their surface structures. Environmental conditions have major influences on the pathogenicity, because the epiphytic phase of pathogens is strongly influenced by water availability, temperature and surface wetness (Agrios 1997). Thus, to cope with this problem, farmers frequently use chemical pesticides and fungicides. Unfortunately, continuous and increasing application of chemical substances generated pathogen resistance to antimicrobial agents as well as environment pollution (Compant et al. 2005). In addition, more and more chemical pesticides are no longer allowed to be in use and therefore alternatives are badly needed (Maheshwari 2013).

As an alternative, specific salt-tolerant or salt-sensitive PGPB, so-called biological control agents, are getting in use to protect plants from pathogen attack. In these cases, plant growth stimulation occurs through suppression of phytopathogens by the production of siderophores, antibiotics or other substances. Other mechanisms are also involved in the antifungal activity such as allelochemicals, biocidal volatiles, hydrolytic and detoxification enzymes (Sturz and Christie 2003; Chakraborty et al. 2011).

Furthermore, plants are capable to acquire an induced systemic resistance (ISR) towards phytopathogens after inoculation with PGPB. When these bacteria are associated with plant roots, they stimulate a protection of the plant host and confer a resistance against a large spectrum of different pathogenic agents (Van Hulten et al. 2006). The importance of bacteria in growth promotion and their ability to elicit the induced systemic tolerance also against abiotic stress conditions, such as e.g. UV-radiation and desiccation stress, has been documented in detail (Damodaran et al. 2013). Many salt-tolerant PGPB were studied for their capability to eliminate phyto-pathogenic fungi in number of rhizosphere (Wolf et al. 2002; Belimov et al. 2005). Bhakthavatchalu et al. (2013) demonstrated the potential effects of volatile and diffusible metabolites produced by the salt tolerant *P. aeruginosa* FP6 in the biocontrol. Among biocontrol rhizobacteria, bacteria belonging to the genera *Pseudomonas* and *Bacillus* are the most important ones (Van Peer et al. 1990; Cook 1993; Paul et al. 2006; Bhakthavatchalu et al. 2013; Saravanan et al. 2013).

It is well known that continuous spreading of heavy metal into the environments occurs, e.g. as contaminants of household compost into soils or via air pollution. This constitutes a significant environmental pollution and its negative impact on human health and agriculture. Rhizosphere, as an important interface of soil and plant, plays a significant role in the phytoremediation of soils contaminated with heavy metals and organic xenobiotics. Microbial populations are known to affect heavy metal mobility and availability to the plant through chelating mechanism, acidification, phosphate solubilization and redox changes. Therefore, they can enhance phytoremediation processes (Jing et al. 2007).

The most common heavy metal contaminants are Cd, Cr, Cu, Hg, Pb and Ni. The level of these toxic elements increases daily in soils and cause a considerable and accelerated pollution. This is the backside consequence of industrialization with its huge production of gas, fuel, fertilizer, sewage and pesticide (Kabata-Pendias 2011). However, phytoremediation is an alternative biological approach of treating the polluted areas and reducing the phytotoxicity in polluted soils (Weyens et al. 2010; Nanda and Abraham 2013). Phytoremediation is defined as the utilization of plants in the elimination of toxic metal and organics. To improve these activities, PGPR that facilitate the proliferation of various plants especially under environmentally stressful conditions may be used to lower the level of growth inhibiting stress ethylene within the plant and also to provide the plant with iron from the soil (Glick 2003).

Few studies showing the utilization of PGPR to reduce the negative effects of Lead (Pb), Chromium (Cr) and other toxic heavy metals on plant growth were reported (Janhmohammadi et al. 2013; Khan et al. 2013; Nakbanpote et al. 2014). Field and laboratory experiments have also been performed to show the positive effect of bacteria of the genus *Pseudomonas* in the bioremediation system for decontamination of petroleum and salt-affected soils (Greenberg et al. 2007). It is important to mention the relationship between the mechanisms involved in mitigation of salt stress and bioremediation, such as the reduction of ethylene, which is increased by several stress conditions – high salinity/osmolarity and contamination by heavy metals.

9.5 Conclusion

In summary, the successful restoration of plant growth under salinity conditions after inoculation with halotolerant or halophilic PGPB provides the basis for a suitable alternative of a successful formulation to improve crop growth and yield in saline soils. Additionally, halotolerant PGPB are able to enhance plant growth under saline conditions by various mechanisms. These mechanisms are similar to those of salt-sensitive PGPB plant interactions. The major difference resides in the osmoregulatory adaptation of halotolerant PGPB and some specific traits to relief deficiencies in salt-stressed plants.

References

Aamir M, Aslem A, Khan MY, Jamshaid MU, Ahmed M, Asghar HN, Zahir ZA (2013) Co-inoculation with *Rhizobium* and plant growth promoting Rhizobacteria (PGPR) for inducing salinity tolerance in mung bean under field condition of semi-arid climate. Asian J Biol 1:17–22

Abdel-Ghaffar BA, El-Shourbagy MN, Basha EM (1998) Responses of NaCl stressed wheat to IAA. In: Proceedings of the 6th Egypt botany conference, Cairo University, Giza, Egypt, vol 6, pp 79–88

Abolhasani M, Lakzian A, Tajabadipour A, Haghnia G (2010) The study of salt and drought tolerance of *Sinorhizobium bacteriua* to the adaptation to alkaline condition. Aust J Basic Appl Sci 4:882–886

Abu-Sharar TM, Bingham FT, Rhoades JD (1987) Stability of soil aggregates as affected by electrolyte concentration and composition. Soil Sci Soc Am J 51:309–314

Agrios GN (1997) Plant physiology. Academic, New York

Ahmad F, Husain FM, Iqbal A et al (2011) Rhizosphere and root colonization by bacterial inoculants and their monitoring methods: a critical area in PGPR research. In: Ahmad F (ed) Microbes and microbial technology: agricultural and environmental applications. Springer, New York, pp 363–391

Ahmed M, Sandhu GR (1988) Response of legumes to salt stress: effect on growth and nitrogen status of soybean. Pak J Agric Res 9:463–468

Ahmed M, Zahir AZ, Nazli F, Akram F, Arshad M, Khalid M (2013) Effectiveness of halo-tolerant, auxin producing *Pseudomonas* and *Rhizobium* strains to improve osmotic stress tolerance in mung bean (*Vigna radiata* L). Braz J Microbiol 44:1341–1348

Alavi P, Starcher MR, Zachow C, Müller H, Berg G (2013) Root-microbe systems: the effect and mode of interaction of Stress Protecting Agent (SPA) *Stenotrophomonas rhizophila* DSM 14405 T. Front Plant Sci 4:1–10. doi:10.3389/fpls.2013.00141

Ali RM, Abbas HM (2003) Response of salt stressed barely seedlings to phenylurea. Plant Soil Environ 49:158–162

Alizadeh O, Parsaeimehr A (2011) The influence of plant growth promoting rhizobacteria (PGPR) on the reduction of abiotic stresses in crops. Extrem Life Biospeol Astrol 3:93–99

Allotey DFK, Asiamah RD, Dedzoe CD, Nyamekye AL (2008) Physico-chemical properties of three salt-affected soils in the lower volta basin and management strategies for their sustainable utilization. West Afr J Appl Ecol 12:1–14

Aly MM, El-Sabbagh, El-Shouny WA, Ebrahim MKH (2003) Physiological response of Zea mays to NaCl stress with respect to *Azotobacter chroococcum* and *Streptomyces niveus*. Pak J Biol Sci 6:2073–2080

Arora NK, Singhal V, Maheshwari DK (2006) Salinity-induced accumulation of poly-β-hydroxybutyrate in rhizobia indicating its role in cell protection. World J Microbiol Biotechnol 22(6):603–606

Ashraf M, Hasnain S, Berge O (2006) Effect of exo-polysaccharides producing bacterial inoculation on growth of roots of wheat (*Triticum aestivum* L) plants in a salt-affected soil. Int J Environ Sci Technol 3:43–51

Asim M, Aslam M, Bano A, Munir M, Majeed A, Abbas SH (2013) Role of phytohormones in root nodulation and yield of soybean under salt stress. Am J Res Commun 1:191–208

Awad NM, Turky AS, Abdelhamid MT, Attia M (2012) Amelioration of environmental salt stress on the growth of Zea mays (L.) plants by exo-polysaccharide producing bacteria. J Appl Sci Res 8:2033–2044

Azevedo NAD, Prisco JT, Filho J, De Lacerda CF, Silva JV, Da Costa PHA, Gomes-Filho E (2004) Effects of salt stress on plant growth, stomatal response and solute accumulation of different maize genotypes. Braz J Plant Physiol 16:31–38

Bakker EP, Booth IR, Dinnbier U, Epstein W, Gajewska A (1987) Evidence for multiple K+ export systems in Escherichia coli. J Bacteriol 169:3743–3749

Bartels D, Sunkar R (2005) Drought and salt tolerance in plants. Crit Rev Plant Sci 24:23–58

Bashan Y, Moreno M, Troyo E (2000) Growth promotion of the Sea water-irrigated oil seed halophyte Salicornia bigelovi inoculated with mangrove rhizosphere bacteria and halotolerant Azospirillum spp. Biol Fertil Soil 32:265–272

Belimov AA, Hontzeas N, Safronova VI, Demchinskaya SV, Piluzza G, Bullitta S, Glick BR (2005) Cadmium-tolerant plant growth-promoting bacteria associated with the roots of Indian mustard (Brassica juncea L Czern). Soil Biol Biochem 37:241–250

Bernard T, Pocard J-A, Perroud B, Le Rudulier D (1986) Variations in the response of salt-stressed Rhizobium strains to bétaïnes. Arch Microbiol 143:359–364

Bhakthavatchalu S, Shivakumar S, Sullia SB (2013) Characterization of multiple plant growth promotion traits of Pseudomonas aeruginosa FP6, a potential stress tolerant biocontrol agent. Ann Biol Res 4:214–223

Bianco C, Defez R (2011) Soil bacteria support and protect plants against abiotic stress. In: Shanker A (ed) Abiotic stress in plants – mechanisms and adaptation. Inntech Europe, Rijeka, pp 143–169

Biro B, Villanyi I, Koves-Péchy K (2002) Abundance and adaptation level of some soil microbes in salt-affected soils. Agrokem Talajtan 51:99–106

Blaha D, Prigent-Combaret C, Mirza MS, Moënne-Loccoz Y (2006) Phylogeny of the 1-aminocyclopropane-1-carboxylic acid deaminase-encoding gene acdS in phytobeneficial and pathogenic Proteobacteria and relation with strain biogeography. FEMS Microbiol Ecol 56:455–470

Boncompagni E, Qsteras M, Poggl M-C, Le Rudulier D (1999) Occurrence of choline and glycine betaine uptake and metabolism in the family Rhizobiaceae and their roles in osmoprotection. Appl Environ Microbiol 65:2072–2077

Bouhmouch I, Brahada F, Filali-Maltouf A, Aurag J (2001) Selection of osmotolerant and effective strains of Rhizobiaceae for inoculation of common bean (Phaseolus vulgaris) in Maroccan saline soils. Agronomie 21:591599

Broek AV, Lambrecht M, Vanderleyden J (1998) Bacterial chemotactic motility is important for initiation of wheat root colonization by Azospirillum brasilense. Microbiology 144:2599–2606

Cardinale M, Ratering S, Suarez C, Montoya AMZ, Geissler-Plaum R, Schnell S (2014) Modulation rhizosphere microbiota to mitigate salt stress of barley plants. In: Microbial ecology and application of inoculants in biocontrol – Indo-German Workshop. Book of Abstracts, IARI, New Delhi, pp. 35–37

Chakraborty U, Roy S, Chakraborty AP, Dey P, Chakraborty B (2011) Plant growth promotion and amelioration of salinity stress in crop plants by salt-tolerant bacterium. Recent Res Sci Technol 3:67–70

Chookietwattana K, Maneewan K (2012) Screening of efficient halotolerant solubilizing bacterium and its effect on promoting plant growth under saline conditions. World Appl Sci J 16:1110–1117
Compant S, Duffy B, Nowak J, Clément C, Ait Barka E (2005) Use of plant growth promoting bacteria for biocontrol of plant diseases: principles, mechanisms of action, and future prospects. Appl Environ Microbiol 71:4951–4959
Cook RJ (1993) Making greater use of introduced microorganisms for biological control of plant pathogens. Annu Rev Phytopathol 31:53–80
Cordovilla MP, Ligero F, Lluch C (1994) The effect of salinity on N2-fixation on growth, symbiotic performance and nitrogen assimilation in Faba bean (Vicia faba L) under salt stress. Plant Soil 172:289–297
Creus CM, Sueldo RJ, Barassi CA (1998) Water relations in Azospirillum-inoculated wheat seedlings under osmotic stress. Can J Bot 76:238–244
Damodaran T, Sah V, Rai RB, Sharma DK, Mishra VK, Jha SK, Kannan R (2013) Isolation of salt tolerant endophytic and rhizospheric bacteria by natural selection and screening for promising plant growth promotion rhizobacteria (PGPR) and growth vigour in tomato under sodic environment. Afr J Microbiol Res 7:5082–5089
Das S, De M, Ray R, Ganguly D, Jana TK, De TK (2011) Salt tolerant culturable microbes accessible in the soil of the Sundarbans mangrove forest, India. Open J Ecol 1:35–40
Davis PJ (2004) Nature, occurrence and functions. In: Davis PJ (ed) Plant hormones biosynthesis, signal transduction, action, vol 1. Kluwer Academic, Dordrecht, pp 1–15
Del Rio LA, Corpas FJ, Sandalio LM, Palma JM, Barroso JB (2003) Plant peroxisomes, reactive oxygen metabolism and nitric oxide. IUBMB Life 55:71–81
Dodd IC, Pérez-Alfocea F (2012) Microbial amelioration of crop salinity stress. J Exp Bot 8:1–14
Egamberdieva D (2011) Survival of Pseudomonas extremorientalis TSAU20 and P chlororaphis TSAU13 in the rhizosphere of common bean (Phaseolus vulgaris) under saline conditions. Plant Soil Environ 57:122–127
Egamberdiyeva D (2005) Characterization of Pseudomonas species isolated from the rhizosphere of plants grown in szernozem soil, semi-arid region of Uzbekistan. Sci World J 5:501–509
Epstein W (1986) Osmoregulation by potassium transport in Escherichia coli. FEMS Microbiol Rev 39:73–78
Fougere F, Le Rudulier D, Streeter JG (1991) Effects of salts on amino acid, organic acid, and carbohydrates composition of roots, bacteroids, and cytosol of alfalfa (Medicago sativa L.). Plant Physiol 96:1228–1236
François LE, Maas EV, Donovan TJ, Youngs VL (1986) Effect of salinity on grain yield and quality, vegetative growth, and germination of semi-dwarf and durum wheat. Agron J 78:1053–1058
Frenkel H, Goertzen JO, Rhoades JD (1978) Effect of clay type and content, exchangeable sodium percentage, and electrolyte concentration on clay dispersion and soil hydraulic conductivity. Soil Sci Soc Am J 142:32–39
Galinski EA (1995) Osmoadaptation in bacteria. Adv Microb Physiol 37:274–325
Galinski EA, Trüper HG (1982) Betaine, a compatible solute in the extremely halophilic phototrophic bacterium Ectothiorhodospira halochloris. FEMS Microbiol Lett 13:357–360
Galinski EA, Trüper HG (1994) Microbial behaviour in salt-stressed ecosystems. FEMS Microbiol Rev 15:95–108
Gerhardt KE, Greenberg BM, Glick BR (2006) The role of ACC deaminase in facilitating the phytoremediation of organics, metals and salt. Curr Trends Microbiol 2:61–73
Ghadiri H, Hussein J, Dordipour E, Rose C (2004) The effect of soil salinity and sodicity on soil erodibility, sediment transport and downstream water quality. In: 13th international soil conservation organization conference – Brisbane, July 2004 Conserving Soil and Water for Society, Sharing Solutions 631, pp 1–6
Ghorai S, Pal KK, Dey R, Jasrai YT (2013) A comparative study of effect Rhizobacteria onto seedling vigour of groundnut in in vitro condition isolated from Kutch region. Sch Acad J Biosci 1:179–182

Gill PK, Sharma AD, Singh P, Bhullar SS (2002) Osmotic stress-induced changes in germination, growth and soluble sugar content of Sorghum bicolour (L.) moench seeds. Bulg J Physiol 28:12–25

Glick BR (2003) Phytoremediation: synergistic use of plants and bacteria to clean up the environment. Biotechnol Adv 21:383–393

Glick BR (2005) Modulation of plant ethylene levels by the bacterial enzyme ACC deaminase. FEMS Microbiol Lett 251:1–7

Glick BR, Holguin G (2003) Transformation of Azospirillum brasilense Cd with ACC deaminase gene from Enterobacter cloacae UW4 fused to the tetr gene promoter improves its fitness and plant growth promoting ability. Microb Ecol 46:122–133

Glick BR, Penrose DM, Li J (1998) A model for the lowering of plant ethylene concentrations by plant growth promoting bacteria. J Theor Biol 190:63–68

Golezani KG, Yengabad FM (2012) Physiological responses of lentil (*Lensculinaris Medik*) to salinity. Int J Agric Crop Sci 4:1531–1535

Gontia I, Kavita K, Schmid M, Hartmann A (2011) *Brachybacterium saurashtrense* sp. nov., a halotolerant root-associated bacterium with plant growth promoting potential. Int J Syst Evol Microbiol 61:2799–2804

Gouffi K, Blanco C (2000) Is the accumulation of osmoprotectant the unique mechanism involved in bacterial osmoprotection? Int J Food Microbiol 55:171–174

Greenberg BM, Huang X-D, Gerhardt K, Glick BR, Gurska J, Wang W, Lampi M, Khalid A, Isherwood D, Chang P, Wang H, Wu SS, Yu X-M, Dixon DG, Gerwing P (2007) Field and laboratory tests of a multi-process phytoremediation system for decontamination of petroleum and salt impacted soils. In: Gavaskar AR, Silver CF (eds) Proceedings of the ninth international in situ and on-site remediation symposium, Chapter B-04. Batelle Press, Columbus

Guasch-Vidal B, Estévez J, Dardanelli MS, Soria-Díaz ME, Fernández de Córdoba F, Balog CIA, Manyani H, Gil-Serrano A, Thomas- Oates J, Hensbergen PJ, Deelder AM, Megías M, Van Brussel AAN (2013) High NaCl concentrations induce the nod genes of *Rhizobium tropici* CIAT899 in the absence of flavonoid inducers. Mol Plant Microbe Interact 26:451–460

Gunes A, Inal A, Alpaslan M (1996) Effect of salinity on stomatal resistance, proline, and mineral composition of pepper. J Plant Nutr 19:389–396

Halverso LJ, Jones TM, Firestone MK (2000) Release of intracellular solutes by four soil bacteria exposed to dilution stress. Soil Sci Soc Am J 64:1630–1637

Hamaoui B, Abbadi JM, Burdman S, Rashid A, Sarig S, Okon Y (2001) Effects of inoculation with Azospirillum brasilense of chickpeas (Cicer arietinum) and faba beans (Vicia faba) under different growth conditions. Agronomie 21:553–560

Han HS, Lee KD (2005) Plant growth promoting rhizobacteria effect on antioxidant status, photosynthesis, mineral uptake and growth of lettuce under soil salinity. Res J Agric Biol Sci 1:210–215

Hartmann A (1988) Osmoregulatory properties of *Azospirillum* spp. In: Klingmüller W (ed) Azospirillum IV: genetics, physiology, and ecology. Springer, Berlin, pp 122–130

Hartmann A, Prabhu SR, Galinski EA (1991) Osmotolerance of diazotrophic rhizosphere bacteria. Plant Soil 137:105–109

Hartmann A, Guendisch C, Bode W (1992) *Azospirillum* mutants improved in iron acquisition and osmotolerance as tools for the investigation of environmental fitness traits. Symbiosis 13:271–279

Honma M, Shimomura T (1978) Metabolism of 1-aminocyclopropane-1-carboxilic acid. Agric Biol Chem 42:1825–1831

Husen E, Wahyudi AT, Suwanto A, Giyanto (2011) Soybean response to 1-aminocyclopropane-1-carboxylate deaminase-producing pseudomonas under field soil conditions. Am J Agric Biol Sci 6:273–278

Hussain MI, Asghar HN, Arshad M, Shahbaz M (2013) Screening of multi-traits rhizobacteria to improve maize growth under axenic conditions. J Anim Plant Sci 23:514–520

Ibekwe AM, Poss JA, Grattan SR, Grieve CV, Suarez D (2010) Bacterial diversity in cucumber (Cucumis sativus) rhizosphere in response to salinity, soil pH, and boron. Soil Biol Biochem 42:567–575

Jan NE, Ud-Din J, Kawabata S (2014) Impact of saline-alkali stress on the accumulation of solids in tomato fruits. Pak J Bot 46:161–166

Janhmohammadi M, Bihmata MR, Ghasemzadeh F (2013) Influence of rhizobacteria inoculation and lead stress on the physiological and biochemical attributes of wheat genotypes. Cercet Agron Mold 1:49–67

Jha Y, Subramanian RB (2013) Paddy plants inoculated with PGPR show better growth physiology and nutrient content under saline conditions. Chil J Agric Res 73:213–219

Jha B, Gontia I, Hartmann A (2012) The roots of the halophyte Salicornia brachiata are source of new halotolerant diazotrophic bacteria with plant growth promoting potential. Plant Soil 356:265–277

Jha PN, Gupta G, Jha P, Mehrotra R (2013) Association of rhizospheric /endophytic bacteria with plants; a potential gateway to sustainable agriculture. GJAS 3:73–84

Jing YD, He Z-L, Yang XE (2007) Role of soil rhizobacteria in phytoremediation of heavy metal contaminated soils. J Zhejiang Univ Sci B 8:192–207

Jofré E, Fischer S, Rivarola V, Balegno H, Mori G (1998) Saline stress affects the attachment of Azospirillum brasilense Cd to maize and wheat roots. Can J Microbiol 44:416–422

Jouyban Z (2012) The effects of salt stress on plant growth. Tech J Eng App Sci 2:7–10

Kabata-Pendias A (2011) Trace elements in soils and plants. CRC Press, Taylor and Francis group, Boca Raton

Kadouri D, Jurkevitch E, Okon Y (2003) Involvement of the reserve material poly-ß-hydroxybutyrate in Azospirillum brasilense stress endurance and root colonization. Appl Environ Microbiol 69:3244–3250

Kahlaoui B, Hachicha M, Rejeb S, Rejeb MN, Hanchi B, Misle E (2011) Effects of saline water on tomato under subsurface drip irrigation: nutritional and foliar aspects. J Soil Sci Plant Nutr 11:69–86

Kang S-M, Khan AL, Waqas M, You Y-H, Kim J-H, Kim J-G, Hamayun M, Lee I-J (2014) Plant growth-promoting rhizobacteria reduce adverse effects of salinity and osmotic stress by regulating phytohormones and antioxidants in Cucumis sativus. J Plant Interact 9:673–682

Kannan R, Damodaran T, Pandey BK, Umamaheswari S, Rai RB, Jha SK, Mishra V, Sharma DK, Sah V (2014) Isolation and characterization of endophytic plant growth-promoting bacteria (PGPB) associated to the sodicity tolerant polyembryonic mango (Mangifera indica L.) root stock and growth vigour in rice under saline sodic environment. Afr J Microbiol Res 8:628–636

Karlidag H, Yildirim E, Turan M, Pehluvan M, Donmez F (2013) Plant growth-promoting rhizobacteria mitigate deleterious effects of salt stress on strawberry plants (*Fragaria* x *ananassa*). Hortic Sci 48:563–567

Kausar R, Shahzad M (2006) Effect of ACC-deaminase containing rhizobacteria on growth promotion of maize under salinity stress. J Agric Soc Sci 2:216–218

Kay BD (1990) Rates of change of soil structure under different cropping systems. Adv Soil Sci 12:1–52

Kaya C, Tuna AL, Okant AM (2010) Effect of foliar applied kinetin and indole acetic acid on maize plants grown under saline conditions. Turk J Agric For 34:529–538

Kaya C, Ashraf M, Dikilitas M, Tuna AL (2013) Alleviation of salt stress induced adverse effects on maize plants by exogenous application of indole acetic acid (IAA) and inorganic nutrients – A field trial. Aust J Crop Sci 7:249–254

Kempf B, Bremer E (1998) Uptake and synthesis of compatible solutes as microbial stress responses to high-osmolality environments. Arch Microbiol 170:319–330

Khalid S, Parvaiz M, Nawaz K, Hussain H, Arshad A, Shawaka S, Sarfaraz ZN, Waheed T (2013) Effect of Indole Acetic Acid (IAA) on morphological, biochemical and chemical attributes of Two varieties of maize (Zea mays L.) under salt stress. World Appl Sci J 26:1150–1159

Khan MY, Asghar HN, Jamshaid MU, Akhtar MJ, Zahir AZ (2013) Effect of microbial inoculation on wheat growth and phyto-stabilization of chromium contaminated soil. Pak J Bot 45:27–34

Kim K, Jang Y-J, Lee S-M, Oh B-T, Chae J-C, Lee K-J (2014) Alleviation of salt stress by Enterobacter sp EJ01 in tomato and Arabidopsis is accompanied by Up-regulation of conserved salinity responsive factors in plants. Mol Cells 37:109–117

Kogut M, Russell NJ (1987) Life at the limits: considerations on how bacteria can grow at extremes of temperature and pressure or with high concentrations of ions and solutes. Sci Prog 71:381–400

Kohler J, Caravaca F, Alguacil MM, Roldan A (2009) Elevated CO2 increases the effect of an arbuscular mycorrhizal fungus and a plant-growth promoting rhizobacterium on structural stability of a semiarid agricultural soil under drought conditions. Soil Biol Biochem 41:1710–1716

Le Rudulier D, Bernard T (1986) Salt tolerance in *Rhizobium*: a possible role for betaines. FEMS Microbiol Rev 39:67–72

Le Rudulier D, Bernard T, Pocard J-A, Goas G (1983) Accroissement de l'osmotolerance chez Rhizobium meliloti par la glycine betaine et la proline betaine. CR Acad Sci Paris 297:155–160

Le Rudulier D, Strom AR, Dandekar AM, Smith LT, Valentine RC (1984) Molecular biology of osmoregulation. Science 224:1064–1068

Maheshwari DK (2011) Bacteria in agrobiology: plant nutrient management. Springer, Berlin/Heidelberg, p 345

Maheshwari DK (2012) Bacteria in agrobiology: stress management. Springer, Berlin/Heidelberg, p 333

Maheshwari DK (2013) Bacteria in agrobiology: disease management. Springer, Berlin/Heidelberg, p 495

Mahmoodzadeh H, Khorasani FM, Besharat H (2013) Impact of salt stress on seed germination indices of five wheat cultivars. Ann Biol Res 4:93–96

Mapelli F, Marasco R, Rolli E, Barbato M, Cherif H, Guesmi A, Ouzari I, Daffonchio D, Borin S (2013) Potential for plant growth promotion of rhizobacteria associated with *Salicornia* growing in Tunisian hypersaline soils. Biomed Res Int. doi:10.1155/2013/248078

Marschner H (1995) Functions of mineral nutrients. In: Marschner H (ed) Mineral nutrition of higher plants, vol 2. Academic, London, pp 229–312

Mayak S, Tirosh T, Glick BR (2004) Plant growth-promoting bacteria that confer resistance in tomato to salt stress. Plant Physiol Biochem 42:565–572

Maziah M, Zuraida AR, Halimi M, Zulkifli S, Sreeramanan S (2009) Responses of banana plantlets to rhizobacteria inoculation under salt stress condition. Am-Euras J Sustain Agric 3:290–305

Miller KJ, Wood JM (1996) Osmoadaptation by rhizosphere bacteria. Annu Rev Microbiol 50:101–136

Montero-Calasanz MC, Santamaría C, Albareda M, Daza A, Duan J, Glick BR, Camacho M (2013) Alternative rooting induction of semi-hardwood olive cuttings by several auxin-producing bacteria for organic agriculture systems. Span J Agric Res 11:146–154

Nabti E, Sahnoune M, Adjrad S, Van Dommelen A, Ghoul M, Schmid M, Hartmann A (2007) A halophilic and osmotolerant Azospirillum brasilense strain from Algerian soil restores wheat growth under saline conditions. Eng Life Sci 7:354–360

Nabti E, Sahnoune M, Ghoul M, Fischer D, Hofmann A, Rothballer M, Schmid M, Hartmann A (2010) Restoration of growth of durum wheat (*Triticum durum var. Waha*) under saline conditions Due to inoculation with the rhizosphere bacterium Azospirillum brasilense NH and extracts of the marine alga Ulva lactuca. J Plant Growth Regul 29:6–22

Nabti E, Sahnoune M, Ghoul M, Fischer D, Rothballer M, Schmid M, Hartmann A (2012) Enhancement and restoration of growth of durum wheat (*Triticum durum, var waha*) on saline soil by using Azospirillum brasilense NH and marine alga Ulva lactuca. In: Krueger D, Meyer II (eds) Algae, ecology, economic uses and environmental impact marine biology. Nova Sciences, New York, pp 29–52

Nadeem SM, Hussain I, Naveed M, Asghar HN, Zahir ZA, Arshad M (2006a) Performance of plant growth promoting rhizobacteria containing ACC-deaminase activity for improving growth of maize under salt-stressed conditions. Pak J Sci 43:114–120

Nadeem SM, Zahir AZ, Naweed M, Arshad M, Shahzad SM (2006b) Variation in growth and ion uptake of maize due to the inoculation with plant growth promoting rhizobacteria under salt stress. Soil Environ 25:78–84

Nadeem SM, Zahir ZA, Naveed M, Arshad M (2007) Preliminary investigations on inducing salt tolerance in maize through inoculation with rhizobacteria containing ACC deaminase activity. Can J Microbiol 53:1141–1149

Nakbanpote W, Panitlurtumpai N, Sangdee A, Sakulpone N, Sirison P, Pimthong A (2014) Salt-tolerant and plant growth promoting bacteria isolated from Zn/Cd contaminated soil : identification and effect on rice under saline conditions. J Plant Interact 9:397–387

Nanda S, Abraham J (2013) Remediation of heavy metal contaminated soil. Afr J Biotechnol 12:3099–3109

Naz R, Bano A (2013) Influence of exogenously applied salicylic acid and plant growth promoting rhizobacteria inoculation on the growth and physiology of sunflower (*Helianthus annuus* L) under salt stress. Pak J Bot 45:367–373

Naz I, Bano A, Ul-Hassan T (2009) Isolation of phytohormones producing plant growth promoting rhizobacteria from weeds growing in Khewra salt range, Pakistan, and their implication in providing salt tolerance to Glycine max (L.). Afr J Biotechnol 8:5762–5766

Pandey P, Maheshwari DK (2007a) Bioformulation of Burkholderia sp. MSSP with a multispecies consortium for growth promotion of *Cajanus cajan*. Can J Microbiol 53(2):213–222

Pandey P, Maheshwari DK (2007b) Two-species microbial consortium for growth promotion of *Cajanus cajan*. Curr Sci 92(8):1137–1142

Paul D, Dineshkumar N, Nair S (2006) Proteomics of a plant growth-promoting rhizobacterium *Peudomonas fluorescens* MSP-393, subjected to salt shock. World J Microbiol Biotechnol. doi:10.1007/s11274-005-9043-ycbrs

Perrig D, Boiero ML, Masciarelli O, Penna C, Ruiz OA, Cassan F, Luna V (2007) Plant growth promoting compounds produced by two strains of *Azospirillum brasilense*, and implications for inoculants formation. Appl Microbiol Biotechnol 75:1143–1150

Pichereau V, Pocard J-A, Hamelin J, Blanco C, Bernard T (1998) Differential effects of Dimethylsulfoniopropionate, Dimethylsulfonioacetate and other S-Methylated compounds on the growth of *Sinorhizobium meliloti* at low and high osmolarities. Appl Environ Microbiol 64:1420–1429

Porcel R, Aroca R, Ruiz-Loeano JM (2012) Salinity stress alleviation using arbuscular mycorrhizal fungi. Agron Sustain Dev 32:181–200

Puentea ME, Holguina G, Bernard R, Glick B, Bashan Y (1999) Root-surface colonization of black mangrove seedlings by *Azospirillum halopraeferens* and *Azospirillum brasilense* in seawater. FEMS Microbiol Ecol 29:283–292

Qureshi RH, Ahmed R, Ilyas M, Aslam Z (1980) Screening of wheat (*Triticum aestivum* L) for salt tolerance. Pak J Agric Sci 27:9–26

Rabie GH, Almadini AM (2005) Role of bioinoculants in development of salt-tolerance of *Vicia faba* plants under salinity stress. Afr J Biotechnol 4:210–222

Rabie GH, Aboul Nasr MB, Al-Humiany A (2005) Increased salinity tolerance of Cowpea plants by dual inoculation of an arbuscular mycorrhizal fungus *Glomus clarum* and a nitrogen-fixer *Azospirillum brasilense*. Mycobiology 33:51–60

Rachidai A, Driouch A, Ouassou A, El Hadrami I (1994) Effet du traitement salin sur la germination du blé dur (Triticum durum Desf.). Rev Amélior Prod Agr 6:209–228

Rajput L, Imran A, Mubeen F, Hafeez FY (2013) Salt-tolerant PGPR strain *Planococcus rifietoensis* promotes the growth and yield of wheat (*Triticum aestivum* L) cultivated in saline soil. Pak J Bot 45:1955–1962

Ramadoss D, Lakkineni VK, Bose P, Ali S, Annapurna K (2013) Mitigation of salt stress in wheat seedlings by halotolerant bacteria isolated from saline habitats. Springer Plus 2:1–7

Raza FA, Faisal M (2013) Growth promotion of maize by desiccation tolerant *Micrococcus luteus-chp37* isolated from Cholistan desert, Pakistan. Aust J Crop Sci 7:1693–1698

Rengasamy P (2006) World salinization with emphasis on Australia. J Exp Bot 57:1017–1023

Rengasamy P (2010a) Soil processes affecting crop production in salt-affected soils. Funct Plant Biol 37:613–620

Rengasamy P (2010b) Osmotic and ionic effects of various electrolytes on the growth of wheat. Aust J Soil Res 48:120–124

Rengasamy P, Olsson KA (1991) Sodicity and soil structure. Aust J Soil Res 29:935–952
Riou N, Le Rudulier D (1990) Osmoregulation in *Azospirillum brasilense*: glycine betaine transport enhances growth and nitrogen fixation under salt stress. J Gen Microbiol 136:1455–1461
Riou N, Poggi MC, Le Rudulier D (1991) Characterization of an osmoregulated periplasmic glycine betaine-binding protein in *Azospirillum brasilense* Sp7. Biochemistry 73:1187–1193
Robinson SP, Jones GP (1986) Accumulation of glycine betaine in chloroplasts provides osmotic adjustment during salt stress. Aust J Plant Physiol 13:659–668
Roiston DE, Biggar JW, Nielsen DR (1984) Effect of salt on soils. Calif Agric 38:11–13
Rojas-Tapias D, Moreno-Galvan A, Pardo-Diaz S, Obando M, Rivera D, Bonilla R (2012) Effect of inoculation with plant growth promoting bacteria (PGPB) on amelioration of saline stress in maize (Zea mays). Appl Soil Ecol 61:264–272
Rothballer M, Schmid M, Hartmann A (2003) In situ localization and PGPR-effect of *Azospirillum brasilense* strains colonizing roots of different wheat varieties. Symbiosis 34:261–279
Saghafi K, Ahmadi J, Asgharzadeh A, Bakhtiari S (2013) The effect of microbial inoculants on physiological responses of two wheat cultivars under salt stress. Int J Adv Biol Biom Res 4:421–431
Sairam RK, Tyagi A (2004) Physiology and molecular biology of salinity stress tolerance in plants. Curr Sci 3:407–421
Saleem M, Arshad M, Hussain S, Bhatti AS (2007) Perspective of plant growth promoting rhizobacteria (PGPR) containing ACC deaminase in stress agriculture. J Ind Microbiol Biotechnol 34:635–648
Sapsirisopa S, Chookietwattana K, Maneewan K, khaengkhan P (2009) Effect of salt-tolerant Bacillus inoculum on rice KDML 105 cultivated in saline soil. Asian J Food Ag-Ind. Special Issue S69–S74
Sarathambal C, Ilamurugu K (2013) Saline tolerant plant growth promoting diazotrophs from rhizosphere of Bermuda grass and their effect on rice. Indian J Weed Sci 45:80–85
Saravanakumar D, Samiyappan R (2007) ACC deaminase from *Pseudomonas fluorescens* mediated saline resistance in ground-nut (*Arachis hypogea*) plants. J Appl Microbiol 102:1283–1292
Saravanan S, Muthumanickam P, Saravanan TS, Santhaguru K (2013) Antagonistic potential of fluorescent pseudomonads and its impact on growth of tomato challenged with phytopathogens. Afr Crop Sci J 21:29–36
Sarin MN, Narayanan A (1968) Effects of soil salinity and growth regulators on germination and seedling metabolism of wheat. Physiol Plant 21:1201–1209
Schloter M, Hartmann A (1998) Endophytic and surface colonization of wheat roots (*Triticum aestivum*) by different *Azospirillum brasilense* strains studied with strain-specific monoclonal antibodies. Symbiosis 25:159–179
Selvakumar G, Panneerselvam P, Ganeshamurthy AN (2012) Bacterial mediated alleviation of abiotic stress in crops. In: Maheshwari DK (ed) Bacteria in agrobiology: stress management. Springer, Berlin, pp 205–223
Shamseldin A, Nyalwidhe J, Werner D (2006) A proteomic approach towards the analysis of salt tolerance in *Rhizobium etli* and *Sinorhizobium meliloti* strains. Curr Microbiol 52:333–339
Sharma RC, Gupta NK, Gupta S, Hasegawa H (2005) Effect of NaCl salinity on photosynthetic rate, transpiration rate, and oxidative stress tolerance in contrasting wheat genotypes. Photosynthesis 43:609–613
Shen M, Kang YJ, Wang HL, Zhang XS, Zhao QX (2012) Effect of plant growth promoting rhizobacteria (PGPRs) on plant growth, yield, and quality of tomato (*Lycopersicon esculentum* Mill) under simulated seawater irrigation. J Gen Appl Microbiol 58:253–262
Shirmardi M, Savaghebi GR, Khavazi K, Akbarzadeh A, Farahbakhsh M, Rejali F, Sadat A (2010) Effect of microbial inoculants on uptake of nutrient elements in two cultivars of sunflowers (*Helianthus annuus* L). Not Sci Biol 2:57–66
Shookietwattana K, Maneewan K (2012) Screening of efficient halotolerant phosphate solubilizing bacterium and its effect on promoting plant growth under saline conditions. World Appl Sci J 16:1110–1117

Shukla PS, Agarwal PK, Jha B (2012) Improved salinity tolerance of *Arachis hypogeae* (L.) by the interaction of halotolerant plant-growth promoting rhizobacteria. J Plant Growth Regul 31:195–206

Siddikee MA, Chauhan PS, Anandham R, Gwang-Hyun H, Tongmin S (2010) Isolation, characterization, and use for plant growth promotion under salt stress of ACC deaminase-producing halotolerant bacteria derived from coastal soil. J Microbiol Biotechnol 20:1577–1584

Sleator RD, Hill C (2001) Bacterial osmoadaptation: the role of osmolytes in bacterial stress and virulence. FEMS Microbiol Rev 26:49–71

Sturz AV, Christie BR (2003) Beneficial microbial allelopathies in the root zone: the management of soil quality and plant disease with rhizobacteria. Soil Tillage Res 72:107–123

Suarez C, Ratering S, Kramer I, Schnell S (2014a) *Cellvibrio diazotrophicus* sp. nov., a nitrogen-fixing bacterium isolated from the rhizosphere of salt meadow plants and emended description of the genus Cellvibrio. Int J Syst Evol Microbiol 64:481–486

Suarez C, Ratering S, Geißler-Plaum R, Schnell S (2014b) *Rheinheimera hassiensis* sp. nov. and *Rheinheimera muenzenbergensis* sp. nov., two new species from the rhizosphere of *Hordeum secalinum*. Int J Syst Evol Microbiol. doi:10.1099/ijs.0.061200-0

Talibart R, Jebbar M, Gouesbet G, Himdi-Kabbab S, Wroblewski H, Blanco C, Bernard T (1994) Osmo-adaptation in rhizobia: ectoine-induced salt-tolerance. J Bacteriol 176:5210–5217

Tank N, Saraf M (2010) Salinity-resistant plant growth promoting rhizobacteria ameliorates sodium chloride stress on tomato plants. J Plant Interact 5:51–58

Tilak KVBR, Ranganayaki N, Pal KK, De R, Saxena AK, Nautiyal SC, Mittal S, Tripathi AK, Johri BN (2005) Diversity of plant growth and soil health supporting bacteria. Curr Sci 89:136–150

Tripathi AK, Mishra BM, Tripathi P (1998) Salinity stress responses in the plant growth promoting rhizobacteria, *Azospirillum* spp. J Biosci 23:463–471

Ul-Hassan T, Bano A (2014) Role of plant growth promoting rhizobacteria and L -tryptophan on improvement of growth, nutrient availability and yield of wheat (*Triticum aestivum*) under salt stress. Int J Appl Agr Res 4:30–39

Upadhyay SK, Maurya SK, Singh DP (2012) Salinity tolerance in free-living plant growth promoting rhizobacteria. Indian J Res 3:73–78

Van Hulten M, Pelser M, Van Loon LC, Pieterse CMJ, Ton J (2006) Costs and benefits of priming for the defense in Arabidopsis. Proc Natl Acad Sci U S A 103:5602–5607

Van Peer R, Van Kuik AJ, Rattink H, Schippers B (1990) Control of *Fusarium* wilt in carnation grown on rockwool by *Pseudomonas* sp. strain WCS417r and by Fe-EDDHA. Neth J Plant Pathol 96:119–132

Vinocur B, Altman A (2005) Recent advances in engineering plant tolerance to abiotic stress: achievements and limitations. Curr Opin Biotechnol 16:123–132

Weyens N, Truyens S, Dupae J, Newman L, Taghavi S, van der Lelie D, Carleer R, Vangronsveld J (2010) Potential of the TCE-degrading endophtye *Pseudomonas putida* W619-TCE to improve plant growth and reduce TCE phytotoxicity and evaporation in poplar cuttings. Environ Pollut 158:2915–2919

Wolf A, Fritze A, Hagemann M, Berg G (2002) *Stenotrophomonas rhizophila* sp. nov., a novel plant-associated bacterium with antifungal properties. Int J Syst Evol Microbiol 52:1937–1944

Xiong L, Zhu JK (2002) Molecular and genetic aspects of plant responses to osmotic stress. Plant Cell Environ 25:131–139

Yadav JSP, Agarwal RR (1961) A comparative study on the effectiveness of gyps and dhaincha (*Sesbania actuleata*) in the reclamation of saline-alkaline soil. J Indian Soil Sci 9:151–156

Yildirim E, Taylor AG, Spittler TD (2006) Ameliorative effects of biological treatments on growth of squash plants under salt stress. Sci Hortic 111:1–6

Yildirim E, Karlidag H, Turan M, Dursun A, Goktepe F (2011) Growth, nutrient uptake, and yield promotion of *Broccoli* by plant growth promoting rhizobacteria with manure. Sci Hortic 46:932–936

Younesi O, Baghbani A, Namdari A (2013a) The effects of *Pseudomonas fluorescensce* and *Rhizobium meliloti* co-inoculation on nodulation and mineral nutrient contents in alfalfa (*Medicago sativa*) under salinity stress. Int J Agric Crop Sci 5:1500–1507

Younesi O, Chaichi MR, Postini K (2013b) Salt tolerance in alfalfa following inoculation with *Pseudomonas*. Middle-East J Sci Res 16:101–107

Zafar-ul-Hye M, Ahmed M, Shahzad SM (2013) Synergistic effect of Rhizobia and plant growth promoting rhizobacteria on the growth and nodulation of lentil seedlings under axenic conditions. Soil Environ 32:79–86

Zahran HH (1999) Rhizobium-legume symbiosis and nitrogen fixation under severe conditions and in an arid climate. Microbiol Mol Biol Rev 63:968–989

Zhang H, Kim M-S, Dowd SE, Shi H, Pare PW (2008) Soil bacterial confer plant tolerance by tissue-specific regulation of the sodium transporter HKT1. Mol Plant Microbe Interact 21:737–744

Chapter 10
Beneficial Usages of Halophilic Microorganisms

Safiye Elif Korcan, Muhsin Konuk, and Sevim Feyza Erdoğmuş

Abstract Petroleum hydrocarbons are the most important contaminants in many saline environments. These environments could have remarkably economic, ecological and scientific value. Halophiles are organisms that require more than 0.2 M NaCl for their growth and can resist the effects of osmotic stress. Halophilic microorganisms propose potential applications in various fields of biotechnology. These microorganisms could be used as a source of metabolites, compatible solutes and other compounds of industrial value. In the last two decades, there have been many reports on the biodegradation of hydrocarbons in salinity environments. This chapter focuses on our growing understanding of Archaea responsible for the degradation of hydrocarbons under aerobic conditions in salinity environs.

Keywords Saline environments • Halophilic microorganisms • Biotechnological usage

10.1 Introduction

Halophilic microorganisms are of a group that they can be procaryotic or eukaryotic organisms. They need a high amount of salt for their growth and have developed different adaptations survived in salty environments. Halophiles can be classified either regarding to their resistance to the effects of osmotic stress or according to the degree of their salt requirement: weak halophiles grow optimally at 0.2–0.85M (1–5 %) sodium chloride (NaCl); moderate halophiles grow optimally at 0.85–3.4M

S.E. Korcan
Biology Department, Faculty of Science and Literature, Afyon Kocatepe University, 03200 Afyonkarahisar, Turkey

M. Konuk (✉)
Department of Molecular Biology and Genetics, Faculty of Engineering and Natural Sciences, Üsküdar University, 34662 İstanbul, Turkey
e-mail: mkonuk@gmail.com

S.F. Erdoğmuş
Department of Laboratory and Veterinary Health, Technical Vocational School of Higher Education of Bayat, Afyon Kocatepe University, 03780 Afyonkarahisar, Turkey

D.K. Maheshwari, M. Saraf (eds.), *Halophiles*, Sustainable Development and Biodiversity 6, DOI 10.1007/978-3-319-14595-2_10

(5–20 %) NaCl; and extreme halophiles grow optimally at 3.4–5.1M (20–30 %) of NaCl solution. Halophiles prevent the loss of cellular water in their saline environments either by accumulating osmotically balanced levels of internal salts or by producing compatible solutes or osmolytes that can be used to maintain the stability of biomolecules. These osmolytes are usually amino acids or sugars and polyols (Oren 2002; Dassarma et al. 2010). In recent years, the number of biotechnological usages of halophilic microorganisms has been increasing, and additional applications are under development.

10.2 Some Biotechnological Uses of Halophilic Microorganisms

10.2.1 Production of Bacteriorhodopsin

Halophilic microorganisms have a number of potential applications in various fields of biotechnology. Halophilic biomolecules used for multiform specialized applications (Shivanand and Mugeraya 2011). Approximately 80 US patents belong to bacteriorhodopsin from *Halobacterium* and it is one of the most recognized products derived from halophiles. Bacteriorhodopsin from *Halobacterium* sp. is being considered for use as an erasable photochromic film. Bacteriorhodopsin films also have the potential to be utilized as biochips which can pass electrical signals, thereby replacing the integrated circuits in modern computers. Since bacteriorhodopsin can convert light into electrical impulses, it can also be used as a light sensor. It has even been suggested that bacteriorhodopsin could be used to give sight to industrial robots (Dassarma et al. 2010).

10.2.2 Production of Extremozymes

Extremozymes make a group of proteins of great biotechnological interest due to their stability in extreme conditions, opening new possibilities for different industrial processes that run in severe conditions of temperature, pH and pressure or in presence of organic solvents (Perez et al. 2011). Halophilic enzymes are of particular interest due to their activity in environments with low water action. Many archaeal enzymes (Amylases, pullulanases and α-glucosidases) involve in carbohydrate metabolism, particularly those from the glycosyl hydrolase family, are of special interest to the industrial biotechnology sector like starch processing. In all starch processing units high temperatures are required (Schiraldi et al. 2002). Amylases have been characterized from many halophilic strains (*Halomonas meridiana, Haloarcula hispanica,* and *Natronococcus amylolyticus*). Similarly, a kind of α-amylase is produced extracellularly by the haloalkaliphilic archaeon, *Natronococcus* sp. (Kobayashi et al. 1996). Type II archaeal pullulanase was recently detected in crude cell extracts of *Desulfurococcus mucosus* and was successfully cloned in *Bacillus subtilis* (Duffner et al. 2000). The discovery of this thermostable enzyme could deliver an enormous benefit to the starch bioconversion process in the saccharification step.

Ectoine, a cyclic tetrahydropyrimidine (1,4,5,6-tetrahydro-2-methyl-4-pyrimidinecarboxylic acid), can be considered a marker for halophilic bacteria: this solute was first detected in the phototrophic *Halorhodospora halochloris*. Recently it has also been observed in the halophilic bacteria *Halomonas boliviensis, Methylarcula marina, M. terricola,* and *Methylophaga* sp. (Roberts 2005). Ectoine and its derivatives have been patented as moisturizers in cosmetics (Montitsche et al. 2000), although the most promising application may be as stabilizers in the polymerase chain reaction, (Sauer and Galinski 1998; Dassarma et al. 2010).

Betaines decrease side effects of anti-inflammatory preparations. They are compatible solutes present in halophilic phototrophic bacteria, chemotrophic bacteria and archea. Mannosylglycerate is another novel compatible solute widely found in the halotolerant *Methanothermus fervidus*, *Pyrococcus furiosus* and *Rhodothermus marinus*. This compound plays a major role in thermal adaptation for the producing organisms. Diglycerol phosphate accumulates under salt stress in the hyperthermophile *Archaeoglobus fulgidus*. This is a potently useful protein stabilizer (Shivanand and Mugeraya 2011).

Halophilic proteases are used in the detergent, food, baking, dairy, leather industry, manufacturing of soy products and in aspartame production. Proteolytic activity with potential industrial application has been characterized in *Halobacterium* spp., *Haloferax mediteranei*, *Natrialba asiatica*, *N. magadii*, *Natronococcus occultus*, and *Natronomonas pharaonis* (Dassarma et al. 2010).

Esterases and lipases are widely used as biocatalysts because of their ability to produce optically pure compounds. These types of enzymes have been use in the detergent industry. An esterase, from *Haloarcula marismortui,* exhibits a preference for short chain fatty acids and monoesters (Müller-Santos et al. 2009).

Endonuclease enzymes have also been isolated from *Halococcus acetoinfaciens* (*HacI*) *Halobacterium cutirubrum* (*HcuI*), *H. halobium (HhlI)*, and *Halobacterium salinarium (HsaI)* and they are very important enzymes in DNA investigations (Dassarma et al. 2010).

10.2.3 Fermented Foods

Halobacteria, *Halococci*, and *Natronococci* have been isolated from various food sources including fermented foods and sauces including kimchi. Proteases have also been isolated from halophiles from different sources like Thai fish sauce (Hiraga et al. 2005; Namwong et al. 2006). β-Carotene is used in the food industry as a natural food colorant. An extremely halophilic bacterium exhibited optimum growth and carotenoid production in presence of 25 % (w/v) NaCl.

10.2.4 Archaeal Lipids

Archaeal lipids are an excellent source for obtaining the liposomes. These lipids have been suggested as monomers for bioelectronics. They may offer a superior alternative for several biotechnological applications, including delivery systems for drugs, genes or cancer imaging agents (Patel and Sprott 1999). Ether lipids were obtained from the extreme halophile *Halobacterium cutirubrum* (Gambacorta et al. 1995; Schiraldi et al. 2002) and patented. Archaeal polar lipids, from *H. salinarum*, have been used as an adjuvant. *Halobacterium* sp. Halocin can protect tissues including the myocardium from damage caused when blood returns after ischemic conditions. Currently, halocin-H6 is the only known biological molecule that exerts a specific inhibitory effect on the Na+/H+ exchanger (NHE) in the mammalian cells (Lequerica et al. 2006).

10.2.5 The Use of Halophiles for Medical Applications

Halophilic archaea (*H. marismortui* and *H. mediterranei*) are capable of producing large quantities of Polyhydroxyalkanoates. These compounds have been utilized for their application as biodegradable plastics (Lafferty et al. 1988). Halophiles are also a kind of phytochemicals that mitigate the damage from reactive oxygen species and DNA repairing enzymes. Thus, they protect the cells against damaging of radiation. These chemicals may be translated for medical applications (Dassarma et al. 2010).

10.2.6 The Use of Halophiles for Bioremediation and Biodegradation

Biodegradation is the metabolic ability of microorganisms to transform or mineralize organic contaminants in to less harmful and nonhazardous substances, which are then integrated into natural biogeochemical cycles. The use of halophiles for bioremediation and biodegradation of various materials from industrial effluents to soil contaminants are being widely examined. Bioremediation and biodegradation in salty environments require the application of extremozymes from those which are able to grow under harsh conditions as explained above. These enzymes are good candidates for the bioremediation of hypersaline environments and treatment of saline effluents (Perez et al. 2011; Shivanand and Mugeraya 2011).

In many cases, the textile discharges contain high salinity, thereby causing problems for conventional biological treatments (Mellado and Ventosa 2003). Halotolerant and halophilic bacteria usually tolerate obvious amounts of toxic metals in their environment. Thus they are used in bioremediation of oil (Margesin and Schinner 1999) and oxyanion pollution (Nriagu and Payna 1988) but their effectiveness in decolorization of textile effluent has been reported inadequate. Some studies carried out by halophilic and halotolerant bacteria showed that *Halomonas* sp. isolated from discharges of textile industries have aexeptional ability in decolorizing the azo dyes. Demirci et al. (2011) showed that *Halobacillus* sp. C-22 was also able to decolorize azo dyes.

Haloferax mediterranei from salt ponds in Spain has been described as a denitrifier and was shown to grow using NO^{-3}, NO^{-2} or NH^{+4} as inorganic nitrogen sources. Therefore this could be used for bioremediation of nitrate and nitrite concentrations in the presence of high salt concentration (Bonete et al. 2008).

Many hypersaline environments including natural saline lakes, called thalassohaline environments, originated by evaporation of seawater. Salt flats, saline industrial discharges, oil fields, and salt marshes are often contaminated with high levels of petroleum hydrocarbons, toxic metals and textile effluents. These systems have considerable economic, ecological and scientific values. Bioremediation technology uses microorganisms to degrade toxic pollutants to harmless products (Philip et al. 2000).

10.2.7 Aromatic Hydrocarbons Degradation

Polycyclic aromatic hydrocarbons (PAHs) belong to the group of persistent organic pollutants (POPs). These are organic contaminants which have resistance to degradation, can remain in the environment for long periods, and have the potential to cause adverse environmental effects. PAHs have been listed as priority pollutants by both the US Environmental Protection Agency and the European Union due to their toxicity, carcinogenicity and mutagenicity (Schoeny et al. 1993; Liu et al. 2013). During the last century, industrial development caused a significant increase of PAH concentrations in the natural environment (Maliszewska-Kordybach 1999). One of the most effective and economical ways to remove these aromatic pollutants is microbial degradation (Alexander 1981; Parales et al. 2002; Parales and Haddock 2004). In recent years, the biodegradation of PAHs has received great attention and a variety of microorganisms have been reported to play important roles in remediation process (Saxena and Thakur 2005; Tapilatu et al. 2010; Arulazhagan and Vasudevan 2011; Lease et al. 2011). PAH degradation appears to be associated with certain phylogenetic groups of bacteria, particularly *Sphingomonas, Burkholderia*, and *Pseudomonas* amongst Gram-negative species and *Mycobacterium* and *Rhodococcus* amongst Gram-positive species. Because high and floating salinity promotes the loss of cell wall integrity, protein denaturalization, and changes in osmotic pressure. The biological treatment of industrial hypersaline wastewaters and the bioremediation of polluted hypersaline environments are not possible with conventional microorganism (Oren et al. 1992; Pieper and Reineke 2000; Pernetti and Di Palma 2005). Studies demonstrated that hydrocarbon biodegradation by halophilic prokaryotes seems to be possible (Buchan et al. 2000; Nicholson and Fathepure 2004; Garcia et al. 2005; Arulazhagan and Vasudevan 2009). Particularly, halophilic archaea capable of degrading hydrocarbon have been the subject of growing attention in recent years due to problems encountered by the oil industry in hypersaline waste water removal and oil polluted salt marshes (Tapilatu et al. 2010).

Bertrand et al. (1990) were among the first to report the isolation of a halophilic Archaea, strain EH4, which was recently classified as *Haloarcula vallismortis.* Subsequently, Kulichevskaya et al. (1991) isolated a *Halobacterium* strain from hypersaline oil-contaminated waste water. This strain degrades n-C10 to n-C30 alkanes. Another extremely halophilic archaea was identified as *Haloferax mediterranei* strain M-11 which having to use oil as its sole source of carbon (Zvyagintseva et al. 2001). Ashok et al. (1995) have isolated some bacterial strains belonging to the genus *Micrococcus, Pseudomonas,* and *Alcaligenes* from soil samples that degraded naphthalene and anthracene as the sole sources of carbon at 7.5 % salinity. Reports also exist on the ability of archaea that degrade aliphatic hydrocarbons at high salinity. Fu and Oriel (1999) reported that an isolate, *Haloferax* sp. D1227, from highly saline soil contaminated by petrol, could grow by using aromatic hydrocarbons such as benzoic acid, cinammic acid, and 3-phenylpropionic acid as C source. Zhao et al. (2009) have shown the degradation of phenanthrene in the presence of 5–15 % NaCl by a halophilic *Halomonas*, *Chromohalobacter, Alcanivorax, Marinobacter, Idiomarina,* and *Thalassospira.* Garcia et al. (2005) described a novel moderate halophile, *Halomonas organivorans*, able to degrade different aromatic compounds. The capacity of certain halophilic and halotolerant bacteria to degrade benzene and toluene has also been demonstrated by (Nicholson and Fathepure 2005). Al-Mueini et al. (2007) showed the degradation of pentadecane and eicosane by an actinomycete strain belonging to the genus *Actinopolyspora*, isolated from a wastewater pool on oil production site in Oman. Cuadros-Orellana et al. (2006), showed that 44 novel halophilic Archea from five different extremely salinated environment could use p-hydroxybenzoic acid as sole C source. Among them, the species belonged to Halobacteriaceae family were be able to degrade p-hydroxybenzoic acid higher than the others.

Tao et al. (2007) studied with *Sphingomonas* sp. GY2B strain, isolated from PAH contaminated soil, and they reported that this strain could employ fenantren as C source effectively.

Tapilatu et al. (2010) work clearly showed that Archaea belonging to both *Haloarcula* and *Haloferax* genera could degrade the hydrocarbons (n-alkanes). They reported that their degradation ability in 225 g/L NaCl at 40 °C was optimum. Al-Mailem et al. (2010) showed that strains of *Haloferax, Halobacterium* and *Halococcus* could grow using both PAHs and alkanes (from C8 to C34) as their sole carbon sources in both pristine and hydrocarbon-contaminated hypersaline environments. Under aerobic conditions, some halophilic archaeal isolates, mainly from the *Halobacteria* group have been described as hydrocarbon degraders. Bonfa et al. (2011) showed that halophilic *Haloferax* sp. strains are able to degrade PAHs including naphthalene, anthracene, phenanthrene, pyrene and benzo[a]anthracene at high salinity (20 % NaCl).

Al-Mailem et al. (2010) reported the ability of *Marinobacter sedimentalis* and *Marinobacter falvimaris* isolated from hypersaline Sabkh as to degrade biphenyl, phenanthrene, anthracene, and naphthalene as the sole carbon source at 6 % NaCl. Recently, Gao et al. (2006) have isolated a phenanthrene-degrading organism, *Marinobacter nanhaiticus* D15-8W, from South Chinese Sea. This strain degrades naphthalene, phenanthrene or anthracene as the sole source of carbon in the presence of 0.5–15 %. The presence of archaea has been demonstrated in natural marine ecosystems that were previously contaminated with petroleum hydrocarbons (Redmond and Valentine 2012). Additionally, some studies suggested that archaea are the primary component of the microbial community in hypersaline environments (Andrei et al. 2012). In the RNA based analyses suggested the presence of certain sequences related from Euryarchaeota phylum to Crenarchaeota phylum (Jurelevicius et al. 2014a, b). Erdoğmuş et al. (2013) showed the degradation of aromatic hydrocarbons as sole carbon sources in the presence of 20 % NaCl by a number of archaeal strains isolated from brine samples of Çamaltı Saltern in Turkey. Although biodegradation of aromatic compounds has been one of the most intensively studied aspects of microbial catabolism in microorganisms, we know very little about the aromatic compound's metabolism in the domain of Archaea (Redmond and Valentine 2012). Further studies to enlighten both pathways and genes involved in hydrocarbon degradation in halophilic Archaea living in different hypersaline environments are clearly needed (Tapilatu et al. 2010).

10.3 Metabolism of Aromatic Compounds by Domain Archaea

Microorganisms break down PAHs aerobically by use of enzymes called dioxygenases (Fig. 10.1). Dioxygenases add an oxygen to the PAH's aromatic ring and is followed by use of an electron transport chain to final catalysis of the PAH (Chadha et al. 2006).

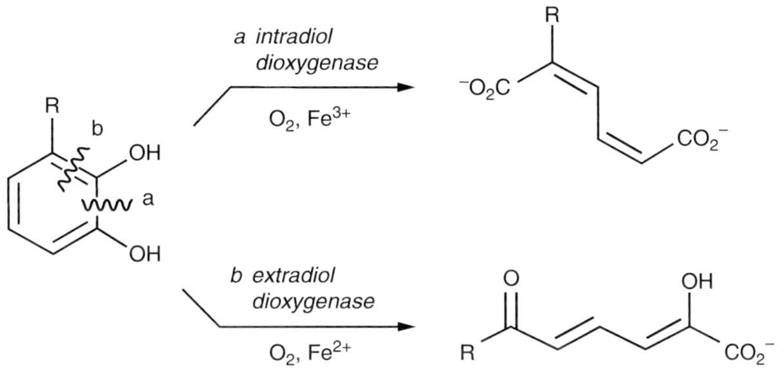

Fig. 10.1 Degradation of aromatic rings by extradiol and intradiol dioxygenases (Bugg 2003)

Degradation pathways of a variety of aromatic hydrocarbons converge at either protocatechuate or catechol. In the meta cleavage pathway, the ring fission occurs adjacent to one of the hydroxyls (extradiol cleavage), and the main enzymes involved in this fission are the catechol 2,3- CTD and protocatechuate 4,5-dioxygenases (Van der Meer et al. 1992). Ring-cleavage dioxygenases constitute central enzymes in the bacterial recycling of aromatic compounds. In the β-ketoadipate pathway, catechol and protocatechuate are cleaved between their two hydroxyl groups by 1,2-CTD or protocatechuate 3,4-PCD. Catechol 2,3-dioxygenases constitute a group of enzymes that are considered crucial for degradation of a wide range of aromatic compounds in contaminated habitats (Moharikar et al. 2003). These pathways (Fig. 10.2) have been identified in the main aromatic compound degrader bacteria, including species of the genera *Acinetobacter, Alcaligenes, Bacillus,* and *Pseudomonas* (Van der Meer et al. 1992; Harwood and Parales 1996).

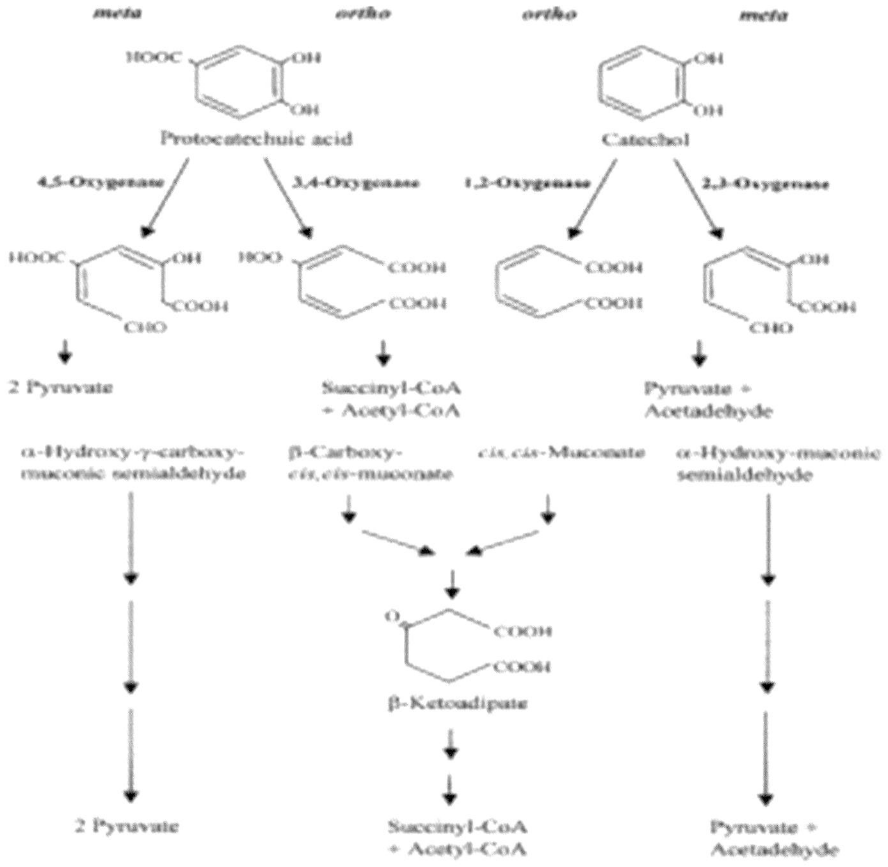

Fig. 10.2 Oxidative aromatic ring degradation (Iwagami 1999)

Aerobic halophilic bacteria degrade phenols mainly by the meta and ortho cleavage pathways. Some bacteria degrade phenols using one of these two pathways, while others may degrade phenols using both pathways (Jiang et al. 2006; Borgne et al. 2008). The accumulation of catechol and *cis, cis*-muconic acid has been observed during the growth of several *Halomonas* strains on phenol, indicating that these bacteria used the ortho pathway (Maskow and Kleinsteuber 2004; Borgne et al. 2008).

Garcia et al. (2005) indicated that catechol 1,2 and protocatechuate 3,4-dioxygenase activities were detected in *H. organivorans*. Dalvi et al. (2012) have analyzed the draft genome sequence of the extremely halophilic benzene and toluene degrading *Arhodomonas* sp. strain Seminole. The analysis predicted 13 putative genes that encode upper and lower pathway enzymes for aromatic compound degradation. These proteins share 44–77 % sequence identity with proteins previously described in non-halophilicorganisms. Bonfa et al. (2011), showed the presence of 1,2-CTD and 3,4-PCD genes in phenol degrading bacteria, *Halomonas organivorans*, *Arhodomona saquaeolei,* and *Modicisalibacter tunisie*. According to the measured enzymatic activities, *H. organivorans* preferentially used the catechol ortho pathway for degrading the benzoic, cinnamic, salicylic, phenylpropionic, and *p*-aminosalicylic acids and the protocatechuate pathway for degrading the *p*-hydroxybenzoic, *p*-coumaric, and ferulic acids (Garcia et al. 2005). A recent study carried out in our laboratory reported the isolation of nine archaeal isolates belonging to various genera that degraded *p*-hydroxybenzoate, naphthalene, phenanthrene, and pyrene as the sole carbon and energy sources in the presence of 20 % NaCl. This study also showed that the isolates possessed genes that encode 1,2-CAT and 3,4-PCA and the expression of these genes was measured spectrophotometrically (Erdogmuş et al. 2013).

10.3.1 Environmental Factors Influence the Rate of PAH Degradation

The effect of temperature is considered as one of the main factors influencing degradation of PAHs. As temperature increases, the solubility of PAH molecule increases.

PAH compounds have low bioavailability because of their hydrophobicity. Also, the water solubility of hydrophobic compounds dramatically decreases with increasing salinity and temperature. These factors limit their bioavailability in hypersaline environments (Dewulf et al. 1995). The first concern to use of hydrocarbons dissolved in an aqueous phase which may be the case for compounds with a relatively elevated aqueous solubility (Goswami et al. 1983). Other two mechanisms contact directly between cells and the particulate or liquid droplet hydrocarbon, and the interaction between pseudosolubilized hydrocarbons in the form of surfactant generated emulsion (Ward 2010). Many hydrocarbonoclastic bacteria synthesize extracellular biosurfactants which help to solubilize hydrocarbons through micellarization and thus enhance their uptake. In the case of a direct contact between bacteria and the hydrocarbon, the pseudosolubilization mechanism may occur at the cell surface. There are still few studies on the mechanisms of hydrocarbon uptake by archaea. Recently, two archaea belonging to the genera *Halovivax* and *Haloarcula* (*Halobacteriaceae*) were shown to produce peptidoglycolipid biosurfactants (Kebbouche-Gana et al. 2009). They both exhibited a high capacity of emulsion stabilization even at high salt concentration (up to 35 % NaCl), and both strains were able to reduce the growth medium surface tension below 29 mN/m. Several species of Halobacteriales also produce exopolysaccharides (EPS) (Nicolaus et al. 1999, 2004) which can be good emulsifiers. For instance, *Haloferax mediterranei* and *Haloarcula japonica* produce sulfated EPS containing hydrophilic anionic sulfate and uronic acid groups combined to the hydrophobic heteropolysaccharide (Anton et al. 1988; Nicolaus et al. 2004). There are also little data available in the literature that describes the cell surface properties of archaea in relation to the solubility of the growth substrates.

Microorganisms, capable of degrading PAHs, normally exhibit high cell surface hydrophobicity (CSH) (Haritash and Kaushik 2009), which enables their adherence to the surface of solid PAHs (Hwang et al. 2008), resulting in a higher biodegradation efficiency (Das and Mukherjee 2007; Abbasnezhad et al. 2011). The biosurfactant production not only leads to the hydrocarbon emulsification, but also plays a role in the change of the cell surface hydrophobicity (CSH). This latter may in turn improve the affinity of microbial cells for a specific substrate by increasing its availability (Zhang and Miller 1995; Kumar et al. 2008). Al-Tahhan et al. (2000) demonstrated that, even at very low concentrations (less than the critical micelle concentration of 0.1 mM), rhamnolipid biosurfactants induced the cell surface of *Pseudomonas spp*. to become hydrophobic through the release of lipopolysaccharides, resulting in increased degradation rates of hexadecane (C16). Interestingly, these authors showed further that the relative CSH was modified similarly by different *P. aeruginosa* strains in the presence of rhamnolipids but with different temporal dynamics. The change of CSH appears to be a function of the bacterial strain, its growth stage, and the type of growth substrate.

Evidences for biosurfactant production by the strain *Haloferax* sp. MNSC14 grown on different hydrocarbons were demonstrated by the simultaneous increase in emulsification activity and decrease in surface tension. The surface hydrophobicity of the cells was also shown to increase in the presence of hydrocarbons (Djeridi et al. 2013).

Several halophiles are emulsifier-producers which could make possible our understanding to access the hydrocarbons instead of the effects of the salts on the hydrocarbons' solubility. In this sense, applications of emulsifiers or halophilic emulsifier-producers could be part of bioremediation strategy in future.

The negative impact of increasing salinity on hydrocarbons biodegradation is also observed in environments where halotolerant microorganisms tend to be dominant, Abed et al. (2006). They reported that almost 100 % of initial phenanthrene and dibenzothiophene were degraded at 35 gL^{-1}, while the best degradation results for pristane (approximately 75 %) and noctadecane (around 85 %) occurred between salinities of 35 and 80 $g.L^{-1}$. Díaz et al. (2002) confirmed that alkanes' biodegradation was the highest, 65 %, at 40 $g.L^{-1}$. Even in typical hypersaline environments a negative impact on hydrocarbon biodegradation induced by increasing salinity has been reported. Ward and Brock (1978) observed that the negative effect of salinity increase hexadecane biodegradation more than on glutamate biodegradation. Nevertheless, Bertrand et al. (1990) isolated a novel halophilic archaeon strain from interface water sediment with salinity of 310 $g.L^{-1}$ (31 %), recently and it was classified as *Haloarcula vallismortis* by Tapilatu et al. (2010).

Al-Mailem et al. (2010) isolated from a sabkha on the coast of Arabian Gulf four strains of Haloarchaea (*Haloferax*, *Halobacterium* and *Halococcus*) able to degrade several aliphatic and aromatic hydrocarbons, with biodegradation rates increasing as salinities increased through 58–175 $g.L^{-1}$ NaCl. Zvyagintseva and Tarasov (1988) evaluated hydrocarbonoclastic activity by *Dietzia maris*, an actinobacterium, at salinities of 5–100 $g.L^{-1}$ NaCl. Optimal biodegradation occurred at 50–100 gL^{-1} NaCl. Among the halophilic fungi, *Fusarium lateritium* and *Drechslera* sp. were shown to metabolize crude oil, with degradation efficiency improved more than thrice if petroleum was an additional source of carbon and energy instead of being the exclusive one (Minai-Tehrani et al. 2009).

Halomonas sp. strain PH2-2 was able to degrade up to 1,100 mg L^{-1} of phenol but cell growth was inhibited at higher phenol concentrations. The strain was able to remove phenol in media containing 18 % NaCl but the removal efficiency decreased from 95 % to 64 % in comparison to media containing 7 % NaCl. The results indicated the potential application of the strain PH2-2 for treatment of hypersaline phenol-containing industrial wastewaters (Haddadi and Shavandi 2013).

Recent excellent reviews by Borgne et al. (2008), Martins and Peixoto (2012), Timmis et al. (2010), and Patzelt (2007) helped us to understanding of the hydrocarbon biodegradation by halophilic and halotolerant microorganisms. Nonetheless, our knowledge on biochemistry, genetics, and pathways of hydrocarbon degradation in halophiles and halotolerants is very few. Such information is crucial for designing novel and more efficient technologies for the remediation of contaminated high salinity environments and for understanding the carbon cycle in such extreme habitats.

10.4 Conclusion

Increasing anthropogenic activities cause pollution in our natural environment. Particularly extremely saline environments are affected from these activities. Our knowledge on organic removing/degrading microorganisms is negligible so little. However halophilic microorganisms are recently being taken consideration by biotechnologists due to their enzymes which work in highly saline conditions; strategies to resist extreme conditions; specific proteins help them to adapt to various harsh environmental conditions. These features provide them an optimal accommodation in their living areas. Therefore, these microorganisms have a strong potential in the area of biotechnology and particularly bioremediation.

References

Abbasnezhad H, Foght JM, Gray MR (2011) Adhesion to the hydrocarbon phase increases phenanthrene degradation by *Pseudomonas fluorescens* LP6a. Biodegradation 22:485–496

Abed RMM, Al-Thukair A, De Beer D (2006) Bacterial diversity of a cyanobacterial mat degrading petroleum compounds at elevated salinities and temperatures. FEMS Microbiol Ecol 57:290–301

Alexander M (1981) Biodegradation of chemicals of environmental concern. Science 211: 132–138

Al-Mailem DM, Sorkhoh NA, Al-Awadhi H, Eliyas M, Radwan SS (2010) Biodegradation of crude oil and pure hydrocarbons by extreme halophilic archaea from hypersaline coasts of the Arabian Gulf. Extremophiles 14(3):321–328

Al-Mueini R, Al-Dalali M, Al-Amri IS, Patzelt H (2007) Hydrocarbon degradation at high salinity by a novel extremely halophilic actinomycete. Environ Chem 4:5–7

Al-Tahhan RA, Sandrin TR, Bodour AA, Maier RM (2000) Rhamnolipid-induced removal of lipopolysaccharide from *Pseudomonas aeruginosa*: effect on cell surface properties and interaction with hydrophobic substrates. Appl Environ Microbiol 66:3262–3268

Andrei AŞ, Banciu HL, Oren A (2012) Living with salt: metabolic and phylogenetic diversity of archaea inhabiting saline ecosystems. FEMS Microbiol Lett 330(1):1–9

Anton J, Meseguer I, Rodrıguez-Valera F (1988) Production of an extracellular polysaccharide by *Haloferax mediterranei*. Appl Environ Microbiol 54:2381–2386

Arulazhagan P, Vasudevan N (2009) Role of a moderately halophilic bacterial consortium in the biodegradation of polyaromatic hydrocarbons. Mar Pollut Bull 58(2):256–262

Arulazhagan P, Vasudevan N (2011) Role of nutrients in the utilization polycyclic aromatic hydrocarbons by halotolerant bacterial strain. J Environ Sci (China) 23(2):282–287

Ashok BT, Saxena S, Musarrat J (1995) Isolation and characterization of four polycyclic aromatic hydrocarbon degrading bacteria from soil near an oil refinery. Lett Appl Microbiol 21(4): 246–248

Bertrand JC, Almallah M, Acquaviva M, Mille G (1990) Biodegradation of hydrocarbons by an extremely halophilic archaebacterium. Lett Appl Microbiol 11:260–263

Bonete MJ, Martınez-Espinosa RM, Pire C, Zafrilla B, Richardson DJ (2008) Nitrogen metabolism in haloarchaea. Saline Syst 4:9. doi:10.1186/1746-1448-4-9

Bonfa MR, Grossman MJ, Mellado E, Durrant LR (2011) Biodegradation of aromatic hydrocarbons by haloarchaea and their use for the reduction of the chemical oxygen demand of hypersaline petroleum produced water. Chemosphere 84(11):1671–1676

Borgne SL, Paniagua D, Vazquez-Duhalt R (2008) Biodegradation of organic pollutants by halophilic bacteria and archaea. J Mol Microbiol Biotechnol 15:74–92

Buchan A, Collier LS, Neidle EI, Moran MA (2000) Key aromatic ring cleaving enzyme, protocatechuate 3,4 dioxygenase, in the ecologically important marine roseobacter lineage. Appl Environ Microbiol 66:4662–4672

Bugg TDH (2003) Dioxygenase enzymes: catalytic mechanisms and chemical models. Tetrahedron 59:7075–7101

Chadha M, Feldman SM, Boolbol S et al (2006) Phase1/2 study of partial breast brachytherapy in patients who have previously received external beam radiation therapy (RT) to the breast. Int J Radiat Oncol Biol Phys 66:S31

Cuadros-Orellana S, Pohlschroder M, Durrant LR (2006) Isolation and characterization of halophilic archaea able to grow in aromatic compounds. Int Biodeterior Biodegrad 57:151–154

Dalvi S, Azetsu S, Patrauchan MA, Aktaş DF, Fathepurea BZ (2012) Proteogenomic elucidation of the initial steps in the benzene degradation pathway of a novel halophile, *Arhodomonas* sp. strain rozel, isolated from a hypersaline environment. Appl Environ Microbiol 78(20):7309–7316

Das K, Mukherjee AK (2007) Differential utilization of pyrene as the sole source of carbon by *Bacillus subtilis* and *Pseudomonas aeruginosa* strains: role of biosurfactants in enhancing bioavailability. J Appl Microbiol 102:195–203

Dassarma P, Coker JA, Huse V, Dassarma S (2010) Halophiles, industrial applications, encyclopedia of industrial biotechnology: bioprocess, bioseparation, and cell technology. University of Maryland, Baltimore, Flickinger MC (eds)

Demirci A, Mutlu MB, Güven A, Korcan E, Güven K (2011) Decolorization of textile azo-metal complex dyes by a halophilic bacterium isolated from Çamaltı Saltern in Turkey. Clean Soil, Air, Water 39(2):177–184

Dewulf J, Drijvers D, Langenhove HV (1995) Measurement of Henry's law constant as function of temperature and salinity for the low temperature range. Atmos Environ 29:323–333

Diaz MP, Boyd KG, Grigson SGW, Burgess JG (2002) Biodegradation of crude oil across a wide range of salinities by an extremely halotolerant bacterial consortium MPD-M, immobilized onto polypropylene fibers. Biotechnol Bioeng 79(2):145–153

Djeridi I, Militon C, Grossi V, Cuny P (2013) Evidence for surfactant production by the haloarchaeon *Haloferax* sp. MSNC14 in hydrocarbon-containing media. Extremophiles 17:669–675

Duffner F, Bertoldo C, Andersen JT, Wagner K, Antranikian G (2000) A new thermoactive pullulanase from *Desulfurococcusmucosus*: cloning, sequencing, purification, and characterization of the recombinant enzyme after expression in *Bacillus subtilis*. J Bacteriol 182:6331–6338

Erdoğmuş SF, Mutlu B, Korcan SE, Güven K, Konuk M (2013) Aromatic hydrocarbon degradation by halophilic Archaea isolated from Çamalt Saltern, Turkey. Water Air Soil Pollut 224:1449. doi:10.1007/s11270-013-1449-9

Fu W, Oriel P (1999) Degradation of 3-phenylpropionic acid by *Haloferax* sp. D1227. Extremophiles 3:45–53

Gambacorta A, Ghiozzi A, De Rosa M (1995) Archaeal lipids and their biotechnological applications. World J Microbiol Biotechnol 11:115–132

Gao Y, Yu XZ, Wu SC, Cheung KC, Tam NF, Qian PY, Wong MH (2006) Interactions of rice (*Oryza sativa* L.) and PAH-degrading bacteria (*Acinetobacter* sp.) on enhanced dissipation of spiked phenanthrene and pyrene in waterlogged soil. Sci Total Environ 372(1):1–11

Garcia MT, Ventosa A, Mellado E (2005) Catabolic versatility of aromatic compound degrading halophilic bacteria. FEMS Microbiol Ecol 54:97–109

Goswami PC, Singh HD, Bhagat SD, Baruah JN (1983) Mode of uptake of insoluble solid substrates by microrganisms. 1. Sterol uptake by an arthtobacter species. Biotechnol Bioeng 25:2929–2943

Haddadi A, Shavandi M (2013) Biodegradation of phenol in hypersaline conditions by *Halomonas* sp. strain PH2-2 isolated from saline soil. Int Biodeterior Biodegrad 85:29–34

Haritash AK, Kaushik CP (2009) Biodegradation aspects of polycyclic aromatic hydrocarbons (PAHs): a review. J Hazard Mater 169:1–15

Harwood CS, Parales RE (1996) The β-ketoadipate pathway and the biology of self-identity. Annu Rev Microbiol 50:553–590

Hiraga K, Nishikata Y, Namwong S, Tanasupawat S, Takada K, Oda K (2005) Purification and characterization of serine proteinase from a halophilic bacterium, Filobacillus sp. RF2-5. Biosci Biotech Biochem 69:38–44

Hwang G, Ban YM, Lee CH, Chung CH, Ahn IS (2008) Adhesion of *Pseudomonas putida* NCIB 9816–4 to a naphthalene-contaminated soil. Colloids Surf B 62:91–96

Iwagami SG (1999) Protein purification and genetic characterization of a streptomycete protocatechuate 3,4 dioxygenase. Master of Science Thesis, The University of British Colombia, Microbiology and Immunology, Columbia (http://circle.ubc.ca/handle/2429/9166)

Jiang HL, Tay ST, Maszenan AM, Tay JH (2006) Physiological traits of bacterial strains isolated from phenol-degrading aerobic granules. FEMS Microbiol Ecol 57:182–191
Jurelevicius D, Almeida-Couto CR, Alvarez VM, Vollú RE, Almeida- Dias F, Seldin L (2014a) Response of the archaeal community to simulated petroleum hydrocarbon contamination in marine and hypersaline ecosystems. Water Air Soil Pollut 225:1871
Jurelevicius D, de Almeida Couto CR, Alvarez VM, Vollú RE, de Almeida Dias F, Seldin L (2014b) Response of the archaeal community to simulated petroleum hydrocarbon contamination in marine and hypersaline. Ecosyst Water Air Soil Pollut 225:1871
Kebbouche-Gana S, Gana ML, Khemili S, Fazouane-Naimi F, Bouanane NA, Penninckx M, Hacene H (2009) Isolation and characterization of halophilic archaea able to produce biosurfactants. J Ind Microbiol Biotechnol 36:727–738
Kobayashi K, Kato M, Miura Y, Komeda KT, Iwamatsu A (1996) Gene cloning and expression of new trehalose-producing enzymes from the hyperthermophilic archaeon *Sulfolobussolfataricus* KM1. Biosci Biotechnol Biochem 60:1882–1885
Kulichevskaya IS, Milekhina EI, Borzenkov IA, Zvyagintseva IS, Belyaev SS (1991) Oxidation of petroleum hydrocarbons by extremely halophilic archeobacteria. Microbiology 60:596–601
Kumar M, Vladimir L, Materano ADS, Ilzins OA, Luis L (2008) Biosurfactant production and hydrocarbon-degradation by halotolerant and thermotolerant *Pseudomonas* sp. World J Microbiol Biotechnol 24:1047–1057
Lafferty RM, Korsatko B, Korsatko W (1988) Microbial production of poly-β-hydroxybutyric acid. In: Rehm HJ, Reed G (eds) Biotechnology, vol 6b. VCH, Weinheim, pp 135–176
Lease CWM, Bentham RH, Gaskin SE, Juhasz AL (2011) Isolation and identification of pyrene mineralizing *Mycobacterium* spp. from contaminated and uncontaminated sources. Appl Environ Soil Sci. doi:10.1155/2011/409643
Lequerica JL, O'Connor JE, Such L, Alberola A, Meseguer I, Dolz M, Torreblanca M, Moya A, Colom F, Soria B (2006) J Physiol Biochem 62(4):253–262
Liu F, Liu JL, Chen QY, Wang B, Cao ZG (2013) Pollution characteristics and ecological risk of polycyclic aromatichydrocarbons (PAHs) in surface sediments of the southern part of the Haihe River system in China. Chin Sci Bull 58(27):3348–3356
Maliszewska-Kordybach B (1999) Sources, concentrations, fate and effects of polycyclic aromatic hydrocarbons (PAHs) in the environment. Part A: PAHs in air. Pol J Environ Stud 8(3):131–136
Margesin R, Schinner F (1999) Biological decontamination of oil spills in cold environments. J Chem Technol Biotechnol 74:381–389
Martins LF, Peixoto RS (2012) Biodegradation of petroleum hydrocarbons in hypersaline environments. Braz J Microbiol 43(3):865–872
Maskow T, Kleinsteuber S (2004) Carbon and energy fluxes during haloadaptation of *Halomonas* sp. EF11 growing on phenol. Extremophiles 8:133–141
Mellado E, Ventosa A (2003) In: Barredo JL (ed) Microorganisms for health care, food and enzyme production. Research Signpost, Kerala, p 233
Minai-Tehrani D, Minuoi S, Herfatmanesh A (2009) Effect of salinity on biodegradation of polycyclic aromatic hydrocarbons (PAHs) of heavy crude oil in soil. Bull Environ Contam Toxicol 82:179–184
Moharikar A, Kapley A, Purohit HJ (2003) Detection of dioxygenase genes present in various activated sludge. Environ Sci Pollut Res Int 10:373–378
Montitsche L, Driller H, Galinski E (2000) Ectoine and ectoine derivatives as moisturizers in cosmetics. Patent US060071, May 9
Müller-Santos M, de Souza EM, Pedrosa FD, Mitchell DA, Longhi S, Carriere F, Canaan S, Krieger N (2009) First evidence for the salt-dependent folding and activity of an esterase from the halophilic archaea haloarculamarismortui. Biochim Biophys Acta 1791(8):719–729
Namwong S, Takada K, Tsunemi M, Tanasupawat S, Oda K (2006) A halophilic serine proteinase from halobacillus sp. SR5-3 isolated from fish sauce: purification and characterization. Biosci Biotechnol Biochem 70:1395–1401
Nicholson CA, Fathepure BZ (2004) Biodegradation of benzene by halophilic and halotolerant bacteria under aerobic conditions. Appl Environ Microbiol 70:1222–1225

Nicholson CA, Fathepure BZ (2005) Aerobic biodegradation of benzene and toluene under hypersaline conditions at the Great Salt Plains, Oklahoma. FEMS Microbiol Lett 245:257–262

Nicolaus B, Lama L, Esposito E, Manca MC, Improta R, Bellitti MR, Duckworth AW, Grant WD, Gambacorta A (1999) *Haloarcula* spp. able to biosynthesize exo-endopolymers. J Ind Microbiol Biotechnol 23:489–496

Nicolaus B, Moriello VS, Lama L, Poli A, Gambacorta A (2004) Polysaccharides from extremophilic microorganisms. Orig Life Evol Biosph 34:159–169

Nriagu JO, Payna JM (1988) Quantitative assessment of worldwide contamination of air, water and soils by trace metals. Nature 333:134–139

Oren A (2002) Diversity of halophilic microorganisms: environments, phylogeny, physiology, and applications. J Ind Microbiol Biotechnol 28:56–63

Oren A, Gurevich P, Azachi M, Henis Y (1992) Microbial degradation of pollutants at high salt concentrations. Biodegradation 3:387–398

Parales RE, Haddock JD (2004) Biocatalytic degradation of pollutants. Curr Opin Biotechnol 15:374–379

Parales RE, Bruce NC, Schmid A, Wackett LP (2002) Biodegradation, biotransformation, and biocatalysis (B3). Appl Environ Microbiol 68:4699–4709

Patel GB, Sprott GD (1999) Archaeobacterial ether lipid liposomes (archaeosomes) as novel vaccine and drug delivery systems. Crit Rev Biotechnol 19:317–357

Patzelt H (2007) Hydrocarbon metabolism, citable reviews in the life sciences. Sultan Qaboos University, Al-Khod, Sultanate of Oman. doi:10.1002/9780470015902.a0000473

Perez D, Martín S, Fernández-Lorente G, Filice M, Guisán JM, Ventosa A, Garcia MT, Mellado E (2011) A novel halophilic lipase, LipBL, showing high efficiency in the production of eicosapentaenoic acid (EPA). PLoS ONE 6(8):e23325

Pernetti M, Di Palma L (2005) Experimental evaluation of inhibition effects of saline wastewater on activated sludge. Environ Technol 26(6):695–703

Philip L, Iyengar L, Venkobacher L (2000) Site of interaction of copper on *Bacillus polymyxa*. Water Air Soil Pollut 119:11–21

Pieper DH, Reineke W (2000) Engineering bacteria for bioremediation. Curr Opin Biotechnol 11(3):262–270

Redmond MC, Valentine DL (2012) Natural gas and temperature structured a microbial community response to the deepwater horizon oil spill. PNAS 109(50):20292–20297

Roberts MF (2005) Organic compatible solutes of halotolerant and halophilic microorganisms. Saline Syst 1(5):1–30

Sauer T, Galinski EA (1998) Bacterial milking: a novel bioprocess for production of compatible solutes. Biotechnol Bioeng 57:306–313

Saxena P, Thakur IS (2005) Purification and characterization of catechol 1,2 dioxygenase of pseudomonas fluorescens for degradation of 4-chlorobenzoic acid. Ind J Biotechnol 4:134–138

Schiraldi C, Di Lernia I, De Rosa M (2002) Trehalose production: exploiting novel approaches. Trends Biotechnol 20(10):420–425

Schoeny R, Poirier K (1993) Provisional guidance for quantitative risk assessment of polycyclic aromatic hydrocarbons, EPA, U.S. Environmental Protection Agency, Cincinnati, Ohio, USA, EPA/600/R-93/089; ECAO-CIN-842

Shivanand P, Mugeraya G (2011) Halophilic bacteria and their compatible solutes osmoregulation and potential applications. Curr Sci 100(10):1516–1521

Tao XQ, Lu GN, Dang Z, Yang C, Yi XY (2007) A phenanthrene-degrading strain *Sphingomonas* sp. GY2B isolated from contaminated soils. Process Biochem 42:401–408

Tapilatu YH, Grossi V, Acquaviva M, Militon C, Bertrand JC, Cuny P (2010) Isolation of hydrocarbon-degrading extremely halophilic archaea from an uncontaminated hypersaline pond (Camargue, France). Extremophiles 14(2):225–231

Timmis KN, McGenity TJ, Meer JR, de Lorenzo V (eds.). (2010). Handbook of Hydrocarbon and Lipid Microbiology. Berlin; Heidelberg: Springer-Verlag; 10.1007/978-3-540-77587-4, pp 670

Van der Meer JR, de Vos WM, Harayama S, Zehnder AJB (1992) Molecular mechanisms of genetic adaptation to xenobiotic compounds. Microbiol Rev 56:677–694

Ward OP (2010) Microbial biosurfactants and biodegradation. Adv Exp Med Biol 672:65–74, In: Sen R (eds) Biosurfactants
Ward DM, Brock TD (1978) Hydrocarbon degradation in hypersaline environments. Appl Environ Microbiol 35(2):353–359
Zhang Y, Miller RM (1995) Effect of rhamnolipid (biosurfactant) structure on solubilization and biodegradation of n-alkanes. Appl Environ Microbiol 61:2247–2251
Zhao B, Wang H, Mao X, Li R (2009) A rapid screening method for bacteria degrading polycyclic aromatic hydrocarbons. Lett Appl Microbiol 49:408–410
Zvyagintseva IS, Tarasov AL (1988) Extreme halophilic bacteria from saline soils. Microbiology 56:664–668
Zvyagintseva IS, Poglazova MN, Gotoeva MT, Belyaev SS (2001) Effect on the medium salinity on oil degradation by nocardioform bacteria. Microbiology 70(6):652–656

Chapter 11
Halophilic Bacteria: Potentials and Applications in Biotechnology

Fatemeh Mohammadipanah, Javad Hamedi, and Mona Dehhaghi

Abstract Halophilic bacteria are categorized based on their requirements to salt. They apply some strategies in adaptation to the high saline environment. Their low nutritional requirements and resistance to high concentrations of salt, introduce as potent agents in wide range of biotechnological applications. Halophilic bacteria are very divergent and more than 150 species introduce in 70 genera of halophilic bacteria are reported. Use of resistant microorganisms and their metabolites under harsh industrially conditions is a strikingly important subject. Various halophilic enzymes in different enzymatically processes, compatible solutes as macromolecule stabilizers, biopolymers, biofertilizers and pharmaceutically active molecules from halophilic bacteria are among the important applications of these group. Additionally, they have many potential in bioremediation of various organic and inorganic pollutants from environment. This chapter covers the different aspects of potential applications of halophilic bacteria in biotechnology.

Keywords Halophilic bacteria • Physiology • Taxonomy • Ecology • Biotechnological application

11.1 Halophilic Bacteria, Physiology, Ecology and Taxonomy

11.1.1 Introduction

Microbial life exists in hypersaline habitat reflects the adaption of microorganisms to high concentration of salts. Halophilic and halotolerant organisms are found in all three domains of life: *Archaea*, *Bacteria*, and *Eukarya*. According to the requirement

F. Mohammadipanah (✉) • J. Hamedi • M. Dehhaghi
Department of Microbial Biotechnology, School of Biology and Center of Excellence in Phylogeny of Living Organisms, College of Science, University of Tehran, Tehran 14155-6455, Iran

Microbial Technology and Products Research Center, University of Tehran, Tehran 14155-6455, Iran
e-mail: fmohammadipanah@ut.ac.ir

D.K. Maheshwari, M. Saraf (eds.), *Halophiles*, Sustainable Development and Biodiversity 6, DOI 10.1007/978-3-319-14595-2_11

for salt components, halophiles are classified into the groups including slightly halophiles (2–5 % NaCl), moderately halophiles (5–20 % NaCl) and extremely halophiles (20–30 % NaCl) while strains grow in 0–5 % salinity are considered halotolerant microorganisms (Ventosa et al. 1998).

The halophilic bacteria can be isolated from salt lakes, sea, saline soil and salted foods and in high saline environments such as Dead Sea, soda lakes and saltern crystallizer ponds. Most of the physiological groups including anaerobic chemoheterotrophs, photoautotrophs, photoheterotrophs and chemolithotrophs can be found within halophilic bacteria. They posses special features in their cell structures for adaptation to osmotic pressure. For example, cell wall of some halophilic bacteria become more hydrophilic in high salt concentration than that in the low salinity. This mechanism facilitates more contact with water molecules at high salinity media (Oren 2002b). Also, cell membrane is affected under high salt condition. Changes can be found in both phospholipids and fatty acid profile. It has been recognized that the negatively charged phospholipids content enhance with increasing the salinity (Russell 1993). Proteins of halophiles have specific features to remain active in high concentration of salt. An increase in number of acidic amino acids in protein composition occurs by thriving in saline condition (Rao and Argos 1981; Benachenhou and Baldacci 1991). The halophilic bacterial species described since 2006 are presented in Table 11.1.

Table 11.1 Introduced halophilic bacterial species since 2006

Halophilic bacterial species	Salt concentration range for growth	References
Alcanivoracaceae		
Kangiella		Ahn et al. (2011)
Kangiella spongicola	2–15 %	
Alteromonadaceae		
Marinimicrobium	2–22 %	Moller et al. (2010)
Marinimicrobium haloxylanilyticum		
Marinobacter	0.5–20 %	Kaeppel et al. (2012)
Marinobacter adhaerens	1–15 %	Montes et al. (2008)
Marinobacter guineae	1–5 %	Gao et al. (2013)
Marinobacter nanhaiticus	5–7.5 %	Kharroub et al. (2011)
Marinobacter oulmenensis	0.5–15 % (5 %)	Xu et al. (2008)
Marinobacter pelagius	1–15 % (4–8 %)	Guo et al. (2007)
Marinobacter segnicrescens	6 %	Liebgott et al. (2006)
Marinobacter vinifirmus		
Marinobacterium	1–10 % (2–5 %)	Kim et al. (2010b)
Marinobacterium lutimaris		
Microbulbifer	1–20 % (10 %)	Tang et al. (2008)
Microbulbifer halophilus		
Bacillaceae		
Alkalibacillus		
Alkalibacillus halophilus	1–20 %	Tian et al. (2007)

(continued)

Table 11.1 (continued)

Halophilic bacterial species	Salt concentration range for growth	References
Alteribacillus		
Alteribacillus bidgolensis	0.5–12.5 % (5–7.5 %)	Didari et al. (2012)
Amphibacillus		
Amphibacillus cookie	6–26 % (5 %)	Pugin et al. (2012)
Aquibacillus		
Aquibacillus halophilus	0.5–20 % (10 %)	Amoozegar et al. (2014a)
Aquisalibacillus		
Aquisalibacillus elongatus	3–20 % (10 %)	Márquez et al. (2008)
Bacillus		
Bacillus aidingensis	8–33 % (12 %)	Xue et al. (2008)
Bacillus aurantiacus	Up to 15 % (3–7 %)	Borsodi et al. (2008)
Bacillus chagannorensis	3–20 % (7 %)	Carrasco et al. (2007b)
Bacillus coahuilensis	0.5–10 % (5 %)	Cerritos et al. (2008)
Bacillus daqingensis	0–16 % (3 %)	Wang et al. (2014a)
Bacillus salaries	3–20 % (10–12 %)	Lim et al. (2006b)
Bacillus halochares	6–23 % (15 %)	Pappa et al. (2010)
Bacillus hunanensis	0.5–15 % (2–4 %)	Chen et al. (2011)
Bacillus iranensis	2.5–15 % (5–7.5 %)	Bagheri et al. (2012)
Bacillus neizhouensis	0.5–10 % (2–4 %)	Chen et al. (2009c)
Bacillus oceani	0–10 %	Liu et al. (2013)
Bacillus oryzaecorticis	9 %	Hong et al. (2014)
Bacillus persicus	0.5–10 % (2.5 %)	Didari et al. (2013)
Bacillus salsus	0.5–7.5 % (3 %)	Amoozegar et al. (2013b)
Bacillus taeanensis	0–12 % (2–10 %)	Lim et al. (2006a)
Gracilibacillus		
Gracilibacillus orientalis	1–20 % (10 %)	Carrasco et al. (2006)
Gracilibacillus halophilus	7–30 % (15 %)	Chen et al. (2008c)
Gracilibacillus saliphilus	1–22 % (10–15 %)	Tang et al. (2009)
Halalkalibacillus		
Halalkalibacillus halophilus	5–25 % (10–15 %)	Echigo et al. (2007)
Halobacillus		
Halobacillus alkaliphilus	0.5–20 % (10 %)	Romano et al. (2008)
Halobacillus kuroshimensis	0.5–22 % (6 %)	Hua et al. (2007)
Halobacillus mangrove	5–20 % (10 %)	Soto-Ramírez et al. (2008)
Halobacillus profundi	0.5–30 % (10 %)	Hua et al. (2007)
Halolactibacillus		
Halolactibacillus alkaliphilus	0.5–30 % (2.5 %)	Cao et al. (2008)
Natribacillus		
Natribacillus halophilus	7–23 % (10–15 %)	Echigo et al. (2012)
Oceanobacillus		
Oceanobacillus aidingensis	0–21 % (6–8 %)	Liu and Yang (2014)

(continued)

Table 11.1 (continued)

Halophilic bacterial species	Salt concentration range for growth	References
Oceanobacillus limi	2.5–10 % (7.5 %)	Amoozegar et al. (2014b)
Oceanobacillus manasiensis	5–10 %	Wang et al. (2010b)
Oceanobacillus neutriphilus	0–17 % (3–5 %)	Yang et al. (2010)
Ornithinibacillus		
Ornithinibacillus halophilus	0.5–12.5 % (5–7.5 %)	Bagheri et al. (2013)
Piscibacillus		
Piscibacillus salipiscarius	2–3 % (10–20 %)	Tanasupawat et al. (2007)
Pontibacillus		
Pontibacillus halophilus	2–25 % (5–10 %)	Chen et al. (2009d)
Salisediminibacterium		
Salisediminibacterium halotolerans	3–30 %	Jiang et al. (2011)
Saliterribacillus		
Saliterribacillus persicus	0.5–22.5 % (7.5 %)	Amoozegar et al. (2013a)
Salsuginibacillus		
Salsuginibacillus kocurii	3–20 % (10 %)	Carrasco et al. (2007a)
Sediminibacillus		
Sediminibacillus halophilus	0–20 % (5–7.5 %)	Carrasco et al. (2008)
Streptohalobacillus		
Streptohalobacillus salinus	0–20 % (7 %)	Wang et al. (2011b)
Virgibacillus		
Virgibacillus albus	1–17 % (5–10 %)	Zhang et al. (2012)
Virgibacillus byunsanensis	8 %	Yoon et al. (2010)
Virgibacillus chiguensis	5–10 % (0–30 %)	Wang et al. (2008)
Virgibacillus kekensis	0–25 % (10 %)	Chen et al. (2008b)
Virgibacillus salarius	0.5–25 % (7–10 %)	Hua et al. (2008)
Virgibacillus sediminis	1–20 % (5–10 %)	Chen et al. (2009a)
Carnobacteriaceae		
Alkalibacterium		
Alkalibacterium gilvum	0–17.5 % (2–5 %)	Ishikawa et al. (2013)
Alkalibacterium subtropicum	0–17 % (1–3 %)	Ishikawa et al. (2011)
Chitinophagaceae		
Fodinibius		
Fodinibius salinus	10–15 %	Wang et al. (2012b)
Gracilimonas		
Gracilimonas mengyeensis	2–15% (5–9 %)	Wang et al. (2013)
Thermoflavifilum		
Thermoflavifilum aggregans	Up to 5 % (0.1–0.25 %)	Anders et al. (2014)
Desulfobacteraceae		
Desulfosalsimonas		
Desulfosalsimonas propionicica		
Desulfohalobiaceae	Up to 20 % (6 %)	Kjeldsen et al. (2010)

(continued)

Table 11.1 (continued)

Halophilic bacterial species	Salt concentration range for growth	References
Desulfohalobium		
Desulfohalobium utahense	0.2–24 % (8–10 %)	Jakobsen et al. (2006)
Desulfohalophilus		
Desulfohalophilus alkaliarsenatis	12.5–33 %	Blum et al. (2012)
Ectothiorhodospiraceae		
Aquisalimonas		
Aquisalimonas asiatica	1–20 % (7–10 %)	Márquez et al. (2007)
Arhodomonas		
Arhodomonas recens	2–25 % (10–12 %)	Saralov et al. (2012)
Spiribacter		
Spiribacter salinus	10–25 % (15 %)	León et al. (2014)
Flammeovirgaceae		
Fabibacter		
Fabibacter pacificus	0.5–10 % (2 %)	Huo et al. (2013)
Flavobacteriaceae		
Psychroflexus		
Psychroflexus lacisalsi	0.5–15 % (5.8–7.0 %)	Zhang et al. (2010)
Halanaerobiaceae		
Halanaerobium		
Halanaerobium sehlinense	5–30 % (20 %)	Abdeljabbar et al. (2013)
Halobacteriaceae		
Fuchsiella		
Fuchsiella alkaliacetigena	0–14 % (7–8.5 %)	Zhilina et al. (2012)
Halanaerobacter		
Halanaerobacter jeridensis	6–30 % (15 %)	Mezghani et al. (2012)
Halanaerocella		
Halanaerocella petrolearia	6–26 % (15 %)	Gales et al. (2011)
Halobellus		
Halobellus clavatus	10–30 % (15 %)	Cui et al. (2011)
Halobellus limi	8–30 % (15–23 %)	Cui et al. (2012)
Halobellus salinus	15–30 % (20 %)	Cui et al. (2012)
Halomicroarcula		
Halomicroarcula pellucida	20–30 % (25 %)	Echigo et al. (2013)
Halorussus		
Halorussus rarus	8–25 % (12 %)	Cui et al. (2010)
Halomonadaceae		
Aidingimonas		
Aidingimonas halophila	1–25 % (5–10 %)	Wang et al. (2009)
Cobetia		
Cobetia crustatorum	6.5 %	Kim et al. (2010c)
Halomonas		
Halomonas aidingensis	0.5–25 %	Liu et al. (2011)

(continued)

Table 11.1 (continued)

Halophilic bacterial species	Salt concentration range for growth	References
Halomonas alkalitolerans	1–23 % (7 %)	Wang et al. (2011a)
Halomonas andesensis	0.5–20 % (1–3 %)	Guzmán et al. (2010)
Halomonas beimenensis	0–15 % (5–10 %)	Wang et al. (2012a)
Halomonas cerina	7.5–20 % (7.5–10 %)	González-Domenech et al. (2008b)
Halomonas cibimaris	3–15 % (5–10 %),	Jeong et al. (2013)
Halomonas daqiaonensis	7–8 %	Qu et al. (2011)
Halomonas daqingensis	1–15 % (5–10 %)	Wu et al. (2008)
Halomonas denitrificans	2–20 % (8–10 %)	Kim et al. (2007)
Halomonas fontilapidosi	3–25 % (5–7.5 %)	González-Domenech et al. (2009)
Halomonas gomseomensis	1–20 % (8–12 %)	Kim et al. (2007)
Halomonas huangheensis	1–20 % (7–10 %)	Miao et al. (2014)
Halomonas janggokensis	1–20 % (10–15 %)	Kim et al. (2007)
Halomonas nitroreducens	3–20 % (5–7.5 %)	González-Domenech et al. (2008a)
Halomonas qiaohouensis	0.5–20 % (2–6 %)	Wang et al. (2014b)
Halomonas ramblicola	1–30 % (5–7.5 %)	Luque et al. (2012)
Halomonas rifensis	0.5–20 % (5–7.5 %)	Amjres et al. (2011)
Halomonas salaria	0–25 % (10–20 %)	Kim et al. (2007)
Halomonas shengliensis	0–15 % (5–15 %)	Wang et al. (2007)
Halomonas stenophila	3–15 % (5–10 %)	Llamas et al. (2011)
Halomonas smyrnensis	(10 %)	Poli et al. (2013)
Halomonas titanicae	0.5–20 % (8–10 %)	Sánchez-Porro et al. (2010)
Halomonas zhanjiangensis	1–20 %	Chen et al. (2009b)
Hyphomonadaceae		
Henriciella		
Henriciella litoralis	1–10 %	Lee et al. (2011)
Litorimonas		
Litorimonas taeanensis	1–6 % (2–3 %)	Jung et al. (2011)
Idiomarinaceae		
Aliidiomarina		
Aliidiomarina taiwanensis	1.5–5 %	Huang et al. (2012)
Methanosarcinaceae		
Methanohalophilus		
Methanohalophilus levihalophilus	1–8 % (2–2.3 %)	Katayama et al. (2014)
Methylococcaceae		
Methylomarinum		
Methylomarinum vadi	1–8 %	Hirayama et al. (2013)
Natranaerobiaceae		
Natranaerobaculum		
Natranaerobaculum magadiense	0–9 % (7–8 %)	Zavarzina et al. (2013)

(continued)

Table 11.1 (continued)

Halophilic bacterial species	Salt concentration range for growth	References
Piscirickettsiaceae		
Thiomicrospira		
Thiomicrospira halophila	2.9–20 % (9 %)	Sorokin et al. (2006)
Pseudomonadaceae		
Pseudomonas		
Pseudomonas salegens	1–10 % (3 %)	Amoozegar et al. (2014c)
Rhodobacteraceae		
Hwanghaeicola		
Hwanghaeicola aestuarii	1.5–6 % (2–3 %)	Kim et al. (2010a)
Oceanicola		
Oceanicola antarcticus	0.5–7.5 % (2 %)	Huo et al. (2014)
Oceanicola flagellatus	0–21 % (6–8 %)	Liu and Yang (2014)
Roseivivax		
Roseivivax sediminis	1–15 % (5–10 %)	Xiao et al. (2012)
Rhodospirillaceae		
Limimonas		
Limimonas halophila	15–30 % (20 %)	Amoozegar et al. (2013c)
Rhodothermaceae		
Salinibacter		
Salinibacter iranicus	12–30 % (18 %)	Makhdoumi-Kakhki et al. (2012)
Salinibacter luteus	12–30 % (18 %)	Makhdoumi-Kakhki et al. (2012)
Salisaeta		
Salisaeta longa	5–20 % (10 %)	Vaisman and Oren (2009)
Salinisphaeraceae		
Salinisphaera		
Salinisphaera japonica	1–30 % (7.5–10 %)	Shimane et al. (2013)
Staphylococcaceae		
Aliicoccus		
Aliicoccus persicus	0.5–12.5 % (7.5 %)	Amoozegar et al. (2014a)
Salinicoccus		
Salinicoccus kunmingensis	0.5–25 % (8–10 %)	Chen et al. (2007)
Thermotogaceae		
Petrotoga		
Petrotoga halophila	0.5–9 % (4–6 %)	Miranda-Tello et al. (2007)
Unclassified Bacillales		
Geomicrobium halophilum	5–25 % (10–15 %)	Echigo et al. (2010)
Unclassified Gammaproteobacteria		
Thiohalomonas denitrificans	6–17 % (9–12 %)	Sorokin et al. (2007)
Thiohalomonas nitratireducens	6–15 (6 %)	Sorokin et al. (2007)

11.1.2 Physiology of Halophilicity in Bacteria

The principle parameter in preliminary classification of moderately halophilic bacteria is their requirement for salt. Most halophilic bacteria are moderately halophiles rather than extreme halophiles. Temperature and nutrients content are the critical factors to determine the requirement for salt, as many true halophiles may behave as halotolerant in complex media (Kushner 1993). In some halophilic archaea such as *Haloferax volcanii* optimum salt tolerance increased to higher values with increasing temperature (Mullakhanbhai and Larsen 1975). Similar observation was reported for *Deleya halophila* in which optimum salt concentration at 32 or 42 °C was 7.5 %, while this value decreased to 5 % with decreasing temperature up to 22 °C (Quesada et al. 1987). Halophiles apply two below major mechanisms to adapt to osmotic stress.

11.1.2.1 Salt-In Mechanism

Accumulation of salts in cytoplasm is a strategy for osmoadaptation. This mechanism is applied by (1) extremely halophilic archaea of genera such as *Haloquadratum*, *Halorhabdus*, *Natronobacterium*, *Natronococcus* and *Halobacterium* (2) and halophilic anaerobic bacteria (order of *Haloanaerobiales*) (3) *Salinobacter ruber* (Oren 2008). In this way, halophiles accumulate potassium and chloride ions in cytoplasm that leads to higher concentration of these ions in cells opposed to the saline environment. The role of anionic regulation in osmoadaptation has been demonstrated. Chloride ion has an important role in osmoregulation in different groups of halophiles especially in *Haloanaerobiales* and *Salinobacter ruber* (Kanekar et al. 2012).

11.1.2.2 Compatible Solutes

Production of organic low molecular weight compounds is a major mechanism to cope with the high saline condition. This mechanism has been demonstrated in prokaryotes and eukarya (Roberts et al. 1992). Compatible solutes are highly soluble in water that either have been synthesized *de novo* by halophiles or accumulated by uptake from surrounding medium. These compounds harbor different classes of molecules including sugars, amino acids, polyoles and their derivatives and comparison with the salt-in strategy, this mode of osmotic adaptation costs high energy (Oren et al. 2006).

11.1.3 Ecology of Halophillic Bacteria

Abundance of moderately halophilic bacteria in saline ecosystems such as saltern crystallizer ponds, brines, alkaline saline habitats, saline soils, Dead Sea, salt flats, evaporated ponds, deep-sea hypersaline basins and alkaline saline habitats, has been slightly investigated. Evaporation of seawater creates the thalassohaline environments with the ionic composition similar to the sea. The main ions in these environments include Na^+ and Cl^- whereas ecosystems in which the ionic composition of hypersaline environment is different from the seawater is called athalassohaline environment. They varies in ionic composition and pH. Alkaline soda lakes have a high concentration of carbonate/bicarbonate ions, while the Mg^{+2} and Ca^{+2} are abundant in the Dead Sea (Sass et al. 2001; Oren 2002a).

11.1.4 Taxonomy of Halophilic Bacteria

Moderately halophilic bacteria constitute a diverse and heterogeneous group of different genera. These bacteria belong to both Gram-negative or Gram-positive and aerobic or facultative anaerobic species. Although, Gram-negative halophilic bacteria have been considered in different genera (*Halomonas*, *Pseudomonas*, *Flavobacterium*, *Halovibrio*, *Deleya and Chromobacterium*), their phylogenetic studies revealed that most of them belong to two genera of *Halomonas* and *Chromohalobacter*. The Gram-positive moderately halophilic aerobic bacteria are mainly included in genera *Halobacillus*, *Salinicoccus*, *Nesterenkonia*, *Marinococcus*, and *Tetragenococcus* (except of two *Bacillus* species) as outlined by Ventosa et al. (1998).

The family Halomonadaceae contains a large number of halophilic bacteria. The genus *Halomonas* contained the species of *Deleya* and *Halovibrio* genera, while further rearrangements based on 16S rRNA similarities and phylogenetic studies in this genus were carried out and a numbers of species were reclassified within the genus *Chromohalobacter*. The family *Halomonadaceae* has been studied using 23S rRNA analysis for the first time by Arahal et al. (2002). According to this study, two phylogenetic groups were characterized in this family. In group 1, *Halomonas elongata*, *Halomonas halmophila*, *Halomonas halophila*, *Halomonas eurihalina*, and *Halomonas salina* were classified and group 2 constituted of *Halomonas aquamarina*, *Halomonas magadiensis*, *Halomonas variabilis*, *Halomonas meridiana*, *Halomonas venusta*, *Halomonas halodurans*, and *Halomonas subglaciescola*. Recently, Rafael et al. (2010) studied the phylogenetic relationships in family *Halomonadaceae* using 16S and 23S rRNA sequences comparison and concluded that the family *Halomonadaceae* was not a monophyletic group (Fig. 11.1).

Further studies have been carried out using protein-encoding genes to reach multilocus sequence analysis (MLSA) in order to compare the members of the family *Halomonadaceae*. In an evaluation of the species relationships in family *Halomonadaceae* based on MLSA approach, it was suggested that heterogeneity in genus *Halomonas* may lead to dividing the genus *Halomonas* to two or more genera, however chemotaxonomic and phenotypic features did not confirm this separation (de la Haba et al. 2011).

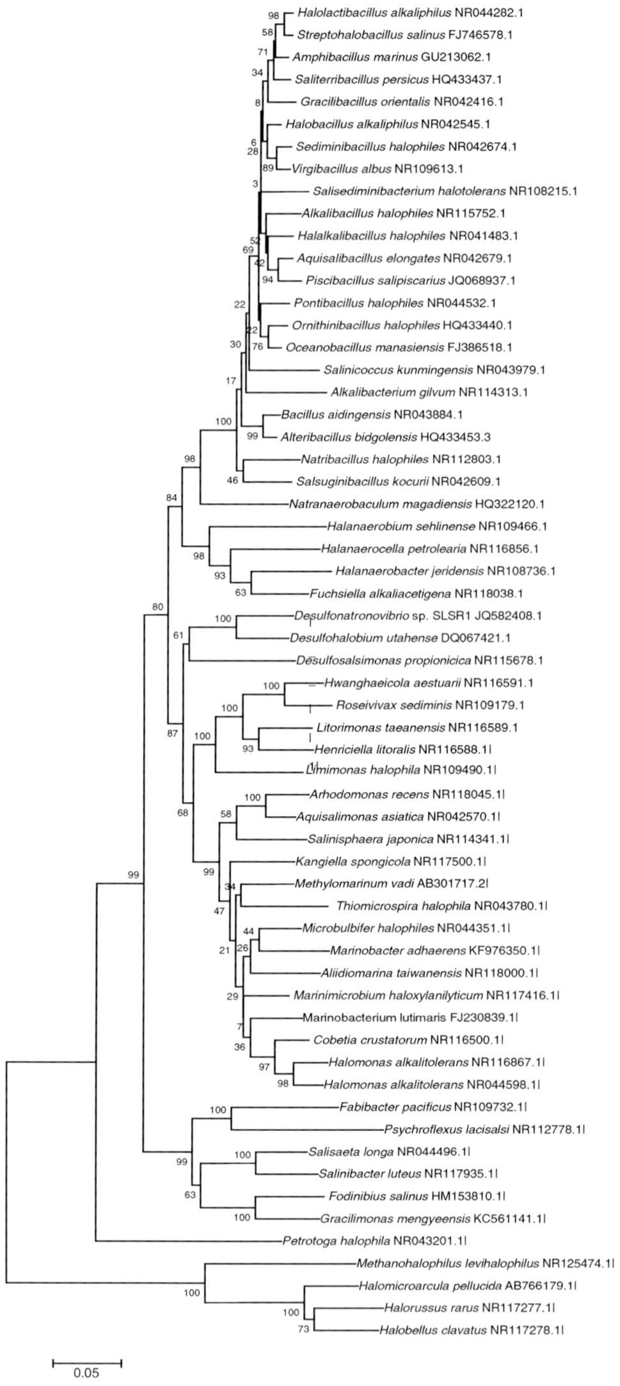

Fig. 11.1 Neighbour-joining phylogenetic tree, based on 16S rRNA gene sequence comparison, showing the phylogenetic relationship among 60 genera of halophilic bacteria

11.2 Potential Industrial Exploitation of Halophilic Bacteria

The potential application of halophilic bacteria can be considered in several fields. As regards, the halophilic bacteria can tolerate high concentrations of salt and have low nutritional requirement, that confer them a significant potential in harsh industrial processes. This section focuses on different aspects of applications or potential application of halophilic bacteria in industry.

11.2.1 Production of Enzymes

There are many microbial enzymes that have been used in various industries, but stability and characterization under the extreme temperatures or salt concentrations is a significant issue. Some enzymes of halophilic bacteria make as desired candidates for industry because of their stability in high ionic circumstance of process. Furthermore, most of halobacterial enzymes are also relatively stable at high temperatures. Thus, halobacteria have attracted much attention in recent years due to beneficial production of halophilic exoenzymes that can be used in diverse fields of biotechnology. For example, halophilic hydrolytic enzymes are mixed in detergents or treatment of saline waste waters and food industry, proved as efficient approach. To prevent precipitation, the halophilic enzymes have acquired lots of negatively charged amino acids on their surfaces (Madern et al. 2000). The charge of surface area of halophilic enzymes plays an important role in solvent-enzyme interaction. A halophilic α-amylase isolated from *Alicyclobacillus acidocaldarius* contained a negative charge on surface area that leads to the stability enhancement in salty circumstance (Matzke et al. 1997). The amino acid composition of enzymes revealed that in addition to the higher acidic amino acids residues, they have higher hydrophobic and lower aliphatic residues compared to their nonhalophilic counterparts (Madern and Zaccai 1997). The halophilic members in genera *Bacillus*, *Halomonas*, *Salinovibrio*, *Chromohalobacter*, and *Salinicoccus* have identified with potential in production of a variety of exoenzymes under hypersaline habitats (Sánchez-Porro et al. 2003). In other studies, many moderately halophilic bacteria belonged to genera *Marinobacter*, *Virgibacillus*, *Halobacillus*, *Geomicrobium*, *Chromohalobacter*, *Oceanobacillus*, *Bacillus*, *Halomonas* and *Staphylococcus* were identified with ability of industrially important enzymes production and salt requirement of 3–20 % (w/v) NaCl. Also, some halophilic bacteria with combined extracellular hydrolytic activities have been identified (Neagu et al. 2014; Kumar et al. 2012). Enzymes identified in halophilic bacteria are listed in Table 11.2.

Table 11.2 Enzymes isolated from halophilic bacteria

Category of enzyme	Enzyme	Halophilic bacterial strains	References
Hydrolases	Nuclease	*Micrococcus varians* subsp. *halophilus*	Kamekura (1986)
	Lipase	*Salinivibrio* sp.	Amoozegar et al. (2008)
		Marinobacter lipolyticus	Martín et al. (2003)
		Salinicoccus sp. JAS4	Jayachandra et al. (2012)
	Amylase	*Micrococcus varians* subsp. *Halophilus*	Kobayashi et al. (1986)
		Halomonas meridian	Coronado et al. (2000)
		Psychromonas antarcticus	Mountfort et al. (1998)
		Halobacillus sp.	Amoozegar et al. (2003)
	Chitinase	*Virgibacillus marismortui*	Essghaier et al. (2012)
	Protease	*Pseudomonas* sp.	Van Qua et al. (1981)
		Chromohalobacter sp.	Vidyasagar et al. (2007)
		Salicola sp.	de Lourdes Moreno et al. (2009)
		Salinicoccus sp. JAS4	Jayachandra et al. (2012)
	Beta-lactamase	*Chromohalobacter* sp.	Tokunaga et al. (2004)
	Cellulase	*Halomonas sp.*	Shivanand et al. (2013)
	Endoglucanase	*Vibrio* sp.	Gao et al. (2010), Huang et al. (2010)
		Halomonas sp.	
Oxidoreductases	Malate dehaydrogenase	*Bacillus haloalkaliphilus*	Weisser and Trüper (1985)
	Malic enzyme	*Salinovibrio costicola*	Salvarrey and Cazzulo (1980)
	Isocytrate dehydrogenase	*Halomonas halodenitrificans*	Weisser and Trüper (1985)
	Alanine dehydrogenase	*Halomonas elongata*	Bylund et al. (1991)
Transferases	Aspartate transcarbamylase	*Salinovibrio costicola*	Ahonkhai et al. (1989)
	Nucleoside diphosphate kinase	*Halomonas* sp.	Yonezawa et al. (2001)
Lyases	Phosphoenolpyruvate carboxykinase	*Salinovibrio costicola*	Salvarrey et al. (1989)

11.2.1.1 Nucleases

Nuclease H of *Micrococcus varians* subsp. *halophilus*, is applied in industrial production of 5′-guanylic acid (5′-GMP) for degradation of RNA. This enzyme is active at 12 % salt and 60 °C (Kamekura et al. 1982). Another halophilic *Bacillus* sp. strain N23-2 producing nuclease has been identified. The enzyme hydrolyzed DNA and RNA with optimal activity at 5–17 % M NaCl or KCl (Onishi et al. 1983).

11.2.1.2 Agarases

One of the most important glycosyl hydrolase is agarase that catalyzes the hydrolysis of agar. Agarases are used in molecular biology experiments and in extraction of the bioactive compounds from algae that widely are used in food industry as food additives. *Alteromonas* sp. ATCC 43961 can grow and produce agarase in a wide range of salinity. The enzyme could hydrolyse agar to oligosaccharides such as neoagarohexaose and neoagarobiose. Also, it has been found that this enzyme can be considered as an effective agent for controling of algae bloom in water bodies (Coyne et al. 1995).

11.2.1.3 Hydantoinase

Production of D-amino acids are essential in biosynthesis of peptide hormones, antibiotics and pesticide and are conducted by activation of D specific hydantoinase. This process is followed by chemically conversion under acidic condition. Use of hydantoinase from halophilic *Pseudomonas* sp. ATCC 55940 in production of D(-) *N*-carbamoylphenylglycine (CPG) has been shown as an effective alternative. This halophilic bacterium produces 5–6 % CPG (% w/v) within 10–15 h at salt concentrations above 7 % (w/v) (Bastawade et al. 2000).

11.2.1.4 Amylases

Amylases are one of the most commonly used industrial enzymes that catalyze the hydrolysis of starch and are used widely in several fields of biotechnology. Halophilic α-amylases have received more attention due to their ability to remain active in the presence of high salt concentrations. Certain halophilic bacteria such as *Micrococcus varians* subsp. *halophilus* produces amylase with optimal activity at 4.5–6 % NaCl and pH 6–7. The enzyme has two components with 86 and 60 kDa molecular mass (Kamekura 1986). The extracellular amylase gene, *AmyH* has also been isolated from the halophile *Halomonas meridian* (Coronado et al. 2000). This amylase exhibited maximal activity at pH 7.0 and the highest amylase production was obtained with 5 % of salt content. These amylases have significant applications in treatment of saline waste water. *Psychromonas antarcticus* a psychrophilic, halophilic, anaerobic bacterium isolated from a high saline pond in Antarctica, showed amylolytic ability in processing starch at low temperatures (Mountfort et al. 1998). A moderately halophilic bacterium of spore forming *Halobacillus* sp. strain MA-2 having amylase production ability showed maximum activity at 5 % (w/v) NaCl in presence of 100 mM sodium arsenate (Amoozegar et al. 2003).

11.2.1.5 Lipases

Lipases are among the most important enzymes with several applications in various fields of pharmaceutical, detergent and food industries. An extracellular lipase was characterized from moderately halophilic *Salinivibrio* sp. strain SA-2. The maximum activity of the enzyme reported at pH 7.5 and 50 °C. The enzyme remained active in presence of 17 % NaCl (Amoozegar et al. 2008). *Marinobacter lipolyticus* a novel moderate halophile was isolated from a hypersaline habitat with lipolytic activity that grow optimally at 7.5 % NaCl (Martín et al. 2003).

11.2.1.6 Xylanases

Xylanases degrade xylan part of the hemicellulose that is considered as the second most natural abundant renewable resource. Xylanases used in removal of residual lignin from pulp (Oksanen et al. 2000). The complete degradation of xylan involved endo-1,4-β-D-xylanases, β-D-xylosidases, α-L-arabinofuranosidases, α-D-glucuronidases and acetylxylanesterases (Collins et al. 2005). Pioneer work on characterization and purification of halotolerant endo-xylanases was done by Wejse et al. (2003) from a novel halophilic bacterium, strain CL8 with the highest sequence similarity value of 91.6 % to *Oceanospirillum linum* and *Marinobacter* sp. str. CAB. The major products of these enzymes activation are xylobiose and xylotriose. The activity of enzymes remained stable at 5 M NaCl, but the optimal activity occurred at 1 M NaCl. β-xylanase and β-xylosidase produced by an aerobic xylanolytic *Gracilibacillus* sp. TSCPVG growing at moderate to extreme salinity (1–30 %) and neutral to alkaline pH (6.5–10.5) were characterized. The highest growth and enzyme activity of 3.5 U/ml was obtained in 3.5 % NaCl. It was the first report of extremely halo-alkali-thermo-stable xylanase (Giridhar and Chandra 2010). A novel mannanase gene (*A021*) was isolated from marine bacterium *Pantoea agglomerans* and was cloned in *E. coli*. Maximum activity of enzyme was observed at pH 6 at 55 °C that enhanced in correlation to the increase of NaCl concentration up to 0.75 M (Wang et al. 2010a).

11.2.1.7 Proteases

Proteases are among important group of enzymes with diverse applications in industry. Using of proteases in detergents, laundry and food industries are the most common application of these enzymes. Recently, proteases in pharmaceutical industry and bioremediation process has attracted more attention. There has been extensive research on the extremophile proteases because of their utility in detergents and food industries. An extracellular protease produced by *Pseudomonas* sp. strain A-14 was identified that its optimal activity was at 18 % NaCl and pH 8 (Van Qua et al. 1981). Halothermophilic protease production by *Chromohalobacter* sp. strain TVSP101 with maximum activity at pH 8 and 4.5 M NaCl is also reported

(Vidyasagar et al. 2007). On the other hand, *Salicola* sp. IC10 showed lipase and protease activities with optimum growth at 15–20 % (w/v) NaCl, pH 8.0 and at 40 °C. This protease hydrolyses casein, gelatin, bovine serum albumin and egg albumin, while the most specific substrate to the enzyme was egg albumin (de Lourdes Moreno et al. 2009).

11.2.1.8 Chitinase

Chitinase is proved as a significant enzyme of biocontrol tools in agriculture or environmental investigations (Jung et al. 2005; Duo-Chuan 2006). A moderately halophilic bacterium *Virgibacillus marismortui* was isolated from shallow salt lakes with ability to produce chitinase (in absence of salt as well as in presence of high salinity (25–30 % NaCl w/v)). Such strains can be significant for biocontrol purposes (Essghaier et al. 2012).

11.2.1.9 Cellulases

Cellulases have been known as industrially significant enzymes in paper, agriculture, food and laundry industries. A halophilic bacterium, *Halomonas* sp. strain PS47 has been known to produce cellulase under saline condition of 6 % NaCl. The maximum activity was at pH 7.1 and 50 °C (Shivanand et al. 2013). A novel salt-tolerant endo-β-1,4-glucanase Cel5A was also identified from *Vibrio* sp. G21. The enzyme obtained from this bacterium has a catalytic domain of glycosyl hydrolase and a cellulose binding domain. The gene *cel5A* was cloned in *E. coli* for the over expression of enzyme (Gao et al. 2010). A novel endoglucanase was identified from halophilic bacterium *Halomonas* sp. S66-4. The enzyme was cloned in *E. coli* and purified recombinant enzyme obtained with the highest activity (4.9 U/mg) at pH 5 and 6 % NaCl (Huang et al. 2010).

11.2.2 Production of Protective Biomolecules

Halophilic microorganisms apply two main strategies in response to hyperosmolarity. Some halophiles use accumulation of salts in the cytoplasm, called "salt in" mechanism. This strategy can be found in halophilic archaea in the order *Halanaerobiales* (Ventosa et al. 1998) and the *Salinobacter ruber* (Antón et al. 2002; Oren 2002a). Other halophilic microorganisms synthesize low molecular weight osmoregulatory compounds that are known as compatible solutes (Da Costa et al. 1998). There are different classes of compatible solutes including amino acids and their derivatives (proline, glutamate, glutamine, ectoine and taurine), sugars (sucrose, trehalose), methylamines (glycine betaine) and polyoles (mannitol, erythritol, and glycerol) (Oren 2002a; Yancey 2005). Halophilic bacteria either synthesize the compatible solutes *de novo* or take up from surrounding environment. In addition to the involvement in osmoregulation, they allow the microorganism to maintain the biomolecules native structure to cope with high salinity fluctuations in their habitats. In fact, compatible solutes stabilize proteins by effect on proper folding of especially polypeptides (Arakawa and Timasheff 1985).

As compatible solutes have protective properties; they can be utilized as protective and stabilizer agents of macromolecules particularly proteins. Compatible solutes might be used as anti-aggregation agents during protein refolding in protein crystalization (Lentzen and Schwarz 2006). This feature of compatible solutes has been recruited for therapeutic purposes. For example, inhibition of protein aggregation aiming at treatment of some amyloidosis disease such as Alzheimer's and Parkinson's (Luley-Goedl and Nidetzky 2011). There are several reports that compatible solutes can be considered as protectant of proteins against heat denaturation (Ramos et al. 1997; Borges et al. 2002; Sawangwan et al. 2010) by alteration in thermodynamics of unfolding phenomenon (Ryu et al. 2008).

It has been shown that voglibiose the analogue of glucosyl glycerol (GG) may be considered as a therapeutic agent in treatment of diabetes mellitus (Ishida et al. 1998). Additionally, investigations showed that GG can be regarded as prebiotic food additives (Luley-Goedl and Nidetzky 2011). The sake contains some osmolytes such as glycosylglycerol (GG), (2R)-1-O-α-D-glucosylglycerol (R-GG-I) and (2S)-1-O-α-D-glucosylglycerol (S-GG-I) which exist in sake can be applied in treatment of skin (Nakahara et al. 2007). Glucosylglycerols are able to stimulate sensory neurons and increase the levels of insulin-like growth factor-I in skin fibroblasts (Harada et al. 2010; Luley-Goedl and Nidetzky 2011).

Among all compatible solute obtained from halophilic bacteria, ectoine is among the most applicable compound implemented extensively in cosmetic production, pharmaceutical industry, macromolecules protectant and as enhancer of PCR and DNA microarray processes (Schnoor et al. 2004; Mascellani et al. 2007; Graf et al. 2008).

11.2.2.1 Proline

Some Gram-positive bacteria namely, *Salinicoccus roseus* and *Salinicoccus hispanicus* accumulate amino acids under high salinity condition. Accumulation of amino acids are not considered as an important solute in osmoregulation (Ventosa et al. 1998). *Halobacillus halophilus* is a moderately halophilic bacterium that grows at NaCl concentrations from 0.5 to 2 M. Production of proline as a main solute at high salinity concentrations (2–3 M NaCl) is known. However, proline is the predominant solute in exponential growth stage and accumulation of ectoine increases during the stationary phase (Köcher et al. 2011). Moreover, *H. halophilus* changes the production of osmolytes mechanism under different salt concentrations, as at intermediate salinity (1 M NaCl) produces glutamine and glutamate, while at higher concentration of NaCl (2–3 M NaCl) produces proline, as the predominant solute (Saum and Müller 2007).

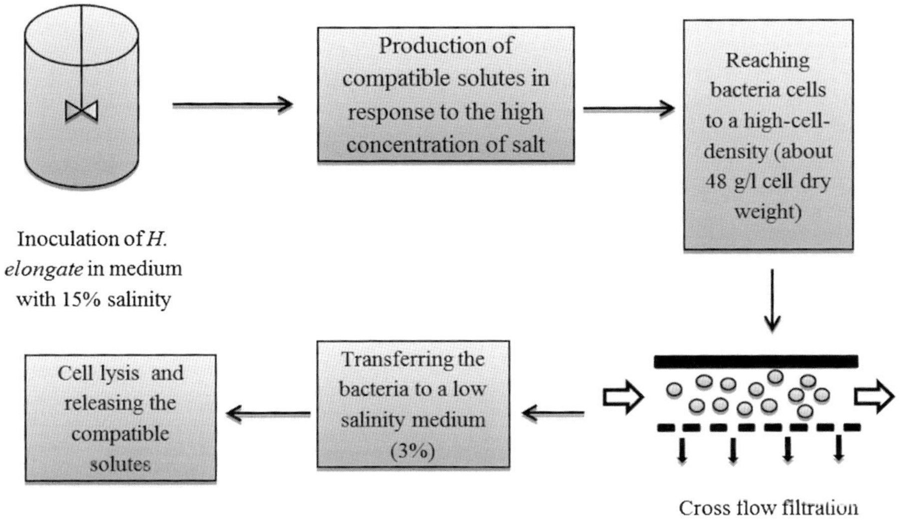

Fig. 11.2 Bacterial milking process to production of ectoine

11.2.2.2 Ectoine

Ectoine (1,4,5,6-tetrahydro-2-methyl-4-pyrimidinecarboxylic acid) is widespread among halophiles that was first discovered in *Ectothiorhodospira halochloris* a halophilic bacterium. Ectoine and betaine can also be obtained using biotechnological process called "bacterial milking" from *Halomonas elongate* as shown in Fig. 11.2 (Sauer and Galinski 1998). Salinity and temperature are two major factors that affect the production profile of ectoine and hydroxyectoine. At high salinity and temperature (20 % NaCl and 40 °C) production of hydroxyectoine increases, while at lower salinity and temperature (15 % salinity and 25 °C) only production of ectoine occurs (Margesin and Schinner 2001).

11.2.2.3 Trehalose

Trehalose is a non-reducing disaccharide consists of two glucose molecules with a glycosidicα-(1-1) bond. Trehalose is significantly stable as it is not readily broken to its ingredients. Bacteria accumulate the trehalose as a osmoprotectant to cope with stress conditions. In addition to the response to salinity, trehalose can accumulate under another stress conditions such as rehydration or heat shock. Trehalose has biotechnological applications and regarded as protectant of the macromolecules and enzymes activity, food additives, marker in transgenic plants in pharmaceutical and cosmetic industry (Iturriaga et al. 2009). The structural diversity of compatible solutes have been shown in Fig. 11.3.

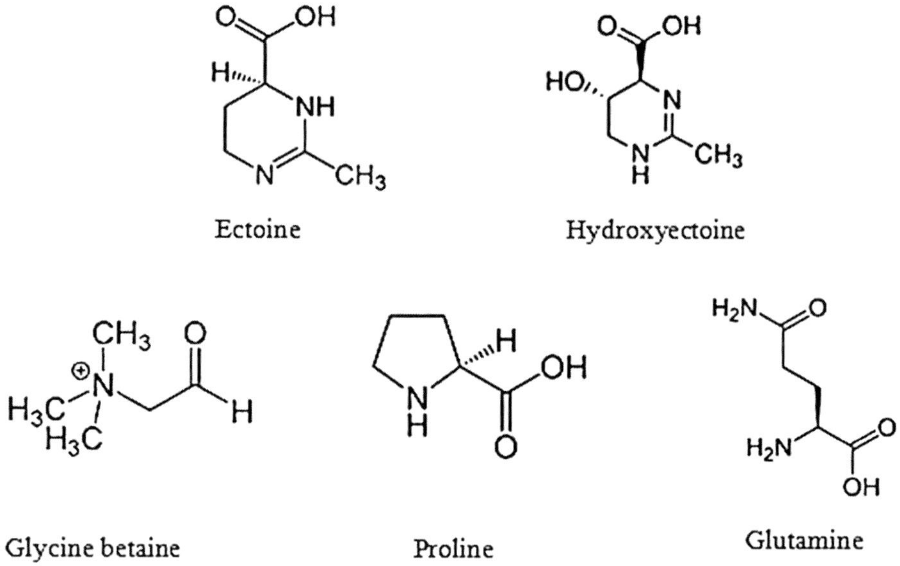

Fig. 11.3 Examples of compatible solutes in halophilic bacteria

11.2.3 Bioremediation of Polluted Environments

In recent decades the growth of numerous industries along with human activities increase the pollution of environment with dangerous pollutants including synthetic dyes, metals and petroleum derivatives. As the most produced waste waters have high concentration of salt, use of halophiles offers a promise alternative in treatment of these waste.

11.2.3.1 Treatment of Saline Wastewaters

Various industrial processes such as oil extraction, tannery or seafood processing, lead to large accumulation of saline wastewater that affect the efficiency of treatment process. It has been accounted that 5 % of total world liquid wastes are highly saline in nature (Lefebvre and Moletta 2006). Conventional treatments have low efficiency in chemical oxygen demand (COD) removal due to the detrimental effects of salt on microbial communities. The halophilic microorganisms expose to plasmolysis in cope with high salinity of wastewaters. Applying halophilic and halotolerant bacteria in treatment of saline wastewater can improve the efficiency of process and can be considered as a remarkable solution for COD removal from saline wastes.

There are studies upon use of aerobic or anaerobic halophiles in removal of saline wastewaters. Ability of *Halanaerobium lacusrosei* (an anaerobic halophilic bacterium) in COD removal from saline (synthetic) wastewater as investigated in an anaerobic packed-bed reactor. It was revealed that *H. lacusrosei* was able to reduce the COD concentration from 3,2 to 1,5 mg/L in salt free wastewater while the maximum removal efficiency was observed at 3 % salt concentration (Kapdan and Erten 2007). Due to occurrence of plasmolysis in bacteria under high salinity, conventional activated sludge for treatment of saline wastewaters did not decrease the COD effectively (Burnett 1974). Thus, halotolerant strains such as *Halobacter halobium* in improve the COD removal in activated sludge (Kargi 2002).

Tannery saline wastewater has large quantities of salts that make it difficult to be treated by conventional methods. Salt tolerant bacteria isolated from tannery wastewater, characterized as *Pseudomonas aeruginosa*, *Bacillus flexus*, *Exiguobacterium homiense* and *Staphylococcus aureus* showed 80 % decrease in COD at 8 % salinity in the biological treatment systems (Sivaprakasam et al. 2008). In addition, a salt tolerant protease from *P. aeruginosa* isolated from tannery saline effluent was also used for wastewater treatment. Protease activity was observed at salt concentrations from 4 to 7 % (Sivaprakasam et al. 2011).

Treatment of phenolic wastewater by a moderately halophilic species of *Halomonas* sp. under saline conditions was reported (Hinteregger and Streichsbier 1997). Interestingly treatment of synthetic wastewater with halotolerant bacterium *Halobacter halobium* supplemented activated sludge in an aerated rotating biodisc contactor is also observed (Kargi and Dinçer 1998). Moreover, *Staphylococcus xylosus* mix with activated sludge has also been proved effective in treatment of saline wastewaters (Sohair et al. 2010). Earlier, Yoshie et al. (2006) reported a denitrification system for saline wastewater treatment using halophilic denitrifying bacteria. γ-Proteobacteria especially *Halomonas* spp. was responsible for denitrification activity under saline condition.

Production of pickled plums, a traditional Japanese food, generates 40,000 t saline wastewaters annually (Kubo and Morakami 1999) which contain 15 % NaCl. Activated sludge cannot process the undiluted saline wastewaters because of high salt concentrations and low pH that prevent the microorganisms' activity. However, dilution with large quantities of water diminish the salts concentration is uneconomical. The employment of halotolerant bacteria in COD reduction of such wastewaters has been described by Kubo et al. (2001).

11.2.3.2 Treatment of Effluent Containing Synthetic Dyes

Industrial effluents that release to the environment contain synthetic dyes which are widely used in textile, paper printing, food, leather, pharmaceutical and many other sector of industries (Kuhad et al. 2004; Couto 2009). Synthetic dyes can be categorized according to their main chemical structures including triphenylmethane, nitro, nitroso, phthalein, polyene, triarylmethine, anthraquinone, and diazo (Fig. 11.4) (Doble and Kumar 2005). Dyes are toxic and have carcinogenic effects on living systems. The effluent containing dyes has high COD and low BOD. There are several physical and chemical methods for removal of dyes from effluents including flocculation, electroflotation, membrane filtration, oxidation, and adsorption and so on.

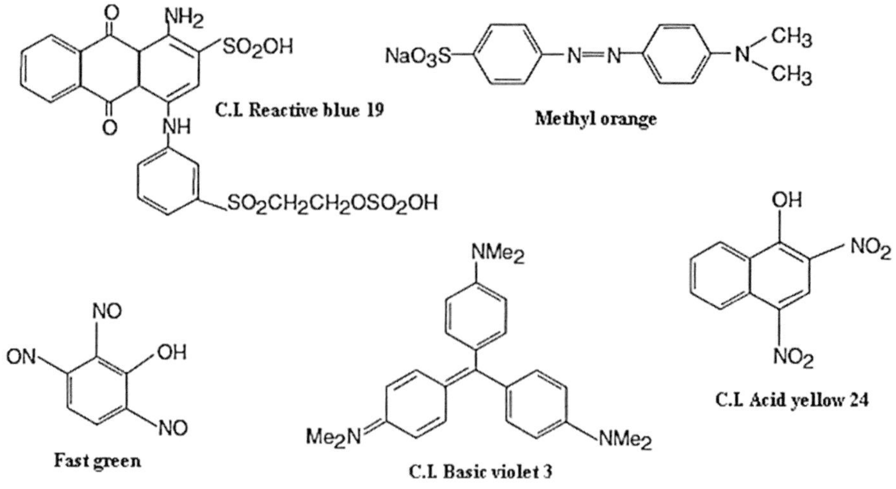

Fig. 11.4 Examples of industrial synthetic dyes

Some microorganisms are known to be capable to degrade the dyes while the high concentration of salt in textile effluents limits the activity of such microorganism. *Halomonas* sp. strain IP8 have been isolated from textile effluents with the ability to decolorize some textile dyes. The bacterium grew well at range of NaCl 1–20 % (w/v) with the optimum growth at 3–5 % (w/v), whereas the strain was not able to grow in absence of NaCl. The concentration of dye decreased from 50 to 20 mg/L during 16–24 h by *Halomonas* sp. strain IP8. Decolorization activity was occurred at temperature range of 25–45 °C, while the optimum temperature was found at 35 °C. Maximum decolorization activity was observed at 1–1.5 M NaCl (Pourbabaee et al. 2011). Earlier, Guo et al. (2008) described decolorization of azo dye by *Halomonas* sp. strain GTW under saline concentration of 10–20 % NaCl. The decolorization of different azo dyes (reactive brilliant red, acid black, acid scarlet and acid red) occurred effectively above 90 % in 24 h of treatment incubation.

11.2.3.3 Bioremediation of Heavy Metals

Presence of elevated heavy metals concentration in the environment leads to serious ecological consequences. Presence of high concentrations of salt prevent the treatment of these effluents. Therefore, identifying the metal resistant halophilic microorganisms biotreatment of saline wastewaters is essentially demanded. A moderately halophilic chromium tolerant bacterium, *Vigribacillus* sp. strain H4 was isolated from saline soil. The strain was resistant to 1,000 mg/L chromium and could reduce 99.2 % of 100 mg/L Cr (VI) at 6 % NaCl (Mishra et al. 2012).

11.2.3.4 Biodegradation of Organic Pollutant

Petroleum is a complex compound including different hydrocarbons such as asphaltenes, cycloalkanes, aliphatics, aromatics and resins having high stability and toxicity (Philp et al. 2005; Yemashova et al. 2007). Organic or hydrocarbon containing pollutants produced during the extraction, refining and transport of petroleum can affect the terrestrial and marine environment. Hydrocarbon contaminations resulting from petrochemical industry arise great concerns for environmental protection. Microorganisms have shown ability to degrade the hydrocarbons such as toluene, ethylbenzene, xylenes (BTEX) and benzene. Identification of halophilic microorganisms that degrade petroleum hydrocarbons is critical for improving the treatment processes. Importantly, bioremediation of polluted hypersaline wastewaters by indigenous halophilic and halotolerant bacteria can be considered as ecologically safe efficient and low cost technology (Philp et al. 2005).

Although, microbial diversity and their metabolic activity reduce with increasing the salinity (Riis et al. 2003; Kleinsteuber et al. 2006), after certain period of times, biodegradation occur due to the adaption of microbial communities to high salinity condition (Kleinsteuber et al. 2006). However, there are numerous reports on enhancement of hydrocarbon biodegradation by increasing the salinity (Díaz et al. 2000; Yang et al. 2000). Bacterial consortia including *Marinobacter* sp., *Erwinia ananas* and *Bacillus* spp. showed ability to degrade petroleum hydrocarbons at range of salinity from 0 to 22 % of NaCl (Díaz et al. 2000, 2002).

Aromatic compounds are categorized to several groups such as aromatic alcohols, benzoates, phenols, nitrobenzenes, pargyline and etc. Some aromatic acids such as vanillic acid, cinnamic acid, and syringic acids are also produced naturally by plants. In addition, decomposition of some polymers and pollutants can release these aromatic acids to the environment (Le Borgne et al. 2008). García et al. (2004) have described the ability of *Halomonas organivorans*, to degrade several aromatic acids such as benzoic, *p*-hydroxybenzoic, salicylic, phenylacetic, *p*-coumaric, ferulic and phenylpropionic acids. *Halomonas campisalis* used phenol as carbon and energy source and converted it to CO_2 at 0–15 % NaCl (Alva and Peyton 2003).

Chlorophenols are organic chemicals that have been used as bactericide, fungicides, wood preservative and solvent in synthetic chemistry. Chlorophenols is found in water, soil and air. These compounds have mutagenic and carcinogenic effects on organisms. Veenagayathri and Vasudevan (2013) described the degradation of 4-chlorophenol by a moderately halophilic bacterial consortium under saline conditions. The optimum chlorophenol degradation was observed at 5 % of NaCl. Employment of haloalkaliphile bacterium *Halomonas campisalis* in degradation of phenol and catechol is also reported.

11.2.3.5 Bioflocculants

Flocculants are defined as agents used in aggregation of particles together to form flocs. These compounds are widely used in wastewaters treatment, drinking water purification, fermentation and food industries. Bioflocculants are macromolecules produced by different microorganisms such as fungi, bacteria and algae that mediate bioflocculation process (Gao et al. 2006). Several studies exhibit the ability of microorganisms in production of bioflocculants. Bioflocculant producing *Alcaligenes latus* B-16, *Bacillus* sp. and *Enterobacter* sp. are reported (Kurane and Nohata 1994; Yokoi et al. 1997; Suh et al. 1997). Also, bioflocculant produced by *Klebsiella* sp. isolated from activated-sludge was investigated (Cheng et al. 2004; Sheng et al. 2006). Sam et al. (2011) described the bioflocculating activity of exopolymeric substances (EPSs) produced by halophilic bacterium *Halomonas* sp. AAD6. Furthermore, *Halobacillus* sp. was introduced as a bioflocculant producing halophile. Its flocculant comprising polysaccharides and protein showed the highest activity at pH 7 (Cosa et al. 2012).

11.2.4 Applications of Halophilic Bacteria in Agricultural Fields

Halophilic bacteria can be recruited in several aspects of agriculture. This group are applied in biocontrol of phytopathogens, solubilization of essential elements and promoting the plants by production of growth factors that are discussed below.

11.2.4.1 Biological Control

Several pathogens affect the plants' health and cause lose or reduction of products. Applying conventional pesticides or fungicides can pose a real risk to the future of mankind and environment. Introduction of the microbial agents to restrain and control of plants diseases referred as "biological control" can be considered as an eco-friendly and efficient alternative.

Production of Fungicides

Several fungicides are used to control plant pathogens in agriculture, but the side effect of these chemical agents have generated lots of environmental and health issues due to enrichment of chemical residues in the environment and development of fungal resistance to fungicides (Chen et al. 2008a). Stem canker of tomato is a serious disease caused by *Botrytis cinerea. B. cinerea* is a necrotrophic fungus that causes gray mould on plants such as tomato, strawberry, apple and other crops (Williamson et al. 2007). Sadfi-Zouaoui et al. (2007) have described the use of moderately halophilic bacteria *Bacillus subtilis* J9 and *Halomonas* sp. K2-5 in control of stem canker of greenhouse tomatoes. The plants treated with halophilic bacteria showed smaller lesions. Essghaier et al. (2008) reported the employment of moderately halophilic bacteria *Virgibacillus marismortui*, *B. subtilis*, *Bacillus pumilus*, *Bacillus licheniformis*, *Terribacillus halophilus*, *Halomonas elongata*, *Planococcus rifietoensis*, *Staphylococcus equorum* and *Staphylococcus* sp. in biocontrol against *B. cinerea*. All of these strains (except of *H. elongata*) were able to produce chitinase and ß-1,3-glucanase. *Halomonas subglaciescola*, *Halobacillus litoralis*, *Marinococcus halophilus*, *Salinococcus roseus*, *Halovibrio variabilis* and *Halobacillus halophilus* were also introduced as inhibitor agents of *B. cinerea* (Sadfi-Zouaoui et al. 2008).

Isolation and identification of plant growth promoting bacteria from saline rhizospheric soil have shown their ability to control some phytopathogenic fungi. Strain of *Cellulosimicrobium* sp. have shown antifungal activity against *B. cinerea*, *Fusarium oxysporum* and *Verticillium dahlia* (Nabti et al. 2014).

Biofertilizers

Phosphate solubilizing bacteria accelerate mobilization of phosphorus by production of organic acids, by which improve the growth of plants as biofertilizers (Deubel and Merbach 2005). Capability of halophilic phosphobacteria on growth of *Avicennia officinalis* seedlings are described (Ravikumar et al. 2010). It is reported that halophilic phosphobacteria increased the level of photosynthetic pigments. Also, the function of halotolerant bacteria *Bacillus megatherium* var. *phosphaticum* and *Bacillus megatherium* ATCC 14581 insolubilizing the phosphorus in saline soils has been investigated (Xiang et al. 2011). *Azotobacter* strains can produce growth promoting hormones including gibberellins, cytokinin and auxins that improve the plants growth. Species of *Azotobacter chroococcum*, *Azotobacter vinelandii* and *Azotobacter beijerinckii* indicate nitrogen fixation and oxine hormone (IAA) production at salinity of 3 % (Ravikumar et al. 2004). A halophilic bacterium *Azospirillum brasilense* isolated from saline soils exhibited production of IAA that caused enhanced growth of wheat under saline condition. In absence of osmoprotectants *A. brasilense* showed resistant to 300 mmol/L NaCl, while in presence of glycine betaine resistance to salt increased to 600 mmol/L (Nabti et al. 2007).

11.2.5 The Role of Halophilic Bacteria in Production of Biopolymers

A wide range of biopolymer structures are produced by microorganisms. Microbial polymers and exopolysaccharides are used for several applications especially in food and pharmaceutical technology or oil exploration. Some prevalent biopolymers produced by halophilic bacteria are explained in this section.

11.2.5.1 Biosurfactant Production

Biosurfactant compounds diminish the surface tension that is produced by several agents. Biosurfactants have been employed in bioremediation, petroleum industry, textiles, food, pharmaceuticals and cosmetics (Cameotra et al. 2010). High production of biosurfactant is detected in a halotolerant strain of *Nocardiopsis* sp. which occurred when olive oil and phenylalanine were used as the carbon and nitrogen sources, respectively. The maximum biosurfactant production was observed at 3 % (w/v) NaCl, while the 80 % of compound activity remained in presence of 12 % (w/v) of NaCl (Khopade et al. 2012). Production of biosurfactant by marine *Streptomyces* species B3 is also described that could reduce the tension of water to 29 mN/m (Khopade et al. 2012).

11.2.5.2 Exopolysaccharide

Bacterial exopolysaccharides (EPS) have wide spread applications in biotechnology. They are used as emulsifiers, gelling agents, in situ microbially enhanced oil recovery, food and cosmetic industries. Extreme habitats have been studied to identify the potential of biopolymers production in their indigenous bacteria (Nicolaus et al. 2010; Poli et al. 2010).

Capability of *Halomonas* species to produce EPS was investigated (Bouchotroch et al. 2000). EPS produced by *Halomonas* species showed pseudoplastic behaviour and can be considered as an emulsifying agent (Calvo et al. 2002). Mauran is a viscous EPS produced by halophilic bacterium *Halomonas maura*. The bacterium grows at range of salt concentrations from 1 to 15 % w/v and the optimum growth occurred in medium containing 3–15 % NaCl. Several properties such as high viscosifying capacity, pseudoplastic behaviour and solubility under a wide range of pH and salinity make it a promising agent in different fields of biotechnology (Llamas et al. 2011). Strains of *Halomonas ventosae* and *Halomonas anticariensis* have been found as polysaccharide producers with 50 kDa molecular weight and composed of glucose, mannose and galactose units. The quantities of produced EPS by these novel strains were reported 28.35 and 28.95 mg per 100 ml, respectively (Mata et al. 2006). Recently, Sam et al. (2011) reported the ability of a halophilic bacterium, *Halomonas* sp. AAD6 in production of EPS with bioflocculating activity. The large scale EPS production by these bacteria can be carried out using sugar beet and starch molasses as carbon sources.

11.2.5.3 Bioplastics

Poly-β-hydroxyalkanoate (PHA) have found in prokaryotes as a storage polymer. PHA is used to produce biodegradable plastics. PHAs have other applications including artifacts and implants of medicine, tissue engineering and drug delivery systems (Quillaguamán et al. 2010). PHAs are synthesized by various microorganisms under excess of carbon and energy sources. Some halophilic bacteria are capable to produce PHA. The ability of *Halomonas boliviensis* to produce PHB has been described using several carbon sources such as sucrose, maltooligosaccharides, glucose, xylose, sodium acetate and butyric acid (Quillaguaman et al. 2005, 2006, 2007). Biswas et al. (2009) described the production of PHA by *Cobetia marina* by consuming glucose and valerate as carbon sources.

11.2.6 Application of Halophilic Bacteria in Food Biotechnology

Halophilic enzymes are functional at high concentration of salt, while salt suppress the nonhalophilic microorganisms' activity thereby prevent the contamination of foods (Margesin and Schinner 2001). For this reason, use of halophilic bacteria in food processes have been attracted more attention.

11.2.6.1 Fermented Foods

Fermented fish sauce is a traditional condiment used widely in East Asia (Longfils et al. 2008). This product is produced by mixing the fish with high concentration of salt (20–30 %) so as to preserve it at ambient temperature for 6–18 months. Several studies have been carried out to lower down the procedure time and decreasing the pH and salt concentration, but some disadvantages such as denaturing the peptides, growth of pathogen microorganisms and loss of flavor can not be ruled out (Akolkar et al. 2010). Proteases from halophilic microorganisms secreted in fermented product have an important role in lowering the time of fermentation. *Bacillus subtilis* CN2 and *Fillobacillus* sp. RF2-5 with their ability in production of proteases have been found suitable to strengthen the quality of fish sauces (Uchida et al. 2004; Hiraga et al. 2005). Further, acceleration of fermentation process using halophilic bacteria, *Vibriobacillus* spp. SK33 and *Staphylococcus* sp. SK-1-5 (Yongsawatdigul et al. 2007) reduced the procedure time from 18 to 4 months. Thus, salt tolerant enzymes produced halophilic bacteria hydrolyze proteins and can accelerate the fish fermentation period. Myeolchi-jeotgal is also a traditional Korean food obtained by fermentation of anchovy (a type of fish) under 20 % salt. It is considered as a useful food containing biogenic amines resulted from microflora (Mah et al. 2002, 2003). *Paenibacillus tyraminigenes* is a novel species isolated from Myeolchi-jeotgal that was able to produce tyramine. Tyramine is synthesized from tyrosine and can be used as a sympathomimetic agent in medicine. However, high concentration of this compound proved to be toxic for human (Ten Brink et al. 1990). Tyramine also has served as an antimicrobial agent, antioxidant, inhibitor of neurotoxins and bacteriocine-like agent (Yen and Hsieh 1997; Giulian 2000). Mah et al. (2008) has observed the optimum production of tyramine in medium containing 5,000 μg/ml tyrosine.

11.2.7 Role in Production of Therapeutic Compounds

Moderately halophilic bacteria are also used for the production of bioactive compounds. In recent decades marine microorganisms have been known as an important and less explored source for the production of therapeutics such as antibacterial, antiviral and anti-tumor agents. Among exhaustible source of marine bacteria, cyanobacteria have attracted much attention as a potent group in production of pharmaceuticals (Singh et al. 2011). Largazole is an anti-proliferative agent isolated from a cyanobacterium, *Symploca* sp. (Taori et al. 2008). The compound exhibit histone deacetylase inhibitor activity (HDAC) that can be introduced as apro-drug (Ying et al. 2008; Bowers et al. 2008). Apratoxins are other important bioactive compound derived from cyanobacteria with anticancer activity. Apratoxin A was also isolated from *Lyngbya bouillonii* showed activity against 60 cancer cell lines (Shoemaker 2006). *Pseudomonas* has been known as a potential candidate for the production of medically important products with more than 800 bioactive compounds isolated from its strains (Berdy 2005). Therefore, halophilic members of this genus may be a promising sources for drug discovery. The bioactive compounds derived marine bacteria and their biological activities have been listed in Table 11.3.

11.3 Conclusion

Halophilic bacteria possess great characteristics which introduce them a considerable group in industrial processes. Although, halophiles can posses interesting potentials in biotechnology, few studies have been performed to discover their potentials in industry. Using these microorganisms reduces the risk of contamination during the production process. Discovery of enzymatic abilities of halophilic microorganisms can accelerate the industrial processes and meet the increasing demands for biocatalysts. Compared with mesophilic enzymes, halophilic counterparts can withstand industrial reaction conditions far better, that lead to more yields and economic advantages in industries. In 2012, industrial enzymes consisted a global market of $4.5 billion and this value increased to $4.8 billion in 2013. According to the BCC research, the enzymes market is expected to reach $7.1 billion by 2018 (BCC research 2014). In respect to the growing demands for enzymes, identification of halophilic enzymes as adaptable agents against industrially harsh conditions seems to be an alternative approach. In addition to halophilic enzymes, stabilizing agents derived from halophiles have attracted remarkable attention in several aspects of biotechnology. Production of ectoine from *Halomonas elongata* has been scaled up by Bitop GmbH. These compounds can be used as efficient stabilizers of biomolecules such as enzymes, proteins, nucleic acids and membrane. Applying genetic tools in manipulation of halophilic bacteria can be considered as a promising approach in order to take advantage of their unique molecules in other production hosts. As they grow easily under simple conditions due to their low

Table 11.3 Biologically active compounds derived from marine bacteria

Compound	Strain	Reported biological activity	Structure	References
MC21-A	*Pseudoalteromonasphenolica*	Antibacterial	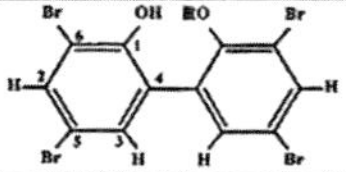	Isnansetyo and Kamei (2003)
- 4hydroxybenzaldehyde	*Pseudomonas* sp.	Antibacterial	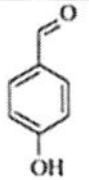	Wratten et al. (1977)
- 2-*n*-heptyl-4-quinolinol,			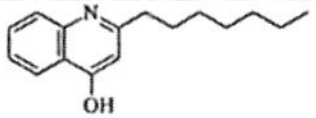	
- 2-*n*-pentyl-4-quinolinol			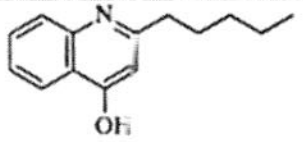	
Moiramides A, B, and C	*Pseudomonas fluorescens*	Antibacterial	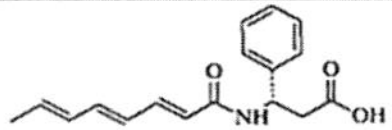 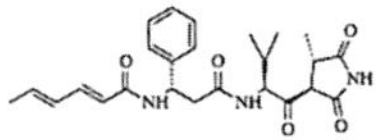 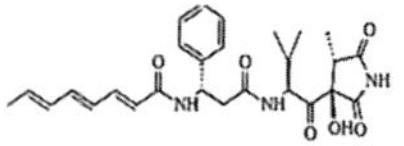	Needham et al. (1994)
2-undecyl-4-quinolone,	*Pseudomonas* sp.	Anti-*Plasmodium falsifarum*	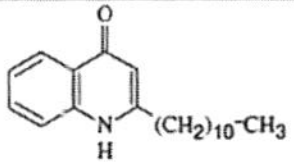	Bultel-Poncé et al. (1999)
2-undecen-1′-yl-4-quinolone,		Antibacterial	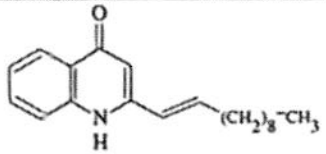	
2-nonyl-4-quinolone,		Antiviral	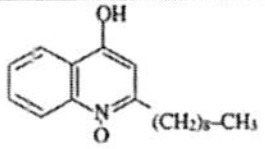	
2-nonyl-4-hydroxyquinoline *N*-oxide				

(continued)

Table 11.3 (continued)

Compound	Strain	Reported biological activity	Structure	References
2,4-diacetylphloroglucinol (DAPG)	*Pseudomonas* sp.	Active against *S. aureus* MRSA and VRSA	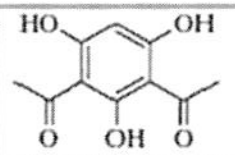	Isnansetyo et al. (2003)
Zafrin (4b-methyl-5,6,7,8 tetrahydro-1 (4b-H)-phenanthrenone)	*Pseudomonas stutzeri*	Active against *S. aureus* and *S. typhi*	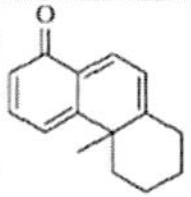	Uzair et al. (2006)
Bushrin (7-(3-furyl)-3,7-dimethyl-7,8-dihydro-1-naphthalenol)	*Pseudomonas stutzeri*	Antibacterial Active against *C. albicans*	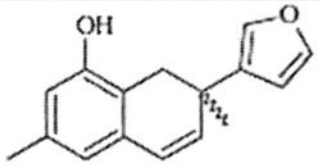	Ahmad et al. (2006)
Miuraenamides	*Paraliomyxa miuraensis*	Antifungal activity	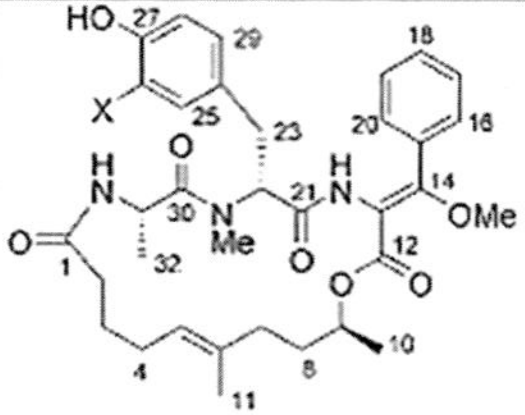	Ojika et al. (2008)
Bromo-alterochromides A and B	*Pseudoalteromonas maricoloris*	Antibacterial	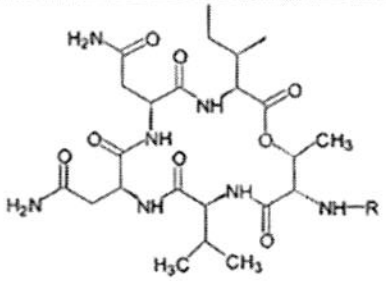	Sobolevskaya et al. (2005)
Thiomarinols A-G	*Alteromonas rava*	Antibacterial	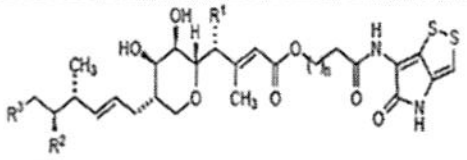	Shiozawa et al. (1993)

Bogorol A	*Bacillus* sp.	Antibacterial	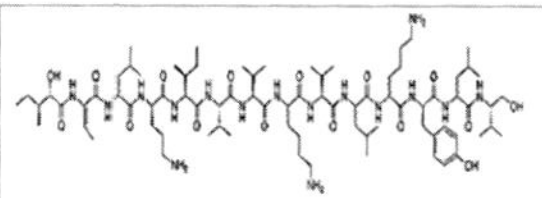	Barsby et al. (2001)
Loloatins A to D	*Bacillus* sp.	Active against MRSA*S. aureus* and vancomycin-resistant enterococci	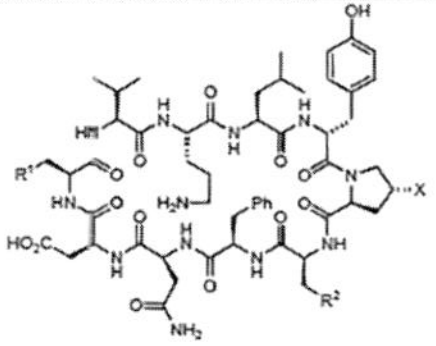	Gerard et al. (1999)
3-phenylpropionic acid	*Bacillus licheniformis*	Antibacterial	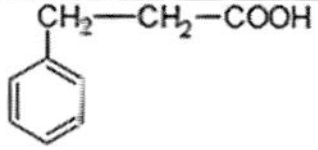	Devi et al. (2010)
4,4′-oxybis[3-phenylpropionic acid]			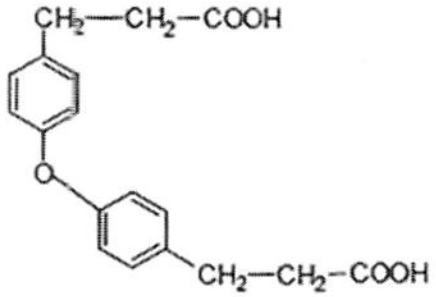	
Largazole	*Symploca* sp.	Histone deacetylase (HDAC) inhibitor	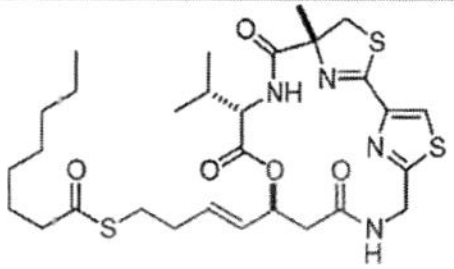	Taori et al. (2008)
Apratoxins A (3) and D	*Lyngbya* spp.	Antiproliferative agents	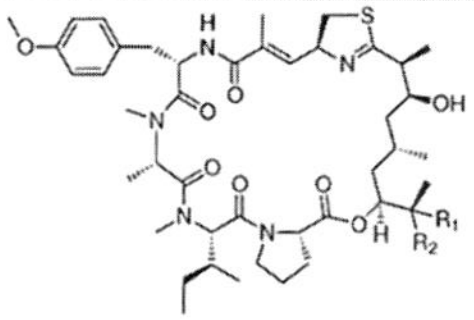	Luesch et al. (2002)

(continued)

Table 11.3 (continued)

Compound	Strain	Reported biological activity	Structure	References
Coibamide A	*Leptolyngbya* sp.	Anticancer	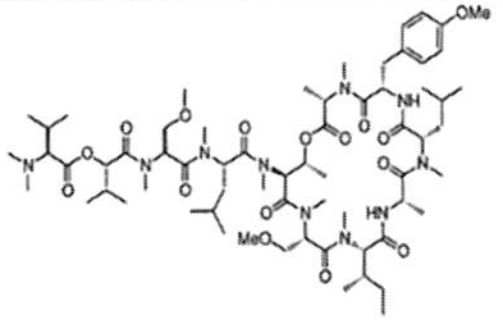	Medina et al. (2008)
CuracinA	*Lyngbya majuscula.*	Antimitotic	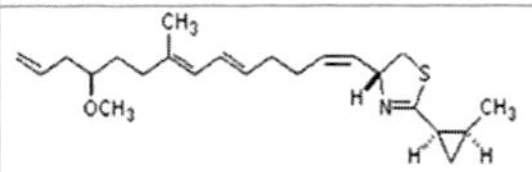	Gerwick et al. (1994)
Cryptophycins	*Nostoc* sp.	Tubulin polymerization	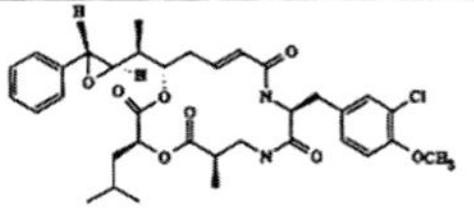	Trimurtulu et al. (1994)
Apratoxin A	*Lyngbya majuscula*	Anticancer	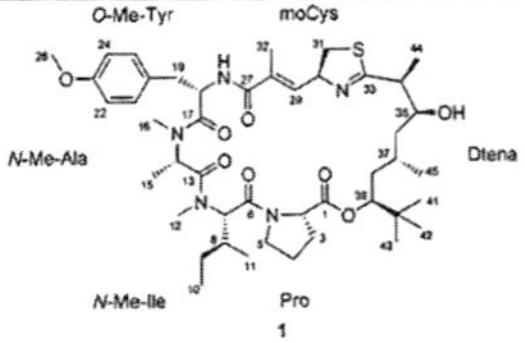	Luesch et al. (2001)
Borophycin	*Nostoc linckia*	Cytotoxicity against human epidermoid carcinoma (LoVo) and human colorectal adenocarcinoma activity	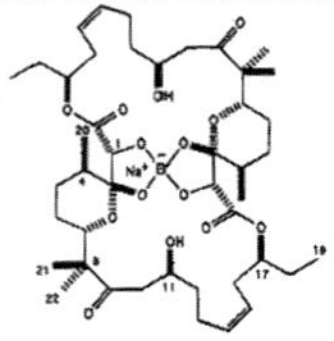	Hemscheidt et al. (1994)

nutritional requirements, they can be assumed as cell factories to produce the recombinant proteins. Despite all advantages of halophilic bacteria, there are few commercially developed examples so far. Implementation of new approaches such as genomics can be considered as an even more powerful tool in discovering the potentials and applications of halophiles.

References

Abdeljabbar H, Cayol JL, Hania WB, Boudabous A, Sadfi N, Fardeau ML (2013) *Halanaerobium sehlinense* sp. nov., an extremely halophilic, fermentative, strictly anaerobic bacterium from sediments of the hypersaline lake Sehline Sebkha. Int J Syst Evol Microbiol 63(6):2069–2074

Ahmad VU, Ahmed N, Kousar F, Uzair B (2006) Antibiotic bushrin. Google Patents

Ahn J, Park JW, McConnell JA, Ahn YB, Häggblom MM (2011) Kangiella spongicola sp. nov., a halophilic marine bacterium isolated from the sponge Chondrilla nucula. Int J Syst Evol Microbiol 61(4):961–964

Ahonkhai I, Kamekura M, Kushner DJ (1989) Effects of salts on the aspartate transcarbamylase of a halophilic eubacterium, *Vibrio costicola*. Biochem Cell Biol 67(9):666–669

Akolkar A, Durai D, Desai A (2010) Halobacterium sp. SP1 (1) as a starter culture for accelerating fish sauce fermentation. J Appl Microbiol 109(1):44–53

Alva VA, Peyton BM (2003) Phenol and catechol biodegradation by the haloalkaliphile Halomonas campisalis: influence of pH and salinity. Environ Sci Technol 37(19):4397–4402

Amjres H, Béjar V, Quesada E, Abrini J, Llamas I (2011) *Halomonas rifensis* sp. nov., an exopolysaccharide-producing, halophilic bacterium isolated from a solar saltern. Int J Syst Evol Microbiol 61(11):2600–2605

Amoozegar M, Malekzadeh F, Malik KA (2003) Production of amylase by newly isolated moderate halophile, *Halobacillus* sp. strain MA-2. J Microbiol Meth 52(3):353–359

Amoozegar MA, Salehghamari E, Khajeh K, Kabiri M, Naddaf S (2008) Production of an extracellular thermohalophilic lipase from a moderately halophilic bacterium, *Salinivibrio* sp. strain SA-2. J Basic Microbiol 48(3):160–167

Amoozegar MA, Bagheri M, Didari M, Fazeli SAS, Schumann P, Sánchez-Porro C, Ventosa A (2013a) *Saliterribacillus persicus* gen. nov., sp. nov., a moderately halophilic bacterium isolated from a hypersaline lake. Int J Syst Evol Microbiol 63(1):345–351

Amoozegar MA, Didari M, Bagheri M, Fazeli SAS, Schumann P, Spröer C, Sánchez-Porro C, Ventosa A (2013b) *Bacillus salsus* sp. nov., a halophilic bacterium from a hypersaline lake. Int J Syst Evol Microbiol 63(9):3324–3329

Amoozegar MA, Makhdoumi-Kakhki A, Ramezani M, Nikou MM, Fazeli SAS, Schumann P, Ventosa A (2013c) *Limimonas halophila* gen. nov., sp. nov., an extremely halophilic bacterium in the family Rhodospirillaceae. Int J Syst Evol Microbiol 63(4):1562–1567

Amoozegar MA, Bagheri M, Makhdoumi-Kakhki A, Didari M, Schumann P, Nikou MM, Sánchez-Porro C, Ventosa A (2014a) *Aliicoccus persicus* gen. nov., sp. nov., a halophilic member of the Firmicutes isolated from a hypersaline lake. Int J Syst Evol Microbiol. doi:ijs.0.058545-058540

Amoozegar MA, Bagheri M, Makhdoumi-Kakhki A, Didari M, Schumann P, Spröer C, Sánchez-Porro C, Ventosa A (2014b) *Oceanobacillus limi* sp. nov., a moderately halophilic bacterium from a salt lake. Int J Syst Evol Microbiol 64(4):1284–1289

Amoozegar MA, Shahinpei A, Sepahy AA, Makhdoumi-Kakhki A, Seyedmahdi SS, Schumann P, Ventosa A (2014c) *Pseudomonas salegens* sp. nov., a halophilic member of the genus Pseudomonas isolated from a wetland. Int J Syst Evol Microbiol. doi:ijs.0.062935-062930

Anders H, Dunfield PF, Lagutin K, Houghton KM, Power JF, MacKenzie AD, Vyssotski M, Ryan JL, Hanssen EG, Moreau JW (2014) *Thermoflavifilum aggregans* gen. nov., sp. nov., a thermophilic and slightly halophilic filamentous bacterium from the phylum Bacteroidetes. Int J Syst Evol Microbiol 64(4):1264–1270

Antón J, Oren A, Benlloch S, Rodríguez-Valera F, Amann R, Rosselló-Mora R (2002) *Salinibacter ruber* gen. nov., sp. nov., a novel, extremely halophilic member of the bacteria from saltern crystallizer ponds. Int J Syst Evol Microbiol 52(2):485–491

Arahal DR, Castillo AM, Ludwig W, Schleifer KH, Ventosa A (2002) Proposal of *Cobetia marina* gen. nov., comb. nov., within the family Halomonadaceae, to include the species *Halomonas marina*. Syst Appl Microbiol 25(2):207–211

Arakawa T, Timasheff S (1985) The stabilization of proteins by osmolytes. Biophys J 47(3):411–414

Bagheri M, Didari M, Amoozegar MA, Schumann P, Sanchez-Porro C, Mehrshad M, Ventosa A (2012) *Bacillus iranensis* sp. nov., a moderate halophile from a hypersaline lake. Int J Syst Evol Microbiol 62(4):811–816

Bagheri M, Amoozegar MA, Schumann P, Didari M, Mehrshad M, Spröer C, Sánchez-Porro C, Ventosa A (2013) *Ornithinibacillus halophilus* sp. nov., a moderately halophilic, Gram-stain-positive, endospore-forming bacterium from a hypersaline lake. Int J Syst Evol Microbiol 63(3):844–848

Barsby T, Kelly MT, Gagné SM, Andersen RJ (2001) Bogorol A produced in culture by a marine Bacillus sp. reveals a novel template for cationic peptide antibiotics. Org Lett 3(3):437–440

Bastawade KB, Gokhale DV, Joshi RR, Kalkote UR, Ravindranathan T, Sudge SS (2000) Halophilic *Pseudomonas* strain having accession No. NCIM 5109 (ATCC 55940) and a process for preparing D (-) N-carbamoylphenylglycine using said strain. Google Patents

BCC research (2014) http://www.bccresearch.com/market-research/biotechnology/enzymes-industrial-applications-markets-bio030g.html

Benachenhou N, Baldacci G (1991) The gene for a halophilic glutamate dehydrogenase sequence, transcription analysis and phylogenetic implications. Mol Gen Genet 230(3):345–352

Berdy J (2005) Bioactive microbial metabolites. J Antibiot 58(1):1–26

Biswas A, Patra A, Paul A (2009) Production of poly-3-hydroxyalkanoic acids by a moderately halophilic bacterium, *Halomonas marina* HMA 103 isolated from solar saltern of Orissa, India. Acta Microbiol Immunol Hung 56(2):125–143

Blum JS, Kulp TR, Han S, Lanoil B, Saltikov CW, Stolz JF, Miller LG, Oremland RS (2012) *Desulfohalophilus alkaliarsenatis* gen. nov., sp. nov., an extremely halophilic sulfate-and arsenate-respiring bacterium from Searles Lake, California. Extremophiles 16(5):727–742

Borges N, Ramos A, Raven ND, Sharp RJ, Santos H (2002) Comparative study of the thermostabilizing properties of mannosylglycerate and other compatible solutes on model enzymes. Extremophiles 6(3):209–216

Borsodi AK, Márialigeti K, Szabó G, Palatinszky M, Pollák B, Kéki Z, Kovács AL, Schumann P, Tóth EM (2008) *Bacillus aurantiacus* sp. nov., an alkaliphilic and moderately halophilic bacterium isolated from Hungarian soda lakes. Int J Syst Evol Microbiol 58(4):845–851

Bouchotroch S, Quesada E, Izquierdo I, Rodríguez M, Béjar V (2000) Bacterial exopolysaccharides produced by newly discovered bacteria belonging to the genus *Halomonas*, isolated from hypersaline habitats in Morocco. J Ind Microbiol Biotechnol 24(6):374–378

Bowers A, West N, Taunton J, Schreiber SL, Bradner JE, Williams RM (2008) Total synthesis and biological mode of action of largazole: a potent class I histone deacetylase inhibitor. J Am Chem Soc 130(33):11219–11222

Bultel-Poncé V, Berge JP, Debitus C, Nicolas JL, Guyot M (1999) Metabolites from the sponge-associated bacterium *Pseudomonas* species. Mar Biotechnol 1(4):384–390

Burnett WE (1974) The effect of salinity variations on the activated sludge process. Water Sew Works 121:37–38

Bylund J, Dyer J, Thompson T, Martin E (1991) Alanine dehydrogenase activity of the extensively halotolerant eubacterium *Halomonas elongata*. Microbios 66(266):45–53

Calvo C, Martínez-Checa F, Toledo F, Porcel J, Quesada E (2002) Characteristics of bioemulsifiers synthesised in crude oil media by *Halomonas eurihalina* and their effectiveness in the isolation of bacteria able to grow in the presence of hydrocarbons. Appl Microbiol Biotechnol 60(3):347–351

Cameotra SS, Makkar RS, Kaur J, Mehta S (2010) Synthesis of biosurfactants and their advantages to microorganisms and mankind. In: Sen R (ed) Biosurfactants. Springer, New York, pp 261–280

Cao SJ, Qu JH, Yang JS, Sun Q, Yuan HL (2008) *Halolactibacillus alkaliphilus* sp. nov., a moderately alkaliphilic and halophilic bacterium isolated from a soda lake in Inner Mongolia, China, and emended description of the genus *Halolactibacillus*. Int J Syst Evol Microbiol 58(9):2169–2173

Carrasco I, Márquez M, Yanfen X, Ma Y, Cowan D, Jones B, Grant W, Ventosa A (2006) *Gracilibacillus orientalis* sp. nov., a novel moderately halophilic bacterium isolated from a salt lake in Inner Mongolia, China. Int J Syst Evol Microbiol 56(3):599–604

Carrasco I, Márquez M, Xue Y, Ma Y, Cowan D, Jones B, Grant W, Ventosa A (2007a) *Salsuginibacillus kocurii* gen. nov., sp. nov., a moderately halophilic bacterium from soda-lake sediment. Int J Syst Evol Microbiol 57(10):2381–2386

Carrasco I, Márquez M, Xue Y, Ma Y, Cowan D, Jones B, Grant WD, Ventosa A (2007b) *Bacillus chagannorensis* sp. nov., a moderate halophile from a soda lake in Inner Mongolia, China. Int J Syst Evol Microbiol 57(9):2084–2088

Carrasco I, Márquez M, Xue Y, Ma Y, Cowan D, Jones B, Grant W, Ventosa A (2008) *Sediminibacillus halophilus* gen. nov., sp. nov., a moderately halophilic, Gram-positive bacterium from a hypersaline lake. Int J Syst Evol Microbiol 58(8):1961 1967

Cerritos R, Vinuesa P, Eguiarte LE, Herrera-Estrella L, Alcaraz-Peraza LD, Arvizu-Gómez JL, Olmedo G, Ramirez E, Siefert JL, Souza V (2008) *Bacillus coahuilensis* sp. nov., a moderately halophilic species from a desiccation lagoon in the Cuatro Cienegas Valley in Coahuila, Mexico. Int J Syst Evol Microbiol 58(4):919–923

Chen YG, Cui XL, Pukall R, Li HM, Yang YL, Xu LH, Wen ML, Peng Q, Jiang CL (2007) *Salinicoccus kunmingensis* sp. nov., a moderately halophilic bacterium isolated from a salt mine in Yunnan, south-west China. Int J Syst Evol Microbiol 57(10):2327–2332

Chen H, Xiao X, Wang J, Wu L, Zheng Z, Yu Z (2008a) Antagonistic effects of volatiles generated by *Bacillus subtilis* on spore germination and hyphal growth of the plant pathogen, Botrytis cinerea. Biotechnol Lett 30(5):919–923

Chen YG, Cui XL, Fritze D, Chai LH, Schumann P, Wen ML, Wang YX, Xu LH, Jiang CL (2008b) *Virgibacillus kekensis* sp. nov., a moderately halophilic bacterium isolated from a salt lake in China. Int J Syst Evol Microbiol 58(3):647–653

Chen YG, Cui XL, Zhang YQ, Li WJ, Wang YX, Xu LH, Peng Q, Wen ML, Jiang CL (2008c) *Gracilibacillus halophilus* sp. nov., a moderately halophilic bacterium isolated from saline soil. Int J Syst Evol Microbiol 58(10):2403–2408

Chen YG, Cui XL, Wang YX, Zhang YQ, Tang SK, Li WJ, Liu ZX, Wen ML, Peng Q (2009a) *Virgibacillus sediminis* sp. nov., a moderately halophilic bacterium isolated from a salt lake in China. Int J Syst Evol Microbiol 59(8):2058–2063

Chen YG, Zhang YQ, Huang HY, Klenk HP, Tang SK, Huang K, Chen QH, Cui XL, Li WJ (2009b) *Halomonas zhanjiangensis* sp. nov., a halophilic bacterium isolated from a sea urchin. Int J Syst Evol Microbiol 59(11):2888–2893

Chen YG, Zhang YQ, Wang YX, Liu ZX, Klenk HP, Xiao HD, Tang SK, Cui XL, Li WJ (2009c) *Bacillus neizhouensis* sp. nov., a halophilic marine bacterium isolated from a sea anemone. Int J Syst Evol Microbiol 59(12):3035–3039

Chen YG, Zhang YQ, Xiao HD, Liu ZX, Yi LB, Shi JX, Zhi XY, Cui XL, Li WJ (2009d) *Pontibacillus halophilus* sp. nov., a moderately halophilic bacterium isolated from a sea urchin. Int J Syst Evol Microbiol 59(7):1635–1639

Chen YG, Hao DF, Chen QH, Zhang YQ, Liu JB, He JW, Tang SK, Li WJ (2011) *Bacillus hunanensis* sp. nov., a slightly halophilic bacterium isolated from non-saline forest soil. Antonie Van Leeuwenhoek 99(3):481–488

Cheng JP, Zhang LY, Wang WH, Yang YC, Zheng M, Ju SW (2004) Screening of flocculant-producing microorganisms and flocculating activity. J Environ Sci 16(6):894–897

Collins T, Gerday C, Feller G (2005) Xylanases, xylanase families and extremophilic xylanases. FEMS Microbiol Rev 29(1):3–23

Coronado MJ, Vargas C, Mellado E, Tegos G, Drainas C, Nieto J, Ventosa A (2000) The α-amylase gene amyH of the moderate halophile *Halomonas meridiana*: cloning and molecular characterization. Microbiology 146(4):861–868

Cosa S, Mabinya LV, Olaniran AO, Okoh AI (2012) Production and characterization of bioflocculant produced by *Halobacillus* sp. Mvuyo isolated from bottom sediment of Algoa Bay. Environ Technol 33(9):967–973

Couto SR (2009) Dye removal by immobilized fungi. Biotechnol Adv 27:227–235

Coyne VE, Stosz SK, Weiner RM (1995) Agarase enzyme system from alteromonas strain 2–40. Google Patents

Cui HL, Gao X, Yang X, Xu XW (2010) *Halorussus rarus* gen. nov., sp. nov., a new member of the family Halobacteriaceae isolated from a marine solar saltern. Extremophiles 14(6):493–499

Cui HL, Yang X, Gao X, Xu XW (2011) *Halobellus clavatus* gen. nov., sp. nov. and Halorientalis regularis gen. nov., sp. nov., two new members of the family Halobacteriaceae. Int J Syst Evol Microbiol 61(11):2682–2689

Cui HL, Yang X, Zhou YG, Liu HC, Zhou PJ, Dyall-Smith ML (2012) *Halobellus limi* sp. nov. and *Halobellus salinus* sp. nov., isolated from two marine solar salterns. Int J Syst Evol Microbiol 62(6):1307–1313

Da Costa M, Santos H, Galinski E (1998) An overview of the role and diversity of compatible solutes in bacteria and archaea. In: Antranikian G (ed) Biotechnology of extremophiles, 61. Springer, New York, pp 117–153

de la Haba RR, del Carmen Marquez M, Papke RT, Ventosa A (2011) Multilocus sequence analysis (MLSA) of the family Halomonadaceae. Int J Syst Evol Microbiol. doi:ijs.0.032938-032930

de Lourdes Moreno M, García MT, Ventosa A, Mellado E (2009) Characterization of *Salicola* sp. IC10, a lipase-and protease-producing extreme halophile. FEMS Microbiol Ecol 68(1):59–71

Deubel A, Merbach W (2005) Influence of microorganisms on phosphorus bioavailability in soils. In: Varma A, Buscot F (eds) Microorganisms in soils: roles in genesis and functions, vol 3. Springer, New York, pp 177–191

Devi P, Wahidullah S, Rodrigues C, Souza LD (2010) The sponge-associated bacterium *Bacillus licheniformis* SAB1: a source of antimicrobial compounds. Mar Drugs 8(4):1203–1212

Díaz MP, Grigson SJ, Peppiatt CJ, Burgess JG (2000) Isolation and characterization of novel hydrocarbon-degrading euryhaline consortia from crude oil and mangrove sediments. Mar Biotechnol 2(6):522–532

Díaz MP, Boyd KG, Grigson SJ, Burgess JG (2002) Biodegradation of crude oil across a wide range of salinities by an extremely halotolerant bacterial consortium MPD-M, immobilized onto polypropylene fibers. Biotechnol Bioeng 79(2):145–153

Didari M, Amoozegar MA, Bagheri M, Schumann P, Spröer C, Sánchez-Porro C, Ventosa A (2012) *Alteribacillus bidgolensis* gen. nov., sp. nov., a moderately halophilic bacterium from a hypersaline lake, and reclassification of Bacillus persepolensis as *Alteribacillus persepolensis* comb. nov. Int J Syst Evol Microbiol 62(11):2691–2697

Didari M, Amoozegar MA, Bagheri M, Mehrshad M, Schumann P, Spröer C, Sánchez-Porro C, Ventosa A (2013) *Bacillus persicus* sp. nov., a halophilic bacterium from a hypersaline lake. Int J Syst Evol Microbiol 63(4):1229–1234

Doble M, Kumar A (2005) Biotreatment of industrial effluents. Elsevier Butterworth–Heinemann, Oxford, pp 111–122

Duo-Chuan L (2006) Review of fungal chitinases. Mycopathologia 161(6):345–360

Echigo A, Fukushima T, Mizuki T, Kamekura M, Usami R (2007) *Halalkalibacillus halophilus* gen. nov., sp. nov., a novel moderately halophilic and alkaliphilic bacterium isolated from a non-saline soil sample in Japan. Int J Syst Evol Microbiol 57(5):1081–1085

Echigo A, Minegishi H, Mizuki T, Kamekura M, Usami R (2010) *Geomicrobium halophilum* gen. nov., sp. nov., a moderately halophilic and alkaliphilic bacterium isolated from soil. Int J Syst Evol Microbiol 60(4):990–995

Echigo A, Minegishi H, Shimane Y, Kamekura M, Usami R (2012) *Natribacillus halophilus* gen. nov., sp. nov., a moderately halophilic and alkalitolerant bacterium isolated from soil. Int J Syst Evol Microbiol 62(2):289–294

Echigo A, Minegishi H, Shimane Y, Kamekura M, Itoh T, Usami R (2013) *Halomicroarcula pellucida* gen. nov., sp. nov., a non-pigmented, transparent-colony-forming, halophilic archaeon isolated from solar salt. Int J Syst Evol Microbiol 63(10):3556–3562

Essghaier B, Fardeau ML, Cayol JL, Hajlaoui MR, Boudabous A, Jijakli H, Sadfi-Zouaoui N (2008) Biological control of grey mould in strawberry fruits by halophilic bacteria. J Appl Microbiol 106:833–846

Essghaier B, Hedi A, Bejji M, Jijakli H, Boudabous A, Sadfi-Zouaoui N (2012) Characterization of a novel chitinase from a moderately halophilic bacterium, *Virgibacillus marismortui* strain M3-23. Ann Microbiol 62(2):835–841

Gales G, Chehider N, Joulian C, Battaglia-Brunet F, Cayol JL, Postec A, Borgomano J, Neria-Gonzalez I, Lomans B, Ollivier B (2011) Characterization of *Halanaerocella petrolearia* gen. nov., sp. nov., a new anaerobic moderately halophilic fermentative bacterium isolated from a deep subsurface hypersaline oil reservoir. Extremophiles 15(5):565–571

Gao J, Bao HY, Xin MX, Liu YX, Li Q, Zhang YF (2006) Characterization of a bioflocculant from a newly isolated *Vagococcus* sp. W31. J Zheijang Univ Sci B 7(3):186–192

Gao Z, Ruan L, Chen X, Zhang Y, Xu X (2010) A novel salt-tolerant endo-β-1, 4-glucanase Cel5A in Vibrio sp. G21 isolated from mangrove soil. Appl Microbiol Biotechnol 87(4):1373–1382

Gao W, Cui Z, Li Q, Xu G, Jia X, Zheng L (2013) *Marinobacter nanhaiticus* sp. nov., polycyclic aromatic hydrocarbon-degrading bacterium isolated from the sediment of the South China Sea. Antonie Van Leeuwenhoek 103(3):485–491

García MT, Mellado E, Ostos JC, Ventosa A (2004) *Halomonas organivorans* sp. nov., a moderate halophile able to degrade aromatic compounds. Int J Syst Evol Microbiol 54(5):1723–1728

Gerard JM, Haden P, Kelly MT, Andersen RJ (1999) Loloatins AD, cyclic decapeptide antibiotics produced in culture by a tropical marine bacterium. J Nat Prod 62(1):80–85

Gerwick WH, Proteau PJ, Nagle DG, Hamel E, Blokhin A, Slate DL (1994) Structure of curacin A, a novel antimitotic, antiproliferative and brine shrimp toxic natural product from the marine cyanobacterium *Lyngbya majuscula*. J Org Chem 59(6):1243–1245

Giridhar PV, Chandra T (2010) Production of novel halo-alkali-thermo-stable xylanase by a newly isolated moderately halophilic and alkali-tolerant *Gracilibacillus* sp. TSCPVG. Process Biochem 45(10):1730–1737

Giulian DJ (2000) Tyramine compounds and their neuronal effects. Google Patents

González-Domenech CM, Béjar V, Martínez-Checa F, Quesada E (2008a) *Halomonas nitroreducens* sp. nov., a novel nitrate-and nitrite-reducing species. Int J Syst Evol Microbiol 58(4):872–876

González-Domenech CM, Martínez-Checa F, Quesada E, Béjar V (2008b) *Halomonas cerina* sp. nov., a moderately halophilic, denitrifying, exopolysaccharide-producing bacterium. Int J Syst Evol Microbiol 58(4):803–809

González-Domenech CM, Martínez-Checa F, Quesada E, Béjar V (2009) *Halomonas fontilapidosi* sp. nov., a moderately halophilic, denitrifying bacterium. Int J Syst Evol Microbiol 59(6):1290–1296

Graf R, Anzali S, Bünger J, Pflücker F, Driller H (2008) The multifunctional role of ectoine as a natural cell protectant. Clin Dermatol 26:326–333

Guo B, Gu J, Ye YG, Tang YQ, Kida K, Wu XL (2007) *Marinobacter segnicrescens* sp. nov., a moderate halophile isolated from benthic sediment of the South China Sea. Int J Syst Evol Microbiol 57(9):1970–1974

Guo J, Zhou J, Wang D, Tian C, Wang P, Uddin MS (2008) A novel moderately halophilic bacterium for decolorizing azo dye under high salt condition. Biodegradation 19(1):15–19

Guzmán D, Quillaguamán J, Muñoz M, Hatti-Kaul R (2010) *Halomonas andesensis* sp. nov., a moderate halophile isolated from the saline lake Laguna Colorada in Bolivia. Int J Syst Evol Microbiol 60(4):749–753

Harada N, Zhao J, Kurihara H, Nakagata N, Okajima K (2010) Effects of topical application of α-D-Glucosylglycerol on dermal levels of insulin-like growth factor-I in mice and on facial skin elasticity in humans. Biosci Biotechnol Biochem 74(4):759–765

Hemscheidt T, Puglisi MP, Larsen LK, Patterson GM, Moore RE, Rios JL, Clardy J (1994) Structure and biosynthesis of borophycin, a new boeseken complex of boric acid from a marine strain of the blue-green alga Nostoc linckia. J Org Chem 59(12):3467–3471

Hinteregger C, Streichsbier F (1997) *Halomonas* sp., a moderately halophilic strain, for biotreatment of saline phenolic waste-water. Biotechnol Lett 19(11):1099–1102

Hiraga K, Nishikata Y, Namwong S, Tanasupawat S, Takada K, Oda K (2005) Purification and characterization of serine proteinase from a halophilic bacterium, *Filobacillus* sp. RF2-5. Biosci Biotechnol Biochem 69(1):38–44

Hirayama H, Fuse H, Abe M, Miyazaki M, Nakamura T, Nunoura T, Furushima Y, Yamamoto H, Takai K (2013) *Methylomarinum vadi* gen. nov., sp. nov., a methanotroph isolated from two distinct marine environments. Int J Syst Evol Microbiol 63(3):1073–1082

Hong SW, Kwon S-W, Kim SJ, Kim SY, Kim JJ, Lee JS, Oh MH, Kim AJ, Chung KS (2014) *Bacillus oryzaecorticis* sp. nov., a moderately halophilic bacterium isolated from the rice-husk. Int J Syst Evol Microbiol. doi:ijs.0.058768-058760

Hua NP, Kanekiyo A, Fujikura K, Yasuda H, Naganuma T (2007) *Halobacillus profundi* sp. nov. and *Halobacillus kuroshimensis* sp. nov., moderately halophilic bacteria isolated from a deep-sea methane cold seep. Int J Syst Evol Microbiol 57(6):1243–1249

Hua NP, Hamza-Chaffai A, Vreeland RH, Isoda H, Naganuma T (2008) *Virgibacillus salarius* sp. nov., a halophilic bacterium isolated from a Saharan salt lake. Int J Syst Evol Microbiol 58(10):2409–2414

Huang X, Shao Z, Hong Y, Lin L, Li C, Huang F, Wang H, Liu Z (2010) Cel8H, a novel endoglucanase from the halophilic bacterium *Halomonas* sp. S66-4: molecular cloning, heterogonous expression, and biochemical characterization. J Microbiol 48(3):318–324

Huang SP, Chang HY, Chen JS, Jean WD, Shieh WY (2012) *Aliidiomarina taiwanensis* gen. nov., sp. nov., isolated from shallow coastal water. Int J Syst Evol Microbiol 62(1):155–161

Huo YY, Xu L, Wang CS, Yang JY, You H, Oren A, Xu XW (2013) *Fabibacter pacificus* sp. nov., a moderately halophilic bacterium isolated from seawater. Int J Syst Evol Microbiol 63(10):3710–3714

Huo YY, Li ZY, You H, Wang CS, Post AF, Oren A, Xu XW (2014) *Oceanicola antarcticus* sp. nov. and *Oceanicola flagellatus* sp. nov., moderately halophilic bacteria isolated from seawater. Int J Syst Evol Microbiol. doi:ijs.0.062588-062580

Ishida H, Kato S, Nishimura M, Mizuno N, Fujimoto S, Mukai E, Kajikawa M, Yamada Y, Odaka H, Ikeda H (1998) Beneficial effect of long-term combined treatment with voglibose and pioglitazone on pancreatic islet function of genetically diabetic GK rats. Horm Metab Res 30(11):673–678

Ishikawa M, Nakajima K, Ishizaki S, Kodama K, Okamoto-Kainuma A, Koizumi Y, Yamamoto Y, Yamasato K (2011) *Alkalibacterium subtropicum* sp. nov., a slightly halophilic and alkaliphilic marine lactic acid bacterium isolated from decaying marine algae collected at a subtropical area of Japan. Int J Syst Evol Microbiol. doi:ijs.0.027953-027950

Ishikawa M, Yamasato K, Kodama K, Yasuda H, Matsuyama M, Okamoto-Kainuma A, Koizumi Y (2013) *Alkalibacterium gilvum* sp. nov., slightly halophilic and alkaliphilic lactic acid bacterium isolated from soft and semi-hard cheeses. Int J Syst Evol Microbiol 63(4):1471–1478

Isnansetyo A, Kamei Y (2003) MC21-A, a bactericidal antibiotic produced by a new marine bacterium, Pseudoalteromonas phenolica sp. nov. O-BC30T, against methicillin-resistant *Staphylococcus aureus*. Antimicrob Agents Chemother 47(2):480–488

Isnansetyo A, Cui L, Hiramatsu K, Kamei Y (2003) Antibacterial activity of 2, 4-diacetylphloroglucinol produced by *Pseudomonas* sp. AMSN isolated from a marine alga, against vancomycin-resistant *Staphylococcus aureus*. Int J Antimicrob Agents 22(5):545–547

Iturriaga G, Suárez R, Nova-Franco B (2009) Trehalose metabolism: from osmoprotection to signaling. Int J Mol Sci 10(9):3793–3810

Jakobsen TF, Kjeldsen KU, Ingvorsen K (2006) *Desulfohalobium utahense* sp. nov., a moderately halophilic, sulfate-reducing bacterium isolated from Great Salt Lake. Int J Syst Evol Microbiol 56(9):2063–2069

Jayachandra S, Kumar A, Merley D, Sulochana M (2012) Isolation and characterization of extreme halophilic bacterium Salinicoccus sp. JAS4 producing extracellular hydrolytic enzymes. Recent Res Sci Technol 4(4):46–49

Jeong SH, Lee JH, Jung JY, Lee SH, Park MS, Jeon CO (2013) *Halomonas cibimaris* sp. nov., isolated from jeotgal, a traditional Korean fermented seafood. Antonie Van Leeuwenhoek 103(3):503–512

Jiang F, Cao S-J, Li ZH, Fan H, Li HF, Liu WJ, Yuan HL (2011) *Salisediminibacterium halotolerans* gen. nov., sp. nov., a halophilic bacterium isolated from Xiarinaoer soda lake sediment in Inner Mongolia, China. Int J Syst Evol Microbiol. doi:ijs.0.034488-034480

Jung WJ, Kuk JH, Kim KY, Kim TH, Park RD (2005) Purification and characterization of chitinase from *Paenibacillus illinoisensis* KJA-424. J Microbiol Biotechnol 15(2):274–280

Jung JY, Kim JM, Jin HM, Kim SY, Park W, Jeon CO (2011) *Litorimonas taeanensis* gen. nov., sp. nov., isolated from a sandy beach. Int J Syst Evol Microbiol 61(7):1534–1538

Kaeppel EC, Gärdes A, Seebah S, Grossart HP, Ullrich MS (2012) *Marinobacter adhaerens* sp. nov., isolated from marine aggregates formed with the diatom *Thalassiosira weissflogii*. Int J Syst Evol Microbiol 62(1):124–128

Kamekura M (1986) Production and function of enzymes of eubacterial halophiles. FEMS Microbiol Lett 39(1):145–150

Kamekura M, Hamakawa T, Onishi H (1982) Application of halophilic nuclease H of *Micrococcus varians* subsp. *halophilus* to commercial production of flavoring agent 5′-GMP. Appl Environ Microbiol 44(4):994–995

Kanekar PP, Kanekar SP, Kelkar AS, Dhakephalkarv PK (2012) Halophiles – taxonomy, diversity, physiology and applications. In: Satyanarayana T, Johri BN, Parkash A (eds) Microorganisms in environmental management: microbes and environment. Springer, Dordrecht, pp 1–34

Kapdan IK, Erten B (2007) Anaerobic treatment of saline wastewater by *Halanaerobium lacusrosei*. Process Biochem 42:449–453

Kargi F (2002) Enhanced biological treatment of saline wastewater by using halophilic bacteria. Biotechnol Lett 24:1569–1572

Kargi F, Dinçer AR (1998) Saline wastewater treatment by halophile-supplemented activated sludge culture in an aerated rotating biodisc contactor. Enzym Microb Technol 22(6):427–433

Katayama T, Yoshioka H, Mochimaru H, Meng XY, Muramoto Y, Usami J, Ikeda H, Kamagata Y, Sakata S (2014) *Methanohalophilus levihalophilus* sp. nov., a slightly halophilic, methylotrophic methanogen isolated from natural gas-bearing deep aquifers, and an emended description of the genus *Methanohalophilus*. Int J Syst Evol Microbiol. doi:ijs.0.063677-063670

Kharroub K, Aguilera M, Jiménez-Pranteda ML, González-Paredes A, Ramos-Cormenzana A, Monteoliva-Sánchez M (2011) *Marinobacter oulmenensis* sp. nov., a moderately halophilic bacterium isolated from brine of a salt concentrator. Int J Syst Evol Microbiol 61(9):2210–2214

Khopade A, Biao R, Liu X, Mahadik K, Zhang L, Kokare C (2012) Production and stability studies of the biosurfactant isolated from marine *Nocardiopsis* sp. B4. Desalination 285:198–204

Kim KK, Jin L, Yang HC, Lee S-T (2007) *Halomonas gomseomensis* sp. nov., *Halomonas janggokensis* sp. nov., *Halomonas salaria* sp. nov. and *Halomonas denitrificans* sp. nov., moderately halophilic bacteria isolated from saline water. Int J Syst Evol Microbiol 57(4):675–681

Kim JM, Jung JY, Chae HB, Park W, Jeon CO (2010a) *Hwanghaeicola aestuarii* gen. nov., sp. nov., a moderately halophilic bacterium isolated from a tidal flat of the Yellow Sea. Int J Syst Evol Microbiol 60(12):2877–2881

Kim JM, Lee SH, Jung JY, Jeon CO (2010b) *Marinobacterium lutimaris* sp. nov., isolated from a tidal flat. Int J Syst Evol Microbiol 60(8):1828–1831

Kim MS, Roh SW, Bae JW (2010c) *Cobetia crustatorum* sp. nov., a novel slightly halophilic bacterium isolated from traditional fermented seafood in Korea. Int J Syst Evol Microbiol 60(3):620–626

Kjeldsen KU, Jakobsen TF, Glastrup J, Ingvorsen K (2010) *Desulfosalsimonas propionicica* gen. nov., sp. nov., a halophilic, sulfate-reducing member of the family Desulfobacteraceae isolated from a salt-lake sediment. Int J Syst Evol Microbiol 60(5):1060–1065

Kleinsteuber S, Riis V, Fetzer I, Harms H, Müller S (2006) Population dynamics within a microbial consortium during growth on diesel fuel in saline environments. Appl Environ Microbiol 72(5):3531–3542

Kobayashi T, Kamekura M, Kanlayakrit W, Onishi H (1986) Production, purification, and characterization of an amylase from the moderate halophile, *Micrococcus varians* subspecies *halophilus*. Microbios 46(188–189):165–177

Köcher S, Tausendschön M, Thompson M, Saum SH, Müller V (2011) Proline metabolism in the moderately halophilic bacterium *Halobacillus halophilus*: differential regulation of isogenes in proline utilization. Environ Microbiol Rep 3(4):443–448

Kubo M, Morakami H (1999) Development of a high-salt concentration wastewaters treatment system. Syokuhin Kikai Souchi 36:94–100

Kubo M, Hiroe J, Murakami M, Fukami H, Tachiki T (2001) Treatment of hypersaline-containing wastewater with salt-tolerant microorganisms. J Biosci Bioeng 91(2):222–224

Kuhad RC, Sood N, Tripathi KK, Singh A, Ward OP (2004) Developments in microbial methods for the treatment of dye effluents. Adv Appl Microbiol 56:185–213

Kumar S, Karan R, Kapoor S, Singh S, Khare S (2012) Screening and isolation of halophilic bacteria producing industrially important enzymes. Braz J Microbiol 43(4):15951603

Kurane R, Nohata Y (1994) A new water-absorbing polysaccharide from *Alcaligenes latus*. Biosci Biotechnol Biochem 58(2):235–238

Kushner D (1993) Growth and nutrition of halophilic bacteria. In: Russell HV, Hechstein LI (eds) The biology of halophilic bacteria, vol 1. CRC Press, Boca Raton, pp 87–103

Le Borgne S, Paniagua D, Vazquez-Duhalt R (2008) Biodegradation of organic pollutants by halophilic bacteria and archaea. J Mol Microbiol Biotechnol 15(2–3):74–92

Lee SH, Shim JK, Kim JM, Choi HK, Jeon CO (2011) *Henriciella litoralis* sp. nov., isolated from a tidal flat, transfer of *Maribaculum marinum* Lai et al. 2009 to the genus *Henriciella* as *Henriciella aquimarina* nom. nov. and emended description of the genus *Henriciella*. Int J Syst Evol Microbiol 61(4):722–727

Lefebvre O, Moletta MR (2006) Treatment of organic pollution in industrial saline wastewater: a literature review. Water Res 40:3671–3682

Lentzen G, Schwarz T (2006) Extremolytes: natural compounds from extremophiles for versatile applications. Appl Microbiol Biotechnol 72(4):623–634

León MJ, Fernández AB, Ghai R, Sánchez-Porro C, Rodriguez-Valera F, Ventosa A (2014) From metagenomics to pure culture: isolation and characterization of the moderately Halophilic bacterium *Spiribacter salinus* gen. nov., sp. nov. Appl Environ Microbiol 80(13):3850–3857

Liebgott PP, Casalot L, Paillard S, Lorquin J, Labat M (2006) *Marinobacter vinifirmus* sp. nov., a moderately halophilic bacterium isolated from a wine-barrel-decalcification wastewater. Int J Syst Evol Microbiol 56(11):2511–2516

Lim JM, Jeon CO, Kim CJ (2006a) *Bacillus taeanensis* sp. nov., a halophilic Gram-positive bacterium from a solar saltern in Korea. Int J Syst Evol Microbiol 56(12):2903–2908

Lim JM, Jeon CO, Lee SM, Lee JC, Xu LH, Jiang CL, Kim CJ (2006b) *Bacillus salarius* sp. nov., a halophilic, spore-forming bacterium isolated from a salt lake in China. Int J Syst Evol Microbiol 56(2):373–377

Liu W, Yang SS (2014) *Oceanobacillus aidingensis* sp. nov., a moderately halophilic bacterium. Antonie Van Leeuwenhoek 105(5):801–808

Liu WY, Wang J, Yuan M (2011) *Halomonas aidingensis* sp. nov., a moderately halophilic bacterium isolated from Aiding salt lake in Xinjiang, China. Antonie Van Leeuwenhoek 99(3):663–670

Liu YJ, Long LJ, Huang XF, You ZQ, Wang FZ, Li J, Kim CJ, Tian XP, Zhang S (2013) *Bacillus oceani* sp. nov., a new slightly halophilic bacterium, isolated from a deep sea sediment environment. Antonie Van Leeuwenhoek 104(5):829–836

Llamas I, Béjar V, Martínez-Checa F, Martínez-Cánovas MJ, Molina I, Quesada E (2011) *Halomonas stenophila* sp. nov., a halophilic bacterium that produces sulphate exopolysaccharides with biological activity. Int J Syst Evol Microbiol 61(10):2508–2514

Longfils P, Monchy D, Weinheimer H, Chavasit V, Nakanishi Y, Schumann K (2008) A comparative intervention trial on fish sauce fortified with NaFe-EDTA and FeSO4+ citrate in iron deficiency anemic school children in Kampot, Cambodia. Asia Pac J Clin Nutr 17(2):250–257

Luesch H, Yoshida WY, Moore RE, Paul VJ, Corbett TH (2001) Total structure determination of Apratoxin A, a potent novel cytotoxin from the marine cyanobacterium *Lyngbya majuscula*. J Am Chem Soc 123(23):5418–5423

Luesch H, Yoshida WY, Moore RE, Paul VJ (2002) New apratoxins of marine cyanobacterial origin from Guam and Palau. Bioorg Med Chem 10(6):1973–1978

Luley-Goedl C, Nidetzky B (2011) Glycosides as compatible solutes: biosynthesis and applications. Nat Prod Rep 28(5):875–896

Luque R, Béjar V, Quesada E, Martínez-Checa F, Llamas I (2012) *Halomonas ramblicola* sp. nov., a moderately halophilic bacterium from Rambla Salada, a Mediterranean hypersaline rambla. Int J Syst Evol Microbiol 62(12):2903–2909

Madern D, Zaccai G (1997) Stabilisation of halophilic malate dehydrogenase from Haloarcula marismortui by divalent cations. Eur J Biochem 249(2):607–611

Madern D, Ebel C, Zaccai G (2000) Halophilic adaptation of enzymes. Extremophiles 4(2):91–98

Mah JH, Han HK, Oh YJ, Kim MG, Hwang HJ (2002) Biogenic amines in Jeotkals, Korean salted and fermented fish products. Food Chem 79(2):239–243

Mah JH, Ahn JB, Park JH, Sung HC, Hwang HJ (2003) Characterization of biogenic amine-producing microorganisms isolated from Myeolchi-Jeot, Korean salted and fermented anchovy. J Microbiol Biotechnol 13(5):692–699

Mah JH, Chang YH, Hwang HJ (2008) *Paenibacillus tyraminigenes* sp. nov. isolated from Myeolchi-jeotgal, a traditional Korean salted and fermented anchovy. Int J Food Microbiol 127:209–214

Makhdoumi-Kakhki A, Amoozegar MA, Ventosa A (2012) *Salinibacter iranicus* sp. nov. and *Salinibacter luteus* sp. nov., isolated from a salt lake, and emended descriptions of the genus *Salinibacter* and of *Salinibacter ruber*. Int J Syst Evol Microbiol 62(7):1521–1527

Margesin R, Schinner F (2001) Potential of halotolerant and halophilic microorganisms for biotechnology. Extremophiles 5(2):73–83

Márquez M, Carrasco I, Xue Y, Ma Y, Cowan D, Jones B, Grant W, Ventosa A (2007) *Aquisalimonas asiatica* gen. nov., sp. nov., a moderately halophilic bacterium isolated from an alkaline, saline lake in Inner Mongolia, China. Int J Syst Evol Microbiol 57(5):1137–1142

Márquez M, Carrasco I, Xue Y, Ma Y, Cowan D, Jones B, Grant W, Ventosa A (2008) *Aquisalibacillus elongatus* gen. nov., sp. nov., a moderately halophilic bacterium of the family Bacillaceae isolated from a saline lake. Int J Syst Evol Microbiol 58(8):1922–1926

Martín S, Márquez M, Sánchez-Porro C, Mellado E, Arahal D, Ventosa A (2003) *Marinobacter lipolyticus* sp. nov., a novel moderate halophile with lipolytic activity. Int J Syst Evol Microbiol 53(5):1383–1387

Mascellani N, Liu X, Rossi S, Marchesini J, Valentini D, Arcelli D, Taccioli C, Citterich MH, Liu CG, Evangelisti R, Russo G, Santos JM, Croce CM, Volinia S (2007) Compatible solutes from hyperthermophiles improve the quality of DNA microarrays. BMC Biotechnol 7:82

Mata JA, Béjar V, Llamas I, Arias S, Bressollier P, Tallon R, Urdaci MC, Quesada E (2006) Exopolysaccharides produced by the recently described halophilic bacteria *Halomonas ventosae* and *Halomonas anticariensis*. Res Microbiol 157(9):827–835

Matzke J, Schwermann B, Bakker EP (1997) Acidostable and acidophilic proteins: the example of the α-amylase from *Alicyclobacillus acidocaldarius*. Comp Biochem Physiol A Physiol 118(3):475–479

Medina RA, Goeger DE, Hills P, Mooberry SL, Huang N, Romero LI, Ortega-Barría E, Gerwick WH, McPhail KL (2008) Coibamide A, a potent antiproliferative cyclic depsipeptide from the Panamanian marine cyanobacterium *Leptolyngbya* sp. J Am Chem Soc 130(20):6324–6325

Mezghani M, Alazard D, Karray F, Cayol JL, Joseph M, Postec A, Fardeau ML, Tholozan JL, Sayadi S (2012) *Halanaerobacter jeridensis* sp. nov., isolated from a hypersaline lake. Int J Syst Evol Microbiol 62(8):1970–1973

Miao C, Jia F, Wan Y, Zhang W, Lin M, Jin W (2014) *Halomonas huangheensis* sp. nov., a moderately halophilic bacterium isolated from a saline–alkali soil. Int J Syst Evol Microbiol 64(3):915–920

Miranda-Tello E, Fardeau ML, Joulian C, Magot M, Thomas P, Tholozan JL, Ollivier B (2007) *Petrotoga halophila* sp. nov., a thermophilic, moderately halophilic, fermentative bacterium isolated from an offshore oil well in Congo. Int J Syst Evol Microbiol 57(1):40–44

Mishra BD, Dutta SK, Dangar TK, Das NN, Thatoi HN (2012) Optimization and characterization of chromium (VI) reduction in saline condition by moderately halophilic *Vigribacillus* sp. isolated from mangrove soil of Bhitarkanika, India. J Hazard Mater 227–228:219–226

Moller MF, Kjeldsen KU, Ingvorsen K (2010) *Marinimicrobium haloxylanilyticum* sp. nov., a new moderately halophilic, polysaccharide-degrading bacterium isolated from Great Salt Lake, Utah. Antonie Van Leeuwenhoek 98(4):553–565

Montes MJ, Bozal N, Mercadé E (2008) *Marinobacter guineae* sp. nov., a novel moderately halophilic bacterium from an Antarctic environment. Int J Syst Evol Microbiol 58(6):1346–1349

Mountfort DO, Rainey FA, Burghardt J, Kaspar HF, Stackebrandt E (1998) *Psychromonas antarcticus* gen. nov., sp. nov., a new aerotolerant anaerobic, halophilic psychrophile isolated from pond sediment of the McMurdo Ice Shelf, Antarctica. Arch Microbiol 169(3):231–238

Mullakhanbhai MF, Larsen H (1975) *Halobacterium volcanii* spec. nov., a Dead Sea halobacterium with a moderate salt requirement. Arch Microbiol 104(1):207–214

Nabti E, Sahnoune M, Adjrad S, Van Dommelen A, Ghoul M, Schmid M, Hartmann A (2007) A halophilic and osmotolerant *Azospirillum brasilense* strain from Algerian soil restores wheat growth under saline conditions. Eng Life Sci 7(4):354–360

Nabti E, Bensidhoum L, Tabli N, Dahel D, Weiss A, Rothballer M, Schmid M, Hartmann A (2014) Growth stimulation of barley and biocontrol effect on plant pathogenic fungi by a *Cellulosimicrobium* sp. strain isolated from salt-affected rhizosphere soil in northwestern Algeria. Eur J Soil Biol 61:20–26

Nakahara M, Mishima T, Hayakawa T (2007) Effect of a sake concentrate on the epidermis of aged mice and confirmation of ethyl α-D-glucoside as its active component. Biosci Biotechnol Biochem 71(2):427–434

Neagu S, Enache M, Cojoc R (2014) Extracellular hydrolytic activities of halophilic microorganisms isolated from Balta Albă salt lake. Roman Biotechnol Lett 19(1):8951–8958

Needham J, Kelly MT, Ishige M, Andersen RJ (1994) Andrimid and moiramides AC, metabolites produced in culture by a marine isolate of the bacterium *Pseudomonas fluorescens*: structure elucidation and biosynthesis. J Org Chem 59(8):2058–2063

Nicolaus B, Kambourova M, Oner ET (2010) Exopolysaccharides from extremophiles: from fundamentals to biotechnology. Environ Technol 31(10):1145–1158

Ojika M, Inukai Y, Kito Y, Hirata M, Iizuka T, Fudou R (2008) Miuraenamides: antimicrobial cyclic depsipeptides isolated from a rare and slightly halophilic myxobacterium. Chemistry 3(1):126–133

Oksanen T, Pere J, Paavilainen L, Buchert J, Viikari L (2000) Treatment of recycled kraft pulps with *Trichoderma reesei* hemicellulases and cellulases. J Biotechnol 78(1):39–48

Onishi H, Mori T, Takeuchi S, Tani K, Kobayashi T, Kamekura M (1983) Halophilic nuclease of a moderately halophilic *Bacillus* sp.: production, purification, and characterization. Appl Environ Microbiol 45(1):24–30

Oren A (2002a) Diversity of halophilic microorganisms: environments, phylogeny, physiology, and applications. J Ind Microbiol Biotechnol 28(1):56–63

Oren A (2002b) Halophilic microorganisms and their environments. Springer, New York, p 100

Oren A (2008) Microbial life at high salt concentrations: phylogenetic and metabolic diversity. Saline Syst 4(2):13

Oren A et al (2006) Life at high salt concentrations. In: The Prokaryotes. A handbook on the biology of bacteria: ecophysiology, isolation, identification, applications. Springer, New York, pp 263–282

Pappa A, Sánchez-Porro C, Lazoura P, Kallimanis A, Perisynakis A, Ventosa A, Drainas C, Koukkou A (2010) *Bacillus halochares* sp. nov., a halophilic bacterium isolated from a solar saltern. Int J Syst Evol Microbiol 60(6):1432–1436

Philp JC, Bamforth SM, Singleton I, Atlas RM (2005) Environmental pollution and restoration: a role for bioremediation. ASM Press, Washington, DC, pp 1–48

Poli A, Anzelmo G, Nicolaus B (2010) Bacterial exopolysaccharides from extreme marine habitats: production, characterization and biological activities. Mar Drugs 8(6):1779–1802

Poli A, Nicolaus B, Denizci AA, Yavuzturk B, Kazan D (2013) *Halomonas smyrnensis* sp. nov., a moderately halophilic, exopolysaccharide-producing bacterium. Int J Syst Evol Microbiol 63(1):10–18

Pourbabaee AA, Bostani S, Amozzegar MA, Naddaf R (2011) Decolorization of cibacron black w-55 under Alkaline conditions by new strain of *Halomonas* sp. isolated from textile effluent. Iran J Chem Chem Eng 30(4):63–70

Pugin B, Blamey JM, Baxter BK, Wiegel J (2012) *Amphibacillus cookii* sp. nov., a facultatively aerobic, spore-forming, moderately halophilic, alkalithermotolerant bacterium. Int J Syst Evol Microbiol 62(9):2090–2096

Qu L, Lai Q, Zhu F, Hong X, Zhang J, Shao Z, Sun X (2011) *Halomonas daqiaonensis* sp. nov., a moderately halophilic, denitrifying bacterium isolated from a littoral saltern. Int J Syst Evol Microbiol 61(7):1612–1616

Quesada E, Bejar V, Valderrama MJ, Ramos-Cormenzana A (1987) Growth characteristics and salt requirement of *Deleya halophila* in a defined medium. Curr Microbiol 16(1):21–25

Quillaguaman J, Hashim S, Bento F, Mattiasson B, Hatti-Kaul R (2005) Poly (β-hydroxybutyrate) production by a moderate halophile, Halomonas boliviensis LC1 using starch hydrolysate as substrate. J Appl Microbiol 99(1):151–157

Quillaguaman J, Delgado O, Mattiasson B, Hatti-Kaul R (2006) Poly (β-hydroxybutyrate) production by a moderate halophile, *Halomonas boliviensis* LC1. Enzym Microb Technol 38(1):148–154

Quillaguamán J, Muñoz M, Mattiasson B, Hatti-Kaul R (2007) Optimizing conditions for poly (β-hydroxybutyrate) production by *Halomonas boliviensis* LC1 in batch culture with sucrose as carbon source. Appl Microbiol Biotechnol 74(5):981–986

Quillaguamán J, Guzmán H, Van-Thuoc D, Hatti-Kaul R (2010) Synthesis and production of polyhydroxyalkanoates by halophiles: current potential and future prospects. Appl Microbiol Biotechnol 85(6):1687–1696

Rafael R, Arahal DR, Márquez MC, Ventosa A (2010) Phylogenetic relationships within the family Halomonadaceae based on comparative 23S and 16S rRNA gene sequence analysis. Int J Syst Evol Microbiol 60(4):737–748

Ramos A, Raven N, Sharp RJ, Bartolucci S, Rossi M, Cannio R, Lebbink J, Van Der Oost J, De Vos W, Santos H (1997) Stabilization of enzymes against thermal stress and freeze-drying by mannosylglycerate. Appl Environ Microbiol 63(10):4020–4025

Rao JM, Argos P (1981) Structural stability of halophilic proteins. Biochemistry 20(23):6536–6543

Ravikumar S, Kathiresan K, Ignatiammal S, Babu Selvam M, Shanthy S (2004) Nitrogen-fixing azotobacters from mangrove habitat and their utility as marine biofertilizers. J Exp Mar Biol Ecol 312(1):5–17

Ravikumar S, Shanthy S, Kalaiarasi A, Sumaya S (2010) Effect of halophilic phosphobacteria on *Avicennia officinalis* seedlings. Ann Biol Res 1:254–260

Riis V, Kleinsteuber S, Babel W (2003) Influence of high salinities on the degradation of diesel fuel by bacterial consortia. Can J Microbiol 49(11):713–721

Roberts M, Lai M, Gunsalus R (1992) Biosynthetic pathways of the osmolytes N epsilon-acetyl-beta-lysine, beta-glutamine, and betaine in Methanohalophilus strain FDF1 suggested by nuclear magnetic resonance analyses. J Bacteriol 174(20):6688–6693

Romano I, Finore I, Nicolaus G, Huertas FJ, Lama L, Nicolaus B, Poli A (2008) *Halobacillus alkaliphilus* sp. nov., a halophilic bacterium isolated from a salt lake in Fuente de Piedra, southern Spain. Int J Syst Evol Microbiol 58(4):886–890

Russell NJ (1993) Lipids of halophilic and halotolerant microorganisms. In: Russell HV, Hochstein LI (eds) The biology of halophilic bacteria, vol 1. CRC Press, Boca Raton, pp 163–210
Ryu J, Kanapathipillai M, Lentzen G, Park CB (2008) Inhibition of β-amyloid peptide aggregation and neurotoxicity by α-D-mannosylglycerate, a natural extremolyte. Peptides 29(4):578–584
Sadfi-Zouaoui N, Essghaier B, Hannachi I, Hajlaoui MR, Boudabous A (2007) First report on the use of moderately halophilic bacteria against stem canker of greenhouse tomatoes caused by *Botrytis cinerea*. Ann Microbiol 57(3):337–339
Sadfi-Zouaoui N, Essghaier B, Hajlaoui M, Fardeau ML, Cayaol J, Ollivier B, Boudabous A (2008) Ability of moderately halophilic bacteria to control grey mould disease on tomato fruits. J Phytopathol 156(1):42–52
Salvarrey MS, Cazzulo JJ (1980) Some properties of the NADP-specific malic enzyme from the moderate halophile *Vibrio costicola*. Can J Microbiol 26(1):50–57
Salvarrey MS, Cannata J, Cazzulo JJ (1989) Phosphoenolpyruvate carboxykinase from the moderate halophile *Vibrio costicola*. Purification, physicochemical properties and the effect of univalent-cation salts. Biochem J 260:221–230
Sam S, Kucukasik F, Yenigun O, Nicolaus B, Oner ET, Yukselen MA (2011) Flocculating performances of exopolysaccharides produced by a halophilic bacterial strain cultivated on agro-industrial waste. Bioresour Technol 102:1788–1794
Sánchez-Porro C, Martin S, Mellado E, Ventosa A (2003) Diversity of moderately halophilic bacteria producing extracellular hydrolytic enzymes. J Appl Microbiol 94(2):295–300
Sánchez-Porro C, Kaur B, Mann H, Ventosa A (2010) *Halomonas titanicae* sp. nov., a halophilic bacterium isolated from the RMS Titanic. Int J Syst Evol Microbiol 60(12):2768–2774
Saralov A, Kuznetsov B, Reutskikh E, Baslerov R, Panteleeva A, Suzina N (2012) *Arhodomonas recens* sp. nov., a halophilic alkane-utilizing hydrogen-oxidizing bacterium from the brines of flotation enrichment of potassium minerals. Microbiology 81(5):582–588
Sass AM, Sass H, Coolen MJ, Cypionka H, Overmann J (2001) Microbial communities in the chemocline of a hypersaline deep-sea basin (Urania basin, Mediterranean Sea). Appl Environ Microbiol 67(12):5392–5402
Sauer T, Galinski EA (1998) Bacterial milking: a novel bioprocess for production of compatible solutes. Biotechnol Bioeng 57(3):306–313
Saum SH, Müller V (2007) Salinity-dependent switching of osmolyte strategies in a moderately halophilic bacterium: glutamate induces proline biosynthesis in *Halobacillus halophilus*. J Bacteriol 189(19):6968–6975
Sawangwan T, Goedl C, Nidetzky B (2010) Glucosylglycerol and glucosylglycerate as enzyme stabilizers. Biotechnol J 2:187–191
Schnoor M, Voß P, Cullen P, Bo¨king T, Galla HJ, Galinski EA, Lorkowski S (2004) Characterization of the synthetic compatible solute homoectoine as a potent PCR enhancer. Biochem Biophys Res Commun 322:867–872
Sheng Y, Zhang Q, Sheng Y, Li C, Wang H (2006) Screening and flocculating properties of bioflocculant-producing microorganisms. J Univ Sci Beijing Miner Metall Mater 13(4):289–292
Shimane Y, Tsuruwaka Y, Miyazaki M, Mori K, Minegishi H, Echigo A, Ohta Y, Maruyama T, Grant WD, Hatada Y (2013) *Salinisphaera japonica* sp. nov., a moderately halophilic bacterium isolated from the surface of a deep-sea fish, *Malacocottus gibber*, and emended description of the genus *Salinisphaera*. Int J Syst Evol Microbiol 63(6):2180–2185
Shiozawa H, Kagasaki T, Kinoshita T, Haruyama H, Domon H, Utsui Y, Kodama K, Takahashi S (1993) Thiomarinol, a new hybrid antimicrobial antibiotic produced by a marine bacterium. Fermentation, isolation, structure, and antimicrobial activity. J Antibiot 46(12):1834–1842
Shivanand P, Mugeraya G, Kumar A (2013) Utilization of renewable agricultural residues for the production of extracellular halostable cellulase from newly isolated *Halomonas* sp. strain PS47. Ann Microbiol 63(4):1257–1263
Shoemaker RH (2006) The NCI60 human tumour cell line anticancer drug screen. Nat Rev Cancer 6(10):813–823
Singh RK, Tiwari SP, Rai AK, Mohaparta TM (2011) Cyanobacteria: an emerging source for drug discovery. J Antibiot 64:401–412

Sivaprakasam S, Mahadevan S, Sekar S, Rajakumar S (2008) Biological treatment of tannery wastewater by using salt-tolerant bacterial strains. Microb Cell Fact 7:15

Sivaprakasam S, Dhandapani B, Mahadevan S (2011) Optimization studies on production of a salt-tolerant protease from *Pseudomonas aeruginosa* strain BC1 and its application on tannery saline wastewater treatment. Braz J Microbiol 42(4):1506–1515

Sobolevskaya M, Smetanina O, Speitling M, Shevchenko L, Dmitrenok P, Laatsch H, Kuznetsova T, Ivanova E, Elyakov G (2005) Controlling production of brominated cyclic depsipeptides by *Pseudoalteromonas maricaloris* KMM 636T. Lett Appl Microbiol 40(4):243–248

Sohair I, Abou-Elela MMK, Mariam EF (2010) Biological treatment of saline wastewater using a salt-tolerant microorganism. Desalination 250:1–5

Sorokin DY, Tourova TP, Kolganova TV, Spiridonova EM, Berg IA, Muyzer G (2006) *Thiomicrospira halophila* sp. nov., a moderately halophilic, obligately chemolithoautotrophic, sulfur-oxidizing bacterium from hypersaline lakes. Int J Syst Evol Microbiol 56(10):2375–2380

Sorokin DY, Tourova TP, Braker G, Muyzer G (2007) *Thiohalomonas denitrificans* gen. nov., sp. nov. and *Thiohalomonas nitratireducens* sp. nov., novel obligately chemolithoautotrophic, moderately halophilic, thiodenitrifying Gammaproteobacteria from hypersaline habitats. Int J Syst Evol Microbiol 57(7):1582–1589

Soto-Ramírez N, Sánchez-Porro C, Rosas-Padilla S, Almodóvar K, Jiménez G, Machado-Rodríguez M, Zapata M, Ventosa A, Montalvo-Rodríguez R (2008) *Halobacillus mangrovi* sp. nov., a moderately halophilic bacterium isolated from the black mangrove Avicennia germinans. Int J Syst Evol Microbiol 58(1):125–130

Suh HH, Kwon GS, Lee CH, Kim HS, Oh HM, Yoon BD (1997) Characterization of bioflocculant produced by *Bacillus* sp. DP-152. J Ferment Bioeng 84(2):108–112

Tanasupawat S, Namwong S, Kudo T, Itoh T (2007) *Piscibacillus salipiscarius* gen. nov., sp. nov., a moderately halophilic bacterium from fermented fish (pla-ra) in Thailand. Int J Syst Evol Microbiol 57(7):1413–1417

Tang SK, Wang Y, Cai M, Lou K, Mao PH, Jin X, Jiang CL, Xu LH, Li WJ (2008) *Microbulbifer halophilus* sp. nov., a moderately halophilic bacterium from north-west China. Int J Syst Evol Microbiol 58(9):2036–2040

Tang SK, Wang Y, Lou K, Mao PH, Jin X, Jiang CL, Xu LH, Li WJ (2009) *Gracilibacillus saliphilus* sp. nov., a moderately halophilic bacterium isolated from a salt lake. Int J Syst Evol Microbiol 59(7):1620–1624

Taori K, Paul VJ, Luesch H (2008) Structure and activity of largazole, a potent antiproliferative agent from the Floridian marine cyanobacterium *Symploca* sp. J Am Chem Soc 130(6):1806–1807

Ten Brink B, Damink C, Joosten H, Huis In't Veld J (1990) Occurrence and formation of biologically active amines in foods. Int J Food Microbiol 11(1):73–84

Tian XP, Dastager S, Lee JC, Tang SK, Zhang YQ, Park DJ, Kim CJ, Li WJ (2007) *Alkalibacillus halophilus* sp. nov., a new halophilic species isolated from hypersaline soil in Xin-Jiang province, China. Syst Appl Microbiol 30(4):268–272

Tokunaga H, Ishibashi M, Arakawa T, Tokunaga M (2004) Highly efficient renaturation of β-lactamase isolated from moderately halophilic bacteria. FEBS Lett 558(1):7–12

Trimurtulu G, Ohtani I, Patterson GM, Moore RE, Corbett TH, Valeriote FA, Demchik L (1994) Total structures of cryptophycins, potent antitumor depsipeptides from the blue-green alga Nostoc sp. strain GSV 224. J Am Chem Soc 116(11):4729–4737

Uchida H, Kondo D, Yamashita S, Tanaka T, Tran LH, Nagano H, Uwajima T (2004) Purification and properties of a protease produced by *Bacillus subtilis* CN2 isolated from a Vietnamese fish sauce. World J Microbiol Biotechnol 20(6):579–582

Uzair B, Ahmed N, Ahmad VU, Kousar F (2006) A new antibacterial compound produced by an indigenous marine bacteria—fermentation, isolation, and biological activity. Nat Prod Res 20(14):1326–1331

Vaisman N, Oren A (2009) *Salisaeta longa* gen. nov., sp. nov., a red, halophilic member of the Bacteroidetes. Int J Syst Evol Microbiol 59(10):2571–2574

Van Qua D, Simidu U, Taga N (1981) Purification and some properties of halophilic protease produced by a moderately halophilic marine *Pseudomonas* sp. Can J Microbiol 27(5):505–510

Veenagayathri K, Vasudevan N (2013) Degradation of 4-chlorophenol by a moderately halophilic bacterial consortium under saline conditions. Br Microbiol Res J 3(4):513–524

Ventosa A, Nieto JJ, Oren A (1998) Biology of moderately halophilic aerobic bacteria. Microbiol Mol Biol Rev 62(2):504–544

Vidyasagar M, Prakash S, Jayalakshmi S, Sreeramulu K (2007) Optimization of culture conditions for the production of halothermophilic protease from halophilic bacterium Chromohalobacter sp. TVSP101. World J Microbiol Biotechnol 23(5):655–662

Wang YN, Cai H, Chi CQ, Lu AH, Lin XG, Jiang ZF, Wu XL (2007) *Halomonas shengliensis* sp. nov., a moderately halophilic, denitrifying, crude-oil-utilizing bacterium. Int J Syst Evol Microbiol 57(6):1222–1226

Wang CY, Chang CC, Ng CC, Chen TW, Shyu YT (2008) *Virgibacillus chiguensis* sp. nov., a novel halophilic bacterium isolated from Chigu, a previously commercial saltern located in southern Taiwan. Int J Syst Evol Microbiol 58(2):341–345

Wang Y, Tang SK, Lou K, Lee JC, Jeon CO, Xu LH, Kim CJ, Li WJ (2009) *Aidingimonas halophila* gen. nov., sp. nov., a moderately halophilic bacterium isolated from a salt lake. Int J Syst Evol Microbiol 59(12):3088–3094

Wang J, Shao Z, Hong Y, Li C, Fu X, Liu Z (2010a) A novel β-mannanase from Pantoea agglomerans A021: gene cloning, expression, purification and characterization. World J Microbiol Biotechnol 26(10):1777–1784

Wang L, Liu WY, Gu ZJ, Chen SF, Yang SS (2010b) *Oceanobacillus manasiensis* sp. nov., a moderately halophilic bacterium isolated from the salt lakes of Xinjiang, China. J Microbiol 48(3):312–317

Wang S, Yang Q, Liu ZH, Sun L, Wei D, Zhang JZ, Song JZ, Wang Y, Song J, Fan JX (2011a) *Halomonas alkalitolerans* sp. nov., a novel moderately halophilic bacteriun isolated from soda meadow saline soil in Daqing, China. J Microbiol 49(1):24–28

Wang X, Xue Y, Ma Y (2011b) *Streptohalobacillus salinus* gen. nov., sp. nov., a moderately halophilic, Gram-positive, facultative anaerobe isolated from subsurface saline soil. Int J Syst Evol Microbiol 61(5):1127–1132

Wang CY, Wu SJ, Ng CC, Tzeng WS, Shyu YT (2012a) *Halomonas beimenensis* sp. nov., isolated from an abandoned saltern. Int J Syst Evol Microbiol 62(12):3013–3017

Wang YX, Liu JH, Xiao W, Zhang XX, Li YQ, Lai YH, Ji KY, Wen ML, Cui XL (2012b) *Fodinibius salinus* gen. nov., sp. nov., a moderately halophilic bacterium isolated from a salt mine. Int J Syst Evol Microbiol 62(2):390–396

Wang YX, Li YP, Liu JH, Xiao W, Lai YH, Li ZY, Ding ZG, Wen ML, Cui XL (2013) *Gracilimonas mengyeensis* sp. nov., a moderately halophilic bacterium isolated from a salt mine in Yunnan, south-western China. Int J Syst Evol Microbiol 63(11):3989–3993

Wang S, Sun L, Wei D, Zhou B, Zhang J, Gu X, Zhang L, Liu Y, Li Y, Guo W (2014a) *Bacillus daqingensis* sp. nov., a halophilic, alkaliphilic bacterium isolated from saline-sodic soil in Daqing, China. J Microbiol 52(7):548–553

Wang YX, Xiao W, Dong MH, Zhao Q, Li ZY, Lai YH, Cui XL (2014b) *Halomonas qiaohouensis* sp. nov., isolated from salt mine soil in southwest China. Antonie Van Leeuwenhoek 106(2):253–260

Weisser J, Trüper HG (1985) Osmoregulation in a New Haloalkaliphilic *Bacillus* from the Wadi Natrun (Egypt). Syst Appl Microbiol 6(1):7–11

Wejse PL, Ingvorsen K, Mortensen KK (2003) Purification and characterisation of two extremely halotolerant xylanases from a novel halophilic bacterium. Extremophiles 7(5):423–431

Williamson B, Tudzynski B, Tudzynski P, van Kan JA (2007) *Botrytis cinerea*: the cause of grey mould disease. Mol Plant Pathol 8(5):561–580

Wratten SJ, Wolfe MS, Andersen RJ, Faulkner DJ (1977) Antibiotic metabolites from a marine pseudomonad. Antimicrob Agents Chemother 11(3):411–414

Wu G, Wu XQ, Wang YN, Chi CQ, Tang YQ, Kida K, Wu XL, Luan ZK (2008) *Halomonas daqingensis* sp. nov., a moderately halophilic bacterium isolated from an oilfield soil. Int J Syst Evol Microbiol 58(12):2859–2865

Xiang W, Liang HZ, Liu S, Luo F, Tang J, Li MY, Che ZM (2011) Isolation and performance evaluation of halotolerant phosphate solubilizing bacteria from the rhizospheric soils of historic Dagong Brine Well in China. World J Microbiol Biotechnol 27(11):2629–2637

Xiao W, Wang YX, Liu JH, Wang ZG, Zhang XX, Ji KY, Lai YH, Wen ML, Cui XL (2012) *Roseivivax sediminis* sp. nov., a moderately halophilic bacterium isolated from salt mine sediment. Int J Syst Evol Microbiol 62(8):1890–1895

Xu XW, Wu YH, Wang CS, Yang JY, Oren A, Wu M (2008) *Marinobacter pelagius* sp. nov., a moderately halophilic bacterium. Int J Syst Evol Microbiol 58(3):637–640

Xue Y, Ventosa A, Wang X, Ren P, Zhou P, Ma Y (2008) *Bacillus aidingensis* sp. nov., a moderately halophilic bacterium isolated from Ai-Ding salt lake in China. Int J Syst Evol Microbiol 58(12):2828–2832

Yancey PH (2005) Organic osmolytes as compatible, metabolic and counteracting cytoprotectants in high osmolarity and other stresses. J Exp Biol 208(15):2819–2830

Yang L, Lai CT, Shieh WK (2000) Biodegradation of dispersed diesel fuel under high salinity conditions. Water Res 34(13):3303–3314

Yang JY, Huo YY, Xu XW, Meng FX, Wu M, Wang CS (2010) *Oceanobacillus neutriphilus* sp. nov., isolated from activated sludge in a bioreactor. Int J Syst Evol Microbiol 60(10):2409–2414

Yemashova NA, Murygina VP, Zhukov DV, Zakharyantz AA, Gladchenko MA, Appanna V, Kalyuzhnyi SV (2007) Biodeterioration of crude oil and oil derived products: a review. Rev Environ Sci Biotechnol 6(4):315–337

Yen GC, Hsieh CL (1997) Antioxidant effects of dopamine and related compounds. Biosci Biotechnol Biochem 61:1646–1662

Ying Y, Taori K, Kim H, Hong J, Luesch H (2008) Total synthesis and molecular target of largazole, a histone deacetylase inhibitor. J Am Chem Soc 130(26):8455–8459

Yokoi H, Yoshida T, Mori S, Hirose J, Hayashi S, Takasaki Y (1997) Biopolymer flocculant produced by an *Enterobacter* sp. Biotechnol Lett 19(6):569–573

Yonezawa Y, Tokunaga H, Ishibashi M, Tokunaga M (2001) Characterization of nucleoside diphosphate kinase from moderately halophilic eubacteria. Biosci Biotechnol Biochem 65(10):2343–2346

Yongsawatdigul J, Rodtong S, Raksakulthai N (2007) Acceleration of Thai fish sauce fermentation using proteinases and bacterial starter cultures. J Food Sci 72(9):M382–M390

Yoon JH, Kang SJ, Jung YT, Lee KC, Oh HW, Oh TK (2010) *Virgibacillus byunsanensis* sp. nov., isolated from a marine solar saltern. Int J Syst Evol Microbiol 60(2):291–295

Yoshie S, Makino H, Hirosawa H, Shirotani K, Tsuneda S, Hirata A (2006) Molecular analysis of halophilic bacterial community for high-rate denitrification of saline industrial wastewater. Appl Microbiol Biotechnol 72(1):182–189

Zavarzina DG, Zhilina TN, Kuznetsov BB, Kolganova TV, Osipov GA, Kotelev MS, Zavarzin GA (2013) *Natranaerobaculum magadiense* gen. nov., sp. nov., an anaerobic, alkalithermophilic bacterium from soda lake sediment. Int J Syst Evol Microbiol 63(12):4456–4461

Zhang H, Hosoi-Tanabe S, Nagata S, Ban S, Imura S (2010) *Psychroflexus lacisalsi* sp. nov., a moderate halophilic bacterium isolated from a hypersaline lake (Hunazoko-Ike) in Antarctica. J Microbiol 48(2):160–164

Zhang YJ, Zhou Y, Ja M, Shi R, Chun-Yu WX, Yang LL, Tang SK, Li WJ (2012) *Virgibacillus albus* sp. nov., a novel moderately halophilic bacterium isolated from Lop Nur salt lake in Xinjiang province, China. Antonie Van Leeuwenhoek 102(4):553–560

Zhilina TN, Zavarzina DG, Panteleeva AN, Osipov GA, Kostrikina NA, Tourova TP, Zavarzin GA (2012) *Fuchsiella alkaliacetigena* gen. nov., sp. nov., an alkaliphilic, lithoautotrophic homoacetogen from a soda lake. Int J Syst Evol Microbiol 62(7):1666–1673

Chapter 12
Microbial Hydrocarbon-Removal Under Halostress

Samir S-A. Radwan and Dina M. Al-Mailem

Abstract Oceans which occupy about 70 % of the earth's surface represent a moderately saline environment. In addition, there are all over the world hypersaline areas, and the industry continuously produces huge amounts of liquid saline wastes. Similar to the nonsaline environments, saline ones harbor halophilic/halotolerant microorganisms with versatile physiological activities. Many of them are phototrophs or chemolithotrophs, and consequently serve as primary producers in these environments. Other microorganisms have the potential for biodegradation of xenobiotic compounds. The saline environments are also subjected to hydrocarbon pollution through legal and illegal human activity. Such environments too harbor halophilic/halotolerant, hydrocarbonoclastic microorganism which rid those environments of hydrocarbon pollutants. These microorganism are either planktonic, or more frequently associated with other microbes to form biofilms on animate and inanimates substrates. Seawaters accommodate numerous hydrocarbonoclastic bacteria belonging predominantly to the Gammaproteobacteria subgroup. Most prominent are the members of the group of the "obligate hydrocarbonoclastic bacteria" (OHCB) belonging to the genera *Alcanivorax*, *Marinobacter*, *Thalassolithus*, *Cycloclasticus* and *Oleispira*. Hypersaline areas harbor in addition, hydrocarbonoclastic haloarchaea. Artificially established biofilms comprising halophilic, hydrocarbonclastic microorganisms revealed active in hydrocarbon removal. It may be expected that the future will witness using such man-made biofilms in bioremediation of saline wastes polluted with hydrocarbons in bioreactors.

Keywords Halophilic microorganisms • Hydrocarbonoclastic biofilms • Hydrocarbons • Oil-bioremediation • Oil-pollution

---

S.S-A. Radwan (✉) • D.M. Al-Mailem
Microbiology Program, Department of Biological Sciences, Kuwait University,
P.O. Box 5969, Safat 13060, Kuwait
e-mail: samir.radwan@ku.edu.kw

D.K. Maheshwari, M. Saraf (eds.), *Halophiles*, Sustainable
Development and Biodiversity 6, DOI 10.1007/978-3-319-14595-2_12

12.1 Introduction

Life evolved on our planet, some 3.5×10^9 years back, first as organic biomolecules and subsequently in the form of unicellular microorganism (prokaryotes). At that time, the primitive atmosphere was absolutely anaerobic, and still remained O_2-free through another couple of billions of years. In view of the absence of molecular oxygen in the primitive atmosphere the ozone shield, which protects life from the UV-irradiation, was absent. Therefore, the pioneer primitive "microorganisms" should have evolved and lived through billions of years deep in the ocean, where UV-rays cannot penetrate. Only after the evolution of the oxygenic phototrophic cyanobacteria some 2.0×10^9 years back, and the consequent accumulation of molecular oxygen in the atmosphere, the ozone shield formed, and life started to migrate from the oceans to terrestrial regions. This brief historical overview is meant to demonstrate the intimate and deep relation of today microorganisms, especially the prokaryotes (actually all living creatures), to the marine (moderately saline) conditions. The halophilic/halotolerant organisms of today are the descendants of those pioneer halophiles. It is therefore, not surprising that "modern" halophilic/halotolerant representatives are rather frequent among all life groups including bacteria, archaea, fungi, algae, higher plants and higher animals. After all, living things all evolved from one or a few common ancestors.

Microorganisms on earth play an essential role in the mineralization of organic matter, recycling of elements and degradation/detoxification of hazardous products. In view of the well-known fact that saline environments occupy a major proportion of the earth's surface, halophilic/halotolerant microorganisms inhabiting such environments, and their role in biodegradation merit special consideration. This article highlights the role of halophilic/halotolerant prokaryotes in the bio-degradative removal of hydrocarbon pollutants in saline and hyper-saline environments.

12.2 Hydrocarbons in Saline Environments

12.2.1 Saline Environments

Oceans cover roughly 70 % of the entire earth's surface. Seawater has a NaCl concentration of about 3.5 % (w/v). Ecosystems connected with this water body viz. estuaries, beaches, salt marshes, inland lakes, rock pools, etc. may exhibit even higher salinity. Hypersaline lakes, e.g. the Dead Sea (Oren 2002) and the Arabian Gulf sabkhas (Al-Mailem et al. 2010b) may become almost saturated (or even supersaturated) with salt. Figure 12.1 shows a hypersaline area at the Kuwaiti coast. With excessive salinity, excessive alkalinity usually prevail, e.g. in Lake Magani in Kenya and Wadi Natrun in Egypt. In addition to the "natural" saline environments, there are also "man-made" saline environments. Examples are salted fish, salted meat and other foods (Vilhelmsson et al. 1996), as well as salt waters disposed by many industries.

Fig. 12.1 A hypersaline coastal area in Kuwait. The salt concentration in the soil was about 4.5 M and in the pond-water 3.5 M

12.2.2 Primary Productivity in Saline Environments

Primary producers in all environments are the origin of organic matter there. Such organisms are autotrophs, i.e. capable of assimilating CO_2, producing organic carbon compounds. Most of them are phototrophs, although in specific environments, chemolithotrophs could be the major primary producers. Like non-extreme environments, saline and hypersaline environments accommodate lower and higher phototrophs, starting with anoxygenic phototrophic bacteria and cyanobacteria (oxygenic phototrophs) up to rooted higher plants. All of them are either halophilic or at least halotolerant [for review see (DasSarma and DasSarma 2012)].

12.2.2.1 Anoxygenic, Phototrophic Bacteria

These are major primary producers in anaerobic environments, e.g., in hypersaline microbial mats (Fourcans et al. 2004). These organisms possess Photosystem I only and lack photosystem II, therefore, their photosynthesis is anoxygenic. This group includes the moderately halophilic green *Chlorobium limicola* and *C. phaeobacteriodes*, the halophilic purple sulfur bacteria *Halochromatium glycolicum*, *H. salexigens*, *Thiocystis violescans*, *Thiocapsa roseopersicina*, *Thiohalocapsa halophila* and *Ectothiorhodospira* spp., and the purple nonsulfur bacteria *Rhodothalassium salexigenes* and *Rhodovibrio salinarum*.

12.2.2.2 Chemolithotrophic, Sulfur-Oxidizing Bacteria

These also occur frequently in microbial mats and fix CO_2 making use of the energy released from oxidation of H_2S and sulfur up to sulfate. Examples are *Beggiatoa* spp. and *Halothiobacillus halophilus*.

12.2.2.3 Cyanobacteria

These are widely distributed in the marine environments; the surface 5 cm layer of seawater worldwide contains millions per ml of picocyanobacteria belonging to the genera *Synechococcus* and *Synechocystis* (Al-Hasan et al. 2001). The likewise unicellular cyanobacterium *Aphanothece halophytica* can tolerate up to 5 M NaCl, and has an optimum of 3.5 M. In addition, a great many of filamentous cyanobacteria e.g. *Microcoleus* spp., *Phormidium* spp. and *Oscillatoria* spp. inhabit moderately saline and hypersaline environments.

12.2.2.4 Algae

Also many of these eukaryotes inhabit the marine ecosystem worldwide and are major primary producers there. Many inhabit the moderately saline seawater, and the prominent representatives are diatoms belonging to the genera *Amphora*, *Nitzchia*, *Navicula* and others. Hypersaline environments also have their extremely halophilic algal inhabitants, e.g. the chlorophyte *Dunaliella* which normally tolerates NaCl-saturation. The marine environment harbors in addition to great variety of phytoplankton (microalgae), a great variety of small and huge thalloid algae (macroalgae) belonging mainly to the Phaeophyta (brown algae) and Rhodophyta (red algae), but also to the Chlorophyta (green algae). All these organisms are oxygenic phototrophs and enrich the saline environments with organic materials.

12.2.2.5 Higher Plants

Many higher plant species grow as wild vegetation in moderately saline and hypersaline areas. We may mention here, one representative example, *Halocnemum strobilacium* belonging to the Chenopodiacea, as a wild inhabitant of hypersaline coastal areas of the Arabian/Persian Gulf. Such higher plants are obviously effective organic matter producers in those saline areas.

12.2.3 Sources of Hydrocarbons

Hydrocarbons may occur naturally in saline and nonsaline environments or may be introduced as contaminants. Almost all living creatures, especially higher plants and animals produce varying proportions of hydrocarbons. These compounds occur mostly as protective layers (cuticle, waxes etc.) in these creatures. Most microbial inhabitants also produce rather small proportions of hydrocarbons. In fact, crude oil is nothing but "preserved" hydrocarbons of ancient creatures that were buried, and most of their organic constituents were decomposed anaerobically (Major 1990). Biologically produced hydrocarbons are one source of those preserved compounds in saline and nonsaline environments. The major hydrocarbon portion, however, reaches those environments as pollutants during legal and illegal activities. Oil spills may arise as a result of equipment failure, human error, natural disasters and/or even deliberate acts (Radwan et al. 1995; Anderson and La Belle 2000). Tanker accidents and pipeline breaks are disasters that repeatedly pollute the marine environment (Bartha 1986). The greatest man-made oil disaster in the history of mankind was the catastrophe associated with the Gulf War of 1990/1991 (Hirschmann 2005). Before their withdrawal from Kuwait in 1991, the Iraqi forces released crude oil in the Gulf water body and blew up and set in fire almost 700 oil wells (Heller et al. 2000), from which crude kept gushing through 7 months, and they released into the Arabian/Persian Gulf water body millions of barrels of crude through 3 successive days (McKinnon and Vine 1991; Radwan 2008). Hydrocarbon derivatives, in the form of pesticides may also reach saline environments as constituents of disposed sewage.

12.2.4 Oil Composition and Oil-Spill Impacts

As mentioned, crude oil is a natural product, it comprises hundreds of organic molecules containing mainly carbon and hydrogen atoms. Oil has its origin in microorganisms, plants and animals buried for long ages (Major 1990).

Hydrocarbons are either aliphatic or aromatic compounds, and in addition, oil contains smaller proportions of asphaltenes and resins (Stevenson 1966; Leahy and Colwell 1990). Aliphatic hydrocarbons consist of carbon chains with varying lengths which could be *n*-alkanes, branched alkanes or cycloalkanes (naphthenes). Alkanes are predominant constituents in crude oil, and are commonly referred to as saturated hydrocarbons or paraffins (Brown et al. 1979). The second major constituents are the aromatics, cyclic compounds, either based on a single ring e.g. benzene, toluene and xylene, or on several benzene rings in linear, angular or cluster arrangement in the polycyclic aromatic hydrocarbons (Sims and Overcash 1983). Asphaltenes make <5 % of the crude and consist of very high molecular weight hydrocarbons, hence their high viscosity or solidity. Resins are like asphaltenes, but contain in addition to carbon and hydrogen, other elements viz. sulfur, oxygen and/or others (Leahy and Colwell 1990).

Pollution of saline and nonsaline environments with crude oil is associated with a lot of ecological problems (Burger 1993). Oil spills in the marine environment behave differently compared with their behavior in the terrestrial environment. In the former case, oil forms a slick which becomes partially emulsified with water due to mechanical effects of wind and wave movement. This results in increasing the exposed area of the oil droplets allowing for more effective microbial attacks (Davis and Gibbs 1975; Cooney 1984). In the terrestrial environment on the other hand, oil penetrates into soil vertically. This is usually associated with reduced hydrocarbon volatilization and consequently toxic-hydrocarbon accumulation (Leahy and Colwell 1990). Soils contaminated with crude oil are not satisfactorily suitable for agricultural and recreational purposes, and are potential sources of contamination of surface and underground waters (Nwoko et al. 2007).

Several physico-chemical factors in the environment are modified after oil spills. Thus, a contaminated soil usually acquires a higher temperature than its pristine counterpart, simply due to the darker color associated with oil contamination (Balks et al. 2002). Contamination also reduces the moisture content of the soil, due to the halophilic nature of oil (Aislabie et al. 2004). The chemistry of the soil also changes. The oil spill is of course associated with an increase of the amount of total organic carbon in the impacted soil, and a simultaneous decrease of nitrate content (Aislabie et al. 2004). Reportedly, the soil acidity increases due to the accumulation of acidic metabolites produced microbiologically during hydrocarbon oxidation (see the next section).

Oil spills have also serious effects on higher living organisms inhabiting the environment. Antarctic region inhabitants e.g. mosses and collembolans, which are already stressed by too low temperature, too low humidity, too high salinity and repeated freeze-thow cycles are stressed more by oil contamination (Konlechner 1985). This is also true for higher plants inhabiting other environments (Henner et al. 1999). Health problems among Penguin chicks (Aislabie et al. 2004) were recorded. Humans suffer from thyroid and endocrine disorders, metabolic diseases, altered immune function, reproductive impairment, genotoxicity and cancer (Fox 2001). Hydrocarbon-utilizing bacteria, in Arctic coastal waters on the other hand, increase in number following petroleum contamination (Horowitz and Atlas 1977). In fact, the numbers of hydrocarbon-utilizing microorganisms in contaminated environments reflect the level of hydrocarbon pollutants there (Colwell and Walker 1977; Atlas 1981; Geiselbrecht et al. 1996; Radwan 2009). In other words, the entire ecosystem changes when the environment receives oil spill (Bonney and Jaber 2011).

12.3 Hydrocarbonoclastic, Halophilic Microorganisms

12.3.1 Characteristics

Effective hydrocarbon-mineralization in the various environments is due to the activity of mainly aerobic microorganisms, although anaerobic biodegradation also occurs. Of course, not all aerobes have the potential for hydrocarbon degradation. This potential is characteristic of those aerobes that posses oxygenases, also called hydroxylases. Those enzyme systems catalyze splitting oxygen molecules into oxygen atoms and introducing the latter into hydrocarbon molecules leading to the corresponding alcohols. This is the key step of microbial attack on the hydrocarbon molecules. Mono-oxygenases are involved in the oxidation of aliphatic hydrocarbons, and dioxygenases in the oxidation of aromatic hydrocarbons. The resulting alcohols are subsequently oxidized through the corresponding aldehydes (ketones) to the corresponding fatty acids. The latter are degraded by β-oxidation into Acetyl CoA units, ready for further metabolism producing either cell material or energy stored as ATP. The mechanisms of hydrocarbon metabolism have been repeatedly reviewed (Klug and Markovetz 1971; Rehm and Reiff 1981; Boulton and Ratledge 1984; Leahy and Colwell 1990; Radwan and Sorkhoh 1993). Microbial hydrocarbon-consumption is usually measured by gas-liquid chromatography (GLC), and is calculated as percent total peak area loss in the experimental GLC-profiles based on the total peak area in GLC-profiles of the hydrocarbons at time zero. Typical GLC-profiles are illustrated in Fig. 12.2.

12.3.2 Responses to Halostress

Halophilic microorganisms with various activities occur all over the globe. Oren (2002) realized that life occurs at most salt concentrations encountered in natural saline habitats, ranging from fresh water to hypersaline lakes. Most hypersaline waters arise via evaporation and concentration of seawater (thalassohaline environments) (Oren 2002). That is why the micro-flora in both environments is usually similar (Forsyth et al. 1971; Ventosa et al. 1998; Al-Mailem et al. 2014a). Saline soils characterized by periodical change of soil salinity usually contain halotolerant rather than halophilic microorganisms (Quesada et al. 1982). Halophilic microorganisms are those that actually require NaCl for growth (Kushner 1978, 1993). In comparison, halotolerant microorganisms are those capable of growth over a range of salinity, but grow best in the absence of NaCl. A third group of the so called haloversatile microorganisms are like the halotolerant group but their growth rate optima are in the presence of NaCl. For ease, halophilic microorganisms are designated moderate halophiles when they grow best in seawater (NaCl content of about 3.5 %, w/v), and extreme halophiles when they grow best in hypersaline environments

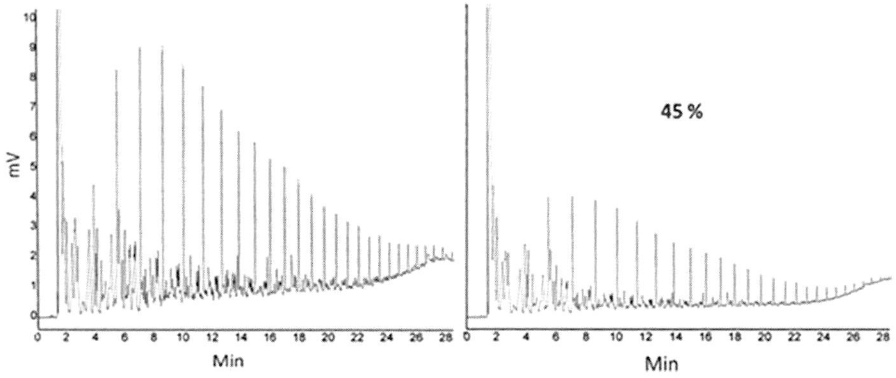

Fig. 12.2 Typical GLC profiles showing oil-consumption by extremely halophilic microorganisms. **Left*: Oil hydrocarbons at time zero. *Right*: Oil hydrocarbons 2 weeks after inoculation with 1 g hypersaline soil. The value on the right profile is the percent of oil consumed as measured in terms of total peak area reduction

(NaCl concentration higher than that of seawater). To keep water flow into cells in saline media, microorganisms must osmoregulate their cytoplasm making use of various compatible solutes, also called osmolytes (Grant et al. 1998; Burg and Ferraris 2008; Shivanand and Mugeraya 2011; DasSarma and DasSarma 2012). Halophiles comprise prokaryotes (bacteria and archaea) as well as eukaryotes (fungi, algae, higher plants, higher animals). Since this article is concerned with halophilic prokaryotes only, mechanisms related to osmoregulation in bacteria and archaea only will be discussed very briefly. It is commonly accepted that haloarchaea accumulate mainly K^+ in order to balance the high Na^+ concentration outside the cells. Consequently, their proteins and ribosomes are stabilized by high K^+ concentrations. In contrast, halophilic bacteria depend on synthesizing or accumulating compatible solutes such as sugars, alcohols, amino acids, betaines, ectoines and others. However, recent results in our laboratory indicate that aerobic, halophilic, hydrocarbonoclastic bacteria from hypersaline environments, like hydrocarbonoclastic haloarchaea from the same habitat use not only the monovalant K^+, but also the divalent Mg^{++} as osmoregulators (Al-Mailem et al. 2013). It has also recently been found that the moderately halohilic, chloride dependent *Halobacillus halophilus* is unique in integrating the concept of compatible solutes with Cl^- accumulation for coping with salt stress.

12.3.3 Identities

For ease, studies related to prokaryotes from environments with moderate salinity and those from hypersaline environments will be reviewed under separate subtitles.

12.3.3.1 Moderate Salinity Environments

Within the past two decades, many studies have been published on moderate halophilic bacteria with hydrocarbonoclastic potential. As mentioned above, the oceans, which represent the moderately saline environment, cover a major part of the earth's surface area. This environment accommodates a great number, of various hydrocarbonoclastic bacterial species, both in pristine and oil-polluted regions (Prince et al. 2003; Head et al. 2006; Yakimov et al. 2007). Before we deal in some detail with individual taxa, it is useful to start this subject by highlighting the unique group of the "obligate hydrocarbonoclastic bacteria" (OHCB) and related taxa. This group, as the name indicates, comprises, a few marine bacterial genera that are obligate hydrocarbon utilizers, *Alcanivorax*, *Marinobacter*, *Thallasolituus*, *Cycloclasticus* and *Oleispira* (Dyksterhouse et al. 1995; Golyshin et al. 2003, 2005; Yakimov et al. 2004, 2007; Liu and Shao 2005). Reportedly, the cell numbers of species belonging to such genera in seawater increase dramatically in response to the introduction of hydrocarbon in this water. Consequently, the rate of hydrocarbon removal increases, especially when limiting nutrients are amended. The biography, ecophysiology, genomics and potential applications of OHCB have been reviewed by Yakimov et al. (2007). Table 12.1 includes lists of halophilic microorganisms with hydrocarbon-utilization potential in the Arabian/Persian Gulf. It is obvious how diverse those microorganisms are.

Alcanivorax spp. most frequently reported as aliphatic and aromatic hydrocarbon-degraders in the marine environments include *A. borkumensis* (Golyshin et al. 2005; Fernandez-Martinez et al. 2003; Al-Awadhi et al. 2012b), *A. dieselolei* (Liu and Shao 2005; Al-Awadhi et al. 2012b) and *A. venustensis* (Fernandez-Martinez et al. 2003; Al-Awadhi et al. 2012b).

A marine genus which merits some more detailed consideration in relation to hydrocarbon-utilization is *Marinobacter*. The genus was first proposed by Guthier et al. (1992); it includes many hydrocarbonoclastic species (Guthier et al. 1992; Huu et al. 1999; Márquez and Ventosa 2005; Shivaji et al. 2005; Green et al. 2006; Al-Mailem et al. 2010a, 2013). The genus is a Gammaproteobacterium, systematically related to *Pseudomonas*. In an earlier attempt to study the hydrocarbon-oxidation mechanisms by *M. hydrocarbonoclasticus*, Lattuati et al. (2002) analyzed the lipids of this species that had been grown in synthetic seawater on a single carbon source, acetate or *n*-icosane. The authors

Table 12.1 Diversity of planktonic, halophilic microorganisms with hydrocarbonclastic potential in the Arabian – Persian Gulf

I. Moderate halophiles in seawater
(i) Culture-dependent analysis (Al-Awadhi et al. 2012a)
Acinetobacter junii, Agarivorans albus, Alcanivorax borkumensis, Alcanivorax dieselolei, Alcanivorax jadensis, Alcanivorax venustensis, Alteromonas addita, Alteromonas macleodii, Alteromonas marina, Arthrobacter globiformis, Cobetia marina, Dietzia maris, Echinicola vietnamensis, Gordonia lacunae, Klebsiella pneumonia, Kocuria rosea, Marinobacter hydrocarbonoclasticus, Marinobacter litoralis, Marinomonas communis, Microbacterium jejuense, Mycobacterium chlorophenolicum, Nesiotobacter exalbescens, Nitratireductor aquibiodomus, Oceanobacillus caeni, Pseudoalteromonas atlantica, Pseudoalteromonas prydzensis, Pseudoalteromonas tetraodonis, Pseudomonas pachastrellae, Pseudomonas stutzeri, Psychrobacter celer, Rhodococcus fascians, Shewanella oneidensis, Stappia kahanamokuae, Thalassospira tepidiphila, Vibrio alginolyticus, Vibrio fortis, Vibrio harveyi, Vibrio nereis, Vibrio sinaloensis
(ii) Culture-independent analysis (Al-Awadhi et al. 2013)
Alteromonas macleodii, Marinobacterium marisflavum, Phaeobacter caeruleus, Pseudoalteromonas phenolica, Thalassolituus oleivorans
II. Extreme halophiles in hypersaline areas
(i) Halophilic bacteria (Al-Mailem et al. 2012, 2013, 2014a, b)
Bacillus oleronius, Chromohalobacter salexigens, Exiguobacterium aurantiacum, Exiguobacterium homiense, Halomonas aquamarina, Halomonas organivorans, Halomonas salina, Halomonas sp., *Marinobacter flavimaris, Marinobacter sedimentalis, Marinobacter sedimentarum, Marinococcus* sp., *Pseudomonas* sp., *Salinicoccus* sp., *Salinivibrio siamensis, Stenotrophomonas maltophalia*
(ii) Haloarchaea (Al-Mailem et al. 2010a, b, 2012, 2014a, b)
Halobacterium piscisalsi, Halobacterium salinarum, Halobacterium sp., *Halococcus* sp., *Haloferax lucentense, Haloferax mucosum, Haloferax* sp., *Haloferax sulfurifontis*

found that *n*-icosane induced the formation of *n*-icosan-1-ol and *n*-icos-11-en-1-ol and a series of β-hydroxy acids ranging from C_{12} to C_{20}. It was suggested that the organism first hydroxylated *n-icosane* to C_{20} primary alcohol, transformed it to the C_{20} β-hydroxy acid which was subsequently degraded into lower homologues. Also the biodegradation of aromatic hydrocarbons by *Marinobacter* species has been studied rather early. Nicholson and Fathepure (2005) established a highly enriched halophilic culture with benzene as the sole source of carbon by using a brine soil. Reportedly, this enrichment completely removed benzene, toluene, ethylbenzene and xylene in culture within 1–2 weeks. The authors also provided experimental evidence that the enrichment produced $^{14}CO_2$ from (^{14}C) benzene. The microbiological analysis revealed that the enrichment was predominated by *Marinobacter* species. Interestingly, hydrocarbonoclastic species of this marine genus were also isolated from saline soils. Gu et al. (2007) described two novel strains which they isolated from an oil-polluted saline soil sample in China; both could grow at 0 and 15 % NaCl. According to their 16S rRNA gene sequences, the two strains were related to *M. lipolytica* and *M. bryozoorum*, they were classified as a novel species, *M. gudaonensis. M. psychrophilus* was isolated from sea ice, it grew at a temperature range of 0–22 °C with the optimum temperature between 15 and 18 °C (Zhang et al. 2008). Lee et al. (2011) isolated a halotolerant *Marinobacter* sp. PY975 capable of aliphatic hydrocarbon biodegradation. The organism grew between 15 and 35 °C with an optimum at 30 °C and in NaCl range of 0–10 % with the optimum at 0 °C. In addition to aliphatic hydrocarbons, this *Marinobacter* sp. could also utilize carbohydrates and organic acids. Gao et al. (2012) isolated *M. nanhaiticus*, an aromatic hydrocarbon utilizing bacterium from the sediments of South China Sea. Reportedly, this species grew best at 25 °C, pH 7.0–8.0 and in the presence of 1–5 % NaCl. In addition to polycyclic aromatic hydrocarbons, this species could also utilizes conventional organic substrates. Moxley and Schmidt (2012) isolated the marine bacterium *Marinobacterium* sp. KM2 which exhibited an absolute growth requirement for methylphenol, this species also utilized aliphatic hydrocarbons, e.g. butan-1-ol and hexadecane. Hedlund et al. (2001) described the *Marinobacter* strain NCE312 capable of degrading the aromatic hydrocarbons 1-methylnaphthalene and 2-methylnaphthalenes. Reportedly, this isolate had a naphthalene dioxygenase phylogenetically similar to that from *Pseudomonas* and *Burkholderia*, suggesting that the oxygenase coding gene may have been transferred horizontally between these lineages of bacteria.

Halomonas is another interesting, moderately halophilic bacterium with hydrocarbonoclastic potential. Garcia et al. (2004) isolated aromatic hydrocarbon utilizing-bacteria whose 16S rRNA gene sequences showed 98 % similarity to those of *Halomonas salina* and *H. halophila*, and classified them under the new species *H. organivorans*. This novel species could use a wide range of organic compounds viz. benzoic acid, *p*-hydroxybenzoic acid, cinnamic acid salicyclic acid, phenylacetic acid, phenylpropionic acid, phenol, *p*-coumaric acid, ferulic acid and *p-aminosalicylic* acid. Wang et al. (2007a) isolated two moderately halophilic bacterial strains from a saline oily soil in China. The 16S rRNA gene sequences of those strains had 96.4 % similarity with the sequence of *H. campisali* and 96.0 % with the sequence of *H. desiderata*. The authors suggested the name *H. gudaonensis* to this novel species. Another novel moderately halophilic species, *H. shenglensis* was also isolated from the above locality in China (Wang et al. 2007b). Reportedly, this novel sp. exhibited 98.1 and 97.8 % 16S rRNA gene sequence similarity to *H. alimentaria* and *H. ventosae*, respectively.

Several recent studies were concerned with hydrocarbon removal in moderately saline environments via microbial consortia. Al-Awadhi et al. (2007) reported that green animate materials from the Arabian/Persian Gulf coast accommodated alkaliphilic and halophilic consortia of hydrocarbonoclastic bacteria. Alkaliphiles comprised species belonging to the genera *Marinobacter*, *Micrococcus*, *Dietzia*, *Bacillus*, *Oceanobacillus* and *Citricoccus* and moderate halophiles were species belonging to the genera *Marinobacter*, *Georgenia*, *Microbacterium*, *Stappia*, *Bacillus*, *Isoptericola* and *Cellulomonas*. Reportedly, those isolates could utilize a wide range of *n*-alkanes and aromatic hydrocarbons. Zhao et al. (2009) established from an oily saline soil a moderately halophilic microbial consortium that degraded phenanthrene at 5–15 % salinity. Molecular analysis revealed that the constituent bacteria, based on the sequences of their electrophoresis-resolved 16S rRNA-genes were affiliated with known halophilic bacteria. Dastgheib et al. (2012) studied biodegradation of polycyclic aromatic hydrocarbons by a halophilic consortium comprising the genera *Marinobacter* and *Halomonas*. This consortium removed phenanthrene in the presence of between 1 and 17 % (w/v) NaCl. Reportedly, none of both bacterial strains alone could metabolize this aromatic hydrocarbon. It was assumed that *Marinobacter* played the central role in phananthrene biodegradation, whereas *Halomonas* played an auxiliary role via removing phenanthrene metabolites which were probably toxic.

Abed et al. (2006) described the bacterial diversity of a hydrocarbonoclastic cyanobacterial mat sample at elevated salinities and temperatures. About 15 % of the 16S rRNA-gene sequences were related to unknown, possibly novel bacteria. The subgroups Betaproteobacteria, Gammaproteobacteria, Deltaproteobacteria, Cytophaga, Flavobacterium, Bacteroidetes and Spirochetes were detected in the mats. Reportedly, *n*-octadecane was optimally biodegraded at salinities between 5 and 12 %; and phenanthrene and dibenzothiophene were utilized optimally at 3.5 % salinity, but considerable proportions of both aromatics were still removed at 8 % salinity. The above compounds were biodegraded within the temperature range of 15 and 40 °C, with optimal removal at 28–40 °C.

Jimenez et al. (2011) described a polyphasic approach for assessing changes in a marine bacterial community in the presence of fuel oil. The bacterial community initially consisted of mainly *Methylophaga* species, but 17 months later the obligate hydrocarbonoclastic genera *Alcanivorax* and *Marinobacter*, also *Thalasssopira* and *Alcaligens* became abundant and exhibited strong hydrocarbon biodegradation potential. Reportedly, the medium salinity affected the numbers of aromatic hydrocarbon degraders, while the identity of the substrate *n*-alkane was a key factor in counting aliphatic hydrocarbon utilizers. The authors concluded that, in addition to

Alcanivorax and *Marinobacter*, *Thalassospira* and *Roseobacter* contributed to removal of aliphatic hydrocarbons, whereas *Mesorhizobium* and *Muricauda* contributed to the degradation of aromatic hydrocarbons.

Lee et al. (2011) used gene-targeted FLX titanium pyrosequencing integrated with stable isotope probing (SIP) applying [^{13}C] biphenyl substrate to reveal that tidal mud flat sediments contain novel aromatic ring hydroxylating dioxygenase (ARHD) genes. Reportedly, more than 80 % of the detected ARHD genes comprised four clades (0.5 distance) with 49–70 % amino acid identity to sequences in public database. The 16S rRNA sequences enriched in the [^{13}C] fraction were from the Betaproteobacteria, bacilli and unclassified phyla. The bacteria belonged to the genera *Pusillimonas*, *Paenibacillus*, *Alcaligenes*, *Variovorax* and *Bacillus*. Al-Awadhi et al. (2012a, c) surveyed Kuwaiti desert soil and marine habitats with at least two-decade history of oil pollution for their inhabitant oil-utilizing bacterioflora. Similar to the moderately saline seawater, the desert soil is characterized by a low water activity. Seawater samples contained predominantly Gammaproteobacteria and desert soil samples mainly Actinobacteria. The Gammaproteobacteria comprised the genera *Psychrobacter*, *Vibrio*, *Agarovorans*, *Alteromonas*, *Marinobacter*, *Marinomonas*, *Alcanivorax*, *Klebsiella*, *Pseudomonas*, *Shevanella*, *Pseudoalteromonas*, *Acinetobacter* and *Cobetia*. The Actinobacteria comprised the genera *Microbacterium*, *Mycobacterium*, *Rhodococcus*, *Kocuria*, *Dietzia*, *Arthrobacter*, *Streptomyces*, *Agrococcus*, *Cellulomonas* and *Nocardia*. In addition, fewer bacterial genera belonging to the Alphaproteobacteria, Fimicutes and Cytophagia were also found. Interestingly, many of the genera had the combined potential for hydrocarbon utilization and nitrogen fixation. Ali et al. (2012) isolated from oil polluted water samples three halotolerant bacterial species belonging to the genera *Bacillus*, *Pseudomonas* and *Micrococcus* with the potential for asphaltene utilization. Reportedly, those bacteria degraded 83–96 % of 2.5 g l^{-1} asphaltene in 21 days at 30 °C and pH 7.

12.3.3.2 Hypersaline Environments

The lists of species in Table 12.1 also show the diversity of extremely halophilic microorganisms with hydrocarbonoclastic potential in the Arabian/Persian Gulf seawater. This microflora comprises halophilic bacteria and haloarchaea.

As should be expected, work on extremely halophilic hydrocarbonoclastic inhabitants of hypersaline environments started much later than that on moderately halophilic/halotolerant organisms. A number of interesting studies have been published on this subject during the past two decades, e.g. Kulichevskaya et al. (1992), Oren et al. (1992), Emerson et al. (1994), Daane et al. (2001), Margesin and Schinner (2001a), Garcia et al. (2004), Nicholson and Fathepure (2005), Al-Awadhi et al. (2007), Zhao et al. (2009), Al-Mailem et al. (2010a, b, 2012), Bonfá et al. (2011). Earlier studies in this field showed some variations in the obtained results. Kerr and Capone (1988) reported that salinity did not affect hydrocarbon biodegradation, whereas Diaz et al. (2000) and Yang et al. (2000) found that salinity enhanced this activity. Riis et al. (2003) and Kleinsteuber et al. (2001) showed that microbial consortia involved in hydrocarbon degradation can adapt to higher salinity.

Hypersaline environments are frequently contaminated with crude oil, heavy metals and/or other toxic products from anthropogenic sources (Hao and Lu 2009). Inoculation of such saline environments with nonextremophilic microorganisms is not effective in their bioremediation, therefore halophiles should be used. Oil reservoirs subject to enhanced oil recovery, usually have extreme conditions, i.e. high temperature (60–90 °C), and the soils and waters around them are saline and harbor halophilic higher plants and microorganisms. Hao and Lu (2009) isolated from one such environments in East China the halophilic bacterial strain TM-1 that had the potential for oil biodegradation. The strain consisted of Gram positive, nonmotile, spherical cells capable of growth at up to 58 °C and 18 % NaCl. Judged by the authors description, this strain should be a *Staphylococcus* sp.. Reportedly, this bacterium degraded various components of different heavy oils, resulting in the removal of aromatics, resins and asphaltenes, and consequent enrichment of the environment with light hydrocarbons. Al-Mailem et al. (2010b) isolated hydrocarbonoclastic haloarchaea from hypersaline coastal areas of the Arabian/Persian Gulf. These comprised two strains of *Haloferax* and one strain each of *Halobacterium* and *Halococcus* that needed at least 1 M NaCl to grow. Optimum salinity was 4 M NaCl and optimum temperature range was 40–45 °C. All strains could grow on a wide range of aliphatic and aromatic hydrocarbons. Depending on the strain identity and medium salinity, those haloarchaea could biodegrade crude oil (13–47 %), *n*-octadecane (28–67 %) and phenanthrene (13–30 %) in culture within 2 weeks. The rates of hydrocarbon biodegradation increased with increasing medium salinity, with an optimum at 3 M NaCl. Even at 4 M NaCl, the biodegradation rates were higher than at 1 and 2 M NaCl. In another study, the same authors (Al-Mailem et al. 2011) reported that those haloarchaea could resist and volatilize Hg^{2+} in culture at extremely high salinity. Individual strains resisted up to between 100 and 200 ppm $HgCl_2$ in hydrocarbon free media with salinities between 1 and 4 M NaCl, but only up to 20 and 30 ppm in the presence of 0.5 % (w/v) crude oil. Reportedly, the individual haloarchaea consumed more crude oil in the presence of 3 M NaCl than in the presence of 2 M NaCl. At both salinities, increasing the $HgCl_2$ concentration in the medium from 0 to 20 ppm resulted in decreasing the oil consumption values, although satisfactory oil consumption still occurred in the presence of 10 ppm $HgCl_2$.

Our group in Kuwait also isolated from the above hypersaline environments two halophilic hydrocarbonoclastic bacteria, *Marinobacter sedimentarum* and *M. flavimaris* with diazotrophic potential (Al-Mailem et al. 2013). Their numbers were in the magnitude of 10^3 cells g^{-1}. Both species failed to grow in the absence of NaCl, showed best growth and hydrocarbon biodegradation activity in the presence of 1–1.5 M NaCl, and still grew and maintained their hydrocarbonoclastic potential at salinities up to 5 M NaCl. They consumed both aliphatic and aromatic hydrocarbons, in addition to Tween 80, as sole sources of carbon and energy. The study provided an experimental evidence for the active nitrogen fixation by both species; both could successfully mineralize crude oil in saline mineral media and hypersaline soil and water microcosms without the addition of any nitrogen fertilizers.

Our group also adopted culture-dependent and culture-independent analysis to study hydrocarbonoclastic microorganisms indigenous to the two hypersaline areas in Kuwait about 100 Km apart (Al-Mailem et al. 2014b). Environmental parameters in both areas were similar, with the exception of the soil organic carbon content, which was in the north considerably higher than in the south. The hydrocarbonocalstic bacterial and haloarchaeal numbers and identities, as analyzed by culture-dependent approach, using media with various salinities, were quite similar. The bacterial species belonged to the genera *Halomonas*, *Chromohalobacter*, *Marinobacter*, *Exiguobacterium*, *Stenotrophomonas*, *Pseudomonas*, *Salinovibrio* and *Bacillus*. The halobacterial species belonged to the genera *Haloferax* and *Halobacterium*. On the other hand, analysis, by fingerprinting of 16S rDNA amplified fractions in the total genomic DNA followed by sequencing of the electrophoresis-resolved bands gave quite different microbial composition for the same samples as compared with the results of the culture-dependent method. Bacterial phylotypes found by the culture-independent approach were affiliated with the genera *Ochrobactrum*, *Sternotrophomonas*, *Rhodococcus* and "*Halomicrobium*", whereas the haloarchaeal genera were affiliated with *Halorussus*, *Halomicrobium* and *Haloorientalis*.

12.3.3.3 Bacteria Associated with Halophilic Plants

There are several publications on hydrocarbonoclastic bacteria associated with roots (rhizospheric bacteria) and leaves (phyllospheric bacteria) of conventional (nonextremophilic) plants, and their role in hydrocarbon removal in soil and air, respectively (Merkl et al. 2005; Radwan 2009; Dashti et al. 2009; Al-Awadhi et al. 2009, 2012b; Sorkhoh et al. 2010b; Ali et al. 2012). Similar, albeit much fewer reports have been published rather recently on bacteria associated with halophilic plants growing in saline environments. The most important findings are reviewed in the following paragraph.

Daane et al. (2001) isolated aromatic hydrocarbon degrading bacteria from contaminated estuarine sediment and salt marsh rhizosphere. The major bacterial isolates were assigned to three main bacterial groups: Gram-negative Pseudomonads; Gram-positive, non-sporing nocardioforms and the Gram-positive, spore forming *Paenibacillus*. The isolates could utilize a wide range of polynuclear aromatic hydrocarbons. It was concluded that the rhizospheres of salt marsh plants contain a variety of aromatic hydrocarbon utilizing bacteria, and that the plant associated microorganisms have the potential for bioremediation of the contaminated sediments. Al-Mailem et al. (2010b) reported that the roots and aerial organs of the halophyte *Halocnemum strobilaceum* (Chenopodiacea) naturally inhabiting hypersaline coastal areas of the Arabian/Persian Gulf accommodate up to 8×10^4

and 3×10^2 g^{-1}, respectively, of extremely halophilic hydrocarbon utilizing microorganisms. Such organisms were 14 to 38-fold more frequent in the rhizosphere than in the plant free soil. They belonged to the haloarchaea *Halobacterium* sp. and *Halococcus* sp., the Firmicutes *Brevibacillus borstenlensis* and the Proteobacteria *Pseudoalteromonas ruthenica* and *Halomonas sinaensis*. The phyllosphere of this plant harbored the Proteobacterium *Ochrobactrum* sp. and the dimorphic yeast *Candida utilis*. All those organisms grew on a wide range of aliphatic and aromatic hydrocarbons, as sole sources of carbon and energy. With the exception of the yeast which could not tolerate salinities more than 2 M NaCl, all the organisms grew successfully in the presence of between 1 and 4 M NaCl, with optimum growth between 1 and 2 M NaCl. The total rhizospheric and phyllospheric microbial consortia could remove oil in nitrogen free and nitrogen containing media. The authors concluded that this wild halophilic plant could be a tool for phytoremediation of oil polluted hypersaline environments. In this context, Sorkhoh et al. (2010b) recorded in the rhizospheres of conventional crops bacterial cells in the magnitude of 10^5 g^{-1} with the combined potential for hydrocarbon utilization and mercury resistance. Reportedly, those nonhalophilic, rhizospheric bacteria were *Citrobacter freundii*, *Enterobacter aerogenes*, *Exiguobacterium aurantiacum*, *Pseudomonas verinii*, *Micrococcus luteus*, *Brevibacillus brevis*, *Arthrobacter* sp. and *Flavobacterium psychrophilum*. This bacterial composition is quite different from that in the rhizospheres of halophilic plants. Furthermore, Sorkhoh et al. (2010a) found that the rhizospheric, hydrocarbonoclastic bacteria had the potential for nitrogen-fixation, which makes them valuable tools in bioremediation.

In addition to the plant-associated halophiles named above, there are also diverse microorganisms associated with other substrates, e.g. the biofouling materials in the marine environment (Table 12.2). Interestingly, many of those halophiles have the combined potential for hydrocarbon-utilization and nitrogen-fixation, which makes them self-independent in their nitrogen-nutrition.

Table 12.2 Diversity of halophilic microorganisms with hydrocarbonoclastic potential on biofouling materials in the Arabian – Persian Gulf (Al-Awadhi et al. 2012b)

Halophilic Microorganisms
Acinetobacter junii, *Agrobacterium tumefaciens*, *Alcanivorax borkumensis*, *Alcanivorax dieselolei*, *Alcanivorax jadensis*, *Alcanivorax venustensis*, *Alteromonas addita*, *Alteromonas macleodii*, *Cobetia marina*, *Dietzia cinnamea*, *Dietzia maris*, *Gordonia lacunae*, *Hoyosella altamirensis*, *Klebsiella pneumoniae*, *Kocuria palustris*, *Kocuria rosea*, *Labrenzia aggregate*, *Marinobacter hydrocarbonoclasticus*, *Marinobacter litoralis*, *Marinomonas communis*, *Microbacterium arborescens*, *Microbacterium jejuense*, *Mycobacterium chlorophenolicum*, *Nesiotobacter exalbescens*, *Nitratireductor aquibiodomus*, *Nitratireductor basaltis*, *Planococcus citreus*, *Planococcus maritimus*, *Pseudoalteromonas atlantica*, *Pseudoalteromonas prydzensis*, *Pseudoalteromonas tetraodonis*, *Pseudomonas pachastrellae*, *Pseudomonas mendocina*, *Pseudomonas stutzeri*, *Rhodococcus ruber*, *Vibrio alginolyticus*, *Vibrio communis*, *Vibrio fortis*, *Vibrio nereis*, *Vibrio shilonii*, *Vibrio sinaloensis*

12.4 Hydrocarbon Bioremediation Potential

Microorganisms including halophilic/halotolerant ones have a potential for use in biotechnology. The potential covers a wide range of medical, environmental, agricultural and industrial applications. Needles to say, and for understandable reasons, studies based on nonextremophilic microorganism are currently much more extensive than those based on extremophiles. In this part of the article, the potential of hydrocarbonoclastic, halophilic/halotolerant organisms only for cleaning hydrocarbon polluted, saline environments and wastes will be considered. A limited number of reviews have been published on this subject (Margesin and Schinner 2001a, b; Le Borgne et al. 2008; Zhuang et al. 2010). The biotechnology applied here is termed bioremediation which means the use of microorganisms in removing xenobiotic pollutants (Atlas and Pramer 1990). Although conventional approaches e.g. land filling, incineration and others are known (Kuiper et al. 2004) and have been applied in practice, they are comparatively costly (Rosenberg 1993), and may lead to air and ground-water pollution.

Bioremediation comprises two approaches; bioaugmentation or seeding and/or biostimulation or fertilization (Atlas and Bartha 1998; Radwan 2009).

12.4.1 Bioaugmentation (Seeding)

As both terms indicate, this approach involves the inoculation of contaminated environment with hydrocarbonoclastic bacterial cocktails to accelerate the hydrocarbon removal (Van Limbergen et al. 1998; Kuiper et al. 2004). Domde et al. (2007) reported that bioaugmentation automatically leads to the introduction into the polluted area of additional gene pools complementary to the already existing genes, thus enhancing the biodegradation, Microorganisms used in bioaugmentation may be wild-types or genetically manipulated to suit the proposed biodegradation activity (Dixon 1996; Erb et al. 1997; Yee et al. 1998; Reineke 1998; Dejonghe et al. 2001). In practice, there are only very few attempts of hydrocarbon bioremediation via bioaugmentation. These few attempts were limited to conventional (nonhalophilic) microorganisms in nonsaline environments. Thus, Domde et al. (2007) working on hydrocarbon contaminated waste water, reported that the water, chemical oxygen demand, a parameter reflecting the organic substance content, was reduced by 52.2 %, compared with only 15.1 % reduction in the control without bioaugmentation. This optimistic result has unfortunately neither been confirmed nor consolidated by subsequent workers in this field. Inoculation of microorganisms in a new environment as a means of enhancing a biological activity has rarely been evaluated as effective. Typical examples are the disappointing practices related to diazotrophic bacterial inoculation as a natural means of nitrogen fertilization, alternative to fertilization with chemicals. In most cases, the introduced microorganisms, as expected, faced a strong competition by the already existing indigenous microflora, a stress which the inoculated ones could not stand. According to own experience, inoculation experiments using conventional hydrocarbonoclastic bacteria were, not successful at all as means of oil bioremediation (Radwan 1991; Radwan et al. 1997).

Sometimes the hydrocarbon pollutant may persist although the proper microorganisms may be present, probably due to the too complex nature of the pollutant (El Fantroussi and Agathos 2005). According to Atlas and Bartha (1998), species of the genus *Pseudomonas* are frequently selected for bioaugmentation, based on their potential for utilization of a wide range of pollutants.

12.4.2 Biostimulation (Biofertilization)

This approach seems to be more appropriate than the bioaugmentation (Bento et al. 2005). It involves the improvement of microbial growth conditions viz. aeration, pH, moisture etc. as well as providing the microflora with limiting nutrients such as nitrogenous compounds, phosphorous compounds and others. Studies under controlled laboratory conditions and in the open field indicate that biostimulation is quite promising as bioremediation technique for removing hydrocarbon pollutants in conventional (nonextreme) environments (Bossert and Bartha 1984; Leahy and Colwell 1990; Atlas 1991; Margesin and Schinner 1998; Choi et al. 2002). However, although numerous studies repeatedly revealed enhancing effects of nitrogen and phosphorous fertilization on hydrocarbon removal by microorganisms (Ramadan et al. 1990; Kerry 1993; Mahro et al. 1994; Braddock et al. 1997; Namkoong et al. 2002), a few other studies failed to confirm this enhancing effect (Seklemova et al. 2001; Rahman et al. 2002). This contradiction demonstrates that there is still a gap of information regarding microbial requirements in the field (Head 1998). This is also valid for extreme (including saline) environments which have been much less frequently studied. However, in addition to satisfying the nitrogen and phosphorous requirements, microorganisms generally benefit also from the addition of energy sources such as sugars and peptides (Radwan et al. 2000a) as well as electron acceptors (El Fantroussi and Agathos 2005).

12.4.3 Phytoremediation

Since chemical and physical remediation approaches for oily environments may harm the ecosystem, particularly the indigenous microorganisms (Harvey et al. 1986), vegetation seems to be an acceptable smart alternative fulfilling the bioremediation objectives (Cunningham et al. 1996; Hutchinson et al. 2003; Radwan 2009). Phytoremediation technology has the benefits of being simple, easy, attractive, economical and environmentally friendly (Komives and Gullner 2000; Radwan 2009). In fact, phytoremediation involves bioaugmentation and biostimulation combined together in one environmentally friendly practice. Higher plants hardly take up the hydrophobic hydrocarbons, yet they harbor on their roots (rhizospheres) unique microbial communities comparatively different from those in their nonrhizosphere soil (Tesar et al. 2002). In other words, cropping and vegetation imply real bioaugmentation practice. Furthermore, root exudates and disintegrating root tissues and falling litter provide the rhizospheric microflora with valuable nutrients (vitamins, amino acids, carbon sources, minerals etc.) (Komives and Gulluer 2000). The rhizosphere is much better aerated and its moisture content is higher than that of the nonrhizosphere soil (Curl and Truelove 1986; Vevrek and Campbell 2002; Radwan 2009). Cropping and vegetation imply real biostimulation practice for the microflora. Conventional cropping managements act enhancing on the rhizosphere microflora, which have been early recognized as metabolizers of several xenobiotic compounds (Boyle and Shann 1995). On the other hand, for successful hydrocarbon phytoremediation, the used higher plant and its rhizospheric microorganisms must be capable of survival and growth in the presence of ambient hydrocarbon and oil concentrations (White et al. 2006). Several of such plants, mostly conventional, but rarely halophiles have been found to fulfill this condition and were used successfully in hydrocarbon phytoremediation programs. Examples are grass species (April and Sims 1990; Schwab and Banks 1994; Radwan et al. 2000b), alfalfa (Schwab and Banks 1994; Wiltse et al. 1998); other legume plants (Radwan et al. 1998), alpine bluegrass (Nichols et al. 1997); *Halocnemum strobilaceum* (Al-Mailem et al. 2010a), clover, sunflower and Indian mustard (Dominguez-Rosado and Pichtel 2004) and others.

12.4.4 Bioremediation of Oily Extreme Environments

Most hydrocarbon bioremediation studies have been done on nonextreme environments and nonextremophilic microorganisms (Van Hamme et al. 2003; Rosernberg 2006; Radwan 2009). Comparable studies on extreme environments are still very few although as mentioned before, hypersaline areas are widely distributed all over the globe and, like the nonextreme areas, they receive various contaminants including hydrocarbons and hydrocarbon derivatives (Lefebvre and Moletta 2006). In this context, hydrocarbon solubility in water decreases with increasing water salinity (Whitehouse 1984). An inverse relationship has been observed between medium salinity level and polyaromatic-hydrocarbon biodegradation activity (Minai-Tehrani et al. 2009).

Halophily among prokaryotes is usually associated with some degree of thermophily. Temperature affects the physical properties of hydrocarbons and consequently their rate of biodegradation (Margesin and Schinner 2001a). Higher temperature reduces oil viscosity, therefore, within certain limits of course, more hydrocarbon bioremediation occurs with increasing temperature. There are reports on hydrocarbonoclastic, thermophilic bacteria, e.g. *Bacillus* sp. (*Geobacillus stearothermophilus*) in hot environments (Sorkhoh et al. 1993). Perfumo et al. (2007) compared the rate of hydrocarbon bioremediation of a hot environment sample at low (18 °C) and high (60 °C) temperatures, in the presence and absence of inorganic fertilizers. Best bioremediation rates occurred at 60 °C and following fertilization with nitrogen, phosphorus and potassium compounds. Higher rates of bioremediation were also measured after bioaugmentation with the thermophilic, hydrocarbonoclastic species *Geobacillus thermoleovorans* T80 (which is indigenous to the studied area), combined with biostimulation at high temperature.

Also cold environments, e.g. Arctic and Antarctic soil and seawater may become polluted with hydrocarbons (Siron et al. 1995; Braddock et al. 1997; Aislabie et al. 2000). Such environments harbor indigenous hydrocarbonoclastic bacteria, whose numbers are higher in hydrocarbon contaminated than in pristine sites (Whyte et al. 1999). However, as it should be expected, the microbial hydrocarbonoclastic activity is very low in frozen areas (Siron et al. 1995).

There are also reports on acidophilic hydrocarbon-utilizing microorganisms (Margesin and Schinner 2001a). Stapleton et al. (1998) reported the successful bioremediation of water and soil samples with ambient pH values of 2 using acidophilic microorganisms. The latter comprised both prokaryotes and eukaryotes capable of degrading aromatic hydrocarbons. A few studies were concerned with hydrocarbon bioremediation by alkaliphilic microorganisms at high pH values (Yakimov et al. 2003; Al-Awadhi et al. 2007). These can be obligate or facultative (alkalitolerant) alkaliphiles (Krulwich and Guffanti 1989) which occur in soda lakes and carbonate rich soils (Kleinsteuber et al. 2001). The haloalkaliphile *Halomonas organivorans* could degrade hydrocarbons at pH 9 in the presence of 10 % NaCl (Oie et al. 2007).

12.4.5 Factors Affecting Hydrocarbon Biodegradation at High Salinity

This subject has been repeatedly reviewed, but only with regard to nonextreme environments and nonextremophilic microorganisms (Atlas 1981, 1995; Leahy and Colwell 1990; Radwan et al. 1995; Radwan 2008, 2009). Factors affecting hydrocarbon biodegradation include temperature, nutrient availability, oxygen, pH and hydrocarbon concentration and composition; their effects on microorganisms are well known and need not to be described here. Earlier investigators identified five possible reasons for the failure of bioremediation in nonextreme environments (Goldstein et al. 1985): the contaminant concentration may be too low to support microbial growth, the presence of microbial inhibitors, the presence of natural microbial enemies (e.g. grazing protozoa), the presence of better utilizable carbon sources and unavailability of the contaminant to the degrading microorganisms. There is reason to believe that all those factors may limit hydrocarbon bioremediation in saline environments in more or less the same way. However, recent studies in our laboratory highlighted effects of some additional specific factors on hydrocarbon bioremediation in hypersaline environments. These factors merit some detailed description here. Table 12.3 shows the effect of specific amendments on microbial oil-removal in hypersaline media.

Table 12.3 Enhancing crude-oil-removal in hypersaline media by specific substances

Substances	% oil removal		References
	Without the substance	With the substance	
Vitamin mixture[a]	20	39	Al-Mailem et al. (2014a)
Magnesium ions	16	36	Al-Mailem et al. (2013)
Potassium ions	20	38	Al-Mailem et al. (2013)

[a]Composed of: (mgl^{-1}): 5.0 thiamine; 10 pyridoxine; 0.1 vitamin B12; 2.0 biotin; 5.0 riboflavin; 2.0 folic acid

In one of the recent studies, experimental evidence was offered that oil-removal in hypersaline environments (>4 M NaCl) was enhanced by nitrogen fertilization and by illumination (Al-Mailem et al. 2012). Hypersaline samples, artificially polluted with 3 % crude oil, were incubated at 40 °C under natural day/night cycles. After 4–6 weeks, the indigenous microflora consumed between 63 and 66 % of the initial amount of crude oil. No oil-removal occurred in the autoclaved controls. Samples fertilized with the organic nitrogen source casminoacids (3 %) lost between 86 and 89 % of the available crude oil. Samples incubated under continuous illumination lost almost double as much more oil than samples incubated in the dark. According to the sequences of their 16S rRNA genes, the indigenous hydrocarbonoclastic microorganisms were affiliated to *Halomonas aquamarina*, *Exiguobacterium aurentiacum*, *Haloferax* sp., *Salinococcus* sp., *Marinococcus* sp., and *Halomonas* sp.. The numbers of these organisms increased during oil bioremediation from the magnitude of 10^4 g^{-1} at time zero to the magnitude of 10^7–10^8 g^{-1} after 6 week incubation, and the *Haloferax* sp. (an archaeaon) proportion increased from 20 to 93 % of the whole hydrocarbonoclastic microflora. At the extremely high salinity used, the limited O_2 solubility in water seemed to have been compensated by the haloarchaon via ATP photophosphorylation using its characteristic bacteriorhodopsin.

In another study, it was found that oil bioremediation by halophilic microorganisms from the above hypersaline samples were enhanced by vitamin-fertilization (Al-Mailem et al. 2014a). In this study, ten hydrocarbonoclastic bacterial species viz. *Halomonas salina*, *H. organivorans*, *Chromohalobacter salexigens*, *Marinobacter sedimentalis*, *M. flavimaris*, *Exiguobacterium homiense*, *Stenotrophomonas maltophilia*, *Pseudmonas* sp., *Salinovibrio siamensis* and *Bacillus oleronius*, as well as five haloarchaal species *Haloferax lucentense*, *Halobacterium salinarum*, *Halobacterium piscisalsi*, *Haloferax mucosum*, and *Halobacterium sulfurifontis*, which had been isolated from the hypersaline environment, were used. The oil and pure hydrocarbon-consumption values by individual microorganisms in media provided with 3–4 M NaCl were significantly enhanced by vitamin fertilization. The most effective vitamins were thiamin, pyridoxine and vitamin B12. Especially haloarchaea are known for their vitamin requirement.

Still in another study (Al-Mailem et al. 2013), the ten extremely halophilic bacterial species and the five haloarchaea named above were found to grow better and to consume more crude oil in hypersaline media receiving between 0.50 and 0.75 M

KCl, and between 1.50 and 2.25 M $MgSO_4$ added as osmoregulators. Within certain limits, the higher the KCl and $MgSO_4$ concentrations in the hypersaline media, the more K^+ and Mg^{2+}, respectively were accumulated by cells of all the tested bacteria and archaea. Also when total natural microbial consortia in the hypersaline samples were used as inocula in bioremediation experiments, hydrocarbon -consumption was enhanced in the presence of the above concentrations of KCl and $MgSO_4$. It was concluded that amendment of calculated concentrations of K^+ and Mg^{2+} could be a promising practice for hydrocarbon-bioremediation in hypersaline environments. Such cations seem to play a significant role as osmoregulators, but other roles like protein (enzyme) molecule stabilization cannot be excluded.

12.5 Promising Perspectives

In their natural habitats, hydrocarbonoclastic microorganisms, like all other microorganisms do not occur alone, not even in environments with long history of hydrocarbon pollution. Nonhydrocarbonoclastic microorganisms do also exist and influence the hydrocarbonoclastic group directly and indirectly. The nature of such microbial associations which range from just "loosely coexisting organisms" (commensals) up to "spatially integrated societies" is a major factor controlling the success or failure of any bioremediation program. The integrated microbial associations as they are expressed in "biofilms" appear to be optimally suited for remediation purposes (Singh et al. 2006; Nilanjana et al. 2012) because of their high microbial biomass and ability to immobilize cells and pollutants. So far, only little attenuation has been devoted to hydrocarbonoclastic biofilms and their role in hydrocarbon remediation. This is true for both nonextreme and extreme environments. However, in view of the continuously increasing volume of hydrocarbon contaminated liquid wastes produced by the modern industry, it should be expected that more attention will be focused on bioreactor dependent hydrocarbon bioremediation. Here, the hydrocarbonoclastic biofilms will represent promising perspectives for bioreactor-dependent waste hydrocarbon removal.

12.5.1 Natural Hydrocarbonoclastic Biofilms

The initial step in biofilm formation is the microbial production of extracellular polysaccharides forming the so called glycocalyx, which mediates pioneer cellular adhesion to the inert substrate and protects the developing biomass from antimicrobial agents (Costerton et al. 1987). Microbial biofilm formation appears to be a rule for all microorganisms in nature, but we will limit our discussion here to hydrocarbonoclastic biofilms. This subject has been studied thoroughly in our laboratory, but also by other researchers elsewhere.

Coastal microbial mats are very widely distributed in marine environments, especially in tropical and subtropical areas. They consist of cyanobacterial biomass harboring in their polysaccharide envelopes bacteria including hydrocarbonclastic ones (Sorkhoh et al. 1992). Another group of natural hydrocarbonoclastic biofilms are the epilithic biomass communities coating coastal gravel particles, they also comprise hydrocarbonoclastic bacteria (Radwan and Al-Hasan 2001). There are also hydrocarbonoclastic biofilms occurring naturally on the surfaces of macroalgae (Radwan et al. 2002), marine biofouling materials (Al-Awadhi et al. 2012b) as well as fish (Radwan et al. 2007). Mention has already been made of hydrocarbonoclastic biofilms on roots (Radwan et al. 1995) and leaves (Swaminathen and Kochlar 1989; Schlegel 1995; Al-Awadhi et al. 2009), that are potentially useful tools for hydrocarbon bioremediation through rhizosphere and phyllopshere technology.

12.5.2 Man-Made Hydrocarbonoclastic Biofilms

So far, there is only limited work done on man-made biofilms, based on hydrocarbonoclastic microorganisms. This is in contrast to the extensive research currently available on the application of biofilms in many other fields viz, human diseases, as well as in removal of many nonhydrocarbon pollutants in the environment (Lear and Lewis 2012).

Our laboratory as well as other laboratories produced several publications on man-made, hydrocarbonoclastic biofilms. Al-Awadhi et al. (2003) isolated microbial consortia from coastal gravel particles and used them for biofilm development on naked gravel particles as well as on glass plates. These biofilms were effective in removing hydrocarbons in batch liquid cultures and the same biofilms could be repeatedly used in successive bioremediation cycles. The microbial consortia in the biofilms consisted of phototrophs, hydrocarbonoclastic bacteria and diazotrophic bacteria. De Lorenzo et al. (2005) used man-made hydrocarbonoclastic biofilms in the treatment of ground water contaminated with aromatic hydrocarbons. Major hydrocarbonoclastic species in those biofilms were *Rhodococcus erythropolis* and *Pseudomonas marginalsi*. Andrews et al. (2006) established biofilms rich in the aromatic hydrocarbon degraders, *Acinetobacter* sp. and *Pseudomonas putida*. Gertier et al. (2009) treated a marine oil spill in microcosms with an oil-sorbent which reportedly attracted *Alcanivorax borkumensis* with the consequence of aliphatic and aromatic hydrocarbon removal. Golby et al. (2012) established environmental mixed species biofilm comprising the genera *Pseudomonas*, *Thauera*, *Hydroyenophaga*, *Rhodoferax* and *Acidovorax*, which was effective in removing hydrocarbon pollutants. Al-Bader et al. (2013) established also mixed species biofilm harboring together anoxygenic and oxygenic phototrophic bacteria, diazotrophic bacteria and hydrocarbonoclastic bacteria, which removed hydrocarbon pollutants effectively in batch cultures. Man-made hydrocarbonoclastic biofilms were also described recently by another few investigators (Chandran and Das 2011; Vaysse et al. 2011). Table 12.4 shows the diversity

Table 12.4 Diversity of halophilic microorganisms with hydrocarbon utilization potential in man-made biofilms

I. Biofilms based on seawater microorganisms (Al-Bader et al. 2013)
(i) Culture independent analysis
Azospirillum brasilense, Azospirillum lipoferum, Dichotomicrobium thermohalophilum, Flexibacter tractuosus, Oceanobacterium insulare, Oleibacter marinus, Rhodobacter aestuarii, Rhodobacter maris, Rhodovulum iodosum, Roseovarius aestuarii, Thalassobius aestuarii
II. Biofilms based on sewage microorganisms (Al-Mailem et al. 2014c)
(i) Culture dependent analysis
Achromobacter denitrificans, Achromobacter xylosoxidans, Ancylobacter polymorphus, Bosea thiooxidans, Brachybacterium paraconglomeratum, Chelatococcus asaccharovorans, Gordonia terrae, Kaistia adipata, Microbacterium esteraromaticum, Microvirga subterranean, Mycobacterium peregrinum, Mycobacterium smegmatis, Pseudomonas aeruginosa, Pseudomonas alcaligenes, Pseudomonas mendocina, Stenotrophomonas acidaminiphila, Zavarzinia compransoris
(ii) Culture independent analysis
Acinetobacter beijerinckii, Alkalitalea saponilacus, Cloacibacterium normanense, Flavobacterium defluvii, Lysobacter capsici, Mucilaginibacter daejeonensis, Ochromonas distigma, Planobacterium taklimakanense, Perlucidibaca piscinae, Petrimonas sulfuriphila, Prolixibacter bellariivorans, Pseudomonas panipatensis, Pseudomonas plecoglossicida, Zavarzinia compransoris

of halophilic microorganisms with hydrocarbonoclastic potential in biofilms recently constructed in our laboratory.

Most of the environmental biofilms described above comprised phototrophic microorganisms. Therefore, according to unpublished data in our laboratory, exposure of biofilms during bioremediation to light enhanced their hydrocarbon removal potential by about 33 %. On the other hand, treating the batch cultures with thioglycollate to reduce the redox potential of the medium inhibited the hydrocarbon consumption by up to 90 %, indicating that most of the hydrocarbon removal occurred aerobically.

12.6 Conclusions

Moderately saline and hypersaline environments in the biosphere harbor a wide diversity of halophilic/halotolerant microorganisms capable of removing oil pollutants in those environments naturally. Such organisms are either planktonic or, more frequently, occur in natural associations as biofilms coating inanimate and animate substrates. Such biofilms immobilize and bring together the proper combination of microorganisms and polluting compounds. It is possible to construct such hydrocarbonoclastic biofilms artificially and to use them in bioreactors in liquid waste hydrocarbon removal. For this, much research is still needed to optimize conditions for optimal hydrocarbon biodegradation.

Acknowledgements Thanks are due to Mr. Mohamed Eliyas for the technical help he offered to us during the preparation of this chapter. Much literature information in this text has been collected within our Research projects RS 01/12, and YS 03/04. For this, we have to thank also our earlier co-workers Mrs. Rasha Sulaiman and Mrs. Maiss Marafie, as well as our current co-worker Mrs. Mayada Kansour.

References

Abed RMM, Al-Thukair A, de Beer D (2006) Bacterial diversity of a cyanobacterial mat-degrading petroleum compounds at elevated salinities and temperatures. FEMS Microbiol Ecol 57:290–301

Aislabie J, Foght J, Saul D (2000) Aromatic hydrocarbon-degrading bacteria from soil near Scott Base, Antarctica. Polar Biol 23:183–188

Aislabie M, Balks M, Foght J, Waterhouse E (2004) Hydrocarbon spills on Antarctic soils: effects and management. Environ Sci Technol 38:1265–1275

Al-Awadhi H, Al-Hasan RH, Sorkhoh NA, Salamah S, Radwan SS (2003) Establishing oil-degrading biofilms on gravel particles and glass plates. Int Biodeterior Biodegrad 51:181–185

Al-Awadhi H, Sulaiman RHD, Mahmoud HM, Radwan SS (2007) Alkaliphilic and halophilic hydrocarbon-utilizing bacteria from Kuwaiti coasts of the Arabian Gulf. Appl Microbiol Biotechnol 77:183–186

Al-Awadhi H, El-Nemr I, Mahmoud H, Sorkhoh NA, Radwan SS (2009) Plant-associated bacteria as tools for the phytoremediation of oily nitrogen-poor soils. Int J Phytoremed 11:11–27

Al-Awadhi H, Al-Mailem D, Dashti N, Hakam L, Eliyas M, Radwan SS (2012a) The abundant occurrence of hydrocarbon-utilizing bacteria in the phyllospheres of cultivated and wild plants in Kuwait. Int Biodeterior Biodegrad 73:73–79

Al-Awadhi H, Dashti N, Kansour M, Sorkhoh NA, Radwan SS (2012b) Hydrocarbon-utilizing bacteria associated with biofouling materials from offshore waters of the Arabian Gulf. Int Biodeterior Biodegrad 69:10–16

Al-Bader D, Kansour M, Rayan R, Radwan SS (2013) Biofilm comprising phototrophic, diazotrophic, and hydrocarbon-utilizing bacteria: a promising consortium in the bioremediation of aquatic hydrocarbon pollutants. Environ Sci Pollut Res 20:3252–3262

Al-Hasan RH, Khanafer M, Eliyas M, Radwan SS (2001) Hydrocarbon accumulation by picocyanobacteria from the Arabian Gulf. J Appl Microbiol 91:533–540

Ali N, Sorkhoh N, Salamah S, Eliyas M, Radwan SS (2012) The potential of epiphytic hydrocarbon-utilizing bacteria on legume leaves for attenuation of atmospheric hydrocarbon pollutants. J Environ Manag 93:113–120

Al-Mailem DM, Sorkhoh NA, Al-Awadhi H, Eliyas M, Samir SS (2010a) Biodegradation of crude oil and pure hydrocarbons by extreme halophilic archaea from hypersaline coasts of the Arabian Gulf. Extremophiles 14:321–328

Al-Mailem DM, Sorkhoh NA, Marafie M, Al-Awadhi H, Eliyas M, Radwan SS (2010b) Oil phytoremediation potential of hypersaline coasts of the Arabian Gulf using rhizosphere technology. Bioresour Technol 101:5786–5792

Al-Mailem DM, Al-Awadhi H, Sorkhoh NA, Eliyas M, Radwan SS (2011) Mercury resistance and volatilization by oil utilizing haloarchaea under hypersaline conditions. Extremophiles 15:39–44

Al-Mailem D, Eliyas M, Radwan SS (2012) Enhanced haloarchaeal oil removal in hypersaline environments via organic nitrogen fertilization and illumination. Extremophiles 16:751–758

Al-Mailem DM, Eliyas M, Radwan SS (2013) Bioremediation of oily hypersaline soil and water via potassium and magnesium amendment. Can J Microbiol 59:837–844

Al-Mailem DM, Eliyas M, Radwan SS (2014a) Enhanced bioremediation of oily hypersaline coastal areas in Kuwait via vitamin-fertilization. Environ Sci Pollut Res 21:3386–3394

Al-Mailem DM, Eliyas M, Khanafer M, Radwan SS (2014b) Culture-dependent and culture-independent analysis of hydrocarbonoclastic microorganisms indigenous to hypersaline environments in Kuwait. Microb Ecol 67:857–865
Al-Mailem DM, Kansour MK, Radwan SS (2014c) Hydrocarbonoclastic biofilms based on sewage microorganisms and their application in hydrocarbon removal in liquid wastes. Can J Microbiol 60:477–486
Anderson C, LaBelle R (2000) Update of comparative occurrence rates for offshore oil spills. Spill Sci Technol Bull 6:303–321
Andrews JS, Mason VP, Thompson IP, Stephens GM, Markx GH (2006) Construction of artificially structured microbial consortia (ASMC) using dielectrophoresis: examining bacterial interactions via metabolic intermediates within environmental biofilms. J Microbiol Meth 64:96–106
April W, Sims RC (1990) Evaluation of the use of prairie grasses for stimulating polycyclic aromatic hydrocarbons treatment in soil. Chemosphere 20:253–265
Atlas RM (1981) Microbial degradation of petroleum hydrocarbons: an environmental perspective. Microbiol Rev 45:180–209
Atlas RM (1991) Microbial hydrocarbon degradation – bioremediation of oil spills. J Chem Technol Biotechnol 52:149–156
Atlas RM (1995) Bioremediation of petroleum pollutants. Int Biodeterior Biodegrad 35:317–327
Atlas RM, Bartha R (1998) Microbial ecology: fundamentals and applications, 4th edn. Benjamin/Cummings Publishing Company Inc., Don Mills
Atlas RM, Pramer D (1990) Focus on bioremediation. ASM News 56:352–353
Balks RM, Paetzold RF, Kimble JM, Aislabie J, Campbell IB (2002) Effects of hydrocarbon spills on the temperature and moisture regimes of cryosols in the Ross Sea region. Antar Sci 14:319
Bartha R (1986) Biotechnology of petroleum pollutant biodegradation. Microb Ecol 12:155–172
Bento FM, Camargo FAO, Okeke BC, Frankenberger WT (2005) Comparative bioremediation of soils contaminated with diesel oil by natural attenuation, biostimulation and bioaugmentation. Bioresour Technol 96:1049–1055
Bonfá MR, Grossman MJ, Mellado E, Durrant LR (2011) Biodegradation of aromatic hydrocarbons by Haloarchaea and their use for the reduction of the chemical oxygen demand of hypersaline petroleum produced water. Chemosphere 84:1671–1676
Bonney M, Jaber M (2011) Environmentally responsible inventory models: non-classical models for a non-classical era. Int J Prod Econ 133:43–53
Bossert I, Bartha R (1984) The fat of petroleum in soil ecosystems. In: Atlas RM (ed) Petroleum microbiology. Macmillan Publishing Co., New York, pp 434–476
Boulton CA, Ratledge C (1984) The physiology of hydrocarbon-utilization microorganisms. In: Wiseman A (ed) Topics in fermentation and enzyme technology, vol 9. Ellis Horwood, Chichester, pp 11–77
Boyle J, Shann J (1995) Biodegradation of phenol 2,4-DCP and 2,4,5-T in field-collected rhizosphere and non-rhizosphere soils. J Environ Qual 24:782–785
Braddock J, Ruth M, Catterall P, Walworth J, McCarthy K (1997) Enhancement and inhibition of microbial activity in hydrocarbon-contaminated arctic soils: implications for nutrient amended bioremediation. Environ Sci Technol 31:2078–2084
Brown DW, Romas LS, Friedman AJ, Maclod WD (1979) Analysis of trace levels of petroleum hydrocarbons in marine sediments using a solvent-slurry extraction procedure. In: Trace organic analysis: a new frontier in analytical chemistry, proceedings 9th materials symposium, National Bureau of Standards, Washington, DC, pp 161–167
Burg MB, Ferraris JD (2008) Intracellular organic osmolytes: function and regulation. J Biol Chem 283:7309–7313
Burger AE (1993) Estimating the mortality of seabirds following oil spills: effects of spill volume. Mar Pollut Bull 26:140–143
Chandran P, Das N (2011) Degradation of diesel oil by immobilized *Candida tropicalis* and biofilm formed on gravel. Biodegradation 22:1181–1189

Choi SC, Kwon KK, Sohn JH, Kim SJ (2002) Evaluation of fertilizer additions to stimulate oil biodegradation in sand seashore mesocosms. J Microbiol Biotechnol 12:431–436
Colwell RR, Walker R (1977) Ecological aspects of microbial degradation of petroleum in the marine environment. Crit Rev Microbiol 5:423–445
Cooney JJ (1984) The fate of petroleum pollutants in freshwater ecosystems. In: Atlas RM (ed) Petroleum microbiology. Macmillan Publishing Co., New York, pp 399–434
Costerton JW, Geesey GG, Cheng KJ (1987) How bacteria stick. Sci Am 238:86–95
Cunningham SD, Anderson TA, Schwab AP, Hsu FC (1996) Phytoremediation of soils contaminated with organic pollutants. Adv Agron 56:55–114
Curl EA, Truelove B (1986) The rhizosphere. Springer, Berlin
Daane LL, Harjono I, Zylstra GJ, Haggblom MM (2001) Isolation and characterization of polycyclic aromatic hydrocarbon-degrading bacteria associated with rhizosphere of salt marsh plants. Appl Environ Microbiol 67:2683–2691
Dashti N, Khanafer M, Ali N, El-Nemr I, Sorkhoh NA, Radwan SS (2009) The potential of oil-utilizing bacterial consortia associated with legume root nodules for cleaning oily soils. Chemosphere 74:1354–1359
DasSarma S, DasSarma P (2012) Halophiles. In: eLS. Wiley Ltd: Chichester. doi: 10.1002/9780470015902.a0000394
Dastgheib SM, Amoozegar MA, Khajeh K, Shavandi M, Ventosa A (2012) Biodegradation of polycyclic aromatic hydrocarbons by a halophilic microbial consortium. Appl Microbiol Biotechnol 65:789–798
Davis SJ, Gibbs CF (1975) The effect of weathering on crude oil residue exposed at sea. Water Resour 9:275–285
De Lorenzo AD, Varcamonti M, Parascandola P, Vignola R, Bernardi A, Sacceddu P, Sisto R, Alteriis ED (2005) Characterization and performance of a toluene-degrading biofilm developed on pumice stones. Microbiol Cell Fact 4:4
Dejonghe W, Boon N, Seghers D, Top EM, Verstraete W (2001) Bioaugmentation of soils by increasing microbial richness: missing links. Environ Microbiol 3:649–657
Diaz MP, Grigson SJW, Peppiatt C, Burgess JG (2000) Isolation and characterization of novel hydrocarbon degrading euryhaline consortia from crude oil and mangrove sediments. Mar Biotechnol 2:522–532
Dixon B (1996) Bioremediation is here to stay. ASM News 62:527–528
Domde P, Kapley A, Purohit HJ (2007) Impact of bioaugmentation with a consortium of bacteria on the remediation of wastewater-containing hydrocarbons. Environ Sci Pollut Res 14:7–11
Dominguez-Rosado E, Pichtel J (2004) Phytoremediation of soil contaminated with used motor oil: II. Greenhouse studies. Environ Eng Sci 21:169–180
Dyksterhouse SE, Gray JP, Herwig RP, Lara JC, Staley JT (1995) *Cycloclasticus pugetii* gen. nov., sp. nov., an aromatic hydrocarbon-degrading bacterium from marine sediments. Int J Syst Bacteriol 45:116–123
El Fantroussi S, Agathos SN (2005) Is bioaugmentation a feasible strategy for pollutant removal and site remediation? Curr Opin Microbiol 8:268–275
Emerson D, Chauhan S, Oriel P, Breznak JA (1994) *Haloferax* sp. D1227, a halophilic Archaeon capable of growth on aromatic compounds. Arch Microbiol 161:445–452
Erb RW, Eichner CA, Wagner-Döbler I, Timmis KN (1997) Bioprotection of microbial communities from toxic phenol mixtures by a genetically designed pseudomonad. Nat Biotechnol 15:378–382
Fernandez-Martinez J, Pujalte MJ, Garcia-Martinez J, Mata M, Garay E, Rodriguez-Valera F (2003) Description of *Alcanivorax venustensis* sp. nov. and reclassification of *Fundibacter jadensis* DSM 12178T (Bruns and Berthe-Corti 1999) as *Alcanivorax jadensis* comb. nov., members of the emended genus *Alcanivorax*. Int J Syst Evol Microbiol 53:331–338
Forsyth MP, Shindler DB, Gochnauer MB, Kushner DJ (1971) Salt tolerance of intertidal marine bacteria. Can J Microbiol 17:825–828
Fourcans A, Garc´ıa de Oteyza T, Wieland A, Solé A, Diestra E, Van Bleijswijk J, Grimalt JO, Kühl M, Esteve I, Muyzer G, Caumette P, Duran R (2004) Characterization of functional bacte-

rial groups in a hypersaline microbial mat community (Salins-de-Giraud, Camargue, France). FEMS Microbiol Ecol 51:55–70
Fox G (2001) Wildlife as sentinels of human health effects in the great lakes-st. Lawrence Basin. Environ Heal Perspect 109:853–861
Gao W, Cui Z, Li Q, Xu G, Jia X, Zheng L (2012) *Marinobacter nanhaiticus* sp. nov., polycyclic aromatic hydrocarbon-degrading bacterium isolated from the sediment of the South China Sea. Antonie Van Leeuwenhock 130:485–491
Garcia MR, Mellado E, Ostos JC, Ventosa A (2004) *Halomonas organivorans* sp. nov., a moderate halophile able to degrade aromatic compounds. Int J Syst Evol Microbiol 54:1723–1728
Geiselbrecht AD, Herwig RP, Deming JW, Staley JT (1996) Enumeration and phylogenetic analysis of polycyclic aromatic hydrocarbon-degrading marine bacteria from Pudget sound sediments. Appl Environ Microbiol 62:3344–3349
Gertier C, Gerdts G, Timmins KN, Yakimov MM, Golyshin PN (2009) Populations of heavy fuel oil-degrading marine microbial community in presence of sorbent materials. J Appl Microbiol 107:590–605
Golby S, Ceri H, Gieg LM, Chatterjee I, Margues LL, Turner RJ (2012) Evaluation of microbial biofilm communities from an Alberta oil sands tailings pond. FEMS Microbiol Ecol 79:240–250
Goldstein JF, Mallory LM, Alexander M (1985) Reason for possible failure of inoculation to enhance biodegradation. Appl Environ Microbiol 50:917–983
Golyshin PN, Martins Dos Santos VA, Kaiser O, Ferrer M, Sabirova YS, Lunsdorf H, Chernikova TN, Golyshina OV, Yakimov MM, Puhler A, Timmis KN (2003) Genome sequence completed of *Alcanivorax borkumensis*, a hydrocarbon degrading bacterium that plays a global role in oil removal from marine systems. J Biotechnol 106:215–220
Golyshin PN, Harayama S, Timmis KN, Yakimov MM (2005) Family Alcanivoraceae. In: Garrity G (ed) Bergey's manual of systematic bacteriology, vol 2. Springer, Berlin, pp 295–298
Grant WD, Gemmell RT, McGenity TJ (1998) Halophiles. In: Horikoshi K, Grant WD (eds) Extremophiles: microbial life in extreme environments. Wiley-Liss, New York, pp 93–132
Green DH, Bowman JP, Smith EA, Gutierrez T, Bolch CJS (2006) *Marinobacter algicola* sp. nov., isolated from laboratory cultures of paralytic shellfish toxin-producing dinoflagellates. Int J Syst Evol Microbiol 56:523–527
Gu SH, Kralovec AC, Christensen ER, Van Camp RP (2007) Source apportionment of PAHs in dated sediments from the Black River, Ohio. Water Resour 37:2149–2161
Guthier MJ, Lafay B, Christen R, Fernandez L, Acquaviva M, Bonin P, Bertrand JC (1992) *Marinobacter hydrocarbonoclasticus* gen. nov., sp. nov., a new, extremely halotolerant, hydrocarbon-degrading marine bacterium. Int J Syst Bacteriol 42:568–576
Hao R, Lu A (2009) Biodegradation of heavy oils by halophilic bacterium. Proc Natl Acad Sci U S A 19:997–1001
Harvey HR, Fallon RD, Patton JS (1986) The effect of organic matter and oxygen on the degradation of bacterial membrane lipids in marine sediments. Geochim Cosmochim Acta 50:795–804
Head IM (1998) Bioremediation: a response to gross environmental abuse. Trends Biotechnol 11:599–608
Head IM, Jones DM, Roling WF (2006) Marine microorganisms make a meal of oil. Nat Rev Microbiol 4:173–182
Hedlund BP, Geiselbrecht AD, Staley JT (2001) *Marinobacter* strain NCE312 has a *Pseudomonas-like* naphthalene dioxygenase. FEMS Microbiol Lett 201:47–51
Heller JM, Wortman WJ, Weir JC (2000) Using modeled data and Geographic Informations System (GIS) technology for the investigation of Gulf war veterans' environmental exposures and illnesses. In: 4th international conference on integrating GIS and Environmental Modeling (GIS/EM4): problems, prospects and research needs, Banff, Alberta, Canada, 2–8 September 2000
Henner P, Schiavon M, Druelle V, Lichtfouse E (1999) Phytotoxicity of ancient gaswork soils: effect of polycyclic aromatic hydrocarbons (PAHs) on plant germentation. Organ Geochem 30:963–969

Hischmann K (2005) The Kuwait oil fires (Environmental disasters). In: Levine M (ed) Facts on File Inc, New York
Horowitz A, Atlas RM (1977) Response of microorganisms to an accidental gasoline spillage in an arctic freshwater ecosystem. Appl Environ Microbiol 33:1252–1258
Hutchinson SL, Schwab AP, Banks MK (2003) Bioremediation of petroleum hydrocarbons in the rhizosphere. In: McCutcheon, Schnoor JL (eds) Phytoremediation: transformation and control of contaminants. Steven C, pp 355–386
Huu NB, Denner EBM, Ha DTC, Wanner G, Stan-Lotter H (1999) *Marinobacter aquaeolei* sp. nov., a halophilic bacterium isolated from a Vietnamese oil-producing well. Int J Syst Bacteriol 49:367–375
Jimenez N, Vinas M, Guiu-Aragones C, Bayona JM, Albaiges J, Solanas AM (2011) Polyphasic approach for assessing changes in an autochthonous marine bacterial community in the presence of Prestige fuel oil and its biodegradation potential. Appl Microbiol Biotechnol 91:823–834
Kerr RP, Copone DG (1988) The effect of salinity on the microbial mineralization of two polycyclic aromatic hydrocarbons in Estuarine sediments. Marine Environ Res 26:181–198
Kerry E (1993) Bioremediation of experimental petroleum spills on mineral soils in the Vestfold Hills, Antarctica. Polar Biol 13:163–170
Kleinsteuber S, Müller RH, Babel W (2001) Expression of the 2,4-D degradative pathway of pJP4 in an alkaliphilic, moderately halophilic soda lake isolate, *Halomonas* sp. EF43. Extremophiles 5:375–384
Klug MJ, Markovetz AJ (1971) Utilization of aliphatic hydrocarbons by microorganisms. Adv Microb Physiol 5:1–43
Komives T, Gullner G (2000) Phytoremediation. In: Wilkinson RE (ed) Plant-environment interaction. Marcel Dekker, New York, pp 437–452
Konlechner JN (1985) Investigation of the fate of paraffin based crude oil in an Antarctic terrestrial ecosystem. N Z Antarct Rec 6:40
Krulwich TA, Guffanti AA (1989) Alkalophilic bacteria. Annu Rev Microbiol 43:435–463
Kuiper I, Lagendijk EL, Bloemberg GV, Lugtenberg JJ (2004) Rhizoremediation: a beneficial plant-microbe interaction. Mol Plant-Microbe Interact 17:6–15
Kulichevskaya IS, Milekhina EI, Borzenkov IA, Zvyagintseva IS, Belyaev SS (1992) Oxidation of petroleum hydrocarbons by extremely halophilic archaeobacteria. Microbiology 60:596–601
Kushner DJ (1978) Life in high salt and solute concentration. In: Kushner DJ (ed) Microbial life in extreme environments. Academic, London, pp 317–368
Kushner DJ (1993) Growth and nutrition of halophilic bacteria. In: Vreeland RH, Hochstein LI (eds) The biology of halophilic bacteria. CRC Press, Boca Raton, pp 87–103
Lattuati A, Metzger P, Acquaviva M, Bertrand JC, Largeau C (2002) *n*-Alkane degradation by *Marinobacter hydrocarbonoclasticus* strain SP 17: long chain β-hydroxy acids as indicators of bacterial activity. Org Geochem 33:37–45
Le Borgne S, Paniagua D, Vazquez-Duhalt R (2008) Biodegradation of organic pollutants by halophilic bacteria and archaea. J Mol Microbiol Biotechnol 15:74–92
Leahy JG, Colwell RR (1990) Microbial degradation of hydrocarbon in the environment. Microbiol Rev 54:305–315
Lear G, Lewis GD (2012) Microbial biofilms: current research and applications. Caister Academic, Auckland
Lee TK, Lee J, Sul WJ, Iwai S, Chai B, Tiedje JM, Park J (2011) Novel biphenyl-oxidizing bacteria and dioxygenase genes from a Korean tidal mudflat. Appl Environ Microbiol 77:3888–3891
Lefebvre O, Moletta R (2006) Treatment of organic pollution in industrial saline wastewater: a literature review. Water Res 40:3671–3682
Liu C, Shao Z (2005) *Alcanivorax dieselolei* sp. nov., a novel alkane degrading bacterium isolated from sea water and deep-sea sediment. Int J Syst Evol Microbiol 55:1181–1186
Mahro B, Schaefer G, Kästner M (1994) Pathways of microbial degradation of polycyclic aromatic hydrocarbons in soil. In: Hinchee RE, Leeson A, Semprini L, Ong SK (eds) Bioremediation of

chlorinated and polycyclic aromatic hydrocarbon compounds. Lewis Publishers, Boca Raton, pp 203–217
Major JT (1990) Genesis and the origin of coal and oil. Apologetics Press, Montgomery
Margesin R, Schinner F (1998) Oil biodegradation potential in alpine habitats. Arc Alpine Res 30:262–265
Margesin R, Schinner F (2001a) Biodegradation and bioremediation of hydrocarbons in extreme environments. Appl Microbiol Biotechnol 56:650–663
Margesin R, Schinner F (2001b) Potential of halotolerant and halophilic microorganisms for biotechnology. Extremophiles 5:73–83
Márquez MC, Ventosa A (2005) *Marinobacter hydrocarbonoclasticus* Gauthier et al. 1992 and *Marinobacter aquaeolei* Nguyen et al. 1999 are heterotypic synonyms. Int J Syst Evol Microbiol 55:1349–1351
McKinnon M, Vine P (1991) Tides of war: eco-disaster in the Gulf. Boxtree Limited, London
Merkl N, Schultze-Kraft R, Infante C (2005) Assessment of tropical grasses and legumes for phytoremediation of petroleum-contaminated soils. Water Air Soil Pollut 165:195–209
Minai-Tehrani D, Minoui S, Herfatmanesh A (2009) Effect of salinity on bioremediation of Polycyclic Aromatic Hydrocarbon (PAHS) of heavy crude oil in soil. Bull Environ Cont Tax 82:179–184
Moxley K, Schmidt S (2012) Isolation of a phenol-utilizing marine bacterium from Durban Harbour (South Africa) and its preliminary characterization as *Marinobacter* sp. KM2. Water Sci Technol 65:932–939
Namkoong W, Hwang EY, Park JS, Choi JY (2002) Bioremediation of diesel-contaminated soil with composting. Environ Pollut 119:23–31
Nichols TD, Wolf DC, Rogers HB, Beyrouty CA, Reynolds CM (1997) Rhizosphere microbial populations in contaminated soils. Water Sci Technol 95:165–178
Nicholson CA, Fathepure BZ (2005) Biodegradation of benzene by halophilic and halotolerant bacteria under aerobic conditions. Appl Environ Microbiol 70:1222–1225
Nilanjana D, Lakshmi V, Geetanjali B, Jaseetha AS, Evy AAM (2012) Application of biofilms on remediation of pollutants-an overview. J Microbiol Biotechnol Res 2:783–790
Nwoko CO, Okeke PN, Agwu OO, Akpan IE (2007) Performance of Phaseolus vulgaris L. in a soil contaminated with spent-engine oil. Afr J Biotechnol 6:1922–1925
Oie CSI, Albaugh CE, Peyton BM (2007) Benzoate and salicylate degradation by *Halomonas campisalis*, an alkaliphilic and moderately halophilic microorganism. Water Res 41:1235–1242
Oren A (2002) Diversity of halophilic microorganisms: environments, phylogeny, physiology and applications. J Ind Microbiol Biotechnol 28:56–63
Oren A, Gurevich P, Azachi M, Hents Y (1992) Microbial degradation of pollutants at high salt concentrations. Biogeosciences 3:387–398
Perfumo A, Banat IM, Marchant R, Vezzulli L (2007) Thermally enhanced approaches for bioremediation of hydrocarbon-contaminated soils. Chemosphere 66:179–184
Prince RC, Lessard RR, Clark JR (2003) Bioremediation of marine oil spills. Oil Gas Sci Technol 58:463–468
Quesada E, Ventosa A, Rodriguez-Valera F, Ramos-Cermenzana A (1982) Types and properties of some bacteria isolated from hypersaline soils. J Appl Microbiol 53:155–161
Radwan SS (1991) Gulf oil spill. Nature 350:456
Radwan SS (2008) Microbiology of oil-contaminated desert soils and coastal areas in the Arabian Gulf. In: Dion P, Nautiyal CS (eds) Microbiology of extreme soils, vol 13, Soil biology. Springer, Berlin, pp 275–297
Radwan SS (2009) Phytoremediation for oily desert soils. In: Singh A, Kuhad RC, Ward OP (eds) Advances in applied bioremediation. Springer, Berlin, pp 279–298
Radwan SS, Al-Hasan RH (2001) Potential application of coastal biofilm-coated gravel particles for treating oily waste. Aqua Microbiol Ecol 23:113–117
Radwan SS, Sorkhoh N (1993) Lipids of *n*-alkane-utilizing microorganisms and their application potential. Adv Appl Microbiol 39:29–90

Radwan SS, Sorkhoh NA, Al-Hasan RH (1995) Self-cleaning and bioremediation potential of the Arabian Gulf. In: Cheremisinoff P (ed) Encyclopedia of environmental control technology, vol 9. Gulf Publishing, Hasbrouck Heights, pp 901–924

Radwan SS, Sorkhoh NA, El-Nemr I, El-Desouky AF (1997) A feasibility study on seeding as a bioremediation practice for the oily Kuwaiti desert. J Appl Microbiol 83:353–358

Radwan SS, Al-Awadhi H, Sorkhoh NA, El-Nemr I (1998) Rhizospheric hydrocarbon-utilizing microorganisms as potential contributors to phytoremediation of the oily Kuwaiti desert. Microbiol Res 153:247–251

Radwan SS, Al-Awadhi H, El-Nemr IM (2000a) Cropping as a phytoremediation practice for oily soil with reference to crop safety as food. Int J Phytoremed 2:383–396

Radwan SS, Al-Mailem D, El-Nemr I, Salamah S (2000b) Enhanced remediation of hydrocarbon contaminated desert soil fertilized with organic carbons. Int Biodeterior Biodegrad 46:129–132

Radwan SS, Al-Hasan RH, Salamah S, Al-Dabbous S (2002) Bioremediation of oily sea water by bacteria immobilized in biofilms coating macroalgae. Int Biodeterior Biodegrad 50:55–59

Radwan SS, Al-Hasan RH, Mahmoud HM, Eliyas M (2007) Oil-utilizing bacteria associated with fish from the Arabian Gulf. J Appl Microbiol 103:2160–2167

Rahman KSM, Thahira-Rahman J, Lakshmanaperumalsamy P, Banat IM (2002) Towards efficient crude oil degradation by a mixed bacterial consortium. Bioresour Technol 85:257–261

Ramadan MA, Tayeb OM, Alexander M (1990) Inoculation size as a factor limiting success of inoculation for bioremediation. Appl Environ Microbiol 56:1392–1396

Rehm HJ, Reiff I (1981) Mechanisms and occurrence of microbial oxidation of long-chain alkanes. Adv Biochem Eng 19:175–216

Reineke W (1998) Development of hybrid strains for the mineralization of chloroaromatics by patchwork assembly. Annu Rev Microbiol 52:287–331

Riis V, Kleinsteuber S, Babel W (2003) Influence of high salinities on the degradation of diesel fuel by bacteria consortia. Can J Microbiol 49:713–721

Rosenberg E (1993) Exploiting microbial growth on hydrocarbons-new markets. Trend Biotechnol 11:419–424

Rosenberg E (2006) Hydrocarbon-oxidizing bacteria. In: Dworkin M, Falkow S, Rosenberg E, Schleifer KH, Stackebrandt E (eds) The prokaryotes, a handbook on the biology of bacteria, vol 2, 3rd edn. Springer, Berlin, pp 564–577

Schlegel HG (1997) General microbiology, 7th edn. Cambridge University Press, Cambridge

Schwab AP, Banks MK (1994) Biologically mediated dissipation of polyaromatic hydrocarbons in the root zone. In: Anderson TA, Coats JR (eds) Bioremediation through rhizosphere technology. American Chemical Society, Washington, DC, pp 132–141

Seklemova E, Pavlova A, Kovacheva K (2001) Biostimulation-based bioremediation of diesel fuel: field demonstration. Biodegradation 12:311–316

Shivaji S, Gupta P, Chaturvedi P, Suresh K, Delille D (2005) *Marinobacter maritimus* sp. nov., a psychrotolerant strain isolated from sea water off the subantarctic Kerguelen islands. Int J Syst Evol Microbiol 55:1453–1456

Shivanand P, Mugeraya G (2011) Halophilic bacteria and their compatible solutes-osmoregulation and potential applications. Curr Sci 100:1516–1521

Sims RC, Overcash MR (1983) Fate of polynuclear aromatic compounds (PNAs) in soil-plant systems. Residue Rev 88:18–68

Singh R, Paul D, Jain RK (2006) Biofilms: implications and bioremediation. Trends Microbiol 14:389–397

Siron R, Pelletier E, Brochu C (1995) Environmental factors influencing the biodegradation of petroleum hydrocarbons in cold seawater. Arch Environ Cont Toxicol 28:406–416

Sorkhoh N, Al-Hassan R, Hopner T, Radwan S (1992) Self-cleaning of the Gulf. Nature 359:109

Sorkhoh NA, Ibrahim AS, Ghannoum MA, Radwan SS (1993) High temperature hydrocarbon degradation by *Bacillus stearothermophilus* from oil polluted Kuwaiti desert. Appl Microbiol Biotechnol 39:123–126

Sorkhoh NA, Ali N, Al-Awadhi H, Dashti N, Al-Mailem DM, Eliyas M, Radwan SS (2010a) Phytoremediation for mercury in pristine and crude oil contaminated soils: contributions of rhizobacteria and their host plants to mercury removal. Ecotoxicol Environ Saf 73:1998–2003

Sorkhoh NA, Ali N, Salamah S, Eliyas M, Khanafer M, Radwan SS (2010b) Enrichment of rhizospheres of crop plants raised in oily sand with hydrocarbon-utilizing bacteria capable of hydrocarbon consumption in nitrogen free media. Int Biodeterior Biodegrad 64:659–664

Stapleton RD, Savage DC, Sayler GS (1998) Biodegradation of aromatic hydrocarbons in an extremely acidic environment. Appl Environ Microbiol 64:4180–4184

Stevenson JJ (1966) Lipids in soil. J Am Oil Chem Soc 43:203–210

Swaminathan MS, Kochlar SL (1989) Hydrocarbon yielding plants. In: Plants and society. Borin M and Sattin M (ed), Macmillan Publishers Limited, London, pp 577–589

Tesar M, Reichenauer TG, Sessitsch A (2002) Bacterial rhizosphere communities of black poplar and herbal plants to be used for phytoremediation of diesel fuel. Soil Biol Biochem 34:1883–1892

Van Hamme JD, Singh A, Ward O (2003) Recent advances in petroleum microbiology. Microbiol Mol Biol Rev 67:503–549

Van Limbergen H, Top EM, Verstraete W (1998) Bioaugmentation in activated sludge: current features and future perspectives. Appl Microbiol Biotechnol 50:16–23

Vaysse PJ, Sivadon P, Goulas P, Grimaud R (2011) Cells dispersed from *Marinobacter hydrocarbonocasticus* SP17 biofilm exhibit a specific protein profile associated with a higher ability to reinitiate biofilm development at the hexadecane–water interface. Environ Microbiol 13:737–746

Ventosa A, Nieto JJ, Oren A (1998) Biology of aerobic moderately halophilic bacteria. Microbiol Mol Biol Rev 62:504–544

Vevrek MC, Campbell WJ (2002) Identification of plant traits that enhance biodegradation of oil. In: 9th annual international petroleum environmental conference, Albuquerque, 22–25 October

Vilhelmsson O, Hafsteinsson H, Kristjansson JK (1996) Isolation and characterization of moderately halophilic bacteria from fully cured salted cod (Bachdoo). J Appl Bacteriol 81:95–103

Wang YN, Cai H, Chi CQ, Lu AH, Lin XG, Jiang ZF, Wu XL (2007a) Halomonas shengliensis sp., a moderately halophilic, denitrifying, crude-oil utilizing bacterium. Int J Syst Evol Microbiol 57:1222–1226

Wang YN, Cai H, Yu SL, Wang ZY, Liu J, WU XL (2007b) *Halomonas gudaonensis* sp. nov., isolated from a saline soil contaminated by crude oil. Int J Syst Evol Microbiol 57:911–915

White PM, Duane JR, Wolf C, Thoma GJ, Reynolds CM (2006) Phytoremediation of alkylated polycyclic aromatic hydrocarbons in a crude oil-contaminated soil. Water Air Soil Pollut 169:1–4

Whitehouse BG (1984) The effects of temperature and salinity on the aqueous solubility of polynuclear aromatic hydrocarbons. Mar Chem 14:319–332

Wiltse CC, Rooney WL, Chen Z, Schwab AP, Banks MK (1998) Greenhouse evaluation of agronomic and crude oil-phytoremediation potential among alfalfa genotypes. J Environ Qual 27:169–173

Whyte LG, Bourbonnière L, Bellerose C, Greer CW (1999) Bioremediation assessment of hydrocarbon-contaminated soils from the high Arctic. Bioremed J 3:69–79

Yakimov MM, Giuliano L, Gentile G, Crisafi E, Chernikova TN, Abraham WR, Lünsdorf H, Timmis KN, Golyshin PN (2003) *Oleispira antarctica* gen. nov., sp. nov., a novel hydrocarbonoclastic marine bacterium isolated from Antarctic coastal sea water. Int J Syst Evol Microbiol 53:779–785

Yakimov MM, Giuliano L, Denaro R, Crisafi E, Chernikova TN, Giuliano WR, Luensdorf H, Timmis KN, Golyshin PN (2004) *Thalassolituus oleivorans* gen. nov., sp. nov., a novel marine bacterium that obligately utilizes hydrocarbons. Int J Syst Bacteriol 54:141–148

Yakimov MM, Timmis KN, Golyshin PN (2007) Obligate oil-degrading marine bacteria. Curr Opin Biotechnol 18:257–266

Yang L, Lai CT, Shieh WK (2000) Biodegradation of dispersed diesel fuel under high saline conditions. Water Res 34:3303–3314

Yee DC, Maynard JA, Wood TK (1998) Rhizoremediation of trichloroethylene by a recombinant, root-colonizing Pseudomonas fluorescens strain expressing toluene ortho-monooxygenase constitutively. Appl Environ Microbiol 64:112–118

Zhang DC, Li HR, Xin YH, Chi ZM, Zhou PJ, Yu Y (2008) *Marinobacter psychrophilus* sp. nov., a psychrophilic bacterium isolated from the Arctic. Int J Syst Evol Microbiol 58:1463–1466

Zhao B, Wang H, Mao X, Li R (2009) Biodegradation of phenanthrene by a halophilic bacterial consortium under aerobic conditions. Curr Microbiol 58:205–210

Zhuang X, Han Z, Nai Z, Zhuang G, Shim H (2010) Progress in decontamination by halophilic microorganisms in saline wastewater and soil. Environ Pollut 158:1119–1126

Chapter 13
Hydrolytic Enzymes in Halophilic Bacteria, Properties and Biotechnological Potential

Mohammad Ali Amoozegar and Maryam Siroosi

Abstract Halophilic bacteria are one of the most important extremophilic microorganisms that can be found in saline or hypersaline environments which are widely distributed around the world. They are not only adapted to live in saline environments in where other organisms are not able to thrive and grow, but also subjected to other kinds of extreme conditions, like high pH values, high or low temperature, low oxygen availability, pressure, and toxic metals. Because of these abilities, biodiversity and biotechnological applications of halophiles have been studied since then these were introduced to microbiologists. Among different applications of halophilic bacteria, enzymes, especially hydrolases, always have received the most attraction in recent years. Enclosed is a summary review on biology of halophilic bacteria and their general applications with a closer look at the hydrolases produced by these microorganisms (haloenzymes).

Keywords Hydrolase • Halophilic bacteria • Haloenzyme • Biochemical characterization

13.1 Introduction

Halophilic bacteria are one of the important extremophilic microorganism groups that are adapted to live in saline environments such as hypersaline lakes and wetlands, saline and hypersaline soils, cold saline environments, alkaline hypersaline lakes, and salted fermented foods. Some of their products have unique features, which could be very good choices for further analysis in order to apply in biotechnology and industrial fields. The most isolated hydrolases from mesophilic organisms cannot do their function in industrial extreme conditions. This problem forces researchers to find ways for stabilizing enzymes by enzyme engineering or

M.A. Amoozegar (✉) • M. Siroosi
Extremophiles Laboratory, Department of Microbiology, School of Biology and Center of Excellence in Phylogeny of Living Organisms, College of Science, University of Tehran, Tehran, Iran
e-mail: amoozegar@ut.ac.ir

D.K. Maheshwari, M. Saraf (eds.), *Halophiles*, Sustainable Development and Biodiversity 6, DOI 10.1007/978-3-319-14595-2_13

exploring enzymes from extremophiles (extremozymes) with natural stability in harsh conditions in industries. Hydrolases from halophilic bacteria (haloenzymes), as a member of extremozymes, are able to do hydrolytic reactions in presence of salt, in a medium with low water activity, and in some cases at extreme pH values, temperatures, and in presence of organic solvents (Karbalaei-Heidari et al. 2007; Shafiei et al. 2011; Li et al. 2014).

Extremophilic microorganisms live in ecosystems with extreme conditions such as high or low temperatures, high or low pH values, high pressure or high salinity, which cannot be tolerated by other mesophilic microorganisms. Halophilic microorganisms founded in the three domains of life in all over the world, are one of the extremophiles needing salts, especially NaCl for their growth. Halophiles inhabit in saline ecosystems such as salterns and hypersaline environments (Gomes and Steiner 2004), thalassohaline and athalassohaline. In thalassohaline brines, the salt composition is the same as seawater salt content, with NaCl as major salt, while athalassohaline systems have different salt composition from seawater (Litchfield and Gillevet 2002).

Although medium ingredients and temperature can alter amount of salt concentration needed for growth of microorganisms (Oren 2008), according to the optimal growth in different concentration of NaCl, extreme halophiles are able to grow optimally at 15–30 % (w/v) (2.5–5.2 M) NaCl, moderate halophiles at 3–15 % (w/v) (0.5–2.5 M) NaCl, slight halophiles at 1–3 % (w/v) (0.2–0.5 M) NaCl, non-halophiles at less than 1 % (w/v) (0.2 M) NaCl, and halotolerants are non-halophiles tolerating high NaCl concentrations (Kushner and Kamekura 1988).

All living organisms need salt, but high salt concentration leads to harmful osmotic stress for cells (DasSarma et al. 2010). Halophiles use two main mechanisms to adapt with high osmotic pressure due to the presence of high NaCl concentration: extremely halophilic archaea and halophilic, anaerobic bacteria (*Haloanaerobiales*) accumulate NaCl or KCl in their cytoplasm (Galinski and Trüper 1994; Ventosa et al. 1998), but most prokaryotes accumulate compatible solutes or osmolytes in their cytoplasm which are highly soluble and low molecular weight organic molecules without any interference with the normal metabolism of cells even at high concentrations in the cytoplasm, so they are named "compatible solutes" (Brown 1976; Lentzen and Schwarz 2006). Polyols (e.g. glycerol), sugars (e.g. trehalose), amino acids (e.g. glutamic acid), and quaternary amines are the bases of compatible solutes (e.g. glycine betaine) (Karan et al. 2012b). Although many bacterial and archaeal compatible solutes are the same, some of archaeal ones carry a negative charge (Martin et al. 1999).

In the domain *Bacteria* different types of halophilic (especially moderate rather than extreme halophiles) and halotolerant bacteria are founded (Oren 2002a), which can be photosynthetic, lithotrophic, or heterotrophic microorganisms (Jiang et al. 2006). Halophilic bacteria are related to different phyla such as: *Cyanobacteria, Proteobacteria, Firmicutes, Actinobacteria, Spirochaetes and Bacteroidetes*. All members of Halanaerobiales (in phylum *Firmicutes*) are anaerobic, fermentative, and halophiles (Ratnakar 2013). Optimal growth of moderately halophilic bacteria occurs at 3–15 % (w/v) NaCl (Ventosa et al. 1998), and this range for extremely halophilic bacterial growth is in upper limit from 20 % (w/v) to saturation (Ventosa 2006).

Diversity of halobacteria from different hypersaline habitats has been studied. In some cases, alongside exploring biodiversity and taxonomy of halobacteria, biotechnological potential of halobacteria has been examined (Babavalian et al. 2014).

13.2 Biotechnological Potential of Halobacteria

Halophilic bacteria, like other microorganisms, have unique properties leading to specific biotechnological and industrial applications. This group of bacteria using in biotechnology has some distinct advantages. They are usually fast growing with simple nutritional properties, the presence of salt in the culture medium reduces unwanted contaminations, they are capable of growing in exceptionally high concentrations of ions, and finally genetic tools for their manipulation are available (Oren 2002b).

Overall, halophilic bacterial applications could be categorized in some groups. Bioremediation of toxic compounds including organic and inorganic materials, production of haloenzymes, compatible solutes, and other valuable compounds could be valuable in different fields of biotechnology and industries (Oren 2002a; Amoozegar et al. 2011, 2012; Alavi et al. 2014).

As mentioned above, compatible solutes are intracellular organic compounds protecting cell against high osmotic stress in saline environments. Compatible solutes could be used as stabilizers of enzymes, nucleic acids, membranes, and whole cells, or as stress-protective agents (Margesin and Schinner 2001). For example, the protective effect of glycine and ectoine (as compatible solutes) on chymotrypsin inhibitor 2 (as a bio-macromolecule) has been proved (Lentzen and Schwarz 2006). Bacterial milking is a process to produce the large scale of compatible solutes ectoine and hydroxyectoine by *Halomonas elongate*. Broadly speaking, *H. elongate* produces compatible solutes in response to the osmotic pressure, induced by salinity of the medium. Transferring cells to a medium with low salinity leads to release compatible solutes from cells to the medium, resulting in osmotic equilibrium. The compatible solutes are purified and the bacterial biomass is returned to the medium with high salinity for the next round of ectoine and hydroxyectoine production (Sauer and Galinski 1998).

Polyhydroxyalkanoates (PHAs) are one of polyhydroxyesters synthesizing and storing in some microorganisms as reserves of carbon and energy (Steinbüchel and Schlegel 1991) when the growth occurred in the presence of excess carbon source and limitation of nitrogen, phosphorus and other nutrients (Anderson and Dawes 1990). PHAs have received much attention because of their biodegradability, biocompatibility, and resembling some properties of petrochemical derived plastics (Rathi et al. 2013). Some of halophilic and halotolerant bacteria produce PHAs such as *Bacillus megaterium* strain H16 (Salgaonkar et al. 2013), *Halomonas* sp. SK5 (Rathi et al. 2013), *Halomonas boliviensis* LC1 (Quillaguaman et al. 2005), and *Oceanimonas* sp. GK1 (Ramezani et al. 2014). PHAs can be used to produce packaging materials and medical artifacts (Tan et al. 2014).

Production of saline effluent as a result of fossil fuel extracting, producing different chemicals (Lefebvre and Moletta 2006), and pollution of many natural ecosystems such as saline environments by human activities (industrial and urban wastewaters) (García et al. 2004) are going to be eminent problems of the future. Bioremediation (transformation or degradation of hazardous compounds in a contaminated site by microorganisms) (Muñoz et al. 2001) of the contaminated saline water and soils using conventional methods are difficult because of low degradation rate and performance as a result of NaCl presence (García et al. 2005). Some halophilic bacteria have the ability to degrade organic pollutant and are good alternatives for other bioremediation programs. There are reports on the ability of some halophilic bacteria to degradation of organic contamination. A halophilic consortium obtained from a saline soil in Iran (containing a *Halomonas* strain and one unculturable strain from the genus *Marinobacter*) was able to degrade phenanthrene (Dastgheib et al. 2012). A halobacterial strain TM-1 could degrade and alter the chemical properties of oils (Hao and Lu 2009). Similarly, a halotolerant *Alcanivorax* sp. strain could utilize crude oil, diesel fuel, and various pure aliphatic hydrocarbon substrates (Dastgheib et al. 2011). In another reports, it has been proved that *Halomonas campisalis* could consume benzoate and salicylate (aromatic compounds) (Oie et al. 2007) and *Halomonas* sp. could degrade phenol as their sole source of carbon and energy (Hinteregger and Streichsbier 1997).

13.3 Haloenzymes from Halophilic Bacteria

Nordberg and Hofsten are considered as the pioneers on the study of haloenzymes (Norberg and Hofsten 1969) because of reporting extracellular proteinase from *Halobacterium salinarium* in 1969 (Norberg and Hofsten 1969). It opened the new sight in enzymology and biotechnological applications, so attempts to find new haloenzymes from halobacteria have increased from then. Haloenzymes have unique features rarely finding in other enzymes. The optimum activity and stability of haloenzymes occur at high NaCl concentration and they need NaCl to maintain their structure. They show high resistance against denaturation and can do the enzymatic function in low water or non-aqueous medium (Madern et al. 2000; Setati 2010). High salt concentration limits availability of water in a microenvironment around the protein inducing some changes in structural and functional dynamics of most enzymes. Enzymes in this condition must adapt to compete with salts for hydration (Delgado-García et al. 2012; Karan et al. 2012a) in order to prevent protein aggregation by reducing the number of hydrophobic residues and increasing acidic residues on the surface of proteins (Rao and Argos 1981). On exteriors of haloenzymes acidic residues give negative charges to enzymes and bind to hydrated ions, which reduce enzyme surface hydrophobicity and aggregation at high salt concentration (Klibanov 2001; Marhuenda-Egea and Bonete 2002). In fact, haloenzymes have multilayered hydration shells to maintain their functional conformation in the presence of high ionic concentration (Delgado-García et al. 2012; Karan et al. 2012a).

The effort to find organic solvent tolerant enzymes, especially from halophilic bacteria has increased. These enzymes could be good choices for biotechnological applications in which organic solvents are used (Ogino et al. 2000), because dissolved salt in water, like organic solvents, reduces water activity (Adams et al. 1995). Using enzymes in organic solvents or aqueous solutions containing organic solvents leads to increased solubility of nonpolar substrates and removal of microbial contamination in the reaction mixture (Doukyu and Ogino 2010). In some reports on production or purification of haloenzymes from halophilic bacteria, resistance of the enzymes toward different organic solvents has been examined. Table 13.1 is a summary of organic solvent tolerant hydrolytic enzymes from different halophilic bacterial strains which their activity or stability against the organic solvent have been studied.

Table 13.1 Organic solvent tolerant hydrolases from different halophilic bacteria

Halophilic strains	Enzyme	Organic solvents	References
Thalassobacillus sp. LY18	α-Amylase	DMSO, DMF, Methanol, Acetonitrile, Ethanol, Acetone, Benzene, *n*-Hexane, Isooctane, Dodecane	Li and Yu (2012b)
Nesterenkonia sp. strain F	α-Amylase	Benzene, Chloroform, Toluene, Cyclohexane, 1-Decanol	Shafiei et al. (2011)
B. agaradhaerens Mi-10-6_2	α-Amylase	Dodecane, Decane, Heptane, *n*-Hexane, Methanol, Propanol	Pandey and Singh (2012)
Salinivibrio sp. strain AF-2004	Protease	Ethanol, 1,4-Dioxane, Benzene, Cyclohexane, Toluene	Karbalaei-Heidari et al. (2007)
		Cyclohexanol, 1-Butanol, 1-Hexanol, 1-Octanol, 1-Decanol	
Halobacillus blutaparonensis strain M9	Protease	Ether, Isooctane, Cyclohexane	Santos et al. (2013)
Geomicrobium sp. EMB2	Protease	Ethanol, 1-Butanol, Benzene, Toluene, Cyclohexane, Heptane	Karan et al. (2011)
		Isooctane, *n*-Decane, *n*-Dodecane	
Thalassobacillus sp. LY18	Cellulase	1-Butanol, DMSO, DMF, Methanol	Li et al. (2012a)
Bacillus sp. L1	Cellulase	DMSO, DMF, Methanol, Acetonitrile, Ethanol, Acetone, 1-Butanol, Chloroform, Benzene, Toluene, Cyclohexane, *n*-Hexane, 1-Decanol, Isooctane	Li and Yu (2012b)
Idiomarina sp. W33	Lipase	DMSO, DMF, Methanol, Acetonitrile, Ethanol, Acetone, *tert*-Butanol, Chloroform, Benzene, Toluene, Cyclohexane, *n*-Hexane, 1-Decanol, Isooctane	Li et al. (2014)
M. lipolyticus SM19	Recombinant lipase	DMSO, DMF, Methanol, Acetonitrile, Ethanol, Acetone, 2-Propanol, 1-Propanol, Diethylether, Toluene, Hexane	Pérez et al. (2011)

13.3.1 Proteases

Proteases distribute among different species of the three domains of life. Many reports have been published on screening, production and purification of proteases especially from microorganisms. Among them, bacterial proteases are well-known types of hydrolases which are important in different biological processes with various applications in industrial and biotechnological fields (Rao et al. 1998) such as bioremediation, waste treatment, wool quality improvement, meat tenderization, and in leather, pharmaceutical and detergent industries (Najafi et al. 2005; Vijayaraghavan et al. 2012; Rao et al. 1998). Many halophilic proteases are active and stable in high temperature, high ionic strength, and in presence of organic solvents, which make them good candidates to use for processing of food, leather, and detergent industries (Lanyi 1974). Proteolytic activity of bacteria could be screened qualitatively by culturing them on a solid culture medium containing skim milk. After appropriate incubation time, clear zone around the colony is taken as evidence of proteolytic activity (Rohban et al. 2009). The clear zone produced by a protease producing halophilic bacterium is shown in Fig. 13.1a. Another way to screen protease producing bacteria is culturing them on a solid medium containing gelatin and treating it with acidic mercuric chloride to appearance zones of gelatin clearance around colonies after incubation time (Kumar et al. 2012). Some reports are published on screening protease producing halophilic bacterial isolates from different

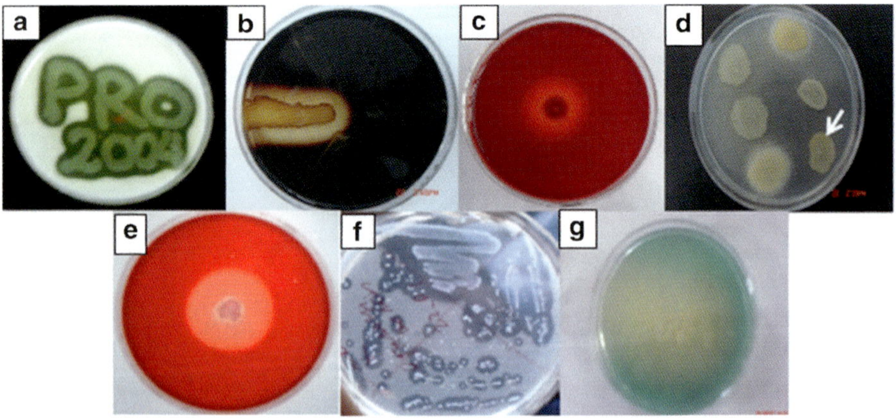

Fig. 13.1 (**a**) Proteolytic activity of *Salinivibrio* strain AF-2004 on skim milk media containing 10 % NaCl (Ali Amoozegar et al. 2007). (**b**) Amylolytic activity of *Nesterenkonia* sp. strain F on starch agar, after treatment with iodine (with the permission of Dr. Amoozegar, unpublished data) (**c**) Cellulolytic activity of *Marinimicrobium* sp. LS-A18 on a medium containing CMC, after staining with Congo red (Zhao et al. 2012) (**d**) Lipolytic activity of *Halomonas* sp. on a medium containing Tween 80 (with the permission of Dr. Amoozegar, unpublished data) (**e**) Xylanolytic activity of *Chromohalobacter* sp. TPSV 101 on a medium containing xylan, after treating with Congo red (Prakash et al. 2009b) (**f**) Chitinolytic activity of *Saccharospirillum* sp. strain HCh1 on a medium containing amorphous chitin (Sorokin and Kolganova 2013) (**g**) Nucleolytic activity of *Halobacillus* sp. on a DNase test agar medium (with the permission of Dr. Amoozegar, unpublished data)

saline ecosystems (Sánchez-Porro et al. 2003b; Cojoc et al. 2009; Rohban et al. 2009; Kumar et al. 2012). There are few reports on production of proteases by halophilic bacteria without further biochemical analysis of these enzymes. Haloalkaliphilic *Bacillus* sp. Po2 (Patel et al. 2006) and haloalkaliphilic *Bacillus* sp. Ve1 (Patel et al. 2005) produced extracellular proteases. The optimum production of both enzymes was at pH 8.0 and NaCl concentration of 15 % and 10 % for *Bacillus* sp. Po2 and *Bacillus* sp. Ve1, respectively. An extracellular protease from *Geomicrobium* sp. EMB2 showed high stability in organic solvents, salt, surfactants, detergents, and alkaline pH (Karan et al. 2011). Another extracellular proteases produced by *Halobacillus blutaparonensis* were stable in the presence of salt (up to 20 % NaCl) and organic solvents (Santos et al. 2013).

To identify proteases in details, it is necessary to purify enzymes and then determine their biochemical properties. For this purpose a number of proteases from halophilic bacteria have been purified and characterized. Table 13.2 listed these proteases and some of their features.

In some reports, enzymes from halophilic bacteria have been cloned and over expressed in the other hosts. In some of these cases, biochemical properties, function, or structure of the recombinant enzymes, have been determined and compared to the native enzymes. Two proteases from *Oceanobacillus iheyensis* O.M.$A_1$8 and *Haloalkaliphilic bacterium* O.M.$E_1$2 were cloned and expressed in *Escherichia coli*. The recombinant proteases (rO.M.$A_1$8 enzyme and rO.M.$E_1$2 enzyme) were active in the range of pH 8.0–11.0 and temperature 30–50 °C. K_m values for rO.M.$A_1$8 enzyme and rO.M.$E_1$2 enzyme were determined to be 0.77 and 0.80 mg/ml, respectively for casein as substrate (Purohit and Singh 2014). The recombinant protease from *Salinivibrio* sp. strain AF- 2004 (isolated from a hypersaline lake in Iran), which is overexpressed in *E. coli*, showed optimum pH and temperature for enzyme activity at 8.5 and 65 °C, respectively (Karbalaei-Heidari et al. 2008).

13.3.2 Amylases

Amylases are hydrolases capable of hydrolyzing starch and related saccharides. The best known enzymes of this group are α-amylase, β-amylase and glucoamylase.

α-Amylases could be used in different industrial processes (Kadziola et al. 1998; Machius et al. 1995) such as food, fermentation (fuel alcohol production), textile, paper, detergent, and pharmaceutical industries (Souza 2010). The presence of amylolytic activity of bacteria could be qualitatively screened by culturing them on starch agar medium. After incubation time, the plate is flooded with iodine solution and the zone of clearance around the colony is a sign of starch hydrolysis (Sánchez-Porro et al. 2003b; Cojoc et al. 2009; Rohban et al. 2009; Kumar et al. 2010, 2012; Jayachandra et al. 2012; Neagu et al. 2014). In Fig. 13.1b the clearance zone of an amylolytic halophilic bacterium after iodine solution treatment is shown. *Halobacillus* sp. strain MA-2, isolated from a saline soil in Iran produced an extracellular α-amylase. The pH and temperature optima for enzyme production were 7.8 and 30 °C respectively, while the optimum pH and temperature for enzyme activity

Table 13.2 Purified or partially purified proteases from halophilic bacteria

Bacteria	Purification procedure	K_m value	V_{max} value	Conditions for optimal activity			M_r (kDa)	References
				NaCl concentration M or % (w/v)	T (°C)	pH		
Salinivibrio sp. strain AF-2004	1. Acetone precipitation	1.4 mg/ml	264 U/mg	0–0.5 M	55	8.5	31	Karbalaei-Heidari et al. (2007)
	2. Q-Sepharose anion exchange chromatography							
	3. Sephacryl S-200 gel filtration chromatography							
A haloalkaliphilic bacterium sp. AH-6	1. $(NH_4)_2SO_4$ precipitation	2.5 mg/ml	625 U/min	0.15 M	37	9.0–11.0	40	Dodia et al. (2008)
	2. Phenyl Sepharose hydrophobic interaction chromatography							
H. karajensis strain MA-2	1. Acetone precipitation	–	–	0.5 M	50	9.0	36	Karbalaei-Heidari et al. (2009)
	2. Q-Sepharose anion exchange chromatography							
Pseudomonas sp. (A-14)	1. $(NH_4)_2SO_4$ precipitation	–	–	18 %	40	8.0	120	Van Qua et al. (1981)
	2. DEAE-cellulose anion exchange chromatography							
	3. Sephadex G-200 gel filtration chromatography							
Bacillus sp.	Phenyl Sepharose hydrophobic interaction chromatography	2 mg/ml	289.8 μg/min	0.03 M	37	10.0–11.0	29	Gupta et al. (2005)
Pseudoalteromonas sp. strain CP76	1. Q-Sepharose anion exchange chromatography	7.1 μM	961 U/mg	0–1.0 M	55	8.5	38	Sánchez-Porro et al. (2003a)
	2. Superdex S-200 gel filtration chromatography							

S. costicola subsp. *costicola*	1. $(NH_4)_2SO_4$ precipitation	–	–	2.0 %	60	8.0	38	Lama et al. (2005)
	2. Q-Sepharose anion exchange chromatography							
	3. Superdex 200 gel filtration chromatography							
Alkalibacillus sp. NM-Fa4	1. Ethanol precipitation	1.3 mg/mL	1,111 mg/ml/min	1 M	50–52	9.5	19.7	Collins et al. (2005)
	2. Q-Sepharose anion exchange chromatography							
A halophilic bacterium	Partially purified	–	–	10 %	–	9.0–9.5	–	Dodia et al. (2006)
Chromohalobacter sp. TVSP101	Partially purified	–	–	4.5 M	80	7.0	–	Vidyasagar et al. (2007)
B. cereus strain CS1	Partially purified	–	–	2 %	37	9.0	–	D'Costa et al. (2013)

were 7.5–8.5 and 50 °C, respectively (Amoozegar et al. 2003). An extracellular α-amylase from *Halomonas meridiana* exhibiting maximal activity at pH 7.0, 37 °C and 10 % (w/v) NaCl, respectively (Coronado et al. 2000). Another extracellular α-amylase isolated from a haloalkaliphilic bacterium, *B. agaradhaerens* Mi-10-6_2 was active up to 4 M salt, with optimal activity at 2 M salt, pH 10.0–11.0 and 50 °C (Pandey and Singh 2012). In Table 13.3, purified or partially purified α-amylases from different halophilic bacterial strains and their biochemical properties are listed.

13.3.3 Cellulases

Cellulases hydrolyze the β-1,4-D-glucosidic linkages in cellulose, lichenin, and cereal β-D-glucans (ShaoMin and Guang 2013).

There are three main groups of cellulolytic enzymes: endocellulase, exocellulase, and β-glucosidase (Karnchanatat et al. 2008), which act together to completely hydrolyze the β-1,4-D-glycosidic bonds of cellulose to form glucose (Bhat and Bhat 1997; Bhat 2000). Cellulases could be applied in the food and chemical industries, in biomedical science, and as feed additives (Zhang et al. 2012).

CMC/Congo red method is a qualitative method to screening cellulase producing bacteria. In this method bacterial strains are cultured on a solid medium contains carboxy methyl cellulose (CMC). After incubation time, 0.1 % Congo red solution is flooded on plates. The clear zone around the colony is a sign of cellulolytic activity (Aunpad and Panbangred 2003; Rohban et al. 2009; Chen and Liu 2013). In Fig. 13.1c the hydrolysis zone of CMC by this method is shown.

Cellulases being obtained from non-halophilic microorganisms are used in various fields, such as food, textiles and laundry, paper, and agriculture (Percival Zhang et al. 2006). Halotolerant cellulases could be used for treating agricultural waste and the bioremediation of cellulose materials (Wang et al. 2009). In Table 13.4 purified or partially purified cellulases isolated from halobacteria are listed. The cellulase from *Marinimicrobium* sp. LS-A18, isolated from a marine solar saltern in China is a halotolerant, thermostable, and alkali-stable enzyme and a good candidate to biotechnological and industrial applications, with optimum activity in absence of NaCl, at 55 °C and pH 7.0 (Zhao et al. 2012).

The gene *celB*, encoding a cellulase from *Bacillus* sp. BGCS10 was cloned and overexpressed in *E. coli*. The recombinant enzyme with the molecular mass of 62 kDa showed its optimal activity at 55 °C with 2.5 M NaCl (Zhang et al. 2012).

13.3.4 Lipases and Esterases

Lipases catalyze the hydrolysis of triglycerides to fatty acids and glycerol and the reverse reaction in non-aqueous systems (Teo et al. 2003). Lipases are used for fat hydrolysis, esterification, inter-esterification, trans-esterification and organic

Table 13.3 Purified or partially purified α-amylases from halophilic bacteria

Bacteria	Purification procedure	K_m value	V_{max} value	Conditions for optimal activity			M_r (kDa)	References
				NaCl concentration M or % (w/v)	T (°C)	pH		
Thalassobacillus sp. LY18	1. $(NH_4)_2SO_4$ precipitation	–	–	10 %	70	9.0	31	Li and Yu (2012b)
	2. Q-Sepharose anion exchange chromatography							
	3. Sephacryl S-100 gel filtration chromatography							
Nesterenkonia sp. strain F	1. Ethanol precipitation	5.8 mg/ml	1.07 mg/ml/min	0.75–1 M	45	6.5	57	Shafiei et al. (2011)
	2. Q-Sepharose anion exchange chromatography							
	3. Sephacryl S-200 gel filtration chromatography	4.5 mg/ml	1.18 mg/ml/min	0.5 M	45	7.5	106	Shafiei et al. (2011)
Chromohalobacter sp. TVSP 101	1. Ethanol precipitation	125 mM	5.88 U/mg	0–20 %	65	9.0	72	Prakash et al. (2009a)
	2. Butyl Sepharose hydrophobic interaction chromatography							
	3. Sephacryl S-200 gel filtration chromatography	166 mM	5.0 U/mg	0–20 %	65	9.0	62	
Micrococcus halobius ATCC 21727	1. Ethanol precipitation			0.25 M	50–55	6.0–7.0	89	Onishi and Sonoda (1979)
	2. DEAE-cellulose anion exchange chromatography							
	3. Bio-Gel P-200 gel filtration chromatography							
	4. Formation of the glycogen-amylase complex							

(continued)

Table 13.3 (continued)

Bacteria	Purification procedure	K_m value	V_{max} value	Conditions for optimal activity			M_r (kDa)	References
				NaCl concentration M or % (w/v)	T (°C)	pH		
Nesterenkonia sp. strain F	1. Ethanol precipitation 2. Q-Sepharose anion exchange chromatography	6.6 mg/ml	1.07 mg/ml/min	0.25 M	45	7.0–7.5	110	Shafiei et al. (2012)
Bacillus sp. Strain TSCVKK	Partially purified	–	–	10 %	55	7.5		Kiran and Chandra (2008)

Table 13.4 Purified or partially purified cellulases from halophilic bacteria

Bacteria	Purification procedure	K_m value (mg/ml)	V_{max} value	Conditions for optimal activity			M_r (kDa)	References
				NaCl concentration (%) (w/v)	T (°C)	pH		
Salinivibrio sp. NTU-05	1. Q chromatography 2. Sephadex G-200 gel filtration chromatography	3.03	142.86 mol/min/mg	5	35	7.5	29	Wang et al. (2009)
Thalassobacillus sp. LY18	1. $(NH_4)_2SO_4$ precipitation 2. Q-Sepharose anion exchange chromatography 3. Sephacryl G-100 gel filtration chromatography	–	–	10	60	8.0	61	Li et al. (2012a)
Bacillus sp. L1	1. $(NH_4)_2SO_4$ precipitation 2. Q-Sepharose anion exchange chromatography 3. Sephadex G-100 gel filtration chromatography	–	–	7.5	60	8.0	45	Li et al. (2012b)
Marinimicrobium sp. LS-A18	Partially purified	–	–	0	55	7.0	–	Zhao et al. (2012)

biosynthesis such as production of drugs in the pharmaceutical industry. They also could be used in the dairy industry for the hydrolysis of milk fat, in the leather industry for the removal of subcutaneous fat, in the paper industry for the removal of impurities from raw cotton, and using as additives in detergents (Gomes and Steiner 2004). Esterases hydrolyze water soluble esters (Verger 1997) and could be used for stereospecific hydrolysis, trans-esterification, and biosynthesis of ester and other organic compounds (Hou et al. 2013).

In order to qualitatively screening lipase producing bacteria, the strains are cultured on a solid medium supplemented with Tween 80 or tributyrin and the clear zones around colonies, after incubation time at appropriate condition, are the sign of lipolytic activity (Sierra 1957; Gonzalez and Gutierrez 1970; Mourey and Kilbertus 1976; Kumar et al. 2012). Figure 13.1d shows such clear zone around a lipase producing by a halophilic bacterium. On the other way, the strains are cultured on a solid medium containing olive oil and victoria blue. The colonies with blue color zones after incubation time are those producing lipase (Samad et al. 1989; Martín et al. 2003). Screening could be done on a solid medium containing olive oil and rhodamine B. After growing colonies, those exhibited orange-red halo under UV light have lipolytic activity (Bhatnagar et al. 2005).

Until now, a few halophilic or halotolerant lipases and esterases have been described from halophilic bacteria. *Salinivibrio* sp. strain SA-2, which isolated from a hypersaline lake in south of Iran produced extracellular polyextremophilic lipase. The maximal activity of the thermohalophilic lipase was at pH 7.5, 50 °C, with stability at pH range of 7.5–8.0 and retaining 90 % of its activity at 80 °C for 30 min. The maximal enzyme production was during the early-stationary phase at 35 °C and pH 8.0 (Amoozegar et al. 2008). Another extracellular lipase was produced by *Idiomarina* sp. W33 isolated from a saline soil. Purification steps included ammonium sulphate precipitation, DEAE-Sepharose anion exchange, and Sephacryl S-200 gel filtration chromatography. The molecular mass of the purified lipase was estimated to be 67 kDa by SDS-PAGE. The enzyme displayed maximal activity at 60 °C, pH 7.0–9.0 and 10 % (w/v) NaCl and showed stability and activity in presence of hydrophobic organic solvents. This lipase was used to biodiesel production (Li et al. 2014). A lipase from the halophilic bacterium *M. lipolyticus* SM19 (LipBL), has been expressed in *E. coli* with a molecular mass of 45.3 kDa. Optimum enzyme activity was at temperature of 80 °C and in a reaction mixture without salt. The enzyme maintains 20 % of its activity from 0 to 3 M of NaCl concentrations and in presence of different organic solvents tested (Pérez et al. 2011). A moderately halophilic bacterium, *Halobacillus* sp. strain LY5 (isolated from the saline soil in Yuncheng, China) produced extracellular esterase. The enzyme with the molecular mass of 96 kDa was purified according to this procedure: 80 % ammonium sulphate precipitation, diethylaminoethyl (DEAE)-cellulose ion exchange, and Sephacryl S-100 gel filtration chromatography. The optimum enzyme activity was at 50 °C, pH 10.0 and 10 % (w/v) NaCl (Li et al. 2012c). An esterase gene, from a marine bacterium *Pelagibacterium halotolerans* $B2^T$ has been cloned and overexpressed in *E. coli*. The recombinant esterase, PE8, with the molecular mass of 23.19 kDa was an alkaline esterase showing an optimal activity at pH 9.5 and 45 °C against *p*-nitrophenyl acetate as substrate (Wei et al. 2013). Another esterase gene from *P. halotolerans* B2^T has been overexpressed in *E. coli*. The molecular mass of the recombinant enzyme, PE10, was determined to be 29.91 kDa. The optimal activity of the halotolerant enzyme was at 45 °C and pH 7.5, maintaining its activity under 4 M NaCl (Wei et al. 2013).

13.3.5 Xylanases

Xylanases are hydrolases which randomly cleave the β-1,4 linkage of the xylan (Collins et al. 2005). Xylanases could be used for the bleaching of paper pulp (Hung et al. 2011), processing of animal food to increase feed digestibility, clarification of fruit juices (Beg et al. 2001), and conversion of biomass to bioethanol and biodiesel (Zhong et al. 2009). Xylanase producing bacteria could be screened qualitatively by culturing bacteria on solid medium containing xylan. After appropriate incubation time, the zone of hydrolysis is visualized by flooding the plates with aqueous solution of Congo red (Wejse et al. 2003). The hydrolysis zone of xylan by a xylanolytic bacterium is shown in Fig. 13.1e. On the other way, it is possible to use chromogenic AZCL xylan and detecting clearing zones around the colonies as a sign of xylanolytic activities (Sánchez-Porro et al. 2003b). The authors suggested that halophilic xylanases could be used to treatment of agricultural waste, bioremediation of xylan-containing materials, and bio-bleaching (Menon et al. 2010). In some, Table 13.5 reviewed purified or partially purified xylanases from halophilic bacteria. A marine bacterium *B. subtilis* strain cho40, isolated from Chorao island of mandovi estuary Goa in India, produced an extracellular halotolerant xylanase, Xyn40, with optimal pH, temperature, and NaCl concentration for xylanolytic activity at 6.0, 60 °C, and 0.5 M respectively. The xylanase gene of this enzyme has been cloned and overexpressed in *E. coli* with the molecular mass of 22.9 kDa. The recombinant xylanase showed similar features to xylanases isolated from *B. subtilis* (Khandeparker et al. 2011).

13.3.6 Chitinases

Chitinases are hydrolases catalyzing the hydrolytic degradation of chitin, the second most abundant natural polysaccharide after cellulose. Chitinases are used for the removal of chitin from the cell wall of fungi to produce protoplasts and production of biologically active oligosaccharides (Bhattacharya et al. 2007). Chitinolytic bacteria could be screened on a solid medium containing colloidal chitin as a sole carbon source. After culturing and appropriate incubation time, chitinolytic activity is visualized by a clear zones of chitin hydrolysis around colonies, as shown in Fig. 13.1f (Han et al. 2014). Two chitinolytic bacteria, HCh1 and HCh2 strains identified as genera *Saccharospirillum* and *Arhodomonas*, respectively were

Table 13.5 Purified or partially purified xylanases from halophilic bacteria

Bacteria	Purification procedure	K_m value (mg/ml)	V_{max} value	Conditions for optimal activity: NaCl concentration M or % (w/v)	T (°C)	pH	M_r (kDa)	References
B. pumilus strain GESF1	1. $(NH_4)_2SO_4$ precipitation	5.3	0.42 μmol/min/ml	2.5–7.5 %	40	8.0	39.6	Menon et al. (2010)
	2. DEAE-cellulose anion exchange chromatography							
	3. Sephadex-G-200 gel filtration chromatography							
Strain CL8	1. $(NH_4)_2SO_4$ precipitation	5	125,000 nkat/mg protein	4 M	60	6.0	43	Wejse et al. (2003)
	2. Mono Q anion chromatography	1	143,000 nkat/mg	4 M	65	6.0	62	
Bacillus sp. NTU-06	1. Q-Sepharose anion exchange chromatography	3.45	387.3 μmol/min/mg	5 %	40	8.0	24	Wang et al. (2010)
	2. Superdex 200 gel filtration chromatography							
Gracilibacillus sp. TSCPVG	Partially purified	–	–	3.5 %	60	7.5	–	Giridhar and Chandra (2010)
Chromohalobacter sp. TPSV 101	Partially purified	–	–	20 %	65	9.0	~15	Prakash et al. (2009b)

isolated from hypersaline lakes in Russia. The *Saccharospirillum* sp. strain HCh1 had optimal growth on chitin at 1 M NaCl and growth range 0.5–3.25 M NaCl. *Arhodomonas* sp. strain HCh2 grew optimally at 1.5 M but grew up to 2.5 M NaCl (Sorokin and Kolganova 2013). *Bacillus* sp. strain SCH-1 and *Paenibacillus* sp. strain SCH-2 isolated from a salted and fermented food (Han et al. 2014) and *Haloanaerobacter chitinovorans* isolated from a solar saltern (Liaw and Mah 1992) exhibited chitinolytic activity. A moderately halophilic bacterium, *Virgibacillus marismortui* strain M3-23 that isolated from a Tunisian shallow salt lake produced a chitinase in the absence or presence of salt. The enzyme was partially purified by ammonium sulphate precipitation and characterized, with maximal activity at 70 °C and pH 9.0 (Essghaier et al. 2012). A halophilic chitinolytic bacterium, *S. costicola* strain 5SM-1 was isolated from a salted mud in Thailand. The chitinase gene (*chiC*) from this bacterium was cloned and the recombinant enzyme (ChiC) was produced by *E. coli* with the molecular mass of 92 kDa (Aunpad and Panbangred 2003).

13.3.7 Nucleases

RNases and DNases are categorized as a larger group, nucleases. RNases act on RNA to produce 5′ ribonucleotides or 2′, 3′-nucleotids and DNases hydrolyses DNA and release 5′-deoxyribonucleotides or 2′, 3′-nucleotides. 5′-nucleotides are aroma and give good flavor to foods (Kanlayakrit et al. 2001). A qualitative method to identify nuclease producing bacteria is culturing them on a DNase test agar medium and flooding the plates with 1 N HCl solution after appropriate incubation time. Appearance of clear halos around colonies is a sign of nuclease activity (Kanlayakrit et al. 2001), as shows in Fig. 13.1g. *Oceanobacillus*, *Thalassobacillus, Halomonas* (Rohban et al. 2009), *Chromohalobacter*, *Salinivibrio*, and *Bacillus* species (Sánchez-Porro et al. 2003a) isolated from saline environments showed nuclease activity after qualitative screening. A halotolerant strain of *Pseudomonas* (*Pseudomonas* sp. No. 3241) produced extracellular ribonuclease with maximum bacterial growth and enzyme production in a medium without NaCl. The enzyme was partially purified using ethanol precipitation and optimal condition for enzyme activity was determined to be at pH 10, temperature of 50 °C, and 3 M NaCl (Kanlayakrit et al. 2001). A moderately halophilic bacterium, *Bacillus* sp. produced an extracellular nuclease. The enzyme was purified by ethanol precipitation, DEAE-Sephadex A-50 column chromatography, and Sephadex G-200 gel filtration. Molecular mass was estimated to be 138 kDa by Sephadex G-200 gel filtration. The enzyme showed maximal activities in the presence of 1.4–3.2 M NaCl. The enzyme exhibited maximal activity at pH 8.5 and at 50 °C on DNA and at 60 °C on RNA (Onishi et al. 1983). The moderately halophilic bacterium *Micrococcus varians*, isolated from soy sauce mash, produced extracellular nuclease with both DNase and RNase activities. The enzyme was purified by ethanol precipitation, DEAE-Sephadex A-50 column chromatography, Sephadex G-100, and Sephadex G-200

Table 13.6 Other hydrolytic enzymes from halophilic bacteria

Enzyme	Halobacteria	Site of sample collection	References
Pectinase	*Halomonas* sp., *Salinicoccus* sp.	Howz Soltan Lake, Iran	Rohban et al. (2009)
Inulinase	*Thalassobacillus* sp., *Gracilibacillus* sp.	Howz Soltan Lake, Iran	
Pullulanase	*Salicola* sp., *Halomonas* sp., *Thalassobacillus* sp.	Howz Soltan Lake, Iran	
	Salinivibrio sp., *Halomonas* sp., *Bacillus* sp.	Hypersaline environments, Spain	Sánchez-Porro et al. (2003b)

gel filtration chromatography. The molecular mass of the enzyme was determined to be 105 kDa. The nuclease had maximal activity in the presence of 2.9 M NaCl or 2.1 M KCl at 40 °C (Kamekura and Onishi 1974).

13.3.8 Other Hydrolytic Enzymes

In addition to hydrolytic enzymes mentioned above, production of other hydrolyses by halobacteria has been reported using qualitative screening on specific solid culture media (Table 13.6). Further biochemical studies on these enzymes such as determination the effect of different parameters on enzyme production, purification, and enzyme characterization could lead to better identification of them for biotechnological purposes.

13.4 Conclusion

Halophiles, as a member of extremophiles, have a lot of potential industrial and biotechnological applications. The structure of haloenzymes, the mechanism of action in presence of high salt concentration, even in the presence of solvents, can be attractive fields of study for researchers. These findings can be useful for introducing new enzymes to industries and engineering of present mesophilic industrial enzymes to increase their stability under harsh conditions and improve their function in low water activity as a result of high salinity or presence of different solvents.

References

Adams MW, Perler FB, Kelly RM (1995) Extremozymes: expanding the limits of biocatalysis. Nat Biotechnol 13:662–668

Alavi S, Amoozegar MA, Khajeh K (2014) Enzyme (s) responsible for tellurite reducing activity in a moderately halophilic bacterium, *Salinicoccus iranensis* strain QW6. Extremophiles 18:953–961

Amoozegar MA, Malekzadeh F, Malik KA (2003) Production of amylase by newly isolated moderate halophile, *Halobacillus* sp. strain MA-2. J Microbiol Methods 52:353–359
Amoozegar MA, Fatemi AZ, Karbalaei-Heidari HR, Razavi MR (2007) Production of an extracellular alkaline metalloprotease from a newly isolated, moderately halophile, *Salinivibrio* sp. strain AF-2004. Microbiol Res 162:369–377
Amoozegar MA, Salehghamari E, Khajeh K, Kabiri M, Naddaf S (2008) Production of an extracellular thermohalophilic lipase from a moderately halophilic bacterium, *Salinivibrio* sp. strain SA-2. J Basic Microbiol 48:160–167
Amoozegar MA, Hajighasemi M, Hamedi J, Asad S, Ventosa A (2011) Azo dye decolorization by halophilic and halotolerant microorganisms. Ann Microbiol 61:217–230
Amoozegar MA, Khoshnoodi M, Didari M, Hamedi J, Ventosa A, Baldwin SA (2012) Tellurite removal by a tellurium-tolerant halophilic bacterial strain, *Thermoactinomyces* sp. QS-2006. Ann Microbiol 62:1031–1037
Anderson AJ, Dawes EA (1990) Occurrence, metabolism, metabolic role, and industrial uses of bacterial polyhydroxyalkanoates. Microbiol Rev 54:450–472
Aunpad R, Panbangred W (2003) Cloning and characterization of the constitutively expressed chitinase C gene from a marine bacterium, *Salinivibrio costicola* strain 5SM-1. J Biosci Bioeng 96:529–536
Babavalian H, Amoozegar MA, Zahraei S, Rohban R, Shakeri F, Moghaddam MM (2014) Comparison of bacterial biodiversity and enzyme production in three hypersaline lakes; Urmia, Howz-Soltan and Aran-Bidgol. Indian J Microbiol 54:444–449
Beg Q, Kapoor M, Mahajan L, Hoondal G (2001) Microbial xylanases and their industrial applications: a review. Appl Microbiol Biotechnol 56:326–338
Bhat M (2000) Cellulases and related enzymes in biotechnology. Biotechnol Adv 18:355–383
Bhat M, Bhat S (1997) Cellulose degrading enzymes and their potential industrial applications. Biotechnol Adv 15:583–620
Bhatnagar T, Boutaiba S, Hacene H, Cayol JL, Fardeau ML, Ollivier B, Baratti JC (2005) Lipolytic activity from halobacteria: screening and hydrolase production. FEMS Microbiol Lett 248:133–140
Bhattacharya D, Nagpure A, Gupta RK (2007) Bacterial chitinases: properties and potential. Crit Rev Biotechnol 27:21–28
Brown A (1976) Microbial water stress. Bacteriol Rev 40:803
Chen WQ, Liu YY (2013) Isolation and identification of *Halomonas* sp. ZSCW-10: a moderately halophilic bacteria strain with cellulase activity. Adv Mater Res 749:236–241
Cojoc R, Merciu S, Popescu G, Dumitru L, Kamekura M, Enache M (2009) Extracellular hydrolytic enzymes of halophilic bacteria isolated from a subterranean rock salt crystal. Rom Biotechnol Lett 14:4658–4664
Collins T, Gerday C, Feller G (2005) Xylanases, xylanase families and extremophilic xylanases. FEMS Microbiol Rev 29:3–23
Coronado MJ, Vargas C, Hofemeister J, Ventosa A, Nieto JJ (2000) Production and biochemical characterization of an α-amylase from the moderate halophile *Halomonas meridiana*. FEMS Microbiol Lett 183:67–71
D'Costa B, Khanolkar D, Dubey SK (2013) Partial purification and characterization of metalloprotease of halotolerant alkaliphilic bacterium *Bacillus cereus* from coastal sediment of Goa, India. AJB 12:4905–4914
DasSarma P, Coker JA, Huse V, DasSarma S (2010) Halophiles, industrial applications. In: Flickinger MC (ed) Encyclopedia of industrial biotechnology: bioprocess, bioseparation, and cell technology. Wiley, Hoboken, pp 1–43
Dastgheib SMM, Amoozegar MA, Khajeh K, Ventosa A (2011) A halotolerant *Alcanivorax* sp. strain with potential application in saline soil remediation. Appl Microbiol Biotechnol 90:305–312
Dastgheib SMM, Amoozegar MA, Khajeh K, Shavandi M, Ventosa A (2012) Biodegradation of polycyclic aromatic hydrocarbons by a halophilic microbial consortium. Appl Microbiol Biotechnol 95:789–798

Delgado-García M, Valdivia-Urdiales B, Aguilar-González CN, Contreras-Esquivel JC, Rodríguez-Herrera R (2012) Halophilic hydrolases as a new tool for the biotechnological industries. J Sci Food Agric 92:2575–258

Dodia MS, Joshi RH, Patel RK, Singh SP (2006) Characterization and stability of extracellular alkaline proteases from halophilic and alkaliphilic bacteria isolated from saline habitat of coastal Gujarat, India. Braz J Microbiol 37:276–282

Dodia M, Rawal C, Bhimani H, Joshi R, Khare S, Singh S (2008) Purification and stability characteristics of an alkaline serine protease from a newly isolated Haloalkaliphilic bacterium sp. AH-6. J Ind Microbiol Biotechnol 35:121–131

Doukyu N, Ogino H (2010) Organic solvent-tolerant enzymes. Biochem Eng J 48:270–282

Essghaier B, Hedi A, Bejji M, Jijakli H, Boudabous A, Sadfi-Zouaoui N (2012) Characterization of a novel chitinase from a moderately halophilic bacterium, *Virgibacillus marismortui* strain M3-23. Ann Microbiol 62:835–841

Galinski EA, Trüper HG (1994) Microbial behaviour in salt-stressed ecosystems. FEMS Microbiol Rev 15:95–108

García MT, Mellado E, Ostos JC, Ventosa A (2004) *Halomonas organivorans* sp. nov., a moderate halophile able to degrade aromatic compounds. Int J Syst Evol Microbiol 54:1723–1728

García MT, Ventosa A, Mellado E (2005) Catabolic versatility of aromatic compound-degrading halophilic bacteria. FEMS Microbiol Ecol 54:97–109

Giridhar PV, Chandra T (2010) Production of novel halo-alkali-thermo-stable xylanase by a newly isolated moderately halophilic and alkali-tolerant *Gracilibacillus* sp. TSCPVG. Process Biochem 45:1730–1737

Gomes J, Steiner W (2004) The biocatalytic potential of extremophiles and extremozymes. Food Technol Biotechnol 42:223–235

Gonzalez C, Gutierrez C (1970) Presence of lipase among species of extremely halophilic bacteria. Can J Microbiol 16:1165–1166

Gupta A, Roy I, Patel R, Singh S, Khare S, Gupta M (2005) One-step purification and characterization of an alkaline protease from haloalkaliphilic *Bacillus* sp. J Chromatogr 1075:103–108

Han KI, Patnaik BB, Kim YH, Kwon HJ, Han YS, Han MD (2014) Isolation and characterization of chitinase-producing *Bacillus* and *Paenibacillus* strains from salted and fermented shrimp, *Acetes japonicus*. J Food Sci 79:M665–M674

Hao R, Lu A (2009) Biodegradation of heavy oils by halophilic bacterium. Prog Nat Sci 19:997–1001

Hinteregger C, Streichsbier F (1997) *Halomonas* sp., a moderately halophilic strain, for biotreatment of saline phenolic waste-water. Biotechnol Lett 19:1099–1102

Hou J, Han J, Cai L, Zhou J, Lü Y, Jin C, Liu J, Xiang H (2013) Characterization of genes for chitin catabolism in *Haloferax mediterranei*. Appl Microbiol Biotechnol 98:1185–1194

Hung KS, Liu SM, Tzou WS, Lin FP, Pan CL, Fang TY, Sun KH, Tang SJ (2011) Characterization of a novel GH10 thermostable, halophilic xylanase from the marine bacterium *Thermoanaerobacterium saccharolyticum* NTOU1. Process Biochem 46:1257–1263

Jayachandra S, Kumar A, Merley D, Sulochana M (2012) Isolation and characterization of extreme halophilic bacterium *Salinicoccus* sp. JAS4 producing extracellular hydrolytic enzymes. RRST 4:46–49

Jiang H, Dong H, Zhang G, Yu B, Chapman LR, Fields MW (2006) Microbial diversity in water and sediment of Lake Chaka, an athalassohaline lake in northwestern China. Appl Environ Microbiol 72:3832–3845

Kadziola A, Søgaard M, Svensson B, Haser R (1998) Molecular structure of a barley α-amylase-inhibitor complex: implications for starch binding and catalysis. J Mol Biol 278:205–217

Kamekura M, Onishi H (1974) Halophilic nuclease from a moderately halophilic *Micrococcus varians*. J Bacteriol 119:339–344

Kanlayakrit W, Ikeda T, Tojai S, Rodprapakorn M, Sirisansaneeyakul S (2001) Isolation and characterization of extracellular halophilic ribonuclease from halotolerant *Pseudomonas* species. Kasetsart J 35:179–187

Karan R, Singh S, Kapoor S, Khare S (2011) A novel organic solvent tolerant protease from a newly isolated *Geomicrobium* sp. EMB2 (MTCC 10310): production optimization by response surface methodology. New Biotechnol 28:136–145

Karan R, Capes MD, DasSarma S (2012a) Function and biotechnology of extremophilic enzymes in low water activity. Aquat Biosyst 8:15

Karan R, Kumar S, Sinha R, Khare S (2012b) Halophilic microorganisms as sources of novel enzymes. In: Satyanarayana T, Johri BN (ed) Microorganisms in sustainable agriculture and biotechnology. Springer, New York, pp 555–579

Karbalaei-Heidari HR, Ziaee AA, Amoozegar MA (2007) Purification and biochemical characterization of a protease secreted by the *Salinivibrio* sp. strain AF-2004 and its behavior in organic solvents. Extremophiles 11:237–243

Karbalaei-Heidari HR, Ziaee AA, Amoozegar MA, Cheburkin Y, Budisa N (2008) Molecular cloning and sequence analysis of a novel zinc-metalloprotease gene from the *Salinivibrio* sp. strain AF-2004 and its extracellular expression in *E. coli*. Gene 408:196–203

Karbalaei-Heidari HR, Amoozegar MA, Hajighasemi M, Ziaee AA, Ventosa A (2009) Production, optimization and purification of a novel extracellular protease from the moderately halophilic bacterium *Halobacillus karajensis*. J Ind Microbiol Biotechnol 36:21–27

Karnchanatat A, Petsom A, Sangvanich P, Piapukiew J, Whalley AJ, Reynolds CD, Gadd GM, Sihanonth P (2008) A novel thermostable endoglucanase from the wood-decaying fungus *Daldinia eschscholzii* (Ehrenb.: Fr.) Rehm. Enzym Microb Technol 42:404–413

Khandeparker R, Verma P, Deobagkar D (2011) A novel halotolerant xylanase from marine isolate *Bacillus subtilis* cho40: gene cloning and sequencing. New Biotechnol 28:814–821

Kiran KK, Chandra T (2008) Production of surfactant and detergent-stable, halophilic, and alkalitolerant alpha-amylase by a moderately halophilic *Bacillus* sp. strain TSCVKK. Appl Microbiol Biotechnol 77:1023–1031

Klibanov AM (2001) Improving enzymes by using them in organic solvents. Nature 409:241–246

Kumar SA, Arunasri R, Jayachandra Y, Sulochana M (2010) Screening of extracellular hydrolytic enzymes from *Marinobacter hydrocarbonoclasticus* strain AK5. Bioscan 5:97–99

Kumar S, Karan R, Kapoor S, Singh S, Khare S (2012) Screening and isolation of halophilic bacteria producing industrially important enzymes. Braz J Microbiol 43:1595–1603

Kushner DJ, Kamekura M (1988) Physiology of halophilic eubacteria. In: Rodriguez-Valera F (ed) Halophilic bacteria, vol 1. CRC Press, Inc, Boca Raton, pp 109–140

Lama L, Romano I, Calandrelli V, Nicolaus B, Gambacorta A (2005) Purification and characterization of a protease produced by an aerobic haloalkaliphilic species belonging to the *Salinivibrio* genus. Res Microbiol 156:478–484

Lanyi JK (1974) Salt-dependent properties of proteins from extremely halophilic bacteria. Bacteriol Rev 38:272

Lefebvre O, Moletta R (2006) Treatment of organic pollution in industrial saline wastewater: a literature review. Water Res 40:3671–3682

Lentzen G, Schwarz T (2006) Extremolytes: natural compounds from extremophiles for versatile applications. Appl Microbiol Biotechnol 72:623–634

Li X, Yu HY (2012a) Purification and characterization of an organic-solvent-tolerant cellulase from a halotolerant isolate, *Bacillus* sp. L1. J Ind Microbiol Biotechnol 39:1117–1124

Li X, Yu HY (2012b) Characterization of an organic solvent-tolerant α-amylase from a halophilic isolate, *Thalassobacillus* sp. LY18. Folia Microbiol 57:447–453

Li X, Wang HL, Li T, Yu HY (2012a) Purification and characterization of an organic solvent-tolerant alkaline cellulase from a halophilic isolate of *Thalassobacillus*. Biotechnol Lett 34:1531–1536

Li X, Yu HY, Lin YF (2012b) Purification and characterization of an extracellular esterase from a moderately halophilic bacterium, *Halobacillus* sp. strain LY 5. Afr J Biotechnol 11:6327–6334

Li X, Qian P, Wu SG, Yu HY (2014) Characterization of an organic solvent-tolerant lipase from Idiomarina sp. W33 and its application for biodiesel production using Jatropha oil. Extremophiles 18:171–178

Liaw HJ, Mah RA (1992) Isolation and characterization of *Haloanaerobacter chitinovorans* gen. nov., sp. nov., a halophilic, anaerobic, chitinolytic bacterium from a solar saltern. Appl Environ Microbiol 58:260–266

Litchfield C, Gillevet P (2002) Microbial diversity and complexity in hypersaline environments: a preliminary assessment. J Ind Microbiol Biotechnol 28:48–55

Machius M, Wiegand G, Huber R (1995) Crystal structure of calcium-depleted Bacillus licheniformis α-amylase at 2.2 Å resolution. J Mol Biol 246:545–559

Madern D, Ebel C, Zaccai G (2000) Halophilic adaptation of enzymes. Extremophiles 4:91–98

Margesin R, Schinner F (2001) Potential of halotolerant and halophilic microorganisms for biotechnology. Extremophiles 5:73–83

Marhuenda-Egea FC, Bonete MJ (2002) Extreme halophilic enzymes in organic solvents. Curr Opin Biotechnol 13:385–389

Martin DD, Ciulla RA, Roberts MF (1999) Osmoadaptation in archaea. Appl Environ Microbiol 65:1815–1825

Martín S, Márquez M, Sánchez-Porro C, Mellado E, Arahal D, Ventosa A (2003) *Marinobacter lipolyticus* sp. nov., a novel moderate halophile with lipolytic activity. Int J Syst Evol Microbiol 53:1383–1387

Menon G, Mody K, Keshri J, Jha B (2010) Isolation, purification, and characterization of haloalkaline xylanase from a marine *Bacillus pumilus* strain, GESF-1. Biotechnol Bioprocess Eng 15:998–1005

Mourey A, Kilbertus G (1976) Simple media containing stabilized tributyrin for demonstrating lipolytic bacteria in foods and soils. J Appl Microbiol 40:47–51

Muńoz J, Pérez-Esteban B, Esteban M, De La Escalera S, Gomez M, Martínez-Toledo M, Gonzalez-Lopez J (2001) Growth of moderately halophilic bacteria isolated from sea water using phenol as the sole carbon source. Folia Microbiol 46:297–302

Najafi MF, Deobagkar D, Deobagkar D (2005) Potential application of protease isolated from *Pseudomonas aeruginosa* PD100. Electron J Biotechnol 8:79–85

Neagu S, Enache M, Cojoc R (2014) Extracellular hydrolytic activities of halophilic microorganisms isolated from Balta Albă salt lake. Rom Biotech Lett 19:8951–8958

Norberg P, Hofsten B (1969) Proteolytic enzymes from extremely halophilic bacteria. J Gen Microbiol 55:251–256

Ogino H, Nakagawa S, Shinya K, Muto T, Fujimura N, Yasuda M, Ishikawa H (2000) Purification and characterization of organic solvent-stable lipase from organic solvent-tolerant *Pseudomonas aeruginosa* LST-03. J Biosci Bioeng 89:451–457

Oie CS, Albaugh CE, Peyton BM (2007) Benzoate and salicylate degradation by *Halomonas campisalis*, an alkaliphilic and moderately halophilic microorganism. Water Res 41:1235–1242

Onishi H, Sonoda K (1979) Purification and some properties of an extracellular amylase from a moderate halophile, *Micrococcus halobius*. Appl Environ Microbiol 38:616–620

Onishi H, Mori T, Takeuchi S, Tani K, Kobayashi T, Kamekura M (1983) Halophilic nuclease of a moderately halophilic *Bacillus* sp.: production, purification, and characterization. Appl Environ Microbiol 45:24–30

Oren A (2002a) Diversity of halophilic microorganisms: environments, phylogeny, physiology, and applications. J Ind Microbiol Biotechnol 28:56–63

Oren A (2002b) Biotechnological applications and potentials of halophilic microorganisms. In: Oren A (ed) Halophilic microorganisms and their environments, cellular origin, life in extreme habitats and astrobiology, vol 5. Springer, New York, pp 357–388

Oren A (2008) Microbial life at high salt concentrations: phylogenetic and metabolic diversity. Saline Syst 4:13

Pandey S, Singh S (2012) Organic solvent tolerance of an α-amylase from haloalkaliphilic bacteria as a function of pH, temperature, and salt concentrations. Appl Biochem Biotechnol 166:1747–1757

Patel R, Dodia M, Singh SP (2005) Extracellular alkaline protease from a newly isolated haloalkaliphilic Bacillus sp.: Production and optimization. Process Biochem 40:3569–3575

Patel R, Dodia M, Joshi R, Singh S (2006) Production of extracellular halo-alkaline protease from a newly isolated Haloalkaliphilic *Bacillus* sp. isolated from seawater in Western India. World J Microbiol Biotechnol 22:375–382

Percival Zhang YH, Himmel ME, Mielenz JR (2006) Outlook for cellulase improvement: screening and selection strategies. Biotechnol Adv 24:452–481

Pérez D, Martín S, Fernández-Lorente G, Filice M, Guisán JM, Ventosa A, García MT, Mellado E (2011) A novel halophilic lipase, LipBL, showing high efficiency in the production of eicosapentaenoic acid (EPA). PLoS One 6:e23325

Prakash B, Vidyasagar M, Madhukumar M, Muralikrishna G, Sreeramulu K (2009a) Production, purification, and characterization of two extremely halotolerant, thermostable, and alkali-stable α-amylases from *Chromohalobacter* sp. TVSP 101. Process Biochem 44:210–215

Prakash S, Veeranagouda Y, Kyoung L, Sreeramulu K (2009b) Xylanase production using inexpensive agricultural wastes and its partial characterization from a halophilic *Chromohalobacter* sp. TPSV 101. World J Microbiol Biotechnol 25:197–204

Purohit MK, Singh SP (2014) Cloning, over expression and functional attributes of serine proteases from *Oceanobacillus iheyensis* O.M.$A_1$8 and *Haloalkaliphilic bacterium* O.M.$E_1$2. Process Biochem 49:61–68

Quillaguaman J, Hashim S, Bento F, Mattiasson B, Hatti-Kaul R (2005) Poly (β-hydroxybutyrate) production by a moderate halophile, Halomonas boliviensis LC1 using starch hydrolysate as substrate. J Appl Microbiol 99:151–157

Ramezani M, Amoozegar MA, Ventosa A (2014) Screening and comparative assay of polyhydroxyalkanoates produced by bacteria isolated from the Gavkhooni Wetland in Iran and evaluation of poly-β-hydroxybutyrate production by halotolerant bacterium Oceanimonas sp. GK1. Ann Microbiol. doi:10.1007/s13213-014-0887-y (in press)

Rao JM, Argos P (1981) Structural stability of halophilic proteins. Biochemistry 20:6536–6543

Rao MB, Tanksale AM, Ghatge MS, Deshpande VV (1998) Molecular and biotechnological aspects of microbial proteases. Microbiol Mol Biol Rev 62:597–635

Rathi DN, Amir H, Abed R, Kosugi A, Arai T, Sulaiman O, Hashim R, Sudesh K (2013) Polyhydroxyalkanoate biosynthesis and simplified polymer recovery by a novel moderately halophilic bacterium isolated from hypersaline microbial mats. J Appl Microbiol 114:384–395

Ratnakar D (2013) Use of halophile physiology and adaptations in various industrial applications. Res J Biotechnol 8:2

Rohban R, Amoozegar MA, Ventosa A (2009) Screening and isolation of halophilic bacteria producing extracellular hydrolyses from Howz Soltan Lake, Iran. J Ind Microbiol Biotechnol 36:333–340

Salgaonkar B, Mani K, Braganca J (2013) Characterization of polyhydroxyalkanoates accumulated by a moderately halophilic salt pan isolate *Bacillus megaterium* strain H16. J Appl Microbiol 114:1347–1356

Samad MYA, Razak C, Salleh AB, Zin Wan Yunus W, Ampon K, Basri M (1989) A plate assay for primary screening of lipase activity. J Microbiol Methods 9:51–56

Sánchez-Porro C, Mellado E, Bertoldo C, Antranikian G, Ventosa A (2003a) Screening and characterization of the protease CP1 produced by the moderately halophilic bacterium *Pseudoalteromonas* sp. strain CP76. Extremophiles 7:221–228

Sánchez-Porro C, Martin S, Mellado E, Ventosa A (2003b) Diversity of moderately halophilic bacteria producing extracellular hydrolytic enzymes. J Appl Microbiol 94:295–300

Santos AF, Valle RS, Pacheco CA, Alvarez VM, Seldin L, Santos AL (2013) Extracellular proteases of *Halobacillus blutaparonensis* strain M9, a new moderately halophilic bacterium. Braz J Microbiol 44:1299–1304

Sauer T, Galinski EA (1998) Bacterial milking: a novel bioprocess for production of compatible solutes. Biotechnol Bioeng 57:306–313

Setati ME (2010) Diversity and industrial potential of hydrolaseproducing halophilic/halotolerant eubacteria. AJB 9:1555–1560

Shafiei M, Ziaee AA, Amoozegar MA (2011) Purification and characterization of an organic-solvent-tolerant halophilic α-amylase from the moderately halophilic *Nesterenkonia* sp. strain F. J Ind Microbiol Biotechnol 38:275–281

Shafiei M, Ziaee AA, Amoozegar MA (2012) Purification and characterization of a halophilic α-amylase with increased activity in the presence of organic solvents from the moderately halophilic Nesterenkonia sp. strain F. Extremophiles 16:627–635

ShaoMin Y, Guang W (2013) Secretory pathway of cellulase: a mini-review. Biotechnol Biofuels. doi:10.1186/1754-6834-6-177

Sierra G (1957) A simple method for the detection of lipolytic activity of microorganisms and some observations on the influence of the contact between cells and fatty substrates. Antonie Van Leeuwenhoek 23:15–22

Sorokin D, Kolganova T (2013) Bacterial chitin utilization at halophilic conditions. Extremophiles 18:243–248

Souza PM (2010) Application of microbial α-amylase in industry-a review. Braz J Microbiol 41:850–861

Steinbüchel A, Schlegel H (1991) Physiology and molecular genetics of poly (β-hydroxyalkanoic acid) synthesis in Alcaligenes eutrophus. Mol Microbiol 5:535–542

Tan GYA, Chen CL, Li L, Ge L, Wang L, Razaad IMN, Li Y, Zhao L, Mo Y, Wang JY (2014) Start a research on biopolymer polyhydroxyalkanoate (PHA): a review. Polymers 6:706–754

Teo JW, Zhang LH, Poh CL (2003) Cloning and characterization of a novel lipase from *Vibrio harveyi* strain AP6. Gene 312:181–188

Van Qua D, Simidu U, Taga N (1981) Purification and some properties of halophilic protease produced by a moderately halophilic marine *Pseudomonas* sp. Can J Microbiol 27:505–510

Ventosa A (2006) Unusual micro-organisms from unusual habitats: hypersaline environments. In: Logan NA, Lappin-Scott HM, Oyston PCF (eds) Prokaryotic diversity: mechanisms and significance. Cambridge University Press, Cambridge, pp 223–253

Ventosa A, Nieto JJ, Oren A (1998) Biology of moderately halophilic aerobic bacteria. Microbiol Mol Biol Rev 62:504–544

Verger R (1997) Interfacial activation of lipases: facts and artifacts. Trends Biotechnol 15:32–38

Vidyasagar M, Prakash S, Jayalakshmi S, Sreeramulu K (2007) Optimization of culture conditions for the production of halothermophilic protease from halophilic bacterium Chromohalobacter sp. TVSP101. World J Microbiol Biotechnol 23:655–662

Vijayaraghavan P, Jebamalar TRJ, Vincent SGP (2012) Biosynthesis optimization and purification of a solvent stable alkaline serine protease from *Halobacterium* sp. Ann Microbiol 62:403–410

Wang CY, Hsieh YR, Ng CC, Chan H, Lin HT, Tzeng WS, Shyu YT (2009) Purification and characterization of a novel halostable cellulase from *Salinivibrio* sp. strain NTU-05. Enzym Microb Technol 44:373–379

Wang CY, Chan H, Lin HT, Shyu YT (2010) Production, purification and characterisation of a novel halostable xylanase from Bacillus sp. NTU-06. Ann Appl Biol 156:187–197

Wei X, Jiang X, Ye L, Yuan S, Chen Z, Wu M, Yu H (2013) Cloning, expression and characterization of a new enantioselective esterase from a marine bacterium *Pelagibacterium halotolerans* B2^T. J Mol Catal B Enzym 97:270–277

Wejse PL, Ingvorsen K, Mortensen KK (2003) Purification and characterisation of two extremely halotolerant xylanases from a novel halophilic bacterium. Extremophiles 7:423–431

Zhang G, Li S, Xue Y, Mao L, Ma Y (2012) Effects of salts on activity of halophilic cellulase with glucomannanase activity isolated from alkaliphilic and halophilic *Bacillus* sp. BG-CS10. Extremophiles 16:35–43

Zhao K, Guo LZ, Lu WD (2012) Extracellular production of novel halotolerant, thermostable, and alkali-stable carboxymethyl cellulase by marine bacterium *Marinimicrobium* sp. LS-A18. Appl Biochem Biotechnol 168:550–567

Zhong C, Lau MW, Balan V, Dale BE, Yuan Y-J (2009) Optimization of enzymatic hydrolysis and ethanol fermentation from AFEX-treated rice straw. Appl Microbiol Biotechnol 84:667–676

Chapter 14
Isolation and Screening of Halophilic Bacteria for Production of Hydrolytic Enzymes

M. Delgado-García, B. Nicolaus, A. Poli, C.N. Aguilar, and R. Rodríguez-Herrera

Abstract There is a great interest on study of extreme microorganisms because of their special characteristics in biotechnology. Particularly, halophilic microorganisms which live in saline environments throughout of their different adaptation mechanisms produce metabolites with great potential. Some of its biomolecules has been studied and applied in industrial processes, such as exopolyssaccharides, carotenoid pigments, bacteriorhodopsin etc. beside certain enzymes especially hydrolases (pectinases, amylases, proteases, lipases, etc.) are important. Recent researches on halophilic microorganisms and their biomolecules have increased around the world. Saline environments such as saline lakes or saline soils are excellent sources for isolation of halophilic microorganisms. However, few saline environments have been studied in depth in order to evaluate the special characteristics of halophilic biomolecules. In this review, the importance of halophilic microorganisms for biotechnological industries, methods for their isolation; techniques for physiological, taxonomical and molecular characterization have been highlighted so as to establish them as important source for enzyme production.

Keywords Halophilic bacteria • Hydrolytic enzymes • Biomolecules • Biotechnology • Molecular identification

14.1 Introduction

Because of its salt tolerance, halophilic microorganisms are considered extreme organisms (Meseguer 2004), According to salt requirements, halophilic microorganisms are classified as (i) extreme halophiles which require 3.5–5 M NaCl; (ii)

M. Delgado-García • C.N. Aguilar • R. Rodríguez-Herrera (✉)
Food Research Department, School of Chemistry, Universidad Autónoma de Coahuila, 25280 Coahuila, Mexico
e-mail: raul.rodriguez@uadec.edu.mx

B. Nicolaus • A. Poli
Istituto Di Chimica Biomolecolare, Comprensorio ex Olivetti, via Campi Flegrei 34, 80078 Pozzuoli, Naples, Italy

D.K. Maheshwari, M. Saraf (eds.), *Halophiles*, Sustainable Development and Biodiversity 6, DOI 10.1007/978-3-319-14595-2_14

moderately halophiles which grow at 0.5–3.5 M NaCl; (iii) weakly halophilies: need 0.3–0.5 M of NaCl and (iv) halo-tolerant which not need salt for growth but their growth is supported at high levels of salt in its medium. In addition to its salt requirements, halophilic microorganisms can tolerate alkaline pH and survive in hot and dry places (Madigan et al. 2004; Ramírez et al. 2004). Its capacity to grow under high salt concentration is due to adaptation mechanisms such as accumulations of compatible solutes in the cytoplasm. These substances act as biological structures stabilizing and allowing adaption in front of cold, heat, desiccation, etc. Principally, compatible solutes form a water layer generating a cell hydration in presence of high salt in the medium (Kunte et al. 2002; Ramírez et al. 2004). These special characteristics developed scientifically interest not only for studies of isolation and search for new species, but to explore the possibility of searching for biomolecules which may have potential in biotechnological processes (Table 14.1).

14.1.1 *Enzymes*

The most important biotechnological application is halophilic enzymes. The major characteristics that they exhibit are because of large proportion of acidic amino acids, principally, aspartic and glutamic acid, allowing a hydration of protein surface because of presence of carboxylic groups of these amino acids (Lanyi 1974; Pundak and Eisenberg 1981). It has been observed that sodium chloride has a role to stimulate important biological processes such as transcription, translation and transport activity of halophilic enzymes (Averhoff and Müeller 2010). The principal class of enzyme produced by halophilic microorganism is hydrolases. It has been reported the commercial use of halophilic hydrolases e.g., amylases for starch degradation and detergents formulation while proteases are applied during fish oil hydrolysis. More studies on halophilic enzymes are still in progress (Karan et al. 2012).

Table 14.1 Biotechnological applications of biomolecules from halophilic microorganisms

Biomolecules	Application	Reference
Enzymes	Starch hydrolysis, microalgae saccharification (amylases), chitin conversion (chitinases), fish oil hydrolysis (lipases)	Karan et al. (2012)
Bacteriorhodopsin	Holography, spatial light modulators, artificial retina, neural network optical computing, and volumetric and associative optical memories	Margesin and Schinner (2001)
Biopolymers	Exo-polysaccharides: emulsifiers and mobility controllers, remove dyes from textile effluent used in alkaline conditions, pharmaceutical, food, and petroleum industries	Ventosa and Nieto (1995)
	Liposomes: compound in medicines and cosmetics, drug carrier in food or pharmaceutical industry	
Halocins	Natural antibiotics, inhibition of Na+/H+ antiporter generating a protection against the myocardium ischemia and reperfusion injury	Schiraldi et al. (2002)
Carotenoid pigments	Food colorant, additive in cosmetics, multivitamin preparations, etc.	Margesin and Schinner (2001)

14.1.2 Bacteriorhodopsin

Because of its photochromic property, bacteriorhodpsin has potential to be used as material for optic memories elaboration, holographic storage material, modelers spatial light, computer memories, etc. The studies of bacteriorhodopsin are focus on halophilic Archea particularly *H. salinarum* and *H. halobium* (Oesterhelt et al. 1991; Meseguer 2001).

14.1.3 Biopolymers

This is another attractive application because they can be used as emulsifiers, thickeners etc. The halophilic microorganisms can produce liposomes which are used as transporters of compound in medicine, cosmetology and polyhydroxyalkanoates to generate biodegradable polymers: Specially, halophilic polysaccharides such as sulfated polysaccharides produced from *Halomonas* sp., others with substantial quantity of fucose are produced from *Salipiger mucescens* and have a high potential and value (Ventosa and Nieto 1995; Margesin and Schinner 2001).

14.1.4 Halocins

These compounds have potential to be used as myocardial protector. These have role to play as preserving agents in food and leather industries and in control of infectious bacteria (Torreblanca et al. 1994 Schiraldi et al. 2002; Karthikeyan et al. 2013).

14.1.5 Carotenoid Pigments

Some halophilic microorganisms are source of carotenoids such as β-carotene vitamin A precursor, which have application as food additive, food coloring, cosmetics product and as drug component (Schiraldi et al. 2002). Now-a-days some of these applications are commercialized but others are in the process of development at pilot scale (Meseguer 2004). However, the tremendous diversity of halophilic microorganisms found in nature is still far from being fully exploited (Oren 2002).

14.2 Saline Environments

Alkaline hyper-saline environments are rich source of diverse microorganisms, especially halophile, alkalophile and haloalkalophile microorganisms which are useful for the development of new bioprocess and microbial products of

Table 14.2 Ionic composition of some saline environments (Madigan et al. 2004)

Concentration (g/L)				
Ion	Sea water	Great Salad Lake	Dead Sea	Zugm Lake (Egypt)
Na^{+}	10.5	105	40.1	142
K^{+}	0.38	6.7	7.7	2.3
Mg^{+}	1.27	11	44	<0.1
Ca^{2+}	0.4	0.3	17.2	<0.1
Cl^{-}	19	18.1	225	155
Br^{-}	0.065	0.2	5.3	–
SO^{4-}	2.65	27	0.5	23
HCO^{3-}	0.14	0.7	0.2	67
pH	8	7.7	6.1	11

commercial interest. Around the world, there are many saline environments, majority are located in dry and hot zones (Delgado-García 2011), being most of them originated by evaporation of seawater (called thalassohaline environments) where salt is composed mainly by sodium and chloride. When evaporation proceeds, some changes occur in the ionic composition because of precipitation of $CaSO_4$ and other minerals; these habitats are called athalassohaline (Oren 2002). Predominance of different ions depends on topography, geology and climatic conditions. The principal cation is sodium, while the anion is chloride (Table 14.2) (Madigan et al. 2004). Microbial life has been adapted to environments that combine high salt concentration with extremely high pH values. Alkaline soda lakes in Africa, India, China with pH values of 11 and higher salt concentrations about 300 g/L proved a source of potential halophilic microorganisms (Oren 2002).

Around of the world there are many saline habitats like Dead Sea (Israel), Great Salt Lake (USA) and others environments in Korea, and Spain. However, such studies cover only a limited part of the tremendous diversity among the halophiles. Nowadays, there are more investigations of saline environments around the world; particularly in Mexico exist many saline environments and saline saltern. Specially, in the north of Mexico Cuatro Ciénegas basin is important for its aquatic habitats, which has high salt concentrations; and it has been described as an important biodiversity reservoir within the Chihuahuan desert. This zone has saline soils with high levels specially of sodium chloride and others salts such as magnesium, calcium, potassium, sulphate and carbonates (Castro et al. 2011). However, Cuatro Ciénegas has negligible phosphorus levels (Elser et al. 2005; Castro 2007), and has been described as a location with high levels of species endemism (Escalante et al. 2008).

On the other hand, solar salterns found around the world, are extremely hyper saline habitats containing microbial systems (Sabet et al. 2009). Mexico has some solar salterns for example, Las Coloradas in Yucatan State, and Guerrero Negro in Baja California Sur State. Especially the last has been studied for almost 30 years, and it is described as a place with great diversity of microbial systems, especially halophilic microorganisms (Ley et al. 2006; Escalante et al. 2009). About soda lakes, in Mexico exist the former soda Texcoco Lake which has been suffered many disturbances because of human activities. Disturbances include drainage of the lake since seventeenth century to halt flooding of the city and irrigation of the generated soils with sewage sludge since the 1970s (Ortega-Guerrero et al. 1997). However, some studies revealed presence of bacteria, actinomycetes, denitrifiers and others (Luna-Guido et al. 2000). Around the world, there are many examples of saline environments where halophilic microorganisms have been isolated. But, still there is lack of information about new halophilic species, and biotechnological applications of these microorganisms. In the next paragraphs, we will discuss different techniques for isolation, physiological, taxonomical and molecular characterization of halophilic microorganisms. In addition, screening techniques for enzyme production will also be discussed.

14.3 Culture Media for Isolation of Halophilic Microorganisms

For isolation of halophilic microorganisms and enzyme production, salts-added culture medium is used. There are reported different culture media depending of the halophilic microorganism of interest. In the case of extreme halophiles and halotolerant, eubacteria may interfere in the isolation of the desirable moderate halophilic microorganisms. Specific microorganisms need specific nutrients per example diverse vitamins. For this reason, it is an important criterion for selection of the best culture medium according to the microorganism of interest. One of the best media for isolation of halophilic bacteria is named halophilic medium (HM) specific for moderate halophiles because of its low Mg^{2+} content, which does not support good growth of extreme halophiles (Ventosa et al. 1982). However, there is other culture medium composition that is more specific for moderate halophiles such as *Halobacillus* genera, where in all cases, pH should be adjusted at 7.0 (Amoozegar et al. 2003). Composition of different culture media reported for halophilic microorganisms isolation is shown in the next paragraphs.

(a) **Halophilic medium (HM) (g/L)**: NaCl (178), KCl (Amoozegar et al. 2003), $MgSO_4 \cdot 7H_2O$ (1), $CaCl_2 \cdot 2H_2O$ (0.36), NaBr (0.23) $NaHCO_3$ (0.06), $FeCl_3$ (trace), yeast extract (10), glucose (1), agar (24) and distilled water (1,000 mL) (Ventosa et al. 1982).
(b) **HALO medium (g/L)**: sodium citrate (10), sodium thiosulfate (10), sodium collate (3), saccharose (20), NaCl (25), Iron citrate (1), potassium dihydrogen phosphate (2), magnesium sulfate (5) and distilled water (1,000 mL) (Castro 2011).
(c) **Marine agar (Difco) (g/L)**: peptone (5), yeast extract (1), ferric citrate (0.1), NaCl (8.8), $MgSO_4 \cdot 7H_2O$ (8.8), $CaCl_2 \cdot 2H_2O$ (1.8), sodium sulfate (3.24), KCl (0.55), Na_2HCO_3 (0.16), KBr (0.08), strontium chloride (34), boric acid (22), sodium silicate (4), sodium fluoride (2.4), ammonium nitrate (1.6), disodium phosphate (8), agar (15) and distilled water (1,000 mL). This medium is used for different halophilic microorganisms such as *Halobacillus* sp. specified by DSMZ institute (Leibniz-Institute DSMZ, German Collection of Microorganisms and Cell Cultures).

(d) **Nutritive agar modified adding NaCl as salt source**. The quantity of NaCl depending on the halophilic microorganism to isolate: (1) extreme halophilic: need about 3.5–5 M of salt, (2): moderately halophilic: growth at 0.5–3.5 M of salt, (3) weakly halophilic: need 0.3–0.5 M of salt (Delgado-García et al. 2013; Meseguer 2004).

(e) **Specific culture medium for *H. trueperi* and *H. karajensis*** (**g/L**): NaCl (100), $MgSO_4 \cdot 7H_2O$ (5), casein peptone (5), yeast extract (3) (Amoozegar et al. 2003).

There are other culture media, specifically for moderate halophiles, however in the case of the extreme halophiles, it is possible to use the culture medium for *Halobacterium salinarun* NRC-1 due to is the major extreme halophilic bacteria studied. It is important considered that extreme halophiles will tolerate a fairly generous salinity range and ionic composition (Rainey and Oren 2006).

(f) **Culture medium for extreme halophilic microorganisms**, **specifically for *Halobacterium salinarum* NRC-1**, **basal medium** (**g/L**): NaCl (250), $MgSO_4 \cdot 7H_2O$ (20), trisodium citrate (3), KCl (2), tryptone (5), yeast extract (3), distilled water (1), trace metals (0.1 mL): $ZnSO_4$ $7H_2O$ (1.32), MnSO4 H_2O (0.34), Fe (NH_4) $2SO_4$ $6H_2O$ (0.82), $CuSO_4$ $5H_2O$ (0.14), distilled water (200 mL) (Rainey and Oren 2006).

Specifically, for isolation of halophilic Archea, it is important to perform a concentrated seawater solution. According to Rodriguez-Valera et al. (1980) and Torreblanca et al. (1986) this solution should be contain salts in approximately the same proportions as found in sea water. Seawater composition is described below.

(g) **Seawater solution** (**stock solution**) (**g/L**): NaCl (240), $MgCl_2$ $6H_2O$ (30), $MgSO_4$ $7H_2O$ (35), KCl (7), NaBr (0.8), $NaHCO_3$ (0.2), is important to add double-distilled water to near the final required volume and dissolve the salts completely. Finally, add 5 mL $CaCl_2$ $2H_2O$ (slowly) from a 1 M sterile stock solution and adjust pH up to 7.5 (Rainey and Oren 2006).

14.4 Culturing of Soil Samples on Solid Media

Agar is the most commonly used gelling agent for halophiles. For this reason, there are different methods to optimize culturing and isolate different halophilic microorganisms.

14.4.1 Modified Pour Plate

First 10 g of soil was added into 90 mL of sterile distilled water. The Stirred should be vigorously to dissolve major part of soil sample. This was the first dilution of sample (10^{-1}), then, in other flask with 90 mL of sterile distilled water 10 mL of the first dilution was placed. It is important to maintain in agitation the first dilution. This step was repeated to obtain a dilution approximately of 10^{-8}. Finally, agar medium was poured into Petri plate and when this almost solidifies 1 mL of the diluted soil suspension was stirred to homogenized agar medium. This step was repeated for each dilution. This technique was used principally to avoid loss microbial diversity (Fig. 14.1) (Madigan et al. 2004; Delgado-García et al. 2013).

14.4.2 Soil Granules

First of all, little soil granules should be placed with a sterile spatula into the agar plate. After that, soil granules should be having a distance about 1 cm among themselves (Fig. 14.2). This technique of soil granules allow to obtain more colonies of bacteria, besides this technique is easy and cheap (Aquilanti et al. 2004; Delgado-García 2011).

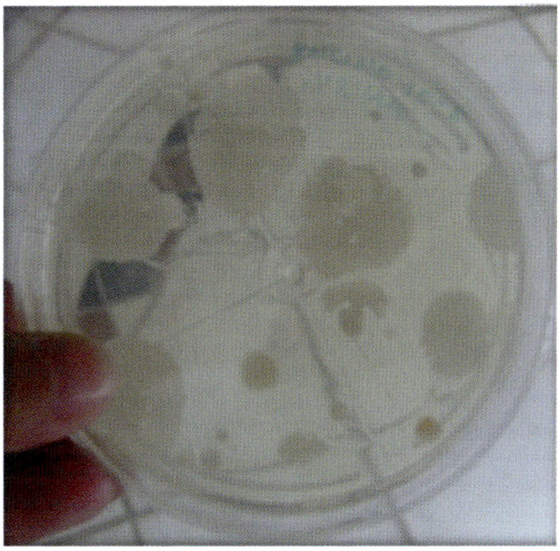

Fig. 14.1 Pour plate technique using a 10^{-5} soil dilution (Delgado-García 2011)

Fig. 14.2 Soil granules technique (Delgado-García 2011)

These types of culturing methods can be used to isolate all kinds of halophilic microorganisms but, in the case of extreme halophilic microorganism, plates should be placed in sealed plastic containers to prevent moisture loss during the incubation period. Incubation time depends on the halophilic microorganism type, per example if objective is to isolate moderate halophilic microorganisms about 2 or 3 days of incubation is sufficient but, in the case of extreme halophilic microorganisms, it could take about 2 or 4 weeks to obtain colonies. The optimum temperature for halophilic bacteria is 37 °C (Rainey and Oren 2006).

14.5 Evaluation of Halophilic Capacity

The salt tolerance evaluation allows selecting the best halophilic strains and determination of the group where each strain belongs.

14.5.1 Growth Kinetics

In order to known adaption of a specific cell to a culture medium with low or high salt concentration, growth kinetics was studied. For this study, it was important to select a range of salt concentrations to establish optimum salt concentration for each strain. First, from a liquid culture with a specific strain is recommended to perform a calibration curve using spectrophotometric methods (λ 600 nm). Then, each strain is grown at different salt concentration, covering the range mentioned for halophilic microorganisms (1) extreme halophilic: need about 3.5–5 M of salt, (2): moderately halophilic: growth at 0.5–3.5 M of salt (3) weakly halophilic: 0.3–0.5 M of salt. After that, it was important to grow the selected strain (s) to specific conditions and to determine bacterial growth in the medium with different salt concentrations at specific time (hours, days, weeks), in this case, it was took 1 mL of the culture and was determinate the absorbance using spectrophotometric methods (λ 600 nm). Later, data was converted and a graphic (cell/mL vs. time) was performed. With this method, it was possible to identify which strain assimilated more rapidly salt from culture medium and optimum salt concentration for microorganism growth (Delgado-García et al. 2013).

14.5.2 Halo Formation in Saline Medium

Formation of a halo by a microorganism in a medium supplemented with salt (NaCl) is an indicative of its halophilic capacity. One of the widely used medium is salt-mannitol agar (g/L): meat extract (10), pluripepetone (10), d-mannitol (10), NaCl (different concentrations), phenol red (0.25), agar (15) (Delgado-García 2013).

After the medium is prepared, place a bacterial colony on the salt-mannitol agar and incubate to specific grow conditions, then, determine halo formation and measure its diameter and quantify days for halo formation. Finally a graphic of halo diameter vs. time is made.

14.5.3 In Vitro Sodium Capture

This technique allows determining halophilic potential of the tested strains, which may use sodium utilization by different metabolic pathways. First, all lab material used in this technique should be washed several times with a nitric acid 10 % (w/v) solution to eliminate sodium and potassium traces, after growing the tested microorganism at a specific salt concentration and grow conditions, sodium capture could be determinate by atomic absorption spectroscopy. It is important to use a sodium standard and control (culture medium with salt concentration without microbial inoculum) (Table 14.3). In addition, it is very important to perform dilutions of the culture medium, approximately 1: 1,000, which will be used for the analysis by atomic absorption spectroscopy. Finally, statistical analysis should be performed to compare the results and deduce which strain is the best sodium scavenger.

The halophilic capacity determination of each microorganisms is a key factor, because it has been mentioned that sodium chloride increase the metabolic activity in halophilic microorganisms (Mudryk and Donderski 1991). Based on these studies, it is possible to have an idea about sodium chloride quantity requirements for halophilic microorganism's growth, because different studies have indicated that halophilic microorganisms require high chloride concentrations to stimulate DNA transcription, translation and transport activity (Lanyi 1974; Pundak and Eisenberg 1981).

14.6 Identification of Halophilic Microorganisms

Phylogeny is the study of evolution and development of species. The comparison of genomic sequences of some macromolecules is the best and reliable technique for inference on the phylogenetic relationships (Herrera 2011). Specially, the study of 16S rDNA gene is the most used to establish phylogenetic and taxonomic

Table 14.3 Specifications for the atomic absorption spectroscopy technique used for study of sodium in vitro capture to determinate the halophilic capacity of the strain (Sánchez-Leal and Arguello 2006; Delgado-García et al. 2013)

Component	Characteristic
Flame	Air-acetylene
Air flux	13.58 L/min
Acetylene flux	2.25 L/min
Lamp current	5 mA
Wavelength	589 nm
Slot width	0.5 nm

relationships among bacteria. Sequence of this gene has been used because is a highly conserved allowing determinate difference among bacteria and construct a dendrogram or phylogenetic tree for their representation (Woese et al. 1990; Delgado-García 2011). Recent advances on mass sequencing will speed a better phylogeny and systematic of organisms. However, still studies on 16S rDNA sequences allow new perspectives on diversity, given to the generation of a new knowledge and classification of bacteria groups and its evolution (Herrera 2011).

14.6.1 DNA Isolation

Although different DNA isolation techniques have been reported in literature, in the next paragraph we will describe one methodology which has offered high DNA quality. First, obtain a cellular pellet by centrifugation (200 g × 10 min) from a liquid culture medium, using per example 200 μL of biomass and 800 μL of sterile distilled water. Then, re-suspend the cellular pellet in a lysozyme solution (5 mg/mL) and incubate at 50 °C for 1 h. After that, apply a thermal shock (1 min in ice and 1 min at 90 °C) for 3 times. Later, add 200 μL of buffer TE 1X and 200 μL of SDS (sodium dodecylsulfate) 10 % and incubate for 1 h in dry bath, then add 100 μL of NaCl (5 M) and 80 μL of CTAB (hexadecyltrimethylammonium bromide) buffer and incubate at 65 °C for 30 min. After that, add an equal volume of chloroform-isoamyl alcohol (24:1) and centrifuge at 1,174 g × 5 min separate the aqueous phase and add 600 μL of isopropyl alcohol, then incubate for 24 h at −20 °C. Later, add 1 mL of alcohol (70 %) and centrifuge at 1,174 g × 5 min. Finally, remove supernatant and add 100 μL of Buffer TE 1X (Maloy 1989; Zavala-Castro 2005; Delgado-García 2011).

14.6.2 Microorganisms Identification by Specific Genes Amplification

One of the most important techniques and most used around the world is the polymerase chain reaction (PCR) (Mullis and Faloona 1987). This technique involves in vitro enzymatic amplification of a specific DNA region located between two DNA regions whose sequence is known. Using a PCR reaction is possible obtain until millions of copies of a gene or DNA region. The PCR technique is indispensable because allow establish new challenges to study and better understand the role of some genes (Mullis and Faloona 1987; Saiki et al. 1988).

There are different PCR variants, for example, RT-PCR used when the sample is a complementary DNA from mRNA. This conversion is due to reverse transcription by an enzyme named reverse transcriptase which converts mRNA in cDNA (Tamay de Dios et al. 2013). Other PCR variant is nested-PCR based in two consecutive rounds of amplification, the first PCR is performed using a pair of external primers. The resulting amplification product is transferred to another tube containing a second primer pair, which is nested. In other words they are internal to the first the initial pair. There are other variants of PCR such multiplex PCR, multiplex with touchdown PCR, etc. however, the basis is the same, amplify a gene to obtain millions of copies (Bartlett and Stirling 2003; Olmos et al. 2003).

14.6.2.1 16S rDNA

It has been reported predominance of bacteria than other microorganisms (fungi, yeast, etc.) in halophilic group. By this reason, there are more studies of isolation, and characterization of bacteria (Vreeland and Hochstein 1992). Also the major representatives of halophilic Archea are prokaryotic (Prescott et al. 2004). Different primers have been used for amplification of the 16S rDNA gene such as those described in Table 14.4 (Guillén-Cruz et al. 2005), which were efficient to identify bacterial strains to specie level (Delgado-García et al. 2013). For PCR reaction is important to add 14.5 μL of sterile distilled water 2.5 μL buffer TE 10X 1 μL of $MgCl_2$ (50 mM), 0.5 μL of dNTP's (20 mM) 2 μL of each primer (10 pM), 0.5 μL of *Taq* polymerase (5 U/mL) and 2 μL of sample DNA. Temperatures used for PCR reactions are: 94 °C (5 min) of initial denaturation followed by 35 cycles (94 °C/1 min, 65 °C/1 min and 72 °C/1 min) and final extension of 72 °C for 8 min (Flores-Gallegos 2009).

There are other useful primers (Table 14.4) for bacteria identification (Muyzer et al. 1995). Specifically GM3 and GM4 universal primers amplify nearly the complete sequence of 16 rDNA of the Bacteria domain, while other primers amplify the V6 and V8 region of 16S rDNA (Table 14.5). Specifically, these primers amplify a short sequence (approximately 58 bp) from the hyper-variable region V6, which is variable enough that allow to distinguish among most closely related bacterial species and also it is useful for differentiation among genera in ecological studies (Lane 1991; Grosskopf et al. 1998).

Table 14.4 Primers used for 16S rDNA amplification (Muyzer et al. 1995; Guillén-Cruz et al. 2005)

Primer	Sequence
16S rDNA forward	5′-AGGAGGTGATCCAACCGCA-3′
16S rDNA reverse	5′-AACTGGAGGAAGGTGGGAT-3′
GM3 forward	5′-AGAGTTTGATCMTGGC-3′
GM4 reverse	5′-TACCTTGTTACGACTT-3

Table 14.5 Primers for amplification of V6 and V8 regions from 16S rDNA (Grosskopf et al. 1998)

Primer	Sequence
A-109	5′-AC(G/T)GCTCAGTAACAGTAACACGT-3′
GC-515	5′-AC(G/T)GCTCAGTAACAGTAACACGT-3′

Table 14.6 Primers used for gene amplification from bacterial strains, during taxonomical studies

Gene	Primers
RNA polymerase subunit β	*rpoβ*F
	5′-GCGAAGTGTTAGAATTACC-3′
	*rpoβ*R
	5′- TCGTATTCTAACCATGCGCC-3′
	Band size = 450 bp
DNA gyrase subunit β topoisomerase II gene	UP-2r
	5′-AGCAGGGTACGGATGTGCGAG-3′
	UP-2Sr
	5′-AGCAGGGTACGGATGTGCGAGCC-3′
	Band size = 1,260 bp

14.6.2.2 Other Genes for Taxonomic Studies

Other gene sequences different to ribosomal ones have been used to differentiate bacterial strains to specie or subspecies level. One of the most used is *rpoβ* gene (RNA polymerase subunit β) sequence because predict similarities among genomes especially in bacteria (Table 14.6). The *rpoβ* gene sequence is more informative than that of 16S rDNA because of its size (4,200 bp). This gene has conserved and alternating variable regions and is used for taxonomic and phylogenetic studies, especially when it is suspected that the tested strain may be new specie (Meintanis et al. 2006). PCR reaction temperatures are: 95 °C/10 min (initial denaturing), 35 cycles (95 °C/1 min, 53 °C/1 min, 72 °C/1 min) and 72 °C/1 min for final extension (Delgado-García 2013).

The *gyrβ* (DNA gyrase subunit β topoisomerase II gene) (Table 14.6) is other gene used for taxonomic studies. This gene allows discriminate highly related species and is possible identify to subspecie level, being an excellent taxonomic marker; the sequence of this gene is about 12,000 bp. PCR reaction temperatures are: 94 °C/5 min (initial denaturing), 35 cycles (94 °C/1 min, 60 °C/1 min, 72 °C/2 min) and 72 °C/5 min for final extension (Yamamoto and Harayama 1995; Li- Tang et al. 2007).

14.6.3 Metagenomics Studies

Metagenomics is defined as the direct genetic analysis of genomes contained with an environmental sample. The analysis start with the cloning of environmental DNA, followed by functional expression screening and then is complemented by direct random shotgun sequencing of the environmental DNA (Tyson et al. 2004; Venter et al. 2004). Direct metagenomic sequencing is an appealing route for investigate microbial community composition, because, it provides simultaneous insight into phylogenetic composition and metabolic capabilities of uncultivated populations (Allen and Banfield 2005; Wilmes et al. 2009).

Metagenomic protocol starts with a soil or water sample, followed by genomic DNA extraction. Then, ligation of heterologous genomic DNA and restriction digested vector is performed; later, the vector is inserting into *E. coli* cells. This procedure allows construct a genomic library and study the DNA sequences in two ways: heterologous gene expression (Function-driven analysis) or by DNA detection by hybridization with specific or by vector-primed insert sequencing (Sequence-driven analysis) (Schloss and Handelsman 2003).

14.6.4 Denaturing Gradient Gel Electrophoresis (DGGE) and Temperature Gradient Gel Electrophoresis (TGGE)

Genetic fingerprints technique provides a pattern of profile of the genetic diversity in a microbial community. One of the fingerprint techniques that have been used in microbial ecology is the electrophoretic separation in high resolution poly-acrylamide gels. Separation of DNA fragments in DGGE and TGGE is based on the decreased electrophoretic mobility of partially melted double-stranded DNA molecules in poly-acrylamide gels containing a linear gradient of DNA denaturants (a mixture of urea and formamide) or a linear temperature gradient, which is created by two water baths attached to a cooling plate under gel (Muyzer and Smalla 1998).

By using DGGE or TGGE, 50 % of the sequence variants can be detected in DNA fragments up to 500 bp (Myers et al. 1985). This percentage can be increased to nearly 100 % by the attachment of a GC-rich sequence, to one side of the DNA fragment. A sequence of guanines and cytosines is added to the 5′-end of one the PCR primers, co-amplified and thus introduced into the amplified DNA fragments (Sheffield et al. 1992). DNA bands in DGGE and TGGE profiles can be visualized using ethidium bromide or SYBR green I (Myers et al. 1985).

14.7 Biochemical and Hybridization Techniques for Taxonomic Studies

Generally, taxonomic studies are used when it is suspected that the tested microorganisms may be new species. Biochemical studies allow knowing cell wall markers for establish relationship and differences among different species of one genus. Polyphasic is one important development for bacterial taxonomy; it has been used to integrate the different kinds of data and information (phenotype, genotypic and phylogenetic) on microorganisms and essentially indicates a consensus type of taxonomy. The principal aim of taxonomy is to give classification to bacteria. Phenotypic methods comprise all those that are not directed toward DNA or RNA; therefore, they also include the chemotaxonomic techniques. The chemotaxonomy is considered one of the essential milestones in development of modern bacterial classification. These types of studies refer to application of analytical methods to collect information on various chemical constituents of cell to classify bacteria (Vandame et al. 1996).

14.7.1 Isoprenic Quinones

Respiratory quinones are lipidic terpenoids constituting the bacterial cytoplasmatic membrane, which play important roles on electron transport, oxidative phosphorylation and possibly, active transport (Ventosa et al. 2004). Two major structural groups, naphtoquinones, which occur less commonly in bacteria and the menaquinones have been reported (Vandame et al. 1996). The study of the respiratory quinones, size and chain saturation allows to determine phylogenetic relations in a bacteria group (Da Costa et al. 2011). In addition, quinones are biomarks of bacterial populations and their study allow understand the respiratory function and electron transport (Irvan 2006).

To determine isoprenic quinone type, the following protocol is used. First, 4 g of lyophilized biomass are added to 50 mL of hexane. Sample is maintained in agitation for 24 h. After that, hexane is evaporated and sample is dried by nitrogen flux. Then, the solvent system of hexane and ethyl acetate (96:4) is prepared and sample is dissolved with chloroform. Later, sample is placed into a TLC (10 × 10 cm) and putting into the chromatographic camera for elution. Finally, TLC is dried and observed under UV light, quinone mark are reveal with cerium-sulfate 0.1 %. Quinones purification is performed from the sample charged in a TLC (20 × 20 cm). The protocol is the same as described above. After the band is obtained, it is cut off from the TLC and purified using methanol-chloroform (2:1) and filtered. Finally, sample is dissolve in chloroform-d_3 and analyzed by H^1 NMR and LC/MS techniques for isoprene units and molecular weight determination (Nicolaus et al. 2001; Romano et al. 2001).

14.7.2 Polar Lipids Composition

The polar lipids are also biomarkers for bacteria identification. Lipid distribution is not identical for all species, for this reason is useful to show evolutionary changes (Barton 2005). A variety of lipids are present in bacteria cells. Polar lipids are the major constituents of the lipid bi-layer of bacterial membranes and have been studied frequently for classification and identification purposes. Phospholipids of cell wall are the most important for their proportion (allow cell or organelles integrity by forming a semi-permeable barrier), also form an active biological material essential for biosynthesis and transport processes (Setlmann and Holst 2002; Barton 2005).

Other types of lipids, such as sphingo-phospholipids, occur in only a restricted number of taxa and are important within these groups of microorganisms (Jones and Krieg 1984; Vandame et al. 1996).

A protocol for determination of polar lipids is as follows: First, 4 g of lyophilized biomass are added to 50 mL of chloroform-methanol-water (65:25:4) and maintained in agitation for 24 h, hexane is evaporated and dried by nitrogen flux. Then, a solvent system chloroform-methanol-water (65:25:4) is prepared and sample is dissolve with chloroform. After that, sample is placed into a TLC (20×20 cm), and then put into the chromatographic camera for elution. Finally, TLC is dried and revealed to know the polar lipid composition with different solutions such as α-naftol (glycolipids), Dittmer (phospholipids), cerium-sulfate 0.1 % (total lipids) and ninhydrin (amino lipids) (Nicolaus et al. 2001). Identification of specific phospholipids is performed by bi-dimensional TLC. The system solvents are chloroform-methanol-water (65:25:4) (one dimension) and chloroform-methanol-acetic acid-water (80:12:15:4) (two dimension) (Fig. 14.3). First, TLC is placed in one system solvent, after that, it is dried and placed in the other system solvent. Finally, TLC is dried and reveal with Dittmer. In this case, for phospholipids identification is important to use standards such as: phosphatidyl serine (PS), phosphatidic acid (AP), inositol phosphatidil (PI), phosphatidil glycerol (PG), phosphoryl ethanol amine (PEA), phosphatidil coline (PC1 and PC3), cardiolipin (CLBH) (Nicolaus et al. 2001).

Phospholipids comprise about 10 % of the bacterial cell dry weight, and each mole of lipids requires about 32 mol of ATP for its synthesis. Thus, phospholipids synthesis requires significant energy investment by the cell, and advantages of maintaining fine control over the pathway are obvious. Bacteria have a pathway where phospholipids are catalyzed by a series of discrete proteins: enzymes are mainly integral inner membrane proteins. The key activated intermediate component during bacterial phospholipids synthesis is CDP-diacylglycerol, which comprises only 0.05 % of the total phospholipids pool.

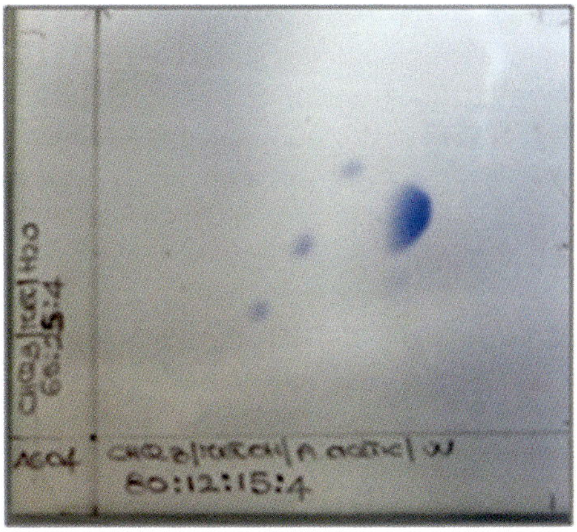

Fig. 14.3 Identification of phospholipids by bidimensional TLC

14.7.3 Fatty Acids Methyl Esters (FAME) Composition

Fatty acids are the major constituents of lipids and lipopolysaccharides and have been used extensively for taxonomic purposes. More than 300 different chemical structures of fatty acids have been identified. The variability in chain length, double-bond position, and substituent groups has proven to be very useful for characterization of bacterial taxa. Mostly, total cellular fatty acid fraction is extracted, but particular fractions such as polar lipids have been analyzed. Cellular fatty acid methyl ester content is a stable parameter, proving that highly standardized culture conditions are used. This method is cheap and rapid and has reached a high degree of automation (Vandame et al. 1996).

Fatty acids between 9 and 20 carbons in length have been used to characterize bacterial genera and species. Simple mutations or plasmid loss or gain do not alter fatty acid composition of an organism. There is an automated system, the MIDI Sherlock Microbial Identification System, which identifies microorganisms based on unique FAME patterns of known strains (Whittaker et al. 2003). The protocol is as follows: lipids are extracted; their hydrolysis is performing by TLC using hexane/Et2O (96:4) and is detected with I_2 vapor. FAME is analyzed by TLC on silica gel using α-naftol as developer. Finally, products are analyzed by GC/MS (Heat et al. 2002; Romano et al. 2007).

14.7.4 DNA-DNA Hybridization

The bacterial specie definition mentioned above is found upon whole genomic DNA-DNA hybridization values (Wayne et al. 1987). Results of this analysis are given in percent of DNA-DNA hybridization and decrease in thermal stability of the hybrid, it is used to delineate species. The percent of DNA binding or the DNA-DNA hybridization value or relative binding ratio is an indirect parameter of sequence similarity between two entire genomes (De Ley et al. 2006). DNA-DNA hybridization is often performed under standard conditions that are not necessarily optimal or stringent for all bacterial DNAs. Generally, optimal conditions for hybridization are preferred, because the optimal temperature curve for hybridization is rather broad (about 5 °C). The hybridization under optimal conditions requires a temperature between 22 and 26 °C, below the melting temperature, measured or calculated at equal salt concentration (Johnson 1991).

First, DNA sample should have a concentration of 1 ng/mL for an efficient DNA-DNA hybridization. One protocol for DNA-DNA hybridization is as follows: DNA is denatured by 10 min at 100 °C followed by quick immersion in water-ice bath. An amount of 50–80 ng/dot of DNA from tested strain is blotted on a positively charged nylon membrane (Roche, Germany). Dots are washed twice by 0.1 × SSC. Then, DNA is cross-linked to nylon by 3 min UV exposure and by 1 h under-vacuum at 120 °C. 1 μg of DNA shared by ultrasonic treatment is dioxigenin-dUTP labeled over night in 20 μL reaction mixture using a nucleotide random priming procedure (Dig DNA Labeling kit, Roche). After that, membranes are pre-hybridized for 3 h at 41 °C in Dig-Easy-Hyb solution and hybridized over night at 41 °C, using a roller tube hybridization incubator, in Dig Easy-Hyb solution containing 20 pg/mL of DIG labeled probe, heat denatured as above described or by 10 min at 68 °C in DIG-Easy-Hyb solution. Then, membrane washes are performed: twice for 5 min at room temperature in 2×SSC solution containing 0.1×SDS, twice for 15 min at 68 °C in 0.1×SSC solution containing 0.1×SDS. Later, immune detection is performing using the anti-dioxigenin AP antibody. Finally, chemi-luminicence is quantified in condition of time-exposure linearity by using a Versa-DOC 400 (BioRad). The DNA-DNA homology percentage is calculated according to Jahnke (1994) (Romano et al. 2007).

14.8 Biotechnological Potential of Halophilic Microorganisms and Enzymes

Enzyme production of halophilic microorganisms is poorly exploited to commercial level. One advantage of these enzymes is its capacity to catalyze reactions under extreme conditions, principally under high salt concentrations, but also are able to be stable at alkaline pH and some of them are thermostable (Enache and Kamekura 2010). These properties allow use them in industrial processes that use hard physicochemical conditions, or use them under low quantity of substrate or in highly concentrated substrates, but the most interesting characteristic of these enzyme is their capacity to catalyze reaction under organic solvents (Setati 2009; Oren 2010).

Hydrolases such as amylases, pectinases, pullulanase, DNases, xylanases, lipases, cellulases, inulinases, etc. is the major enzyme group produced by halophilic microorganisms (Table 14.7) (Cojoc et al. 2009; Rohban et al. 2009; Enache and Kamekura 2010). Halophilic enzymes have other applications per example as flavoring agent, or for pulp and paper industry, detergents formulation, oil fish hydrolysis, etc. (Kamekura and Onishi 1974; Karan et al. 2012).

14.8.1 Detection of Enzyme Production

For enzyme production by halophilic microorganisms is important first to know the enzyme type produced, so it is essential to perform an enzymatic screening. For the screening is crucial to select the substrate to produce the enzyme of interest. There is a simple, rapid and cheap method, which use only a microplate to perform the enzymatic screening. The protocol is as follows: First, culture medium with the specific substrate is prepared 250 μL of each agar is placed in a microplate well (Fig. 14.4). After that, wells are inoculated with the tested strain using the puncture

Table 14.7 Hydrolases produced by diverse halophilic bacteria and their industrial applications (Delgado-García 2012; Karan et al. 2012;)

Enzyme	Microorganisms (examples)	Applications
Amylases	*Haloferax mediterranei*, *Halobacterium salinarum*, *Nesterenkinia halobia*, *Halomonas meridiana*, *Halobacillus* sp.	Saccharification of marine microalgae, starch hydrolysis and textile, food, brewing and distilling industries
Proteases	*Bacillus* sp., *Halobacillus* sp., *Virgibacillus* sp. *Natrialba magadii*	Peptide synthesis, fish sauce preparation, detergents formulations
Lipases	*Salinivibrio* sp. *Natrococcus* sp.	Detergent additives, in the food and paper industries, enantioselective biocatalysis
Xylanases	*Halobacillus* sp., AS-04, *Bacillus pumilus*	Pulp and paper industry, baking industry for increasing loaf volume
DNase	*Microccocus varians*, *Bacillus* sp., N23-2	Acid 5′-guanilic and acid 5′-inosinic as flavor agents
Cellulases and pullulanases	*Bacillus* sp.	Biocatalysis in organic solvents and super critic fluids
Chitinases	*Halobacterium salinarum* NRC-1, *Planococcus riftoensis*	Oligosaccharide synthesis, bioconversion of chitin from fish, crab or shrimp, treatment of chitinous waste

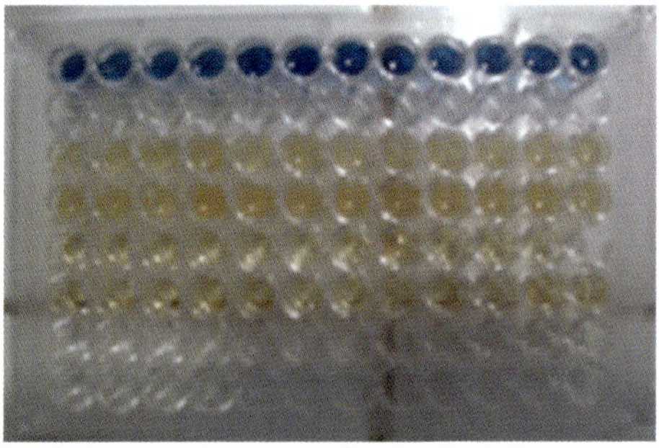

Fig. 14.4 Microplate with agar in each well for enzymatic screening of halophilic microorganism using hydrolysis halos formation. Each well has a substrate specific where the strain will be inoculate (Delgado-García 2013)

technique. Then, microplate is covered and incubated at specific temperature. Finally, each well is reveal and observed with stereoscope, using halo formation as an indicative of the hydrolysis mechanism (Delgado-García et al. 2013).

14.8.2 Genetic Aspects of Halophilic Enzymes

Generally, the halophilic enzymes reported are typically secreted into the extracellular environment throughout the growth cycle of halophilic bacteria in the presence of suitable substrates which would act as direct inducers of respective gene according to the enzyme produced (Setati 2010). It is reported that halophilic proteins are acidic in nature and the increase in acidic amino acids is compensated by decrease in basic amino acids. The frequency of proline and glycine is higher in case of extreme halophiles. Significant increase in the frequency of proline among extreme halophiles is observed in the genes reported as regulators of ion transporters (*aca4*, *esi47*, *hal3*, *hal*), transporters (*ema1*, *gbuA*), osmotic tolerance proteins (*gpd1*, *relA*) and salt toxicity target (*hal2*). On the other hand, high frequency of glycine among extreme halophile is observed in the genes reported to regulators of ion transporters (esi47, hal3, hal5), transporters (*ena1*, *ena2*, *gbuA*, *hal11*), osmotic tolerance protein (*cysK*, *gdp1*, *relaA*) and chaperones (*dnaK*, *groEL*) (Anwar and Chauhan 2012).

There are different studies for gene identification of specific halophilic enzyme per example the amylase produced by *Halomonas meridiana* and codified by *amyH* belongs to the already proposed family of α-amylases composed of the enzymes from *Alteromonas haloplanktis*, *Thermonospora curvata*, *Streptomyces* sp., insects and mammals due to the amino acid homology. The *AmyH* contains the four highly conserved regions in amylase enzymes. The invariant amino acid residues are also conserved in the *AmyH* sequence (Coronado et al. 2000).

A solvent stable protease produced by *Geomicrobium* sp. EMB2 a moderately halophilic bacterium has been characterized. The gene had a coding capacity of 375 amino acids. Also, the amino terminus of the mature extracellular purified protease matched with 65th amino acid onwards of the predicted polypeptide sequence. According to gene bade date, the protease clustered with serine protease marine gamma proteobacterium HTCC2207, *Shewanella loihica* PV-4, *Renibacterium salmoninarum* and other related *Bacillus* sp. and show a 40 % of amino acid hydrophobic residues (Karan et al. 2011).

14.9 Conclusion

In this chapter are discussed diverse methods and techniques for isolation and screening of halophilic bacteria with capacity to produce hydrolytic enzymes. Use of these techniques may help to perform a deep study for characterization of this kind of microorganisms, and generate valuable information for future studies on halophilic microorganisms and enzyme and search for more biotechnological applications. In the near future, halophilic bacteria will be used for their enzymes principally in biocatalysis processes, where the use of organic synthesis is indispensable. Halophilic enzymes may perform a reaction at low aqueous activity and tolerate organic solvents as culture medium which allow use them in pharmaceutical, food and chemical industries. In addition, enantioselectivity and stereoselectivity are very important characteristics of hydrolytic enzymes and are very important in bio-catalysis process. However, studies on halophilic enzymes until now are non significant because there are only few reports about genetic and proteomic studies. For this reason, it will be interesting to know more about genetic of halophilic enzymes and in a future performed studies of cloning, recombinant DNA or over-expression of halophilic enzymes, contributing to generation of new enzymes appropriate for modern biotechnological industries.

Acknowledgment This research was performed at The Food Research Department, School of Chemistry, Universidad Autonoma de Coahuila in Coahuila, Mexico. MDG likes to thank to the National Council of Science and Technology (CONACYT)-Mexico for the financial support during her M.Sc. Degree studies

References

Allen EE, Banfield JF (2005) Community genomics in microbial ecology and evolution. Nat Rev Microbiol 3:489–498

Amoozegar MA, Malekzadeh F, Malik KA, Schumann P, Spöer C (2003) *Halobacillus karajensis* sp. nov., a novel moderate halophile. Int J Syst Evol Microbiol 53:1059–1063

Anwar T, Chauhan RS (2012) Computational analysis of halotolerance gene from halophilic prokaryotes to infer their signature sequences. Int J Adv Biotechnol Bioinforma 1(1):69–78

Aquilanti L, Favilli F, Clementi F (2004) Comparison of different strategies for isolation y preliminary identification of *Azotobacter* from soil samples. Soil Biol Biochem 36:1475–1483

Averhorff B, Müller V (2010) Exploring research frontiers in microbiology: recent advances in halophilic and thermophilic extremophiles. Res Microbiol 161:506–514

Bartlett J, Stirling D (2003) PCR protocols in methods in molecular biology. In: Bartlett J, Stirling D(eds), vol 226, 2nd edn. Humana Press, Totowa, p 153

Barton LL (2005) Relationships in prokaryotes. Springer, New York, p 53

Castro L (2007) Isolation and characterization of halophilic microorganisms in saline soils of Cuatro Cienegas, Coahuila, México. AQM 3(5):33–43

Cojoc R, Merciu S, Popescu G, Dumitru L, Kamekura M, Enache M (2009) Extracellular hydrolytic enzymes of halophilic bacteria isolated from a subterranean rock salt crystal. Rom Biotech Lett 14(5):4658–4664

Coronado MJ, Vargas C, Mellado E, Tegos G, Drainas C, Nieto JJ (2000) The α-amylase gene *amyH* of the moderate halophile *Halomonas meridian* cloning and molecular characterization. Microbiology 146:861–868

Da Costa MS, Albuquerque L, Nobre MF, Wait R (2011) The extraction and identification of respiratory lipoquinones of prokaryotes and their use in taxonomy. In: Rainey F, Oren A (eds) Taxonomy of prokaryotes, vol 38. Academic/Elsevier, Londres, p 197

Delgado-García M (2011) Isolation and identification of halophilic microorganisms in soils of the State of Coahuila. BSc. thesis, Universidad Autónoma de Coahuila, Saltillo, Coahuila

Delgado-García M (2013) Production, characterization and sequencing of xylanase from Halobacillus sp. MSc. thesis, Universidad Autónoma de Coahuila, Saltillo, Coahuila

Delgado-García M, De la Garza-Rodríguez I, Cruz-Hernández MA, Balagurusamy N, Aguilar C, Rodríguez-Herrera R (2013) Characterization and selection of halophilic microorganisms isolated from Mexican soils. J Agric Biol Sci 8(6):457–464

Elser JJ, Schampell JH, García-Pichel F, Wade BD, Souza V, Eguiarte L, Escalante A, Farmer JD (2005) Effects of phosphorus enrichment and grazing snails on modern stromatolitic microbial communities. Freshw Biol 50:1808–1825

Enache M, Kamekura M (2010) Hydrolytic enzymes of halophilic microorganisms and their economic values. Rom Biotech Lett 47(1):47–59

Escalante A, Eguiarte LE, Espinosa-Asuar L, Forney L, Noguez AM, Souza V (2008) Diversity of aquatic prokaryotic communities in the Cuatro Ciénegas basin. FEMS Microbiol Ecol 65:50–60

Flores- Gallegos C (2009) Isolation and identification of native strains of the genus *Azotobacter*. BSc thesis, Universidad Autónoma de Coahuila

Grosskopf R, Janssen PA, Liesack W (1998) Diversity and structure of the methanogenic community in anoxic rice paddy soil microcosms as examined by cultivation and direct 16S rRNA gene sequence retrieval. Appl Environ Microbiol 64:960–969

Guillén-Cruz RF, Hernández-Castillo G, Gallegos-Morales D, Rodríguez-Herrera R, Aguilar C, Padrón-Corral E, Reyes-Valdés H (2005) *Bacillus* sp. as biocontrol in soil with *Fusarim* spp., *Rizotocnia solani*, Kühn and *Phytophtpra capsici* Leonian and their effect in development and crop yield chili (*Capsicum annun* L.). Rev Mex Fitopatol 24(2):105–114

Heat RJ, Jackowski S, Rock CH (2002) Chapter 3: Fatty acid and phospholipid metabolism in prokaryotes. In: Vance DE, Vance JE (eds) Biochemistry of lipids. Lipoproteins and membranes. Elsevier Science, University of Alberta, Edmonton

Herrera C (2011) Filogenia Bacteriana mediante el análisis del ARNr 16S. Carrera de Biología, Facultad de Estudios Superiores Iztacalca. Universidad Nacional Autónoma de México. Available in: http://www.uaem.mx/biologicas/index/progenote/bacteriana.pdf

Irvan (2006) Extraction of bacterial quinines from various different samples by using supercritical CO2. J Technol Process 5(2):112–119

Johnson LL (1991) DNA re-association experiments. In: Stackebrandt E, Goodfellow M (eds) Nucleic acid techniques in bacterial systematics. Wiley, Chichester, pp 21–44

Jones D, Krieg NR (1984) Serology and chemotaxonomy. In: Krieg NR, Holt JG (eds) Bergey's manual of systematic bacteriology, vol 1. The Williams & Wilkins Co., Baltimore, pp 15–18

Kamekura M, Onishi H (1974) Halophilic nuclease from a moderately halophilic *Micrococcus varians*. J Bacteriol 119:339 344

Karan R, Mohan-Singh RK, Kappor S, Khare K (2011) Gene identification and molecular characterization of solvent stable protease from a moderately haloalkaliphilic bacterium, *Geomicrobium* sp. EMB2. J Microbiol Biotechnol 21:129–135

Karan R, Capes MD, DasSarma S (2012) Function and biotechnology of extremophilic enzymes in low water activity. Aquat Biosyst 8(4):1–15

Karthikeyan P, Bhat SG, Chandrasekaran M (2013) Halocin SH10 production by an extreme haloarchaeon *Natrinema* sp. BTSH10 isolated from salt pans of South India. Saudi J Biol Sci 20(2):205–212

Kunte JH, Trüper HG, Stan LH (2002) The quest for the conditions of life. In: Horneck G, Baumstark-Khan C (eds) Physics and astronomy online library. Springer, Berlin, pp 185–200

Lane DJ (1991) 16S/23S rRNA sequencing. In: Stackebrandt ER, Goodfellow M (eds) Nucleic acid techniques in bacterial systematics. Wiley, Chichester, pp 115–175

Lanyi JK (1974) Salt dependent properties of proteins from extremely halophilic bacteria. Bacteriol Rev 38:272–290

Ley RE, Harris JK, Wilcox J, Spear JR, Miller SR, Bebout BM, Maresea JA, Bryant DA, Sogin ML, Pace NR (2006) Unexpected diversity and complexity of the Guerrero Negro hypersaline microbial mat. Appl Environ Microbiol 72:3685–3695

Li-Tang W, Fwu-Ling L, Chun-Ju T, Hiroaki K (2007) Comparison of *gyrB* gene sequences 16S rRNA gene sequences and DNA-DNA hybridization in the *Bacillus subtilis* group. Int J Syst Evol Microbiol 57:1846–1850

Luna-Guido ML, Beltrán-Hernández RI, Solís-Ceballos NA, Hernández-Chávez N, Mercado-García F, Catt JA, Olalde-Portugal V, Dendooven L (2000) Chemical and biological characteristics of alkaline soils form the former Lake Texcoco as affected by artificial drainage. Biol Fertil Soils 32:102–108

Madigan MT, Martinko JM, Parker J (2004) Brock Microbiología de los microorganismos, 10th edn. Pearson Prentice Hall, S.A. de C.V, pp 1, 8, 28, 29, 159, 161, 444–448

Maloy S (1989) Experimental techniques in bacterial genetics. Jones and Barlett. Publishers Boston. Department of Microbiology University of Illinois. Available in: http://books.google.com.mx/books?id=uqXAVLuyQhYCyprintsec=frontcoverydq=Experimental+Techniques+in+Bacterial+Geneticsylr=lang_en

Margesin R, Schinner F (2001) Potential of halotolerant and halophilic microorganisms for biotechnology. Extremophiles 5:73–83

Meintanis C, Chalkou KI, Kormas KA, Lymperopoulou DS, Katsifas EA, Hatzinikolaou DG, Karagouni AD (2006) Application of *rpoB* sequences similarity analysis, REP-PCR and BOX-PCR for the differentiation of species within the genus Geobacillus. Lett Appl Microbiol 46:395–401

Meseguer SI (2004) Los microorganismos halófilos y su potencial aplicado en la biotecnología. Cien Inv 7(2):13–17
Mudryk Z, Donderski W (1991) Effect of sodium chloride on the metabolic activity of halophilic bacteria isolated from the lake Gardno Estuary. Estuar Coast Shelf Sci 14(4):495–498
Mullis KB, Faloona FA (1987) Specific synthesis of DNA in vitro via a polymerase-catalyzed chain reaction. Methods Enzymol 155:335–350
Muyzer G, Smalla K (1998) Application of denaturing gradient gel electrophoresis (DGGE) and temperature gradient gel electrophoresis (TGGE) in microbial ecology. Anto Leeuw 73:127–141
Muyzer G, Teske A, Wirsen CO, Jannasch HW (1995) Phylogenetic relationships of *Thiomicrospira* species and their identification deep-sea hydrothermal vent samples by denaturing gradient gel electrophoresis of 16S rDNA fragments. Arch Microbiol 164:165–172
Myers RM, Fischer SG, Lerman LS, Maniatis T (1985) Nearly all single base substitutions in DNA fragments joined to a GC- clamp can be detected by denaturing gradient gel electrophoresis. Nucleic Acids Res 13:3131–3145
Nicolaus B, Manca MC, Lama L, Esposito E, Gambacorta A (2001) Lipid modulation by environmental stresses in two models of extremophiles isolated from Antarctica. Polar Biols 24:1–8
Oesterhelt D, Bräuchle C, Hampp A (1991) Bacteriorhodopsin a biological material for information processing. Q Rev Biophys 24:425–478
Olmos A, Esteban O, Bertolini E, Cambra M (2003) PCR protocols, chapter 26: Nested RT-PCR in a single closed tube. In: Bartlett J, Stirling D (eds) Methods in molecular biology, vol 226, 2nd edn, Humana Press, Totowa
Oren A (2002) Diversity of halophilic microorganisms: environments, phylogeny, physiology and applications. J Ind Microbiol Biotechnol 28:56–63
Oren A (2010) Industrial and environmental application of halophilic microorganisms. Environ Sci Technol 31(8–9):825–834
Ortega-Guerrero A, Cherry JA, Aravena R (1997) Origin of pore water and salinity in the lacustrine aquitard overlaying the regional aquifer of Mexico City. Can J Hydrol 197:1–4
Prescott LM, John HP y Donald KA (2004) Microbiología. 5ta. Ed. Mc Graw Hill, S.A. España, pp 475, 476, 670–673
Pundak S, Eisenberg H (1981) Structure and activity of malate dehydrogenase from the extreme halophilic bacteria of the Dead Sea. Eur J Biochem 118:463–498
Rainey F, Oren A (2006) Extremophiles: methods in microbiology. Academic/Elsevier, Amsterdam, NED
Ramírez DN, Serrano JA, Sandoval H (2004) Las bacterias halófilas y sus aplicaciones biotecnológicas. Rev Soc Ven Microbiol 24(1–2):1–23
Rodriguez-Valera F, Ruiz-Berraquero F, Ramos-Cormenzana A (1980) Isolation of extremely halophilic bacteria able to grow in defined inorganic media with single carbon sources. J Gen Microbiol 119:535–538
Rohban R, Amoozegar AM, Ventosa A (2009) Screening and isolation of halophilic bacteria producing extracellular hydrolases from Howz Soltan Lake Iran. J Ind Microbiol Biotechnol 36:333–340
Romano I, Nicolaus B, Lama L, Trabasso D, Caracciolo G, Gambacorta A (2001) Accumulation of osmoprotectans and lipid pattern modulation in response to growth conditions by *Halomonas pantelleriense*. Int J Syst Evol Microbiol 24:342–352
Romano I, Lama L, Orlando P, Nicolaus B, Giordano A, Gambacorta A (2007) *Halomonas sinaiensis* sp. nov., a novel halophilic bacterium isolated from a salt lake inside Ras Muhammad Park, Egypt. Extremophiles 11:789–796
Sabet S, Diallo L, Hays L, Jung W, Dillon J (2009) Characterization of halophiles isolated from solar salterns in Baja California, Mexico. Extremophiles 13:643–656
Saiki RK, Gelfand DH, Stoffel S, Scarf SJ, Higuchi R, Horn GT, Mullis K, Erlich HA (1988) Primer-directed enzymatic amplification of DNA with a thermostable DNA polymerase. Science 239:487

Sánchez-Leal L, Arguello AH (2006) Capacidad de las bacterias halófilas para capturar sodio in vitro y su posible aplicación en bio-remediación en suelos salinos sódicos. NOVA-Publicación científica 4:19–32

Schiraldi C, Giuliano M, de Rosa M (2002) Perspectives on biotechnological applications of Archea. Archea 1:75–86

Schloss PD, Handelsman J (2003) Biotechnological prospects from metagenomics. Curr Opin Biotechnol 14:303–310

Seltmann G, Holst O (2002) The bacterial cell wall. Springer, Berlin/New York, p 10

Setati EM (2009) Diversity and industrial potential of hydrolase-producing halophilic/halotolerant eubacteria. Afr J Biotechnol 9(11):1555–1560

Sheffield VC, Beck JS, Stone EM, Myers RM (1992) A simple and efficient method for attachment of a 40-base pair, GC-rich sequence to PCR-amplified DNA. Biotechniques 12:386–387

Tamay de Dios L, Ibarra C, Velasquillo C (2013) Fundamentals chain reaction polymerase (PCR) and real-time PCR. Investig Discapacidad 2(2):70–78

Torreblanca MF, Rodríguez-Valera F, Juez G, Ventosa A, Kamekura M, Kates M (1986) Classification of non-alkaliphilic halobacteria based on numerical taxonomy and polar lipid composition and description and *Halocarcula* gen. nov., and *Haloferax* gen. nov. Syst Appl Microbiol 8:89–99

Torreblanca M, Meseguer I, Ventosa A (1994) Production of halocin is a practically universal feature of archeal halophilic rods. Lett Appl Microbiol 19:201–205

Tyson GW, Chapman J, Hugenholtz P, Allen EE, Ram RJ, Richardson PM, Solovyev VV, Rubin EM, Rokhsar DS, Banfield JF (2004) Community structure and metabolism through reconstruction of microbial genomes from the environment. Nature 428(6978):37–43

Vandame P, Pot B, Gillis M, De Vos P, Kersters K, Swings J (1996) Polyphasic taxonomy, a consensus approach to bacterial systematics. Microbiol Rev 60(2):407–438

Venter JC, Remington K, Heidelberg JF, Halpern AL, Rusch D, Eisen JA, Wu D, Paulsen I, Nelson KE, Nelson W, Fouts DE, Levy S, Knap AH, Lomas MW, Nealson K, White O, Peterson J, Hoffman J, Parsons R, Baden-Tillson H, Pfannkoch C, Rogers YH, Smith HO (2004) Environmental genome shotgun sequencing of the Sargasso Sea. Science 304(5667):66–74

Ventosa A, Nieto JJ (1995) Biotechnological applications and potentialities of halophilic microorganisms. World J Microbiol Biotechnol 11:85–94

Ventosa A, Quesada E, Rodríguez-Valera F, Ruíz-Berraquero F, Ramos-Comenzana A (1982) Numerical taxonomy of moderately halophilic Gram-negative rods. J Gen Microbiol 128:1959–1968

Ventosa A, Márquez CM, Sánchez-Porro C, De la Haba RR (2004) Taxonomy of halophilic Archea and bacteria, pp 73. In: Vreeland RH (ed) Advances in understanding the biology of halophilic microorganisms. Springe (2012). Vestraete W, Environmental biotechnology. ESEB, London, p 48

Vreeland HR., Hochstein LI (1992) The biology of halophilic bacteria, 1st edn. CRC Press, Boca Raton, pp 55–62

Wayne LG, Brenner DJ, Colwell RR, Grimont PA, Kandler O, Krichevsky MI, Moore LH, Moore WEC, Murray RGE, Stackbrandt E, Starr MP, Trüper HG (1987) Report of the ad hoc committee on reconciliation of approaches to bacterial systematics. Int J Syst Bacteriol 37:463–464

Wilmes P, Simmons SL, Denef VJ, Banfield JF (2009) The dynamic genetic repertoire of microbial communities. FEMS Microbiol Rev 33:109–132

Woese CR, Kandler O, Wheelis ML (1990) Towards a natural system of organisms proposal for the domains Archea, Bacteria and Eukarya. Proc Natl Acad Sci U S A 87:4576–4579

Yamamoto S, Harayama S (1995) PCR amplification and direct sequencing of *gyrB* genes with universal primers and their application to the detection and taxonomic analysis of *Pseudomonas putida* strains. Appl Environ Microbiol 61(3):1104–1109

Zavala-Castro JE (2005) Manual of basic techniques in molecular biology. Ediciones de la Universidad de Yucatán, Yucatán, MX

Chapter 15
Perspectives and Application of Halophilic Enzymes

Stuti Patel and Meenu Saraf

Abstract The world of halophilic bacteria is quite diverse. We find representatives of three domains of life, archaea, bacteria and eukarya that are adapted to salt concentration upto saturation. The micro-organisms able to grow upto NaCl concentration (>300 g/l) are found all over the small subunit rRNA based tree of life. Their metabolic diversity is high as well encompassing oxygenic and anoxygenic phototrophs, aerobic heterotrophs, denitrifiers, sulfate reducers, fermenters and methanogens. The proteins of halophilic bacteria are magnificently engineered to function in a milieu containing 2–5 M salt. The proteins and encoding genes of halophiles represent a valuable repository and resource for reconstruction and visualizing processes of habitat selection and adaptive evolution. Search for new enzymes endowed with novel activities and enhanced stability continues to be desirable character for important commercial production. These poly extremophiles are excellent source of enzymes and metabolites possessing inherent ability to function in extreme conditions viz high salt, alkaline pH and facilitating catalysis for biotechnological application in food processing, industrial bioconversion and bioremediation. In brief, we have just begun to realize the great potential and true extent of diversity and suitable industrial applications possible from halophiles.

Keywords Halophilic enzymes • Compatible solutes • Salt-in strategy • Saline environments

15.1 Introduction

Halophiles are the organism that survives in hypersaline areas such as salt or soda lakes, coastal lagoons or man-made salterns etc. Hypersaline environments are those which contain salt concentration in excess approximately 3.5 % total dissolved salts. The two largest and best-studied hypersaline lakes are the Great Salt

S. Patel • M. Saraf (✉)
Department of Microbiology and Biotechnology, University School of Sciences, Gujarat University, Ahmedabad 380 009, Gujarat, India
e-mail: sarafmeenu@gmail.com

D.K. Maheshwari, M. Saraf (eds.), *Halophiles*, Sustainable Development and Biodiversity 6, DOI 10.1007/978-3-319-14595-2_15

Lake, in the western United States, and the Dead Sea, in the Middle East. The Great Salt Lake is larger (3,900 km^2) and shallower (10 m), and contains salts that are close in relative proportion to sea water. The Dead Sea is smaller (800 km^2) and deeper (340 m), and contains a very high concentration of magnesium salts. Both of these lakes are close to neutral pH, although the Great Salt Lake is slightly alkaline while the Dead Sea is slightly acidic (DasSarma and Arora 2001). Several alkaline hypersaline soda brines exist such as Great Basin lakes of USA, Lake Magadi of Kenya, Sambhar Lake of India etc. Soda brines are devoid of divalent cation like Mg^{+2} and Ca^{+2} as they have lower solubility in alkaline environment. Common phenomenon of hypersaline environment is the formation of gradients of salinity due to the seawater evaporation (Satyanarayana et al. 2005). All the three domains of life Archaea, Bacteria and Eukarya possess halophiles which can withstand high ionic concentration in their surroundings. Halophilic microorganism has evolved to exist in such salt saturated environments can be categorized into halotolerant, slightly, moderately and extremely halophilic (Enache and Kamekura 2010) by adapting themselves to flourish in such an extremely saline environment. Halotolerant are those which can tolerate high salt in their surroundings, slightly halophilic can survive in upto 3 % NaCl, moderately halophilic organisms can withstand upto 15 % NaCl and Extreme halophiles are those which can grow in the environment of 15–30 % salt concentration (Moreno et al. 2013).

Eukarya domain possess the sole member ubiquitously present in high-salt environments, is the green alga *Dunaliella* sp. is more a halotolerant than halophilic which can survive broad range of salt concentrations. Some species of *Dunaliella* are found to be the producer of β-carotene is in high demand as an antioxidant, as a source of pro-vitamin A (retinol) and as a food colouring agent. Its antioxidant activities make it popular for use in health food (Oren 2010). Moreover, variety of Diatoms species such as *Amphora coffeaeformis* and *Nitzschia* and *Navicula* are also reported to withstand salt concentration upto 3 mol/L (Das Sarma and Arora 2001). Besides this another rare macro organisms found to be halophilic is the brine shrimp *Artemia* sp. (Oren 2002), halotolerant yeast *Debaromyces hansenii* isolated from sea water (Buchalo et al. 2000). Some representatives of fungi such as meristematic fungus *Trimmmatostroma salinum* and Black yeast *Hortaea Werneckii* (Zalar et al. 1999; Gunde-Cimerman et al. 2000) are also considered as halophiles on the basis of their twin characteristics (i) their absolute requirement of high salt and (ii) their ability to grow in saturated ionic environment.

Haloarchaea in order Halobacterials and family Halobacteriaceae are extreme halophiles requiring at least 1.5 M NaCl for growth. Most of the species grow optimally in media with 3.5–4.5 M NaCl; many are able to grow in saturated NaCl (5.2 M) (Grant et al. 2001). These microorganisms were distributed all over the world in different natural hypersaline lake (Caton et al. 2009) or artificial crystallizer ponds (Oh et al. 2009). Likewise they found in environment with lower salinity (Elshahed et al. 2004). Halophilic archaea use high salt in strategy in order to survive osmotic challenges associated with life in hypersaline environments (Lin et al. 2011). Thus, they have enzymes which are active up to 5 M or more concentration of NaCl or 4 M KCl (Danson and Hough 1997).

Halophilic bacteria includes cyanobacteria, sulfur oxidizing bacteria, phototrophs, *Actinobacteria*, *Spirochetes*, aerobic and anerobic bacteria widely distributed in the saline regions ranging from the moderately halophiles to extremophiles. Filamentous cyanobacteria like *Microcoleus chthonoplastes*, *Phormidium ambiguum*, *Oscillatoria neglecta*, *O. limnetica* and *O. salina* are found to be present from the green mats of hypersaline lakes. Beneath, the layer of Cyanobacteria, Phototrophic bacteria *Chlorobium limnicola* and *Thiocapsa halophile* exist in anaerobic but lighted zones in microbial mats. Third layer in the saline microbial mat is of Sulfur-oxidizing bacteria, *Thiobacillus halophilus* from the western Australian lake, *B. leptiformis*, filamentous *Achromatium volutans* from Solar Lake and *Beggiatoa alba* from Guerrero Negro (Das Sarma and Arora 2001) were described.

Aerobic organotrophic Gram-negative bacteria such as species of *Acinetobacter*, *Alteromonas*, *Deleya*, *Flavobacterium*, *Pseudomonas*, *Vibrio* and *Marinomonas* are abundantly present in brines of medium saline region (Satyanarayana et al. 2005). Aerobic Gram-positive moderately halophilic bacteria of the genera *Halobacillus*, *Bacillus*, *Marinococcus*, *Salinococcus*, *Nesterenkonia*, and *Tetragenococcus* are found to be present. They include cocci such as *Nesterenkonia halobia*, isolated from salterns, producing yellow-red carotenoid pigments; *Tetragenococcus halophilus*, from fermented soy sauces and squid liver sauce, and from brine for curing anchovies, which are capable of lactic acid fermentation; and several *Salinococcus* species from salterns. Other examples include *B. diposauri*, from the nasal cavity of a desert iguana; *B. haloalkaliphilus*, from Wadi Natrun; and *B. halodenitrificans*, from a solar saltern in southern France. *Halobacillus litoralis* and *H. trueperi* are found in the Great Salt Lake (DasSarma and Arora 2001). Besides this, anaerobic fermentative bacteria of the order Halanaerobiales (phyla- Firmicutes) consist of only halophiles as their sole members (Ratnakar 2013).

A fascinating aspect of halophilic organisms is that they can produce some enzymes and metabolites which can maintain their structure and function in near salt-saturating biological niches where normal proteins get precipitated out.

15.2 Diversity of Halophilic Enzyme Producing Microorganisms

Halophilic enzymes producing organisms are widely distributed in the regions ranging from low salt concentration to the region of hypersaline environment. Among them, there are few number of halophilic enzyme producing microorganisms which have been characterized within the niches like salterns, saline deserts, salt lakes etc. While most research performed on hypersaline environments has focused on the microbial diversity and ecology of these environments, there is growing interest in the extracellular hydrolytic enzymes from moderately halophilic bacteria. Most halophilic hydrolase producers have been assigned to the family Halomonadaceae and were shown to produce industrially relevant enzymes such as cellulases, amylases, xylanases, proteases and lipases (Govender et al. 2009). It is generally believed that while these halophilic enzymes perform the similar enzyme function as their non-halophilic counterparts, they can catalyse such reactions under different conditions, such as high salt environments. Genera includes *Bacillus*, *Salibacillus*, *Salinivibrio*, *Halomonas* are moderately halophiles and are known to produce five main enzymes includes amylases, proteases, lipases, pullulanases, and DNases from the solar salterns of South Spain (Moreno et al. 2013). Among the Archaea three genera are known to produce maximum amount of enzymes – *Salicola, Salinibacter* and *Pseudomonas.* Among them *Salicola* is known to produce lipase and protease characterized from crystallizer form of different solar salterns (Ovreas et al. 2003; Maturrano et al. 2006). DNase activity was frequently reported by (Moreno et al. 2013) from the Altacama Desert, an extreme salt environment contaminated with heavy metals besides the activity of other enzymes. In the analyzed community, pullulanase and protease producers were also detected, being xylanases the least represented among the hydrolases tested. It is interesting to emphasize that multiple hydrolytic activity was frequently detected in the isolates reported in this study supporting previous studies in other hypersaline habitats (Moreno et al. 2009) As reported in the study of Rohban et al. (2009) most environmental isolates able to produce hydrolytic enzymes were Gram-positive bacteria, although the isolates were assigned to the family Bacillaceae, comprising species of the genera *Bacillus*, *Halobacillus* and *Thalassobacillus.* Only two isolates were related to the Gram-negative bacteria *Pseudomonas halophila* (Sorokin and Tindall 2006) and *Halomonas organivorans* (García et al. 2004) and the other characterized isolates were related to *Salinicoccus roseus* (Ventosa et al. 1990) Gram-positive non capsulated moderate halophilic *Bacillus* species were found to produce good amount of extracellular amylases from an Indian saline desert (Khunt et al. 2011). Howz Soltan Lake of Iran was reported to be the best site for almost all the hydrolytic enzyme producers. Lipolytic enzyme was found to be the most abundantly found from that site. Various genera such as *Halovibrio*, *Oceanobacillus*, *Thalassobacillus*, *Halobacillus*, *Virgibacillus*, etc. are reported (Dang et al. 2009). Cojoc et al. (2009) reported that the Slanic Prahova salt mine of Romania has the lipase and protease hydrolase producers in ample besides other extracellular enzymes, which includes *Bacillus*, *Halobacillus*, *Pseudomonas*, *Halomonas* and *Staphylococcus*.

Total 14 genera were characterized from the saline deposits of Southern Okinawa Trough (China) belonging to *Halomonas* and *Psychrobacter* groups showing good producers of almost all hydrolytic enzymes. Sambhar Lake (India) was reported to have extreme saline environment ranging from 7 to 30 % or higher salinity and hyper alkaline environment $\geq$9 or 9.5 pH exhibits diverse group of halophilic microorganisms (Upasani and Desai 1990) (Fig. 15.1). Similarly, Rann of Kutch (India) also have hypersaline ecosystem (Fig. 15.2). Pandit et al. (2014) collected soil samples from seven different locations of Kutch region to study the influence of salinity on microbial community. Salinity in different locations of Kutch region ranges from 4.38 to 32.85 % constituting 56–87 % bacteria and 8–40 % archaea. Major prokaryotic phyla observed were *Proteobacteria* (abundance percentages of 18.6–47.7 %), *Euryarchaeota* (7.9–40.1 %), *Bacteroidetes* (9.1–18.7 %), *Firmicutes* (5.8–8.4 %), *Actinobacteria* (2.1–5.8 %), and *Cyanobacteria* (2–4.2 %).

Fig. 15.1 Photograph of Sabhar Lake

Fig. 15.2 Photograph of Rann of Kutch

15.3 Adaptation of Halophilic Proteins with Salts

The stability and properties of proteins are strictly dependent on the interaction with the solvent in which they are dissolved. Proteins are composed of various combinations of hydrophilic and hydrophobic amino acids, and interactions of these residues with water or other solvents determine the folding of the proteins. It has long been known that activities of enzymes of animal and plant origin are strongly inhibited by salts (Warren et al. 1966). Proteins, particularly enzymes, of halophilic microorganisms, however, are extraordinary in that they require high concentration of salts for their stabilities and activities. Two fundamental strategies exist within the microbial world that enables microorganisms to adapt with stress generated from high salt environments are (i) The Salt-in strategy and (ii) The Compatible- Solute strategy.

15.3.1 Salt-In Strategy

Organisms that grow optimally in the presence of extremely high salinities of up to 5 M NaCl, accumulate intracellular KCl in concentrations higher than the external NaCl concentration to maintain a turgor pressure. This is called "Salt-in" strategy found in *Halobacteriales* (archaea) and *Haloanaerobiales* (anaerobic halophilic bacteria) as observed by Hanelt and Muller (2013). While the situation in the anaerobic bacteria is thought to be different; there is evidence that these organisms invest as little as possible in the maintenance of ion gradients. Measurements of the ion composition of exponentially growing cells of *Haloanaerobium praevalens* showed that K^+ is the dominating cation, but that Na^+ levels are also relatively high. Cells entering the stationary phase eventually replace K^+ for Na^+ (Oren et al. 1997). Cellular processes and machineries of organisms following this strategy are adapted to high internal KCl, such that in general these halophiles are restricted to grow at high salt.

15.3.2 Compatible-Solute Strategy

A more flexible strategy is found in halotolerant microorganisms and moderately halophilic bacteria that grow over a wide range of salinities (typically 0.5–3 M NaCl) (Roessler and Muller 2001). This strategy, "the low-salt-in" strategy, relies on the accumulation of high concentrations of organic compatible solutes. Compatible solutes are small, mainly neutral but polar compounds, which are highly soluble in water and do not interfere with the cellular metabolism. Thus, other than for KCl, a broad variation of the intracellular concentration of those compounds is possible without effecting cellular processes. The uptake or synthesis of compatible solutes retains a cytoplasm iso-osmotic with or slightly hyperosmotic compared to its surroundings. Besides this they have protein-stabilizing properties that help in the proper folding of protein chains (Arakawa and Timasheff 1985). Hence they are also known as chemical chaperones (Chattopadhyay et al. 2004). Compatible solutes employ their effect through changes in solvent structure and/or elusive changes in the dynamic properties of the protein and not by structural changes of the protein itself (Lamosa et al. 2003), and also helps in Protein- DNA interaction (Kurz 2008). These compatible solutes include polyols such as glycerol, sugars and their derivatives, amino acids and their derivatives, and quaternary amines such as glycine, betaine and ectoines etc. either by *de novo* synthesis or by uptake from the surroundings.

Ectoine was first discovered in the haloalkaliphilic photosynthetic sulphur bacterium, *Ectothiorhodospira halochloris*. It is also produced in variety of halophilic and halotolerant bacteria (Rothschild and Mancinelli 2001). Ectoine can protect many unstable enzymes and also nucleic acids against the detrimental action of high salinity, thermal denaturation, desiccation and freezing, thereby increasing shelf life and activity of enzyme preparations. Such compatible solutes are sometimes termed 'molecular chaperones' (Kolp et al. 2006). Betaine is the one which is synthesized by *Actinopolyspora halophilus*, *Halomonas elongata* and *Methanohalophilus portucalensis* FDF1 either by oxidation of choline or methylation of glycine while some halophiles have been observed to accumulate betaine when grown in salt rich medium (Roberts 2005). Amino acids like alanine, glutamine, proline are the three main compatible solutes present in several organisms. Some species belonging to *Streptomyces* can accumulate all three amino acids during salt stress; Gram-positive halophiles accumulate low level of glutamine and alanine and high level of proline against high salinity (Empadinhas and da Costa 2008).

15.4 Halophilic Enzymes

The emphasis on enzymes that can tolerate harsh environments has greatly increased over the past decade; halophilic enzymes are those which can work even at hyper saline environment. Extremozymes produced by halophilic microorganisms, have identical enzymatic features like their non-halophilic counterpart, but they exhibit different structural properties. Among these, two main points could be mentioned, (i) a high content in acidic amino-acids located predominantly on the solvent-exposed surfaces of the protein. Negative charges on the amino acids attract water molecules and form the hydrogen bonds to keep the proteins hydrated so that they do not precipitate (Balasubramanian et al. 2002). Increase in negative charges also results in an increase in ion-pair networks in halophilic enzymes (Dym et al. 1995) and (ii) Reduction of hydrophobic surfaces and the unusually high number of ordered side chains (Britton et al. 2006).

Bioinformatics studies of halophilic protein sequences have shown that they also contain different hydrophobic residues. Siglioccolo et al. (2011) determined that the hydrophobic contact in the core of halophilic proteins, exposed to molar concentrations of inorganic salt, is consistently smaller than that in mesophilic proteins. Weaker hydrophobic interactions due to smaller hydrophobic residues can increase the flexibility of protein in high salt, since it prevents the hydrophobic core from becoming too rigid (Mevarech et al. 2000). Halophilic enzymes with organic solvent tolerance are useful for the application in retaining organic solvents as reaction media without any modifications to stabilize the enzyme (Doukyu and Ogino 2010).

Halophilic enzymes are divided into three categories (i) intracellular enzymes-which are not directly in contact with the ionic concentration of the surroundings, (ii) membrane bound enzymes (carrier proteins) – which are in direct contact with the cytoplasmic content as well as the outside medium and (iii) extracellular enzymes- which are directly exposed to the saline medium (Ventosa et al. 1998). It is generally assumed that many halophilic enzymes are polyextremophilic. These enzymes not only remain active and stable in high salt environments but are often also thermotolerant and alkaliphilic (Moreno et al. 2009). A wide variety of bacteria that secrete extracellular hydrolytic enzymes such as amylases, proteases, lipases, DNases, pullulanases and xylanases have been isolated and characterized (Rohban et al. 2009). Certain organisms belonging to *Bacillus* group including *Salibacillus*, *Halobacillus*, *Oceanobacillus*, *Gracilibacillus*, *Virgibacillus*, *Thalassobacillus* and *Piscibacillus* and Gram-negative bacterial species of *Salinivibrio*, *Chromohal*obacter and *Halomonas* (Rohban et al. 2009).

15.4.1 Amylases

Novel amylases with desirable properties of thermal, salt, alkaline and organic solvent stability serve as a necessary approach to replenish the shortage of industrially stable enzymes. This extracellular polymer degrading enzymes from halophiles were proved to be useful in various industrial processes (Mohapatra et al. 1998). Halophilic amylases, (EC: 3.2.1.54), secreted from bacteria such as *Micrococcus halobius* (Onishi and Sonoda 1979), *Halomonas meridian* (Coronado et al. 2000), *Halothermothrix orenii* (Tan et al. 2008), *Streptomyces* sp. (Chakraborty et al. 2009) as well as *Chromohalobacter* sp. (Prakash et al. 2009) has molecular weight that ranges from 50 to 75 kDa and was found to be stable at broad pH, extreme salt environment and can work even at temperatures above 50 °C, makes them an attractive nominee in industrial processes which are commonly performed at low water activity (Margesin and Schinner 2001). For example, the amylase from *Halobacillus* and *Chromohalobacter* spp. were found to be stable at pH 7–10. Some of the enzymes such as the *Chromohalobacter* amylase maintain their stability in the presence and absence of NaCl (Prakash et al. 2009). Culture filtrate of *Acinetobacter* sp. isolated from sea sands found to contain two amylases showing maximum activity at 0.2–0.6 M NaCl at pH 7.0 and 50–55 °C (Onishi and Hidaka 1978). An α- amylase produced by *Halomonas meridian* has been purified and characterized, is a monomeric unit of molecular mass 58 kDa, required 3 M NaCl for its optimal activity and remain stable at 4 M NaCl reaching half-life time of about 83 days (Perez-Pomares et al. 2003).

15.4.2 Proteases

Proteases have been purified and characterized in *Halobacterium* spp., *Haloferax mediteranei*, *Natrialba asiatica*, *Natrialba magadii*, *Natronococcus occultus*, and *Natronomonas pharaonis* (DasSarma et al. 2010). For example; Serine protease from *Halobacterium salinarum* was the first extracellular hydrolase identified from an extremely halophilic archaeon, requires 4 M NaCl for optimal catalytic activity and stability with yields of up to 76 % in an environment with low water activity. This enzyme has potential to be used for peptide synthesis, particularly those containing glycine (Ryu et al. 1994). Serine protease halolysin 172P1 produced by archaeon *Natrialba asiatica* is a thermostable halophilic enzyme showing optimum activity at 75–80 °C, pH 10.7 and 25 % NaCl (Kamekura and Seno 1990). Amino acid sequence of halolysin 172P1 revealed that it consists of 411 amino acids and has high homology with thermitase obtained from *Thermoactinomyces vulgaris.* It has long C-terminal chain of about 120 amino acids which was not found in any extracellular serine protease (Kamekura et al. 1992).

Extracellular protease Haloprotease CP1 isolated from moderately halophilic bacterium *Pseudoalteromonas ruthenica* (Chand and Mishra 2003), showed optimal activity at 55 °C, pH 8.5 and high tolerance to a wide range of NaCl concentrations 0–4 M NaCl (Sanchezz-porro et al. 2003). Protease produced by the *Halobacillus karajensis* strain MA-2, in the presence of gelatin, has maximum activity at pH values ranging from 8.0 to 10.0, with 55 % and 50 % activity at pH 6 and 11, respectively. Moreover, the enzyme activity was strongly inhibited by PMSF, Pefabloc SC and EDTA; indicating that this enzyme probably belongs to the subclass of serine metalloproteases (Karbalaei-Heidari et al. 2009). *Geomicrobium* sp. EMB2, a moderately halophilic strain produces an extracellular protease was stable at 75 % organic solvents, 20 % salts, 2.0 % detergents and 1.0 % surfactants (Karan et al. 2011). Halophilic protease from *Bacillus* sp. EMB9 was uniquely stable in polar solvents like methanol, toluene and n-decane (Sinha and Khare 2014).

15.4.3 Lipases

Isolation and characterization of salt stable lipases from halophilic source has been a growing interest nowadays (Guzman et al. 2008). Availability of such enzymes would facilitate industrial processes that require activity at high salt concentration and low water activity. An intracellular lipase enzymes produced by *Salicola marasensis* LipL, has a maximum activity at 1 M NaCl. It can tolerate upto 4 M NaCl with 6 mM of betaine. This enzyme can act on different substrates such as p-nitrophenyl butyrate, p-nitrophenylvalerate, p-nitrophenylcaprilate and p-nitrophenyldecanoate as well as 4-methylumbelliferone. The lipolytic activity increased in the presence of organic solvents such as 1-butanol, 2-butanol and acetone (5 % and 10 %, v/v), EDTA 1 % (v/v) and metal ions as Ni^{2+} and Ca^{2+}

(Moreno et al. 2013). An esterase from *Haloarcula marismortui* exhibits a preference for short chain fatty acids and monoesters and is dependent on the presence of salt for proper folding and activity (Das Sarma et al. 2010).

15.4.4 Xylanases

Xylans are a heterogeneous group of polysaccharides based on a main chain formed from β-1,4 linked D-xylopyranosyl subunits. Xylanases derived from marine and hypersaline bacteria such as *Glaciecola mesophila* (Guo et al. 2009) and *Chromohalobacter* sp. (Prakash et al. 2009) and *Nesterenkonia* sp. (Govender et al. 2009) exhibit highest stability at wide range of pH from 6.0 to 11.0 and remain active at temperature above 60 °C (Prakash et al. 2009). Xylanases isolated from *Gracibacillus* sp. TSGPVG demonstrated sustainable activity upto 30 % NaCl at temperature 60 °C (Giridhar and Chandra 2010) would be essential for the extracellular activity in such hypersaline environments from where they are isolated.

15.4.5 Nucleases

Kamekura and Onishi (1983) have shown that *Micrococcus varians* can produce Nuclease H enzyme at 1–4 M NaCl and KCl concentration. Maximum activity was obtained at 2.9 M NaCl and 2.1 M KCl at 40 °C, but activity gets lost on dialyzing the enzyme with the buffer containing 3.4 M NaCl (Kamekura and Onishi 1976). An exonucleases enzyme produced from *B. halophilus* which removes nucleotides from 5′ end of the nucleic acids has optimum activity at 1.4–3.2 M NaCl or 2.3–3.2 M KCl, pH 8.5, optimum temperature 50–60 °C and was stimulated by divalent cation Mg^{+2} and Ca^{+2} (Onishi et al. 1983).

15.4.6 Cellulases

Halocella cellulolytica was the first cellulase producing bacterium discovered by (Bolobova et al. 1992) stable at extreme hypersaline environment. Recently, moderately halophilic *Marinobacter* sp. MSI032 has been reported to produce maximum cellulase production in the presence of 2 % NaCl at 27 °C and pH 9.0. Purification using ammonium sulfate precipitation, Sephadex G-200 and DEAE Sepharose chromatography led to 37 % recovery with 12.5 % purification fold, has molecular weight of 68 kDa. It is alkaline in nature with optimum activity at pH 9.0, stable upto pH 12.0 (Wang et al. 2009).

15.4.7 Chitinases

Chitinases are the enzyme that catalyzes the hydrolysis of chitin, an insoluble linear biopolymer with β-1,4 linkage of N-acetyl-D-glucosamine. *Halobacterium salinarum* NRC-1, halophilic archaeon found to acquire a gene *ChiN1* coding for chitinase, a structural homologue of its non-halophilic counterpart, this gene has recombinantly expressed in another archaea *Haloarcula japonica* TR-1 exhibit optimum activity at about 1 M NaCl and was stable upto 5 M NaCl (Hatori et al. 2006). List of organisms with their potential to produce hydrolytic enzymes are shown in Table 15.1.

15.5 Potential Application of the Extremozymes

Halophilic enzymes have customized sizeable interest because of their potential for use in various biotechnological and industrial applications, such as biomedical and chemical sciences, food, leather, laundry detergent, pharmaceutical industries etc. (Raj and Suman 2010), that can be helpful to develop novel products due to their unique properties of withstanding harsh environments.

Table 15.1 Microorganisms producing hydrolytic enzymes

Types of enzymes	Organisms name	References
Amylase	*Streptomyces* sp.	Chakraborty et al. (2009)
	Haloferax mediterranei	Perez-Pomares et al. (2003)
	Halobacterium salinarum	Patel et al. (1993)
	Natronococcus amylolyticus	Kobayashi et al. (1994)
	Haloarcula sp.	Fukushima et al. (2005)
Proteases	*Bacillus* sp.	Shivanand and Jayaraman (2009)
	Salinivibrio sp.	Amoozegar et al. (2007)
	Salicola sp.	Moreno et al. (2009)
	Filobacillus sp.	Hiraga et al. (2005)
	Chromohalobacter sp.	Vidyasagar et al. (2009)
	Nesterenkonia sp.	Bakhtiar et al. (2005)
	Virgibacillus sp.	Sinsuwan et al. (2010)
Xylanase	*Chromohalobacter* sp.	Prakash et al. (2009)
Lipases	*Salinivibrio* sp.	Amoozegar et al. (2008)
	Natronococcus sp.	Boutaiba et al. (2006)
Nucleases	*Halomonas* sp. *Pseudomonas* sp.	Moreno et al. (2013)
Cellulase	*Salinivibrio* sp. NTU05	Wang et al. (2009)
Chitinase	*Virgibacillus* sp., *Terribacillus* sp.	Rohban et al. (2009)

Now-a-days, production of commercially important enzymes like amylases, proteases, lipases etc. has been explored extensively in industries. Amylases constitute a group of interesting enzymes from the industrial as well as biotechnological point of view (Pandey et al. 2000).

Amylases are widely applied in starch saccharification, in the textile, food, brewing and distilling industries. Species like *Halomonas meridian*, *Haloarcula hispanica* and *Natronococcus amylolyticus* producing amylases helps in the production of high fructose corn syrup by hydrolyzing the corn starch. Moreover, it is also used in desizing process in textile industry, and to remove starch from the clothes as detergents (Ratnakar 2013). The enzyme from *Haloarcula* sp. S-1 showed a comparatively high tolerance to many organic solvents (Fukushima et al. 2005) which makes it different from other haloarchaeal α-amylases.

Lipolytic enzymes are valuable biocatalysts due to their broad substrate specificity and high chemo-, regio- and stereo selectivity (Park et al. 2009). These enzymes are currently used as detergent additives, in the food and paper industries, and as enantioselective biocatalysts for the production of fine chemicals (Jaeger and Holliger 2010). However, industrial applications of lipases are often hampered by their low stability in the processes, including low thermostability and loss of activity in presence of the organic solvents, where most of reactions are performed. In this sense, the lipases isolated from extreme microorganisms constitute an excellent alternative in the industrial processes (Pikuta et al. 2007).

Halophilic protease have been widely used in industry for a long time, especially in washing detergent, baking, brewing, cheese and tanning industry (Li and Li 2009). Due to the stability and properties of halophilic proteases, these enzymes are good candidates for use in industrial processes. An interesting extracellular protease, designated haloprotease CP1 has been isolated from the moderately halophilic bacterium *Pseudoalteromonas ruthenica* (Chand and Mishra 2003). The maximal production of the protease CP1 by *P. ruthenica* CP76 was detected at the end of the exponential growth phase at 37 °C, in media containing 7.5 % salt and supplemented with sucrose (50 mM).

Cellulases are generally useful in laundry detergents for softening and shining of the clothes, used in textile industry for polishing of fabrics, for the production of biofuel from cellulosic material etc. (Aygan and Arikan 2008; Wang et al. 2009). Halophilic and halotolerant cellulases derived from *Bacillus* sp. (Aygan et al. 2008), *Salinivibrio* sp. (Wang et al. 2009) have been characterized and were reported to be thermostable, halostable and alkalostable, for various industrial applications (Voget et al. 2006).

Xylanases from *Halorhabdus utahensis* have potential application in the manufacture of coffee, livestock feeds and flour. Xylanases can also be used in place of chlorine bleaching for the removal of residual lignin from pulp (Ratnakar 2013).

Halophilic chitinase producing moderately halophilic bacterium such as *Virgibacillus* sp., *Terribacillus halophilus* and *Planococcus rifitoensis* isolated from shallow salt lakes in Tunisia were reported as good biocontrol agent against the *Botrytis cinerea* causative agent of Grey mold diseases on strawberries and tomatoes (Essghaier et al. 2009).

15.6 Conclusion

Hypersaline environments are widely distributed harboring diverse halophilic microbial communities. Halophiles are salt loving organisms that inhabit hypersaline environment either by accumulating the high solute concentration within the cytoplasm or by maintaining isotonic conditions. Such halophilic microorganisms were known to produce certain valuable salt resistive enzymes containing acidic residues and reduced hydrophobic side chains. These extremozymes were used to design a wide range of novel biocatalytic processes that are more accurate, specific and environment friendly. The stability of these extremophilic enzymes, or extremozymes in the face of adverse conditions has led to their use in a variety of industrial as well as biotechnological applications. Thus the study of halophilic enzymes and their regulation in halophilic microorganisms contributing to understand overall physiology of organisms that they have adapted will bear plenty of fruits in various industrial, agricultural, food, chemical and pharmaceutical processes.

References

Amoozegar MA, Fatemi ZA, Karbalaei-Heidari HR, Razavi MR (2007) Production of an extracellular alkaline metalloprotease from a newly isolated, moderately halophile, *Salinivibrio* sp. strain AF-2004. Microbiol Res 162:369–377

Amoozegar MA, Salehghamari E, Khajeh K, Kabiri M, Naddaf S (2008) Production of an extracellular thermohalophilic lipase from a moderately halophilic bacterium, *Salinivibrio* sp. strain SA-2. J Basic Microbiol 48:160–167

Arakawa T, Timasheff SN (1985) The stabilization of proteins by osmolytes. Biophys J 47:411–414

Aygan A, Arikan B (2008) A new haloalkaliphilic, thermostable endoglucanse from moderately halophilic *Bacillus* sp. C14 isolated from Van Soda Lake. Int J Agric Biol 10:369–374

Aygan A, Arikan B, Korkmas H, Dincer S, Colak O (2008) Highly thermostable and alkaline α-amylase from a halotolerant alkaliphilic *Bacillus* sp. AB68. Braz J Microbiol 39:547–553

Bakhtiar S, Estiveira RJ, Hatti-Kaul R (2005) Substrate specificity of alkaline protease from alkaliphilic feather-degrading *Nesterenkonia* sp. AL20. Enzyme Microb Technol 37:534–540

Balasubramanian S, Pal S, Bagchi B (2002) Dynamics of water molecules at the surface of an aqueous micelle: atomistic molecular dynamics simulation study of a complex system. Curr Sci 82:845–854

Bolobova AV, Simankova MV, Markovitch NA (1992) Cellulase complex of a new halophilic bacterium, *Halocella Cellulolytica*. Microbiology 61:557–562

Boutaiba S, Bhatnagar T, Hacene H, Mitchell DA, Baratti JC (2006) Preliminary characterization of a lipolytic activity from an extremely halophilic archaeon, *Natronococcus* sp. J Mol Catal B Enzyme 41:21–26

Britton KL, Baker PJ, Fisher M, Ruzheinikov S, Gilmour DJ, Bonete MJ et al (2006) Analysis of protein solvent interactions in glucose dehydrogenase from the extreme halophile *Haloferax mediterranei*. Proc Natl Acad Sci 103:4846–4851

Buchalo AS, Nevo E, Wasser SP, Volz PA (2000) Newly discovered halophilic fungi in the Dead Sea (Israel). In: Seckbach J (ed) Journey to diverse microbial worlds. Kluwer, Dordrecht, pp 241–252

Caton TM, Caton IR, Witte LR, Schneegurt MA (2009) Archaeal diversity at the great salt plains of Oklahoma described by cultivation and molecular analyses. Microb Ecol 58(3):519–528

Chakraborty S, Khopade A, Kokare C, Mahadika K, Chopade B (2009) Isolation and characterization of novel α-amylase from marine *Streptomyces* sp. D1. J Mol Catal B Enzyme 58:17–23

Chand S, Mishra P (2003) Research and application of microbial enzymes. India's contribution. Adv Biochem Eng Biotechnol 85:95–124

Chattopadhyay MK, Kern R, Mistou MY, Dandekar AM, Uratsu SL, Richarme G (2004) The chemical chaperone proline relieves the thermosensitivity of a dnaK deletion mutant at 42°C. J Bacteriol 186:8149–8152

Cojoc R, Merciu S, Popescu G, Dumitru L, Kamekura M, Enache M (2009) Extracellular hydrolytic enzymes of halophilic bacteria isolated from a subterranean rock salt crystal. Rom Biotechnol Lett 14:4658–4664

Coronado MJ, Vargas C, Hofemeister J, Ventosa A, Nieto J (2000) Production and biochemical characterization of an α-amylase from the moderate halophile *Halomonas meridian*. FEMS Microbiol Lett 183:67–71

Dang H, Zhu H, Wang J, Li T (2009) Extracellular hydrolytic enzyme screening of culturable heterotrophic bacteria from deep-sea sediments of the Southern Okinawa Trough. World J Microbiol Biotechnol 25:71–79

Danson MJ, Hough DW (1997) The structural basis of protein halophilicity. Comp Biochem Physiol 117(3):307–312

DasSarma S, Arora P (2001) Halophiles. Encyclopedia of life sciences. Macmillan Press, London

DasSarma P, Coker JA, Huse V, DasSarma S (2010) Halophiles, industrial applications. In: Flickinger MC (ed) Encyclopedia of industrial biotechnology: bioprocess, bioseparation, and cell technology. Wiley, New York

Doukyu N, Ogino H (2010) Organic solvent-tolerant enzymes. Biochem Eng J 48:270–282

Dym O, Menarch M, Sussmann JL (1995) Structural features that stabilize halophilic malate dehydrogenase from an archaebacterium. Science 267:1344–1346

Elshahed MS, Najar FZ, Roe BA, Oren A, Dewers TA, Krumholz LR (2004) Survey of archaeal diversity reveals an abundance of halophilic archaea in a low-salt, sulfide- and sulfur-rich spring. Appl Environ Microbiol 70(4):2230–2239

Empadinhas N, da Costa MS (2008) Osmoadaptation mechanisms in prokaryotes: distribution of compatible solutes. Int Microbiol 11:151–161

Enache M, Kamekura M (2010) The halophilic enzyme and their economical values. Rom J Biochem 47(1):47–59

Essghaier B, Fardeau ML, Cayol JL, Hajlaoui MR, Boudabous A, Jijakli H, Sadfi-Zouaoui N (2009) Biological control of grey mould in strawberry fruits by halophilic bacteria. J Appl Microbiol 106:83–846

Fukushima T, Mizuki T, Echigo A, Inoue A, Usami R (2005) Organic solvent tolerance of halophilic α-amylase from a haloarchaeon, *Haloarcula* sp. strain S-1. Extremophiles 9:85–89

García MT, Mellado E, Ostos JC, Ventosa A (2004) *Halomonas organivorans* sp. nov., a moderate halophile able to degrade aromatic compounds. Int J Syst Evol Microbiol 54:1723–1728

Giridhar PV, Chandra TS (2010) Production of novel haloalkali- thermostable xylanase by a newly isolated moderately halophilic and alkali tolerant *Gracibacillus* sp. TSCPVG. Process Biochem 45:1730–1737

Govender L, Naidoo L, Setati ME (2009) Isolation of hydrolase producing bacteria from Sua pan solar salterns and the production of endo-1,4-β-xylanase from a newly isolated haloalkaliphilic *Nesterenkonia sp*. Afr J Biotechnol 8:5458–5466

Grant WD, Kamekura M, Mc Genity TJ, Ventosa A (2001) Class III. *Halobacteria* class. In: Boone DR, Castenholz RW (eds) Bergery's manual of systematic bacteriology. Springer, New York, pp 294–334

Gunde- Cimerman NS, Zalar P, de Hoog GS, Plemenitas A (2000) Hypersaline waters in salterns-natural ecological niches for black yeasts. FEMS Microbiol Ecol 32(3):235–240

Guo B, Chen XL, Sun CY, Zhou BC, Zhang YZ (2009) Gene cloning, expression and characterization of a new cold-active and salt-tolerant endo-β-xylanase from marine *Glaciecola mesophila* KMM 241. Appl Microbiol Biotechnol 84:1107–1115

Guzmán MN, Vargas VA, Antezana H, Svoboda M (2008) Lipolytic enzyme production by halophilic/halotolerant microorganisms isolated from Laguna Verde, Bolivia. Revista Boliviana De Química 25(1):14–23

Hanelt I, Muller V (2013) Molecular mechanisms of adaptation of the moderately halophilic bacterium *Halobacillus halophilus* to its environment. Life 3:234–243

Hatori Y, Sato M, Orishimo K, Yatsunami R, Endo K, Fukui T, Nakamura S (2006) Characterization of recombinant family 18 chitinase from extremely halophilic archaeon *Halobacterium salinarum* strain NRC-1. Chitin Chitosan Res 12:201

Hiraga K, Nishikata Y, Namwong S, Tanasupawat S, Takada K, Oda K (2005) Purification and characterization of serine proteinase from a halophilic bacterium, *Filobacillus* sp.RF2-5. Biosci Biotechnol Biochem 69:38–44

Jaeger KE, Holliger P (2010) Chemical biotechnology a marriage of convenience and necessity. Curr Opin Biotechnol 21:711–712

Kamekura M, Onishi H (1976) Effect of magnesium and some nutrients on the growth and nuclease formation of a moderately halophile *Micrococcus varians* var. *halophilus*. Can J Microbiol 22:1567–1576

Kamekura M, Onishi H (1983) Inactivation of nuclease H of the moderate halophilic *Micrococcus varians* sp. halophiles during cultivation in the presence of salting in type salts. Can J Microbiol 29:46–51

Kamekura M, Seno Y (1990) A halophilic extracellular protease from a halophilic archaebacterium strain 172P1. Biochem Cell Biol 68:352–359

Kamekura M, Seno Y, Holmes ML, Dyall-Smith ML (1992) Molecular cloning and sequencing of the gene for a halophilic alkaline serine protease (halolysin) from an unidentified halophilic archaea strain (172P1) and expression of the gene in *Haloferax volcanii*. J Bacteriol 174:736–742

Karan R, Singh S, Kapoor S, Khare S (2011) A novel organic solvent tolerant protease from a newly isolated *Geomicrobium* sp. EMB2 (MTCC 10310): production optimization by response surface methodology. New Biotechnol 28:136–145

Karbalaei-Heidari HR, Amoozegar MA, Hajighasemi M, Ziaee AA, Ventosa A (2009) Production, optimization and purification of a novel extracellular protease from the moderately halophilic bacterium *Halobacillus karajensis*. J Ind Microbiol Biotechnol 36:21–27

Khunt M, Pandhi N, Rana A (2011) Amylase from moderate halophiles isolated from wild ass excreta. Int J Pharm Biol Sci 1:586–592

Kobayashi T, Kanai H, Aono R, Horikoshi K, Kudo T (1994) Cloning, expression and nucleotide sequencing of α- amylase gene from the haloalkaliphilic archaeon Natronococcus sp. strain Ah-36. J Bacteriol 176:5131–5134

Kolp S, Pietsch M, Galinski EA, Gutschow M (2006) Compatible solutes as protectants for zymogens against proteolysis. Biochim Biophys Acta 1764:1234–1242

Kurz M (2008) Compatible solute influence on nucleic acids: many questions but few answers. Saline Syst 4:6

Lamosa P, Turner DL, Ventura R, Maycock C, Santos H (2003) Protein stabilization by compatible solutes. Effect of diglycerol phosphate on the dynamics of *Desulfovibrio gigas* rubredoxin studied by NMR. Eur J Biochem 270:4606–4614

Li AN, Li DC (2009) Cloning, expression and characterization of the serine protease gene from *Chaetomium thermophilum*. J Appl Microbiol 106:369–380

Lin QS, Chen SH, Hu MY, Rizwan-ul-Haq M, Yang L, Li H (2011) Biodegradation of cypermethrin by a newly isolated actinomycetes HU-S-01 from wastewater sludge. Int J Environ Sci Technol 8(1):45–56

Margesin R, Schinner F (2001) Potential of halotolerant and halophilic microorganisms for biotechnology. Extremophiles 5:73–83

Maturrano L, Valens-Vadell M, Roselló-Mora R, Antón J (2006) *Salicola marasensis* gen. nov., sp. nov., an extremely halophilic bacterium isolated from the Maras solar salterns in Perú. Int J Syst Evol Microbiol 56:1685–1691

Mevarech M, Frolow F, Gloss LM (2000) Halophilic enzymes: proteins with a grain of salt. Biophys Chem 86:155–164

Mohapatra BR, Banerjee UC, Bapuji M (1998) Characterization of a fungal amylase from *Mucor* sp. associated with the marine sponge *Spirastella* sp. J Biotechnol 60:113–117

Moreno ML, García MT, Ventosa A, Mellado E (2009) Characterization of *Salicola* sp. IC10, a lipase- and protease-producing extreme halophile. FEMS Microbiol Ecol 68:59–71

Moreno MDL, Dolores P, María TG, Encarnación M (2013) Halophilic bacteria as a source of novel hydrolytic enzymes. Life 3:38–51

Oh D, Porter K, Russ B, Burns D, Dyall-Smith M (2009) Diversity of Haloquadratum and other haloarchaea in three, geographically distant, Australian saltern crystallizer ponds. Extremophiles 14(2):161–169

Onishi H, Hidaka O (1978) Purification and properties of amylase produced by a moderately halophilic Acinetobacter sp. Can J Microbiol 24:1017–1023

Onishi H, Sonoda K (1979) Purification and some properties of an extracellular amylase from a moderate halophile *Micrococcus halobius*. Appl Environ Microbiol 38:616–620

Onishi H, Mori T, Takeuchi S, Tani L, Kobayashi T, Kamekura M (1983) Halophilic nuclease of a moderately halophilic Bacillus sp.: production, purification and characterization. Appl Environ Microbiol 45:24–30

Oren A (2002) Diversity of halophilic microorganisms: environments, phylogeny, physiology, and applications. J Ind Microbiol Biotechnol 28:58–63

Oren A (2010) Industrial and environmental applications of halophilic microorganisms. Environ Technol 31:825–834

Oren A, Heldal M, Norland S (1997) X-ray microanalysis of intracellular ions in the anaerobic halophilic eubacterium *Haloanaerobium praevalens*. Can J Microbiol 43:588–592

Ovreas L, Bourne D, Sandaa RA, Casamayor EO, Benlloch S, Goddard V (2003) Response of bacterial and viral communities to nutrient manipulations in sea water mesocosms. Aquat Microb Ecol 31:109–121

Pandey A, Nigam P, Soccol CR, Soccol VT, Singh D, Mohan R (2000) Advances in microbial amylases. Biotechnol Appl Biochem 31:135–152

Pandit AS, Joshi MN, Bhargava P, Ayachit GN, Shaikh IM, Saiyed ZM, Saxena AK, Bagatharia SB (2014) Metagenomes from the saline desert of Kutch. Genome A 2(3):1

Park JH, Ha HJ, Lee WK, Généreux-Vincent T, Kazlauskas RJ (2009) Molecular basis for the stereoselective ammoniolysis of N-alkyl aziridine-2-carboxylates catalyzed by *Candida antarctica* lipase B. Chembiochem 10:2213–2222

Patel S, Jain N, Madamwar D (1993) Production of α-amylase from *Halobacterium halobium*. World J Microbiol Biotechnol 9:25–28

Perez-Pomares F, Bautista V, Ferrer J, Pire C, Marhuenda-Egea FC, Bonete MJ (2003) α-Amylase activity from the halophilic archaeon *Haloferax mediterranei*. Extremophiles 7:299–306

Pikuta EV, Hoover RB, Tang J (2007) Microbial extremophiles at the limits of life. Crit Rev Microbiol 33:183–209

Prakash B, Vidyasagar M, Madhukumar MS, Muralikrishna G, Sreeramulu K (2009) Production, purification, and characterization of two extremely halotolerant, thermostable, and alkali-stable α-amylase from *Chromohalobacter* sp. TVSP 101. Process Biochem 44:210–215

Raj E, Suman CE (2010) Purification and characterization of a new hyperthermostable, allosamidin-insensitive and denaturation-resistant chitinase from the hyper thermophilic archaeon *Thermococcus chitonophagus*. Extremophiles 7:43–53

Ratnakar D (2013) Use of halophile physiology and adaptations in various industrial applications. Res J Biotechnol 8(2):1–3

Roberts MF (2005) Organic compatible solutes of halotolerant and halophilic microorganisms. Saline Syst 1:5

Roessler M, Muller V (2001) Chloride dependence of glycine betaine transport in *Halobacillus halophilus*. FEBS Lett 489:125–128

Rohban R, Amoozegar MA, Ventosa A (2009) Screening and isolation of halophilic bacteria producing extracellular hydrolyses from Howz Soltan Lake, Iran. J Ind Microbiol Biotechnol 36:333–340

Rothschild LJ, Mancinelli RL (2001) Life in extreme environments. Nature 409:1092–1101

Ryu K, Kim J, Dordick JS (1994) Catalytic properties and potential of an extracellular protease from an extreme halophile. Enzyme Microb Technol 16:266–275

Sanchez-Porro C, Martin S, Mellado E, Ventosa A (2003) Diversity of moderately halophilic bacteria producing extracellular hydrolytic enzymes. J Appl Microbiol 94:295–300

Satyanarayana T, Raghukumar C, Shivaji S (2005) Extremophilic microbes: diversity and perspectives. Curr Sci 89(1):78–90

Shivanand P, Jayaraman G (2009) Production of extracellular protease from halotolerant bacterium, *Bacillus aquimaris* strain VITP4 isolated from Kumta coast. Process Biochem 44:1088–1094

Siglioccolo A, Paiardini A, Piscitelli M, Pascarella S (2011) Structural adaptation of extreme halophilic proteins through decrease of conserved hydrophobic contact surface. BMC Struct Biol 11:1–12

Sinha R, Khare SK (2014) Effect of organic solvents on the structure and activity of moderately halophilic *Bacillus* sp. EMB9 protease. Extremophiles. doi:10.1007/s00792-014-0683-4

Sinsuwan S, Rodtong S, Yongsawatdigul J (2010) A NaCl-stable serine proteinase from *Virgibacillus* sp. SK33 isolated from Thai fish sauce. Food Chem 119:573–579

Sorokin DY, Tindall BJ (2006) The status of the genus name *Halovibrio fendrich* 1989 and the identity of the strains *Pseudomonas halophila* DSM 3050 and *Halomonas variabilis* DSM 3051. Int J Syst Evol Microbiol 56:487–489

Tan TC, Mijts BN, Swaminathan K, Patel BKC, Divine C (2008) Crystal structure of the polyextremophilic alpha-amylase AmyB from *Halothermothrix orenii*: details of a productive enzyme-substrate complex and an N domain with a role in binding raw starch. J Mol Biol 378:852–870

Upasani V, Desai S (1990) Sambhar Salt Lake: chemical composition of the brines and studies on haloalkaliphilic archaebacteria. Arch Microbiol 154:589–593

Ventosa A, Marquez MC, Ruiz-Berraquero F, Kocur M (1990) *Salinicoccus roseus* gen. nov., sp. nov., a new moderately halophilic Gram-positive coccus. Syst Appl Microbiol 13:29–33

Ventosa A, Nieto JJ, Oren A (1998) Biology of moderately halophilic aerobic bacteria. Microbiol Mol Biol Rev 62:504–544

Vidyasagar M, Prakash S, Mahajan V, Shouche YS, Sreeramulu K (2009) Purification and characterization of an extreme halothermophilic protease from a halophilic bacterium *Chromohalobacter* sp. TVSP101. Braz J Microbiol 40:12–19

Voget S, Steele HL, Streit WR (2006) Characterization of a metagenome-derived halotolerant cellulase. J Biotechnol 126:26–36

Wang CY, Hsieh YR, Ng CC, Chan H, Lin HT, Tzeng WS, Shyu YT (2009) Purification and characterization of a novel halostable cellulase from *Salinivibrio* sp. strain NTU05. Enzyme Microb Technol 44:373–379

Warren JC, Stowring L, Morales MF (1966) The effect of structure-disrupting ions on the activity of myosin and other enzymes. J Biol Chem 241:309–316

Zalar P, de Hoog GS, Gunde- Cimerman NS (1999) Ecology of halotolerant dothideaceous black yeasts. Stud Mycol 43:38–48

Chapter 16
Extracellular Proteases from Halophilic and Haloalkaliphilic Bacteria: Occurrence and Biochemical Properties

Vikram H. Raval, Megha K. Purohit, and Satya P. Singh

Abstract Polyextremophilic organisms, such as haloalkaliphiles have focused attention due to their hydrolytic enzymes. The halophilic microorganisms include heterotrophic and methanogenic archaea; photosynthetic, lithotrophic, and heterotrophic bacteria; and photosynthetic and heterotrophic eukaryotes. Halophiles and haloalkaliphiles are distributed throughout in hypersaline environments, natural hypersaline brines in arid, coastal, and even deep sea locations, as well as in man-made salterns. The microbes possess range of the biocatalysts which allow them to sustain under the prevailing extreme conditions. Among the enzymes, proteases, carbohydrases, and peroxidases are the most cited candidates. The proteases widely occur in microbial world and play significant role in the processing and maintenance of large number of membrane proteins under the cellular conditions. Besides, the proteases are among the commercially most viable enzymes and dominate the worldwide enzyme market. The proteases of halophilic and haloalkaliphilic bacteria from the marine sources appear to have significant role in the detergent industries, ripening of the salted fish, fish sauces and marinades, modifying fish protein concentrations, dehairing and deskinning, non aqueous enzymology and asymmetric catalysis. In the present chapter, the proteases from the halophilic and haloalkaliphilic bacteria have been reviewed with respect to their occurrence and biochemical properties.

Keywords Haloalkaliphiles • Polyextremophiles • Alkaline proteases • Halotolerant enzymes • Enzyme kinetics • Enzyme characterization • Biocatalytic potential • Metagenomics

---

V.H. Raval • M.K. Purohit • S.P. Singh (✉)
Department of Biosciences, Saurashtra University, Rajkot 360 005, India
e-mail: satyapsingh@yahoo.com; satyapsingh125@gmail.com

D.K. Maheshwari, M. Saraf (eds.), *Halophiles*, Sustainable Development and Biodiversity 6, DOI 10.1007/978-3-319-14595-2_16

16.1 Introduction

The biological and chemical diversity of the marine environment has enormous potential for the industrial developments in various fields, such as pharmaceuticals, cosmetics, nutritional supplements, molecular probes, enzymes, fine chemicals, and agrichemicals. Among different value added molecules from the marine life, the enzymes have recently received considerable attention. The enzymes from the marine organisms may have unique features not found in terrestrial organisms. A broader understanding of the marine enzymes will provide opportunity for an integrated biological assessment of the marine ecosystem. Besides microorganisms, many other marine organisms such as fishes, prawns, crabs, snakes, plants, sponges and algae are also explored for their enzymatic potential. Enzymatic properties such as high salt tolerance, thermostability, barophilicity, psycrophilicity, and the large scale cultivation are some of the key targets for the screening of the potential microbes (Forsyth et al. 2005). Many of such properties are reflected by the marine organisms as they thrive in the hydrothermal vents, oceanic caves and areas of high pressure and absence of light (Horikoshi 1999; Niehaus et al. 1999). The marine microbes usually have symbiotic relationship with their vertebrates and invertebrates hosts. Many marine habitats, such as near shore sediments, salt enriched soil and seawater have been investigated for the bacteria diversity and biocatalytic potential (Patel et al. 2005, 2006a, b; Dodia et al. 2006, 2008a; Joshi et al. 2008). Among the enzymes; proteases, amylases and peroxidases are the most explored candidates (Rao et al. 1998; Patel et al. 2005; Dodia et al. 2006). Majority of the extremophiles studied are thermophilic in nature (Ladenstein and Antranikian 1998; Niehaus et al. 1999; Demirjian et al. 2001) and relatively limited attention is paid to halophiles and haloalkaliphiles. The recent advances in the discovery of enzymes from the marine microorganisms would increase the applications of marine enzymes (Rao et al. 1998; Dodia et al. 2008a, b; Joshi et al. 2008; Purohit and Singh 2014; Raval et al. 2014a). Extremophiles mainly include halophiles, alkaliphiles, thermophiles and acidophiles. Certain microbes are able to sustain and adapt to more than one extremity and in this context, thermoacidophiles, thermoalkaliphiles, thermohalophiles and haloalkaliphiles are the key groups. In this chapter, we have focused on the haloalkaliphiles. These organisms live in saline and hypersaline environments. Hypersaline ecosystems represent some of the most unusual and extreme environments (Grant and Mwatha 1998; DasSarma and Arora 2001; Martins et al. 2001). Hyper saline environments are of two types: those arising from seawater with sodium chloride as the predominant salt (thalassohaline), and those from non-seawater sources that contain different ion ratio (athalassohaline) (Upasani and Desai 1990; Litchfield and Gillevet 2002; Litchfield 2004; Baxter et al. 2005; Joshi 2006; Rawal 2012).

16.2 Occurrence

Oceans are the largest saline environment on earth. These environments are categorized as hypersaline habitats and have heavy concentrations of dissolved salts which do not favor normal life forms. Natural evaporation of saline waters lead to environments that are said to be derived environments with predominant sodium chloride concentrations, and are called thalasohaline environments. Generally, the salt composition is similar to sea water, with pH neutral to slightly alkaline. Upon gradual evaporation of the water the ionic composition changes due to gypsum precipitation along with other minerals (Oren 2002a, b).

On gradual evaporation of water from these hypersaline environments, the salinity levels increase and can reach as high as 30 %. A high salinity of this range not only supports growth of halophilic bacteria but also allows growth of true halophilic fungi (black yeast) (Gunde-Cimerman et al. 2000).

Thalasohaline environments harbor a great diversity of microbial life with upto 3–3.5 mol/L NaCl, a threshold value where only few microorganisms can grow. *Halobacterium* and *Dunaliella* are some of the most cited extreme halophiles from the thalassohalic environments (DasSarma and Arora 2001). A summary of the bacteria from thalasohaline environments is depicted in Table 16.1.

The environments with great differences from sea water in ionic strength and resistance for all biological life forms in salinity tolerance are referred as athalassohaline environments. The Dead Seas, alkaline Soda Lakes, carbonate springs, salterns brines and alkaline soil represent athalassohaline environments. These habitats are characterized by high proportion of magnesium and potassium salts after precipitation of NaCl from sea water and leads to non marine proportion. The major components are sulfate, carbonate, chloride and sodium with potassium, calcium and magnesium in trace amount. However, there are unusually high concentrations of organic carbon compounds, nitrogen compounds, and phosphate. Mass develop-

Table 16.1 Chemical diversity of marine environments

Ions (g/l) % dry weights	Typical ocean	Sea water at shores	Dead Sea	Great Salt Lake Utah	Sambhar Lake Rajasthan
Na^+	30.8	10.8	39.2	105.4	37.50
K^+	1.1	0.4	7.3	6.7	0.05
Mg^{2+}	3.7	1.3	40.7	11.1	0.00
Ca^{2+}	1.2	0.4	16.9	0.3	0.00
Cl^-	55.5	19.6	212.4	181.4	21.46
Br^-	0.19	0.1	5.1	0.2	NA
SO_4^{2-}	7.7	2.7	0.5	27	6.00
HCO_3^-/CO_3^{2-}	0.41	0.1	0.2	0.7	1.94/0.60
Salinity	35	35.2	322.6	333.6	164.0
pH	7.9–8.2	8.2	5.9–6.3	7.7	9.0

ments of phototrophic sulfur bacteria, halobacteria, cyanobacteria, and green algae were observed (Imhoff et al. 1979). Compared to smaller hypersaline ponds, the compositions of these lakes are relatively constant due to their size. However, recent human activities have significantly affected the chemistry and biology of these lakes. The size and salinity of the Dead Sea basin are significantly affected due to the incoming freshwater streams for irrigation (Watzman 1997). Hypersaline evaporation ponds are known in Antarctica (e.g. Deep Lake, Organic Lake and Lake Suribati), many of them being stratified with respect to salinity. Some alkaline hypersaline soda brines also exist, such as the Wadi Natrun lakes of Egypt, Lake Magadi in Kenya, and the Great Basin lakes of the western United States (Mono Lake, Owens Lake, Searles Lake and Big Soda Lake). Many hypersaline pools represent dynamic environments, displaying seasonal variations in size, salinity and temperature. In addition to the natural hypersaline lakes, numerous manmade solar salt pans are created for the production of salts. Some deep sea sediments have ancient salt deposits of over 100 million years old, representing salt gradients caused due to evaporation. This leads to the sequential blooms and succession of diverse microbial population comprising of archae, eubacteria, fungi, algae and protozoa (DasSarma and Arora 2001). Table 16.2 represents bacterial diversity of the well known hypersaline environments (Table 16.3).

16.3 Biochemical Properties of Proteases

The biochemical characterization of enzymes is important to evaluate their biotechnological potential. The study of the biochemical properties of proteases, such as the substrate specificity, optimum catalytic pH, temperature, salt requirements, inhibitors, metal ions and the stability profiles is important in assessing their applications prospects. An account of these properties of the protease is being provided in the following sections.

16.3.1 pH

The optimum protease catalysis is generally observed in the range of pH 8–10 and is highly pH dependent (Sanchez-Porro et al. 2003; Gupta et al. 2005; Hiraga et al. 2005; Dodia et al. 2006; Patel et al. 2006b; Purohit and Singh 2011; Raval et al. 2014a). However, proteases with highly alkaline pH for the activity are also reported (Purohit and Singh 2011; Raval et al. 2014a). On the other hand, a low pH optimum of 7.5 for halotolerant extracellular proteases of *Bacillus subtilis* strain FP-133 (Setyorini et al. 2006) and pH 8 for an alkaline protease from haloalkaliphilic *Bacillus* sp. (Patel et al. 2006b) are reported.

Table 16.2 Thallasohaline and other environments (Saline salterns, salt pans, sea waters, saline sediments, evaporation ponds and other sources)

Organism	Site of isolation	NaCl M or % (w/v) or g/l	pH	Remarks	Reference
Haloalkaliphilic bacterium Ve2-20-91	Alkaline sea water Veraval sea coast Gujarat, India	10 %	10	Gram positive, moderately halophilic bacteria isolated for various hydrolytic enzyme production	Raval et al. (2014a, b) and Raval (2013)
Haloalkaliphilic bacterium Ve2-10-10; Oceanobacillus oncorhynchi Ve2-15-91, Oceanobacillus iheyensis Ve2-20-92, Haloalkaliphilic bacterium S-15-9, Haloalkaliphilic bacterium D-10-102		10 %	10		
		15 %	9		
		20 %	9		
		15 %	9		
Halophilic and alkaliphilic bacterium D-15-9 Oceanobacillus oncorhynchi D-20-91, Oceanobacillus oncorhynchi Mi-10-54		10 %	10		
		15 %	10		
		20 %	9		
		5 %	8		
Oceanobacillus ihyensis O.M.A18 and *Haloalkaliphilic bacterium O.M.E12*	Salt panes located in Okha, Gujarat India	20 %	10	Moderately haloalkaliphilic bacteria, isolated for alkaline protease producing	Purohit and Singh (2011, 2014)
		15 %	10		
Haloalkaliphilic bacteria	Muthukur solar salterns from Nellore district, Andhra Pradesh	12 %	12	Haloalkaliphilic, hydrolytic enzyme producing bacteria	Shameer et al. (2013)
Halomonas smyrnensis sp. nov.	Camalti Saltern Area, a wildlife reserve in Sasali, Izmir province located in the Aegean Region of Turkey	10 %	7	Moderately halophilic, exopolysaccharide-producing bacterium	Poli et al. (2013)
			8–10		
Haloalkaliphilic bacteria	Arid saline systems of Southern Tunisia	15–25 %	10 and above	Haloalkaliphilic bacteria	Hidri et al. (2013)
Bacillus agaradhaerens	Salt panes located in Mithapur, Gujarat India	10 %	9	Haloalkaliphilic bacteria, organic solvent tolerance amylase	Pandey and Singh (2012)
Mi-10-6$_2$					

(continued)

Table 16.2 (continued)

Organism	Site of isolation	NaCl M or % (w/v) or g/l	pH	Remarks	Reference
Oceanobacillus sp. *Sj-1*	Saline soil Jodiya, Gujarat India	10 %	9	Haloalkaliphilic bacteria, organic solvent tolerance protease	Pandey et al. (2012)
Halorubrum sfaxense sp. nov.	Solar saltern of Sfax, Tunisia	25 %	7.4	Halophilic archaeon	Trigui et al. (2011)
Geomicrobium halophilum gen. nov. *Geomicrobium halophilum* sp. nov.	Forest soil and garden soil, respectively, Japan	15 %	9	Moderately halophilic and alkaliphilic bacteria	Echigo et al. (2010)
Bacillus marmarensis sp. nov.	Mushroom compost from Yalova, located in the Marmara region of Turkey	0–12 %	11 (8–12.5)	Obligately alkaliphilic bacterium, protease producing	Denizci et al. (2010)
Haloalkaliphilic bacterium S-20-9	Sea water samples of costal Gujarat, Somnath (Western India)	10 %	9	Moderately haloalkaliphilic bacterium	Joshi et al. (2009)
Haloalkaliphilic bacterium sp. AH6	Saline habitats coastal region of Jamnagar, Gujarat, India	10–15 %	10 (8–11)	Moderately haloalkaliphilic bacterium	Dodia et al. (2006, 2008a, b)
Halorubrum californiense sp. nov.	Crystallizer pond, cargill solar salt plant, Newark, California, USA	5 M	7.3	Extremely halophilic archaeon, motile, rod-shaped, pink-pigmented,	Pesenti et al. (2008)
Halomonas daqingensis sp. nov.	Soil sample contaminated with crude oil from Daqing oil field in Heilongjiang Province, China	10 %	9	Gram-negative, moderately halophilic bacteria	Wu et al. (2008)
Halomonas indalinina sp. nov.	Solar saltern in Cabo de Gata, Almeria, Spain	10 %	9	Moderately haloalkaliphilic bacterium	Cabrera et al. (2007)
Bacillus qingdaonensis sp. nov.	Crude sea salt Qingdao, China	12 %	9	Moderately haloalkaliphilic bacterium	Wang et al. (2007)

Halomonas gudaonensis sp. nov.	Saline soil contaminated with crude oil in Gudao in Shengli oilfield, China	15 %	9	Moderately haloalkaliphilic bacterium	Wang et al. (2007)
Halomonas gomseomensis sp. nov. *Halomonas janggokensis* sp. nov. *Halomonas salaria* sp. nov. and *Halomonas denitrificans* sp. nov.	Saline water in Anmyeondo, Korea	12 %	8	Aerobic, Gram-negative, non-spore-forming rodsmotile with peritrichous flagella or polar flagella	Kim et al. (2007)
		15 %	8		
		20 %	8		
		10 %	9		
Chromohalobacter japonicus sp. nov.	Japanese salty food, Japan	12.5 %	8	Moderately halophilic non spore forming bacteria	Sanchez-Porro et al. (2007)
Bacillus lehensis sp. nov.	Soil from Leh, India,	<12 %	8 (7–11)	Alkalitolerant, Gram positive bacteria	Ghosh et al. (2007)
Bacillus okhensis sp. nov.	Saltpan in Okha, India	5 % (0–10 %)	9 (7–10)	Strictly aerobic, rod-shaped halo-alkali tolerant, bacteria	Nowlan et al. (2006) and Dodia (2005)
Alkalibacillus filiformis sp. nov.	Mineral pool in Malvizza, Campania, Italy	10 %	9	Alkalitolerant and halotolerant, Gram-positive bacteria	Romano et al. (2005b)
Thalassobacillusdevorans gen. nov. sp. nov.	Phenol enriched samples in hypersaline habitats, Spain	10 %	7	Halophilic, phenol-degrading, Gram-positive bacterium	Garcia et al. (2005)
Halomonas campaniensis sp. nov.	Algal mat from a mineral pool in Malvizza, Italy	25 %	7–10	Gram-positive, phosphatidylglycerol and diphosphatidyl-glycerol were the predominant polar lipids. Mole G+C; 48.4 %	Romano et al. (2005a)
Virgibacillus dokdonensis sp. nov.	Dokdo, an island located at the edge of the East Sea, Korea	4–5 %	7–8	Slightly halophilic bacterial strain,	Yoon et al. (2005a)

(continued)

Table 16.2 (continued)

Organism	Site of isolation	NaCl M or % (w/v) or g/l	pH	Remarks	Reference
Thalassobacillus devorans gen. nov. sp. nov.	Hypersaline habitats of southern Spain	7.5–10 %	7	Moderately halophilic, phenol-degrading, Gram-positive bacterium	Garcia et al. (2005)
Bacillus pseudofirmus FTU		(17–18 %)	9–10	Halo-tolerant alkaliphilic strain. Mole G+C; 41.3 %	Muntyan et al. (2002)
Chromohalobacter salexigens sp. nov.	Isolated from salterns	(0.9–25 %)	5–10	Gram negative, moderately halophilic bacteria	Arahal et al. (2001)
Halomonas campisalis sp. nov. 4A	Sediment samples taken from the salt plain of Alkali Lake in Washington State (USA)	1.5 M (0.2–4.5 M)	9.5 (6–12)	Moderately halophilic and alkaliphilic, motile with rod-shaped cells	Mormile et al. (1999), Peyton (1999) and Melanie et al. (1999)
Natroniella acetigena	Bottom mud of the soda-depositing Lake Magadi, Kenya	(2–2.56 M)	9.7–10	Extreme halophilic and alkaliphilic	Zhilina et al. (1996)
Halomonas halmophila	Dead Sea	20 %	ND	Gram negative, Halophilic bacteria	Dobson et al. (1990)

Table 16.3 Athalassohaline environments and the microbial diversity

Organism	Site of isolation	NaCl M or % (w/v) or g/l	pH	Remarks	Reference
Oceanobacillus limi sp. nov.	Mud sample of hypersaline lake Aran-Bidgol in Iran	7.5 %	7	Moderately halophilic bacterium,	Amoozegar et al. (2014)
Marinobacter persicus sp. nov.	Hypersaline lake Aran-Bidgol in Iran	7.5–10 %	7	Gram-negative, moderately halophilic bacterium	Bagheri et al. (2013)
Natronococcus roseus sp. nov.	Sediments of soda lake Chagannor in Mongolia, China	30 %	9.5	Halophilic archaeon	Corral et al. (2013)
Desulfohalophilus alkaliarsenatis gen. nov. sp. nov.	Extreme environment of Searles Lake, California	55–330 g/l	9.25	A haloalkaliphilic sulfate-respiring bacterium	Blum et al. (2012)
Alkalibacillus haloalkaliphiles sp.nov.	Saline soil samples Sambhar Lake Rajasthan	10 %	8	Aerobic, Gram-positive, rod-shaped organism	Rawal et al. (2012)
Desulfonatronovibrio halophilus sp. nov.	Aanoxic sediments of hypersaline chloride–sulfate lakes in the Kulunda Steppe Altai, Russia	2 M	7.5	Moderately halophilic sulfate reducing bacteria	Sorokin et al. (2012)
Bacillus iranensis sp. nov.	Hypersaline lake Aran-Bidgol Lake, Iran	5–7.5 % (2.5–15 %)	7.5 7–10	Moderately halophilic bacterium	Bagheri et al. (2012)
Haloarchaeobius iranensis gen. nov., sp. nov.	Aran-Bidgol salt lake, a saline playa in Iran	3.5	7.5 (6–8)	Extremely halophilic archaeon, orange red pigmented	Makhdoumi-Kakhki et al. (2012)
Salinivibrio sharmensis sp. nov.	Saline lake in Ras Mohammed Park, Egypt	10 % (6–16 %)	9 6–10	Haloalkaliphilic bacterium	Romano et al. (2011)
Marinospirillum celere sp. nov.	Mono Lake, USA	2 %	9.5	Haloalkaliphilic bacterium	Namsaraev et al. (2009)
Bacillus persepolensis sp. nov.	Hypersaline lake Howz-Soltan, Iran	10 %	8.5	Moderately halophilic bacterium	Amoozegar et al. (2009)
Bacillus aidingensis sp. nov.	Sediment sample from Ai-Ding salt lake, China	12 %	7.2	Gram-positive, halophilic bacterium	Xue et al. (2008)

(continued)

Table 16.3 (continued)

Organism	Site of isolation	NaCl M or % (w/v) or g/l	pH	Remarks	Reference
Halomonas cerina sp. nov.	Saline soils, Spain	10 %	8	Moderate halophile Gram-negative rods, capsulated and non-motile	Gonzalez-Domenech et al. (2008)
Halomonas saccharevitans sp. nov. *Halomonasarcis* sp. nov. and *Halomonas subterranea* sp. nov.	Salt lake on the Qinghai–Tibet Plateau saline well in the Si-Chuan Basin, China	7.5 % 5 % 5 %	8 8 8	Gram-negative, aerobic, neutrophilic and halophilic bacteria	Xu et al. (2007)
Halomonas mongoliensis sp. *nov. Z-7009* and *Halomonas kenyensis* sp. *nov., AIR-2*	Isolated from Soda Lakes in Mongolia and Kenya	28.0 g/l 28.0 g/l	9.3 (8–10.5) 9.35 (7.5–10.6)	Facultative, moderate halophile new haloalkaliphilic denitrifiers capable of N_2O reduction,	Boltyanskaya et al. (2007)
85 different strains of *Thioalkalivibrio* genus	Soda Lakes in Mongolia (Hotontyn, Dzunundziin, Sharburdiin, Golyntsagan, Gurvany Behiin) Kenya (Nakaru, Magadi, Bogoria, CraterLake, Elmenteita), California, Mono Lake Egypt (Wadi Natrun, Gaara, Umm-risha, Hamra, Beida, Fazdah, Zugm, Ruzita, Khadra) and Siberia (Hadyn, TantarIII, Elongated lake, Narrow Lake, Borzinskoe, Stamp Lake, Tsaidam)	Total salts 50–360 g/l 20–220 g/l 90 g/l 200–380 g/l 10–380 g/l	9.2–10.5 9.5–11.0 9.7 9.5–10.3 9.2–10.6	Haloalkaliphilic, obligately chemolithoautotrophic, sulphur-oxidizing bacteria	Foti et al. (2006)
Haloalkaliphilic strain Z-7026	Soda-Lake sediments isolated on medium I and cellulase production optimized in medium II	2.6 g/l 10 g/l	8–8.5 8.8	Obligate anaerobic cellulase secreting bacterium belongs to the cluster III of *Clostridia* with low G+C content	Zvereva et al. (2006)

Four haloalkaliphilic strains belongs to *Alkalispirillum*-*Alkalilimnicola* group	Soda Lakes Kenyan lake Magadi Mongolia: a mixed sample from two hypersaline lakes and a sample from saline lake Gurvany Nur a mixed sample from eight Wadi Natrun lakes in Egypt	Total salts	10.5	Obligate aerobes, capable of anaerobic growth with acetate using either nitrate or N(2)O as electron acceptors	Sorokin et al. (2006)
		240 g/l	9.5–10.3		
		200–390 g/l	10.2		
		50 g/l	9.2–10.2		
		220–360 g/l			
Haloalkaliphiles belongs to *Thioalkalimicrobium*, *Thioalkalivibrio* and *Thioalkalispira*	Soda Lakes	0.5–1 M	10	Chemolithotrophic, sulfur oxidizing, nitrate reducing	Sorokin and Kuenen (2005)
Halalkalicoccus tibetensis gen. nov. sp. nov.	Lake Zabuye, the Tibetan Plateau, China	3.4 M	10	Haloalkaliphilic archaea	Xue et al. (2005)
Halobacillus dabanensis sp. nov. and *Halobacillus aidingensis* sp. nov.	Salt lakes in Xinjiang, China	10 %	7.5	Gram-positive, moderately halophilic spore-forming bacteria	Liu et al. (2005)
		(0.5–25 %)	(5–11)		
		10 %	7.5		
		(5–20 %)	(6–10)		
Halobiforma lacisalsi sp. nov.	Salt lake in Xinjiang, China	4.3 M (2.6–4.3 M)	7.5	Gram-negative, motile, neutrophilic and extremely halophilic strain	Xu et al. (2005)
Alkalibacillus salilacus	Ai-Ding Lake in Xin-jiang province in China	10–12 %	8 (7–9)	Strictly aerobic, halophilic bacterium, spore-forming, motile rod. Gram-positive.	Jeon et al. (2005)
		(5–20 %)			
H. dabanensis D-8T Halobacillus aidingensis AD6T	Salt lakes of the Xinjiang region of China	10 %	7.5	Spore-forming, moderately halophilic, aerobic, heterotrophic bacteria	Liu et al. (2005)
		(0.5–25 %)			
		10 %	7.5		
		(0.5–20 %)			
Marinococcus halotolerans sp. nov.	Saline soil located in Qinghai, north-west China	10 %	7–7.5	Gram-positive, moderately halophilic bacterium	Li et al. (2005)
Aeromicrobium alkaliterrae	Alkaline soil from Korea	<9 %	11	Gram-positive, Alkaliphilic bacteria	Yoon et al. (2005b)

(continued)

Table 16.3 (continued)

Organism	Site of isolation	NaCl M or % (w/v) or g/l	pH	Remarks	Reference
Thialkalivibrio thiocyanodenitrificans	Soda-lake sediments	0.3–1.8 M	9.6–10	Denitrifying sulfur-oxidizing bacterium	Sorokin et al. (2004)
Halorubrum alkaliphilum sp. nov. *Halorubrum xinjiangense* sp. nov.	Soda lake in Xinjiang, China	20 %	10	Haloalkaliphilic archaeon	Feng et al. (2004)
Thialkalivibrio halophiles sp. nov.	Alkaline hypersaline lake in the Altai Steppe, Russia	0.2–5 M	7.5–9.8	Chemolithoautotrophic, sulfur-oxidizing bacterium. Glycine betaine as key compatible solute	Banciu et al. (2004)
Halorubrum tibetense sp. nov.	Lake Zabuye in Tibet, China	3.4 M	9.5	C20:C20 and C20:C25 major lipids, derivatives of phosphatidylglycerol phosphate and phosphatidylglycerol phosphate methyl ester	Fan et al. (2004)
Halomonas campisalis Z-7398-2	Lake Magadi (Kenya)	1 M	8.8–9.5	Moderately halophilic and alkaliphilic, motile with rod-shaped cells	Boltianskaia et al. (2004)
Methylophaga natronica sp. nov.	Soda lake, Transbaikal	2–3 %	8.5–9	Alkaliphilic and moderately halophilic, restricted-facultatively methylotrophic bacterium	Doronina et al. (2003a)
Methylophaga alcalica sp. nov.	Saline soda lake from an East Mongolia	3–4 %	9.0–9.5	Gram-negative, obligate methylotrophicasporogenous, motile short rods, multiplying by binary fission. Mole G+C; 48.3 %	Doronina et al. (2003b)
Spirochaeta americana sp. nov.	Alkaline, hypersaline Mono Lake, California, USA	2–12 %	8–10.5	Gram-negative, motile, spirochaete-shaped obligately anaerobic. Mole G+C; 58.5 %	Hoover et al. (2003)

Tindallia californiensis sp. nov.	Athalassic, alkaline Mono Lake, California, USA	3–5 % (1–20 %)	9.5 (8–10.5)	Extremely haloalkaliphilic, strictly anaerobic, acetogenic, Gram positive, spore-forming bacterium slightly curved motile rods. Mole G+C; 44.4 %	Pikuta et al. (2003)
Halobacillus salinus sp. nov.	Salt lake on the coast of the East Sea in Korea	2–10 %	7	Gram-positive, halophilic bacterium	Yoon et al. (2003)
Halobacillus karajensis sp. nov.	Saline soil of the Karaj, Iran	1–24 %	6–9.6	Potent amylase producer	Amoozegar et al. (2003)
Ectothiorhodospira	Mono Lake, California, USA	70–90 g/l	9.8	Anaerobic, facultative, arsenite-oxidizing chemoautotroph	Oremland et al. (2002)
Thioalkalivibrio thiocyanoxidans sp. nov. *Thioalkalivibrio paradoxus* sp. nov.	Soda Lakes in South-East Siberia, Kenya and Egypt	4 M	10	Obligately autotrophic, sulfur-oxidizing bacteria able to grow with thiocyanate (SCN-) as the sole energy and nitrogen source	Sorokin et al. (2002a)
Thioalkalimicrobium cyclicum sp. nov. and *Thioalkalivibrio jannaschii* sp. nov.	Alkaline and saline Mono Lake, California, USA	(3–4 %)	10	Obligate autotrophic, sulfur-oxidizing bacteria	Sorokin et al. (2002b)
Salinicoccus alkaliphiles sp. nov.	Soda lake, China	(0–25 %)	6.5–11.5	Novel alkaliphilic and moderately halophilic Gram-positive coccus	Zhang et al. (2002)
Thioalkalispira microaerophila gen. nov. sp. nov.	Soda lakes in South-East Siberia, Kenya and Egypt	(20–25 %)	10	Obligately autotrophic, sulfur oxidizing, capable of growth on thiocyanate	Sorokin et al. (2002c)
Tindallia magadiensis	Soda depositing lake, type strain isolated from Lake Magadi	(0.51–1) M sodium dependent growth	8.5 (7.5–10.5)	Haloalkaliphilic bacteria, Strictly dependent on Na^+ for optimal growth	Kevbrin et al. (1998)
Halomonas meridian sp. nov.	Hypersaline lakes of the Vestfold Hills, Antarctica	ND	ND	Halotolerant, non-pigmented bacteria	James et al. (1990)

16.3.2 Salt

The enzymes from halophiles and haloalkaliphiles can not only withstand higher NaCl concentrations but actually the salt is required for the function and stability of these enzymes. The stability and activity of the enzyme strongly depend on the protein dynamics. Stability is required for the appropriate geometry for ligand binding, as well as to avoid denaturation, while flexibility is necessary to function at a metabolically appropriate rate. The protein folding, stability and solubility is of fundamental importance in the cellular function and molecular biology. High salt affects the conformational stability of the proteins. Halophilic proteins have evolved specific mechanisms that allow them to be stable and soluble at high salt concentrations (Madern et al. 2000). Negative charges on the halophilic proteins bind significant amounts of the hydrated ions, thus reducing the surface hydrophobicity and decreasing the tendency to aggregate at high salt concentration. Halophilic proteins are distinguished from their non-halophilic homologous proteins by exhibiting instability in low salt concentrations, while maintaining soluble and active conformations in high concentrations of salt, for example, up to 5 M NaCl (Madern et al. 2000). The requirement of high salt concentration for the stabilization of the halophilic enzymes, on the other hand, is due to a low affinity binding of the salt to specific sites on the surface of the folded polypeptide, thus stabilizing the active conformation of the protein. It has been proposed that salts exert charge screening, reducing electrostatic repulsion and enhancing hydrophobic interaction, favoring a compact folded structure of halophilic proteins (Karan and Khare 2011).

16.3.3 Stability and Activity of Proteases in NaCl

As discussed above, proteins from halophilic and haloalkaliphilic organisms require salt (NaCl/KCl) for their activity and stability. However, the requirement of salt highly varies. Most of the halophilic proteins are active and stable up to 4 M salt, the optimum being at 1–2 M (Gimenez et al. 2000; Joshi 2006; Dodia et al. 2008a, b; Joshi et al. 2008) and inactivated and denatured at concentrations below 1 M NaCl or loss of the activity in the absence of salt. In general, salt increases activity, solubility, stability and thermal stability of the enzymes from the haloalkaliphilic bacteria. Alkaline proteases from haloalkaliphilic *Bacillus* sp. are active up to 0.2–0.5 M NaCl with decrease in activity with the further increase of NaCl (Gupta et al. 2005; Patel et al. 2006b; Dodia et al. 2008a; Joshi et al. 2008; Purohit and Singh 2011; Raval et al. 2014a).

16.3.4 Organic Solvents

In recent years a class of solvent tolerant microbes has focused attention. Such organisms are attractive for the applications in solvent bioremediation and biotransformation in non-aqueous media and provide rich source of solvent stable enzymes (Sardessai and Bhosle 2004; Gupta et al. 2005; Rahman et al. 2006; Thumar and Singh 2009). The largest role of the solvent tolerant enzymes is in the pharmaceutical sector, where its exquisite regioselective and stereoselective properties enable difficult synthesis. Only limited reports are available in literature on the microorganisms, which produce organic solvent-stable proteases. Particularly, the halophilic and haloalkaliphilic bacteria are least attended for the exploration of the solvent-stable proteases. There are various strategies adopted by the microbial cells, such as cell enlargement, strengthening of the cell membrane, degradation and biotransformation of organic solvents and solvent-efflux pumps.

The level of the toxicity considerably changes among the organic solvents. Solvent toxicity correlates inversely with its $\log P_{ow}$, the logarithm of its partitioning coefficient between defined octanol-water mixture ($\log P_{ow}$) (Sikkema et al. 1995). Organic solvents with lower $\log P_{ow}$ values are more toxic than those with higher $\log P_{ow}$ values. The organic solvent with the lowest $\log P_{ow}$ in which target microorganisms can grow is called the index solvent, and the $\log P_{ow}$ value of the index solvent is referred as the index value. Solvents with a $\log P_{ow}$ below 4.0, such as, benzene ($\log P_{ow}$ 2.13), toluene ($\log P_{ow}$ 2.69), octanol ($\log P_{ow}$ 2.92), xylene ($\log P_{ow}$ 3.12–3.2), and styrene ($\log P_{ow}$ 2.95), are extremely toxic to the microorganisms because they accumulate in the cytoplasmic membrane of bacteria and disrupt it. The solvent toxicity depends not only on the inherent toxicity of the compound but also on the intrinsic tolerance of the target bacteria. Proteases being one of the most explored enzymes have attracted considerable attention over the last few decades. They have variability with respect to origin, mechanism of action and specificity. Proteases from haloalkalilic bacteria catalyze protein hydrolysis under alkaline conditions in the presence of salt and have recently attracted attention due to their vital role in leather, food and detergent industries.

In India, however, the use of industrial enzymes in general and extremozymes in particular is still quite limited. Greater emphasis should be given on efficient production of the enzymes at large scale and vast market potential has to be realized. During the last decade there has been dramatic increase in the need for bioactive compounds with novel activities. Advances in microbiological techniques and enzyme technology in 1960s and 1970s lead to the development of several industrial enzyme applications. The widening spectrum of the applications of the enzymes in varied sectors has necessitated the need for wider sources of the biocatalysts (Gupta and Roy 2004). The enzymes from haloalkaliphilic bacteria and archaea have many specific functions (Margesin and Schinner 2001a, b; Ruiz and DeCastro 2007; Pandey and Singh 2012; Pandey et al. 2012).

16.3.5 Temperature

Thermostable enzymes are of special interest for industrial applications due to their stability under high temperatures and wide pH range. The thermophilic proteases catalyze the reaction and maintain the stability at higher temperatures. In addition, higher temperatures can accelerate the reaction rates, increase the solubility of non-gaseous reactants and products and reduce the incidence of microbial contamination by mesophilic organisms. Thermophiles, such as *Bacillus stearothermophilus, Thermus aquaticus, Bacillus licheniformis, Bacillus pumilus and Thermoanaerobacter yonseiensis*, produce a variety of thermostable extracellular proteases (Carvalho et al. 2008; Ueda et al. 2008; Wang et al. 2008; Zhang et al. 2008; Toyokawa et al. 2010a, b). It has been known that enzymes from thermophilic bacteria are unusually thermostable, while possessing other properties similar to those found in mesophilic bacteria (Battestein and Macedo 2007). It's known that salt enhances the thermostability of alkaline proteases (Dodia et al. 2008a, b). Similarly, Ca^{2+} and Polyethylene glycol also plays important role in enhancing the temperature stability of the enzymes (Ghorbel et al. 2003; Dodia et al. 2008a, b; Manni et al. 2008). The sequencing, structure, and mutagenesis information accumulated during the last 20 years suggest that the molecular cloning and over expression of the genes are key to obtain high quantity of the desired proteins (Shannon et al. 1989; Luke et al. 2007; Berezovsky and Shakhnovich 2008; Guo and Ying 2008; Ni et al. 2009). Protein engineering could be considered as one of the important approaches to obtain improved biocatalysts. As an alternate to these modern but expensive and time consuming techniques, exploration of microbial resources from extreme environments can provide much needed biocatalytic platform (Reza et al. 2008; Rawal et al. 2012; Purohit and Singh 2014; Raval et al. 2014a). While there are number of thermostable proteases reported from thermophilic organisms, non-thermophilic groups of the extremophiles are quite rare (Ramesh et al. 2009). Search for thermostable enzymes from other groups of extremophiles would be quite attractive option to obtain biocatalysts able to function under the multitudes of non-conventional conditions.

16.3.6 Metal Ions

Alkaline proteases require divalent cations, viz. Ca^{2+}, Mg^{2+} and Mn^{2+} or a combination of these cations for their maximum activity. Arguably, these cations protect the enzyme against thermal denaturation and maintain the active conformation of the enzyme at high temperatures (Manni et al. 2008). Activity of an alkaline serine protease from *Bacillus subtilis* is reported to increase in the presence of Ca^{2+}, Mg^{2+} and Mn^{2+} (Bayoudh et al. 2000; Adinarayana et al. 2003). Some alkaline proteases are metal ion-dependent in view of their sensitivity to the chelating agents such as EDTA. Some metal ions such as Ba^{2+}, Mn^{2+}, Mg^{2+}, Co^{2+}, Fe^{3+} and Zn^{2+} are used to stabilize proteases (Rattray et al. 1995; Johnvesly and Naik 2001). The metal ions protect the enzyme against denaturation and play vital roles in maintaining the active confirmation of the enzyme at higher temperatures and salt concentrations. Ca^{+2} is known to activate the protease and increase the thermostability (Kotlova et al. 2007). On the other hand strong inhibitory effects of Hg^{2+} and Cu^{2+} ions on the alkaline proteases of some *Bacillus* sp. are also known (Oberoi et al. 2001; Beg and Gupta 2003). Marginal inhibitory effect of some other metal ions, such as Co^{2+}, Zn^{2+}, Na^{+}, Cd^{2+}, Al^{3+}, Fe^{2+}, Fe^{3+}, have been reported. The metal ion dependence or sensitivity of a particular protease may be advantageous to its application.

16.3.7 Surfactant and Detergents

As stated in above sections, the alkaline proteases are among the most commercially viable enzymes and widely used in the detergent industries. A large number of the alkaline proteases are explored and characterized from varied categories of microbes. Some of these enzymes are stable in with sodium dodecyl sulphate (SDS), sodium linear alkyl benzene sulphonate, surfactants and commercial detergents (Joshi et al. 2008; Raval et al. 2014a). An alkaline protease from *Bacillus clausii* is highly stable in 5 % SDS and 10 % H_2O_2 (Joo et al. 2003). Extracellular alkaline proteases from haloalkaliphilic *Bacillus* sp. are stable in SDS, Triton X-100 and Tween-80 (Patel et al. 2006b; Dodia et al. 2008b; Joshi et al. 2008). The robustness of the alkaline proteases as laundry additives has been assessed with various commercial detergents. Many of these studies suggest catalytic activity and stability of the alkaline proteases in different commercial detergents under harsh laundry conditions (Phadatare et al. 1993; Oberoi et al. 2001; Gupta et al. 2005; Raval et al. 2014a).

16.4 Applications

Besides important roles of the halophiles and haloalkaliphiles in ecology, these prokaryotes have focused attention due to their potential applications in various sectors, such as biomedical and chemical industries, food, leather, laundry detergent and pharmaceutical industries (Rothschild and Manicinelli 2001; Moreno et al. 2013). Some archaeal metabolites, such as extracellular enzymes, osmotically active substances (compatible solutes), exopolysaccharides and special lipids have potential industrial applications (Schiraldi and De Rosa 2002). Among the enzymes, proteases constitute one of the most important groups of the industrial enzymes, accounting for the majority of the worldwide enzyme sales. They have been widely used in various industries for a long time, especially in washing detergent, baking, brewing, cheese industry and tanning industry (Chand and Mishra 2003; Li and Li 2009). They appear to be a good source of various hydrolytic enzymes, proteases in particular. It will open new dimensions for the development of value added products owing to the unique features of such enzymes under the harsh catalytic conditions (Munawar and Engel 2014).

Many proteases are being used as the potential detergent additives. However, some of the lesser known applications of the proteases are also discussed where the proteolytic enzymes of haloalkaliphilic bacteria can be used for varied applications.

16.4.1 Medical Application

Alkaline proteases are also used for developing products of medical importance. Kudrya and Simonenko exploited the elastolytic activity of *B. subtilis* 316 M for the preparation of elastoterase, which was applied for the treatment of burns, purulent wounds, carbuncles, furuncles and deep abscesses (Kudrya and Simonenko 1994). The use of alkaline protease from *Bacillus* sp. strain CK 11-4 as a thrombolytic agent having fibrinolytic activity has been described (Kim et al. 1996).

16.4.2 Photographic Industries

Proteases find potential application in the bioprocessing of the used X-ray films for silver recovery. Used X-ray film contains approximately 1.5–2.0 % (by weight) silver in its gelatin layers (Kumar and Takagi 1999; Sinha et al. 2014). The enzymatic hydrolysis of the gelatin layers on the X-ray film allows the recover of the silver, leading to the recycling of the polyester film base.

16.4.3 Applications in Food Processing and Food Derivatives

Halotolerant microorganisms play important role in various fermentation processes, occurring in the presence of salt and producing various compounds that give characteristic taste, flavor and aroma.

16.4.3.1 Food Formulation and Dietary Products

Alkaline proteases have broad substrate specificity and can hydrolyze proteins from plants, fishes, and animals to produce hydrolysates. The commercial alkaline protease Alcalase, was used in the production of a less bitter hydrolysate and a debittered enzymatic whey protein hydrolysate which can help regulate the blood pressure, in infant food formulations and therapeutic dietary products (Neklyudov et al. 2000).

16.4.3.2 Fish Sauce Fermentation

Fish sauce is a popular seasoning in Southeast Asia, as typified by nampla in Thailand, nuoc mam in Vietnam, and patis in the Philippines. In Thailand, it is produced by mixing fish, such as anchovies with salts and fermenting for 6–12 months at room temperature. The fermentation liquid is rich in fish soluble proteins, peptides, and amino acids with the characteristic Umami taste (Curtis 2009). They are produced during the proteolytic degradation by endogenous proteases in the muscles or digestive tracts of fish, and various microorganisms exist in the fermentation broth (Taira et al. 2007). Hence, the halotolerant proteases of bacteria origin can greatly contribute to fish sauce fermentation in the food industry. A halotolerant proteinase from *B. licheniformis* RKK-04 is a key enzyme for fish sauce fermentation isolated from a fermented Thai fish sauce broth. Its capable to digest the myosin heavy chain of fish protein completely and thus can be used for fish sauce fermentation (Toyokawa et al. 2010a, b).

16.4.3.3 Leather Industry

The traditional methods of bating and dehairing of leather using sodium sulfide treatment leads to the environmental pollution, adding to the sulfide and suspended solids in tannery effluents in the environment. Thus, the bio-treatment of leather using an enzymatic approach would be an attractive option (Boominadhan et al. 2009). Proteases are eco-friendly biocatalysts and can be used as an alternate to chemical processes of pretreatments of leather in tanning industry.

16.5 Genes and Structure and Function Analysis

The discovery of new extremophilic species and the determination of genome sequences provide a route to new enzymes (Purohit 2012; Rawal 2012; Raval 2013). The molecular cloning and over-expression of proteins, protein engineering and directed evolution provide approaches to improve enzyme stability and modify specificity in the manner that may not exist in the natural world (Colquhouna and Sorumb 2002). Knowledge of full nucleotide sequences of the enzyme genes has facilitated the deduction of the primary structure of the encoded enzymes and in many cases, identification of various functional regions. These sequences may serve as the basis for the phylogenetic analysis and prediction of the secondary structure of the proteases (Purohit 2012; Raval et al. 2014b).

It is obvious that halophilic proteases and other enzymes function and are stable in high salt concentrations. The comparison between the sequences of halophilic and non-halophilic homologous protein pairs indicates that there are more number of Asp and Glu, while Lys, Ile and Leu are less frequent in halophilic proteins (Graziano and Merlino 2014). The gene of a highly thermostable protease from a

Bacillus was cloned by PCR and nucleotide sequence was determined (Graziano and Merlino 2014). Similarly, a 1242 base pair DNA fragment from *Bacillus halodurans* isolated from alkaline sediments coding for a potential protease has been cloned and sequenced (Zhang et al. 2008).

The gene encoding the protease *nep* from a haloalkaliphilic archaeon *Natrialba magadii* has been cloned and sequenced. The *nep* gene expressed in *Escherichia coli* and *Haloferax volcanii* yields active form of the enzyme (Kamekura et al. 1992). The *nep* encoded polypeptide with a molecular mass of 56.4 kDa and isoelectric point of 3.77 include 121-amino acids. The primary sequence of *nep* closely relates to serine proteases of the subtilisin family from archaea and bacteria highlighting 50–85 % similarities. *Haloferax volcanii* produces protease which is active in high salt, alkaline pH and high DMSO (Rosana et al. 2008). Similarly, cloning, overexpression, characterization and structural and functional analysis of few alkaline protease genes and proteins from haloalkaliphilic bacteria: *Oceanobacillus iheyensis* O.M.A18, Haloalkaliphilic bacterium O.M.E12 and Haloalkaliphilc bacterium Ve2-20-91 are recently reported (Purohit 2012; Purohit and Singh 2011, 2014; Raval 2013; Raval et al. 2014b). The amino acid sequences of the alkaline proteases show hydrophobic character and stable configurations. The amino acids Asp 141, His 171 and Ser 324 form a catalytic triad with Ile, Leu and Ser being other amino acid moieties present in the active site. The characteristics of the recombinant proteases are similar to the enzymes (Purohit and Singh 2014).

While in another report, the catalytic features of the protease from haloalkaliphilic bacteria are directly linked with the structural features. There is correlation between the biochemical and molecular properties, such as sequence of amino acids and protein structure. The catalytically active site of the proteases generally consists of three residues; Asp (an electrophile), His (base) and Ser (nucleophile). In this triad, the serine residue forms the catalytic site. The molecular structure of the protease comprise of a single chain A, with 17 α-helices and 3 parallel β-strands as predicted by the CLC protein workbench v.5.8.1 (CLC Bio) (Raval et al. 2014b).

16.6 Conclusion

Halophiles and haloalkaliphiles are among the significant groups of the extremophiles. Many novel and unique properties of these microbes have been explored for fermentation, aquaculture, enzyme and antibiotic production. It appears that in coming years, greater potential in biotechnology will be realized for the halophilic and haloalkaliphilic microorganisms. The stability under the multitude of extremities of pH, salt and temperature makes these organisms more attractive. Some of the common traits of these organisms are reflected in their biochemical properties, catalysis and stability of their enzymes that are salt dependent. The enzymes from these organisms primarily include proteases, lipases, amylases, cellulases, chitinase and xylanases. In the recent past, the metagenomic approaches to explore the diversity of non- cultivable microbes have attained significance. The analysis of the population dynamics, diversity, phylogeny and biocatalytic potential of the microbial community of the saline ecosystems would be quite interesting to explore. In this context, the culture independent approaches and in-vitro improvement of the enzymes will lead to the identification and creation of novel biocatalysts.

References

Adinarayana K, Ellaiah P, Prasad DS (2003) Purification and partial characterization of thermostable serine alkaline protease from a newly isolated *Bacillus subtilis* PE-11. AAPS Pharm Scitech 4:1–9

Amoozegar MA, Malekzadeh F, Malik KA, Schumann P, Sproer C (2003) *Halobacillus karajensis* sp. nov., a novel moderate halophile. Int J Syst Evol Microbiol 53:1059–1063

Amoozegar MA, Sanchez-Porro C, Rohban R, Hajighasemi M, Ventosa A (2009) *Bacillus persepolensis* sp. nov., a moderately halophilic bacterium from a hypersaline lake. Int J Syst Evol Microbiol 59:2352–2358

Amoozegar MA, Bagheri M, Makhdoumi-Kakhki A, Didari M, Schumann P, Sproer C, Sanchez-Porro C, Ventosa A (2014) *Oceanobacillus limi* sp. nov., a moderately halophilic bacterium from a salt lake. Int J Syst Evol Microbiol 64:1284–1289

Arahal DR, Garcia MT, Vargas C, Canovas D, Nieto JJ, Ventosa A (2001) *Chromohalobacter salexigens* sp. nov., a moderately halophilic species that includes *Halomonas elongate* DSM 3043 and ATCC 33174. Int J Syst Evol Microbiol 51:1457–1462

Bagheri M, Didari M, Amoozegar M, Schumann P, Sanchez-Porro C, Mehrshad M, Ventosa A (2012) *Bacillus iranensis* sp. nov., a moderately halophilic bacterium from Aran-Bidgol Lake, a hypersaline Iranian lake. Int J Syst Evol Microbiol 62(Pt 4):811–816. doi:10.1099/ijs.0.030874-0

Bagheri M, Amoozegar MA, Didari M, Makhdoumi-Kakhki A, Schumann P, Sproer C, Sanchez-Porro C, Ventosa A (2013) *Marinobacter persicus* sp. nov., a moderately halophilic bacterium from a saline lake in Iran. Antonie Van Leeuwenhoek 104:47–54

Banciu H, Sorokin DY, Galinski EA, Muyzer G, Kleerebezem R, Kuenen JG (2004) *Thialkalivibrio halophilus* sp. nov., a novel obligately chemolithoautotrophic, facultatively alkaliphilic, and extremely salt-tolerant, sulfur-oxidizing bacterium from a hypersaline alkaline lake. Extremophiles 8:325–334

Battestein V, Macedo GA (2007) Effects of temperature, pH and additives on the activity of tannase produced by *Paecilomyces variotii*. Electron J Biotechnol 10(2):191–199

Baxter BK, Litchfield CD, Sowers K, Griffith JD, Dassarma PA, Dassarma S (2005) Microbial diversity of Great Salt Lake. In: Adaptation to life at high salt concentrations in archaea, bacteria, and eukarya cellular origin, life in extreme habitats and astrobiology, vol 9. Springer, Dordrecht, pp 9–25

Bayoudh A, Gharsallah N, Chamkha M, Dhouib A, Ammar S, Nasri M (2000) Purification and characterization of an alkaline protease from *Pseudomonas aeruginosa* MN1. J Ind Microbiol Biotechnol 24:291–295

Beg KQ, Gupta R (2003) Purification and characterization of an oxidation-stable, thiol-dependent serine alkaline protease from *Bacillus mojavensis*. Enzyme Microb Technol 32:294–304

Berezovsky IN, Shakhnovich EI (2008) Physics and evolution of thermophilic adaptation. FEBS J 275(7):1593–1605

Blum JS, Kulp TR, Han S, Lanoil B, Saltikov CW, Stolz JF, Miller LG, Oremland RS (2012) *Desulfohalophilus alkaliarsenatis* gen. nov., sp. nov., an extremely halophilic sulfate- and arsenate-respiring bacterium from Searles Lake, California. Extremophiles 16:727–742

Boltianskaia I, Antipov AN, Kolganova TV, Lysenko AM, Kostrikina NA, Zhilina TN (2004) *Halomonas campisalis*, an obligatorily alkaliphilic, nitrous oxide-reducing denitrifier with a Mo-cofactor-lacking nitrate reductase. Mikrobiologiia 73(3):326–334

Boltyanskaya YV, Kevbrin VV, Lyensko AM, Kolganova TV, Tourova TP, Osipov GA, Zhilina TN (2007) *Holomonas mongoliensis* sp. nov. and Halomonas kenyensis sp. nov., new haloalkaliphilic denitrifiers capable of N_2O reduction, isolated from soda lakes. Microbiology 76:739–747

Boominadhan U, Rajakumar R, Sivakumaar PKV, Joe MM (2009) Optimization of protease enzyme production using Bacillus sp. isolated from different wastes. Bot Res Int 2(2):83–87

Cabrera A, Aguilera M, Fuentes S, Incerti C, Russell NJ, Ramos-Cormenzana A, Monteoliva-Sanchez M (2007) *Halomonas indalinina* sp. nov., a moderately halophilic bacterium isolated from a solar saltern in Cabo de Gata, Almeria, southern Spain. Int J Syst Evol Microbiol 57:376–380

Carvalho RV, Correa TLR, Matos da Silva JC, Mansur LRC, Martins MLL (2008) Properties of an amylase from thermophilic *Bacillus* sp. Braz J Microbiol 39(1):102–107

Chand S, Mishra P (2003) Research and application of microbial enzymes. India's contribution. Adv Biochem Eng Biotechnol 85:95–124

Colquhouna D, Sorumb H (2002) Cloning, characterization and phylogenetic analysis of the fur gene in *Vibrio salmonicida* and *Vibrio logei*. Gene 296:213–220

Corral P, Gutierrez MC, Castillo AM, Dominguez M, Lopalco P, Corcelli A, Ventosa A (2013) *Natronococcus roseus* sp. nov., a haloalkaliphilic archaeon from a hypersaline lake. Int J Syst Evol Microbiol 63(1):104–108

Curtis S (2009) Fermentation of fish soluble proteins, peptides, and amino acids for umami tastes. Food Technol 13:26–30

DasSarma S, Arora P (2001) Halophiles. In: Encyclopedia of life sciences. Macmillan Press, Nature Publishing Group, London, pp 1–9. [WWW document]. http://www.els.net

Demirjian DC, Moris-Varas F, Cassidy CS (2001) Enzymes from extremophiles. Curr Opin Chem Biol 5:144–151

Denizci AA, Kazan D, Erarslan A (2010) *Bacillus marmarensis* sp. nov., an alkaliphilic, protease-producing bacterium isolated from mushroom compost. Int J Syst Evol Microbiol 60:1590–1594. doi:10.1099/ijs. 0.012369-0

Dobson SJ, James SR, Franzmann PD, McMeekin TA (1990) Emended description of *Halomonas halmophila* (NCMB 1971^{T}). Int J Syst Bacteriol 40(4):462–463

Dodia MS (2005) PhD thesis, Saurashtra University, Rajkot

Dodia MS, Joshi RH, Patel RK, Singh SP (2006) Characterization and stability of extracellular alkaline proteases from moderately halophilic and alkaliphilic bacteria isolated from saline habitat of coastal Gujarat, India. Braz J Microbiol 37:244–252

Dodia MS, Bhimani HG, Rawal CM, Joshi RH, Singh SP (2008a) Salt dependent resistance against chemical denaturation of alkaline protease from a newly isolated haloalkaliphilic *Bacillus* sp. Bioresour Technol 99:6223–6227

Dodia MS, Rawal CM, Bhimani HG, Joshi RH, Khare SK, Singh SP (2008b) Purification and stability characteristics of an alkaline serine protease from a newly isolated Haloalkaliphilic bacterium sp. AH-6. J Ind Microbiol Biotechnol 35:121–131

Doronina NV, Darmaeva T, Trotsenko Y (2003a) *Methylophaga natronica* sp. nov., a new alkaliphilic and moderately halophilic, restricted-facultatively methylotrophic bacterium from soda lake of the Southern Transbaikal region. J Syst Appl Microbiol 26:382–389

Doronina NV, Darmaeva TD, Trotsenko YA (2003b) *Methylophaga alcalica* sp. nov., a novel alkaliphilic and moderately halophilic, obligately methylotrophic bacterium from an East Mongolian saline soda lake. J Syst Evol Microbiol 53(Pt 1):223–229

Echigo A, Minegishi H, Mizuki T, Kamekura M, Usami R (2010) *Geomicrobium halophilum* gen. nov., sp. nov., a moderately halophilic and alkaliphilic bacterium isolated from soil. Int J Syst Evol Microbiol 60:990–995

Fan H, Xue Y, Ma Y, Ventosa A, Grant WD (2004) *Halorubrum tibetense* sp. nov., a novel haloalkaliphilic archaeon from Lake Zabuye in Tibet, China. Int J Syst Evol Microbiol 54:1213–1216

Feng J, Zhou P, Liu S (2004) Halorubrum xinjiangense sp. nov., a novel halophile isolated from saline lakes in China. Int J Syst Evol Microbiol 54:1789–1791
Forsyth MP, Shindler DB, Gochnauer MB, Kushner DJ (2005) Salt tolerance of intertidal marine bacteria. Can J Microbiol 17:825–828
Foti M, Ma S, Sorokin DY, Rademaker JLW, Kuenen GJ, Muyzer G (2006) Genetic diversity and biogeography of haloalkaliphilic sulfur-oxidizing bacteria belonging to the genus *Thioalkalivibrio*. FEMS Microbiol Ecol 56:95–101
Garcia MT, Gallego V, Ventosa A, Mellado E (2005) *Thalassobacillus devorans* gen. nov., sp. nov., a moderately halophilic, phenol-degrading, Gram-positive bacterium. Int J Syst Evol Microbiol 55:1789–1795
Ghorbel B, Kamoun AS, Nasri M (2003) Stability studies of protease from *Bacillus cereus* BG1. Enzyme Microb Technol 32:513–518
Ghosh A, Bhardwaj M, Satyanarayana T, Khurana M, Mayilraj S, Jain RK (2007) *Bacillus lehensis* sp. nov., an alkalitolerant bacterium isolated from soil. Int J Syst Evol Microbiol 57:238–242
Gimenez MI, Studdert CA, Sanchez J, De Castro RE (2000) Extra cellular protease of *Natrialba magadii*: purification and biochemical characterization. Extremophiles 4:181–188
Gonzalez-Domenech CM, Martinez-Checa F, Quesada E, Bejar V (2008) *Halomonas cerina* sp. nov., a moderately halophilic, denitrifying, exopolysaccharide-producing bacterium. Int J Syst Evol Microbiol 58:803–809
Grant WD, Mwatha WE (1998) Bacteria from alkaline, saline environments. In: Hattori T, Ishida Y, Maruyama Y, Morita RY, Uchida A (eds) Recent advances in microbial ecology, vol 1. Japan Scientific Societies Press, Tokyo, pp 29–33
Graziano G, Merlino A (2014) Molecular bases of protein halotolerance. Biochim Biophys Acta (BBA) Protein Proteomics. doi:10.1016/j.bbapap.2014.02.018
Gunde-Cimerman N, Zalar P, DeHoog GS, Plementias A (2000) Hypersaline water in salterns. Natural ecological niches for halophilic black yeasts. FEMS Microbiol Ecol 32:235–240
Guo JP, Ying M (2008) High-level expression, purification and characterization of recombinant *Aspergillus oryzae* alkaline protease in *Pichia pastoris*. Protein Expr Purif 58(2):301–308
Gupta MN, Roy I (2004) Enzymes in organic media forms, functions and applications. Eur J Biochem 271(13):2575–2583
Gupta A, Roy I, Patel RK, Singh SP, Khare SK, Gupta MN (2005) One-step purification and characterization of an alkaline protease from haloalkaliphilic *Bacillus* sp. J Chromatogr A 1075:103–108
Hidri DE, Guesmi A, Najjari A, Cherif H, Ettoumi B, Hamdi C, Boudabous A, Cherif A (2013) Cultivation dependant assessment, diversity, and ecology of Haloalkaliphilic bacteria in arid saline systems of southern Tunisia. BioMed Res Int. doi:10.1155/2013/648141, Article ID 648141, 15
Hiraga K, Nishikata Y, Namwong S, Tanasupawat S, Takada K, Oda K (2005) Purification and characterization of serine proteinase from a halophilic bacterium, *Filobacillus* sp. RF2-5. Biosci Biotechnol Biochem 69:38–44
Hoover RB, Pikuta EV, Bej AK, Marsic D, Whitman WB, Tang J, Krader P (2003) *Spirochaeta americana* sp. nov., a new haloalkaliphilic, obligately anaerobic spirochaete isolated from soda Mono Lake in California. Int J Syst Evol Microbiol 53:815–821
Horikoshi K (1999) Alkaliphiles: some applications of their products for biotechnology. Microbiol Mol Biol Rev 63:735–750
Imhoff JF, Sahl HG, Soliman GSH (1979) The Wadi Natrun: chemical composition and microbial mass developments in alkaline brines of Eutrophic Desert Lakes. Geomicrobiol J 1(3):219–234
James SR, Dobson SJ, Franzmann PD, McMeekin TA (1990) *Halomonas meridiana*, a new species of extremely halotolerant bacteria isolated from Antarctic saline lakes. Syst Appl Microbiol 13(3):270–278
Jeon C, Lim J, Lee J, Xu L, Jiang C, Kim C (2005) Reclassification of *Bacillus haloalkaliphilus* Fritze 1996 as *Alkalibacillus haloalkaliphilus* gen. nov., comb. nov. and the description of

Alkalibacillus salilacus sp. nov., a novel halophilic bacterium isolated from a salt lake in China. Int J Syst Evol Microbiol 55:1891–1896

Johnvesly B, Naik GR (2001) Studies on production of thermostable alkaline protease from thermophilic and alkaliphilic *Bacillus* sp. JB-99 in a chemically defined medium. Process Biochem 37:139–144

Joo HS, Kumar CG, Park GC, Paik SR, Chang CS (2003) Oxidant and SDS stable alkaline protease from a halo-tolerant *Bacillus clausii* I-52: production and some properties. J Appl Microbiol 95:267–272

Joshi RH (2006) A PhD thesis, Department of Biosciences, Saurashtra University Rajkot

Joshi RH, Dodia S, Singh SP (2008) Production and optimization of a commercially viable alkaline protease from a haloalkaliphilic bacterium. Biotechnol Bioprocess Eng 13:552–559

Joshi RH, Dodia MS, Singh SP (2009) Production and optimization of a commercially viable alkaline protease from a haloalkaliphilic bacterium. Biotechnol Bioprocess Eng 13:552–559

Kamekura M, Seno Y, Holmes ML, Dyall-Smith ML (1992) Molecular cloning and sequencing of the gene for a halophilic alkaline serine protease (halolysin) from an unidentified halophilic archaea strain (172P1) and expression of the gene in *Haloferax volcanii*. J Bacteriol 174(3):736–742

Karan R, Khare SK (2011) Stability of haloalkaliphilic *Geomicrobium* sp. protease modulated by salt. Biochemistry (Moscow) 76:686–693

Kevbrin VV, Zhilina TN, Rainey FA, Zavarzin GA (1998) *Tindallia magadii* gen. nov., sp. nov.: an alkaliphilic anaerobic ammonifier from Soda Lake deposits. Curr Microbiol 37:94–100

Kim WK, Choi KH, Kim YT, Park HH, Choi JY, Lee YS, Oh HI, Kwon IB, Lee SY (1996) Purification and characterization of a fibrinolytic enzyme produced from *Bacillus* sp. strain CK11-4 screened from Chungkook-Jang. Appl Environ Microbiol 62:2482–2488

Kim K, Jin L, Yang H, Lee S (2007) *Halomonas gomseomensis* sp. nov., *Halomonas janggokensis* sp. nov., *Halomonas salaria* sp. nov. and *Halomonas denitrificans* sp. nov., moderately halophilic bacteria isolated from saline water. Int J Syst Evol Microbiol 57:675–681

Kotlova EK, Ivanova NM, Yusupova MP, Voyushina TL, Ivanushkina NE, Chestukhina GG (2007) Thioldependent serine proteinase from *Paecilomyces lilacinus* purification and catalytic properties. Biochemistry (Moscow) 72:117–123

Kudrya V, Simonenko I (1994) Alkaline serine proteinase and lectin isolation from the culture fluid of *B. subtilis*. Appl Microbiol Biotechnol 41:505–509

Kumar CG, Takagi H (1999) Microbial alkaline proteases: from a bio-industrial view point. Biotechnol Adv 17:561–594

Ladenstein R, Antranikian G (1998) Proteins from hyperthermophiles: stability and enzymatic catalysis close to the boiling point of water. Adv Biochem Eng Biotechnol 61:37–85

Li AN, Li DC (2009) Cloning, expression and characterization of the serine protease gene from *Chaetomium thermophilum*. J Appl Microbiol 106:36–380

Li WJ, Schumann P, Zhang Y, Chen G, Tian X, Xu L, Stackebrandt E, Jiang C (2005) *Marinococcus halotolerans* sp. nov., isolated from Qinghai, north-west China. Int J Syst Evol Microbiol 55:1801–1804

Litchfield CD (2004) Microbial molecular and physiological diversity in hypersaline environments. In: Ventosa A (ed) Halophilic microorganisms. Springer, Berlin/Heidelberg, pp 49–61

Litchfield CD, Gillevet PM (2002) Microbial diversity and complexity in hypersaline environments: a preliminary assessment. J Ind Microbiol Biotechnol 28:48–55

Liu WY, Zeng J, Wang L, Dou YT, Yang SS (2005) *Halobacillus dabanensis* sp. nov. and *Halobacillus aidingensis* sp. nov., isolated from salt lakes in Xinjiang. China Int J Syst Evol Microbiol 55:1991–1996

Luke KA, Higgins CL, Wittung-Stafshede P (2007) Thermodynamic stability and folding of proteins from hyperthermophilic organisms. FEBS J 274(16):4023–4033

Madern D, Ebel C, Zaccai G (2000) Halophilic adaptation of enzymes. Extremophiles 4:91–98

Makhdoumi-Kakhki A, Amoozegar MA, Bagheri M, Ramezani M, Ventosa A (2012) *Haloarchaeobius iranensis* gen. nov., sp. nov., an extremely halophilic archaeon isolated from a saline Lake. Int J Syst Evol Microbiol 62:1021–1026

Manni L, Jellouli K, Agrebi R, Bayoudh A, Nasri M (2008) Biochemical and molecular characterization of a novel calcium-dependent metalloprotease from *Bacillus cereus* SV1. Process Biochem 43(5):522–530
Margesin R, Schinner F (2001a) Biodegradation and bioremediation of hydrocarbons in extreme environments. Appl Microbiol Biotechnol 56:650–663
Margesin R, Schinner F (2001b) Potential of halotolerant and halophilic microorganisms for biotechnology. Extremophiles 5:73–83
Martins RF, Davids W, Abu AW, Levander F, Radstrom P, Hatti-Kaul R (2001) Starch-hydrolyzing bacteria from Ethiopian soda lakes. Extremophiles 2:135–144
Melanie RM, Romine MF, Garcia MT, Antonio V, Bailey TJ, Peyton BM (1999) *Halomonas campisalis* sp. nov., a denitrifying, moderately Haloalkaliphilic bacterium. Syst Appl Microbiol 22:551–558
Moreno ML, Perez D, Garcia MT, Mellado E (2013) Halophilic bacteria as a source of novel hydrolytic enzymes. Life 3:38–51. doi:10.3390/life3010038
Mormile MR, Romine MF, Garcia TM, Ventosa A, Bailey T, Peyton BM (1999) *Halomonas campisalis* sp. nov., a denitrifying, moderately Haloalkaliphilic bacterium system. Appl Microbiol 22:551–558
Munawar N, Engel PC (2014) Halophilic enzymes: characteristics, structural adaptation and potential applications for biocatalysis. Extremophiles 2(4):334–344
Muntyan MS, Tourova TP, Lysenko AM, Kolganova TV, Fritze D, Skulachev VP (2002) Molecular identification of alkaliphilic and halotolerant strain *Bacillus sp*. FTU as *Bacillus pseudofirmus* FTU. Extremophiles 6:195–199
Namsaraev Z, Akimov V, Tsapin A, Barinova E, Nealson K, Gorlenko V (2009) *Marinospirillum celere* sp. nov., a novel haloalkaliphilic, helical bacterium isolated from Mono Lake. Int J Syst Evol Microbiol 59:2329–2332
Neklyudov AD, Ivankin AN, Berdutina AV (2000) Properties and uses of protein hydrolysates (review). Appl Biochem Microbiol 36:452–459
Ni X, Yue L, Chi Z, Li Z, Wang X, Madzak C (2009) Alkaline protease gene cloning from the marine yeast *Aureobasidium pullulans* HN2-3 and the protease surface display on *Yarrowia lipolytica* for bioactive peptide production. Mar Biotechnol 11:81–89
Niehaus F, Bertoldo C, Kahler M, Antranikian G (1999) Extremophiles as a source of novel enzymes for industrial application. Appl Microbiol Biotechnol 51:711–729
Nowlan B, Dodia MS, Singh SP, Patel BKC (2006) *Bacillus okhensis* sp. nov., a halotolerant and alkalitolerant bacterium from an Indian saltpan. Int J Syst Evol Microbiol 56:1073–1077
Oberoi R, Beg QK, Puri S, Sazena RK, Gupta R (2001) Characterization and wash performance analysis of an SDS-Stable alkaline protease from *Bacillus* sp. World J Microbiol Biotechnol 17:493–497
Oremland RS, Hoeft SE, Santini JM, Bano N, Hollibaugh RA, Hollibaugh JT (2002) Anaerobic oxidation of arsenite in Mono Lake water and by a facultative, arsenite-oxidizing chemoautotroph, strain MLHE-1. Appl Environ Microbiol 68(10):4795–4802
Oren A (2002a) Diversity of halophilic microorganisms: environments, phylogeny, physiology, applications. J Ind Microbiol Biotechnol 28:56–63
Oren A (2002b) Halophilic microorganisms and their environments. Kluwer Academic Publishers, Dordrecht. doi:10.1007/0-306-48053-0
Pandey S, Singh SP (2012) Organic solvent tolerance of an α-amylase from haloalkaliphilic bacteria as a function of pH, temperature, and salt concentrations. Appl Biochem Biotechnol 166:1747–1757
Pandey S, Rakholiya KD, Raval VH, Singh SP (2012) Catalysis and stability of an alkaline protease from a haloalkaliphilic bacterium under non-aqueous conditions as a function of pH, salt and temperature. J Biosci Bioeng 114(3):251–256
Patel RK, Dodia MS, Singh SP (2005) Extracellular alkaline protease from a newly isolated haloalkaliphilic *Bacillus* sp.: production and optimization. Process Biochem 40:3569–3575

Patel RK, Dodia MS, Joshi RH, Singh SP (2006a) Production of extracellular haloalkaline protease from a newly isolated haloalkaliphilic *Bacillus* sp. isolated from sea water in Western India. World J Microbiol Biotechnol 22(4):375–382

Patel RK, Dodia MS, Joshi RH, Singh SP (2006b) Purification and characterization of alkaline protease from a newly isolated haloalkaliphilic *Bacillus* sp. Process Biochem 41:2002–2009

Pesenti PT, Sikaroodi M, Gillevet PM, Sanchez-Porro C, Ventosa A, Litchfield CD (2008) *Halorubrum californiense* sp. nov., an extreme archaeal halophile isolated from a crystallizer pond at a solar salt plant in California, USA. Int J Syst Evol Microbiol 58:2710–2715

Peyton BM (1999) *Halomonas campisalis* sp. nov., a denitrifying, moderately haloalkaliphilic bacterium. Syst Appl Microbiol 22:551–558

Phadatare US, Despande VV, Srinivasan MC (1993) High activity alkaline protease from *Conidiobolus coronatus* (NCL 86.2.20) enzyme production and compatibility with commercial detergents. Enzyme Microb Technol 15:72–76

Poli A, Nicolaus B, Denizci A, Yavuzturk B, Kazan D (2013) *Halomonas smyrnensis* sp. nov., a moderately halophilic, exopolysaccharide-producing bacterium from Camalt Saltern area, Turkey. Int J Syst Evol Microbiol 63:10–18

Purohit MK (2012) PhD thesis, Saurashtra University, Rajkot

Purohit MK, Singh SP (2011) Comparative analysis of enzymatic stability and amino acid sequences of thermostable alkaline proteases from two haloalkaliphilic bacteria isolated from Coastal region of Gujarat, India. Int J Biol Macromol 49:103–112

Purohit MK, Singh SP (2014) Cloning, over expression and functional attributes of serine proteases from *Oceanobacillus iheyensis* O.M.A18 and *Haloalkaliphilic bacterium* O.M.E12. Process Biochem 49:61–68

Rahman RNZRA, Geok LP, Basri M, Salleh AB (2006) An organic solvent-stable alkaline protease from *Pseudomonas aeruginosa* strain K: enzyme purification and characterization. Enzyme Microb Technol 39(7):1484–1491

Ramesh S, Rajesh M, Mathivanan N (2009) Characterization of a thermostable alkaline protease produced by marine *Streptomyces fungicidicus* MMLY 614. Bioprocess Biosyst Eng 32:91–800

Rao MB, Tanksale AM, Ghatge MS, Deshpande VV (1998) Molecular and biotechnological aspects of microbial proteases. Microbiol Mol Biol Rev 62:597–635

Rattray FP, Bockelmann W, Fox PF (1995) Purification and characterization of an extracellular proteinase from *Brevibacterium linens* ATCC 9174. Appl Environ Microbiol 61(9):3454–3456

Raval VH (2013) PhD thesis, Saurashtra University Rajkot

Raval VH, Pillai S, Rawal CM, Singh SP (2014a) Biochemical and structural characterization of a detergent stable serine alkaline protease from sea water haloalkaliphilic bacteria. Process Biochem 49(6):955–962

Raval VH, Rawal CM, Bhatt HB, Dahima BR, Pandey S, Singh SP (2014b) Cloning and molecular characterization of a serine protease from sea water haloalkaliphilic bacteria. Ann Microbiol. doi:10.1007/s13213-014-0869-0

Rawal CM (2012) PhD thesis, Saurashtra University Rajkot

Rawal CM, Raval VH, Bhimani HD, Bhensdadiya DV, Kothari CR, Patel AB, Bhatt VD, Parmar NR, Sajnani MR, Koringa PG, Joshi CG, Kothari RK, Singh SP (2012) Whole genome shotgun sequencing of extremophile *Alkalibacillus haloalkaliphilus* C-5 of Indian origin. J Bacteriol 194(17):4775

Reza H, Heidari K, Amoozegar MA, Hajighasemi M, Ziaee A, Ventosa V (2008) Production, optimization and purification of a novel extracellular, protease from the moderately halophilic bacterium *Halobacillus karajensis*. J Ind Microbiol Biotechnol 36:21–27

Romano I, Giordano A, Lama L, Nicolaus B, Gambacorta A (2005a) *Halomonas campaniensis* sp. nov., a haloalkaliphilic bacterium isolated from a mineral pool of Campania Region, Italy. Syst Appl Microbiol 28:610–618

Romano I, Lama L, Nicolaus B, Gambacorta A, Giordano A (2005b) *Alkalibacillus filiformis* sp. nov., isolated from a mineral pool in Campania, Italy. Int J Syst Evol Microbiol 55:2395–2399

Romano I, Orlando P, Gambacorta A, Nicolaus B, Dipasquale L, Pascual J, Giordano A, Lama L (2011) *Salinivibrio sharmensis* sp. nov., a novel haloalkaliphilic bacterium from a saline lake in Ras Mohammed Park (Egypt). Extremophiles 15(2):213–220

Rosana HE, Williams H, Kites S, David K (2008) Purification and characterization of Nep proteases from *H. volcanii*. Prot Exp Purif 43(5):31–37

Rothschild LJ, Manicinelli RL (2001) Life in extreme environments. Nature 409:1092–1101

Ruiz DM, De Castro RE (2007) Effect of organic solvents on the activity and stability of an extracellular protease secreted by the haloalkaliphilic archaeon Natrialba magadii. J Ind Microbiol Biotechnol 34:111–115

Sanchez-Porro C, Mellado E, Bertoldo C, Antranikian G, Ventosa A (2003) Screening and characterization of the protease CP1 produced by the moderately halophilic bacterium *Pseudoalteromonas* sp. strain CP76. Extremophiles 7:221–228

Sanchez-Porro C, Tokunaga H, Tokunaga M, Ventosa A (2007) *Chromohalobacter japonicus* sp. nov., a moderately halophilic bacterium isolated from a Japanese salty food. Int J Syst Evol Microbiol 57:2262–2266

Sardessai YN, Bhosle S (2004) Industrial potential of organic solvent tolerant bacteria. Biotechnol Prog 20(3):655–660

Schiraldi C, De Rosa M (2002) Production of biocatalysts and biomolecules from extremophiles. Trends Biotechnol 20:515–521

Setyorini E, Takenaka S, Murakami S, Aoki K (2006) Purification and characterization of two novel halotolerant extracellular proteases from Bacillus subtilis strain FP-133. Biosci Biotechnol Biochem 70(2):433–440

Shameer S, Prasada BG, Paramageetham C (2013) Isolation of amylase producing bacteria from solar salterns of Nellore district, Andhra Pradesh, India. RRJMB 2(1):13–16

Shannon JD, Baramova EN, Bajarnason JB, Fox JW (1989) Amino acid sequence of a Crotalus atrox venom metalloproteinase which cleaves type IV collagen and gelatin. J Biol Chem 264(20):11575–11583

Sikkema J, De Bont J, Poolman B (1995) Mechanisms of membrane toxicity of hydrocarbons. Microbiol Rev 59:201–222

Sinha R, Srivastava AK, Khare SK (2014) Efficient proteolysis and application of an alkaline protease from halophilic *Bacillus sp.* EMB9. Prep Biochem Biotechnol 44:680–696

Sorokin DY, Kuenen JG (2005) Alkaliphilic chemolithotrophs from soda lakes. FEMS Microbiol Ecol 52:287–295

Sorokin DY, Gorlenko VM, Tourova TP, Tsapin AI, Nealson KH, Kuenen JG (2002a) *Thialkalimicrobium cyclum* sp. nov., and *Thialkalivibrio jannaschii* sp. nov., new species of alkaliphilic, obligately chemolithoautotrophic sulfur-oxidizing bacteria from a hypersaline alkaline Mono Lake (California). Int J Syst Evol Microbiol 52:657–664

Sorokin DY, Tourova TP, Kolganova TV, Sjollema KA, Kuenen G (2002b) *Thioalkalispira microaerophila* gen. nov., sp. nov., a novel litho autotrophic, sulfur oxidizing bacterium from a soda Lake. Int J Syst Evol Microbiol 52:2175–2182

Sorokin DY, Tourova TP, Lysenko AM, Mityushina LL, Kuenen JG (2002c) *Thioalkalivibrio thiocyanoxidans* sp. nov., and *Thioalkalivibrio paradoxus* sp. nov., novel alkaliphilic, obligately autotrophic, sulfur oxidizing bacteria capable of growth on thiocyanate, from soda lakes. Int J Syst Evol Microbiol 52:657–664

Sorokin DY, Tourova TP, Antipov AN, Muyzer G, Kuenen JG (2004) Anaerobic growth of the haloalkaliphilic denitrifying sulfur-oxidizing bacterium *Thialkalivibrio thiocyanodenitrificans* sp. nov., with thiocyanate. Microbiology 150(7):2435–2442

Sorokin DY, Zhilina TN, Lysenko AM, Tourova TP, Spiridonova EM (2006) Metabolic versatility of haloalkaliphilic bacteria from soda lakes belonging to the *Alkalispirillum-Alkalilimnicola* group. Extremophiles 10:213–220

Sorokin DY, Tourova TP, Abbas B, Suhacheva MV, Muyzer G (2012) *Desulfonatronovibrio halophilus* sp. nov., a novel moderately halophilic sulfate-reducing bacterium from hypersaline chloride–sulfate lakes in Central Asia. Extremophiles 16(3):411–417

Taira W, Funatsu Y, Satomi M, Takano T, Abe H (2007) Changes in extractive components and microbial proliferation during fermentation of fish sauce from underutilized fish species and quality of final products. Fish Sci 73:913–923

Thumar JT, Singh SP (2009) Organic solvent tolerance of an alkaline protease from salt-tolerant alkaliphilic *Streptomyces clavuligerus* strain Mit-1. J Ind Microbiol Biotechnol 36:211–218

Toyokawa Y, Takahara H, Reungsang A, Masakazu F, Yuki H, Tachibana S, Masaaki Y (2010a) Purification and characterization of a halotolerant serine proteinase from thermotolerant *Bacillus licheniformis* RKK-04 isolated from Thai fish sauce. Appl Microbiol Biotechnol 9:2434–2435

Toyokawa Y, Takahara H, Reungsang A, Masakazu F, Yuki H, Tachibana S, Masaaki Y (2010b) Purification and characterization of a halotolerant serine proteinase from thermotolerant *Bacillus licheniformis* RKK-04 isolated from Thai fish sauce. Appl Microbiol Biotechnol 86:1867–1875

Trigui H, Masmoudi S, Brochier-Armanet C, Maalej S, Dukan S (2011) Characterization of *Halorubrum sfaxense* sp. nov., a new halophilic archaeon isolated from the solar saltern of Sfax in Tunisia. Int J Microbiol Article ID 240191, 8 pages. doi:10.1155/2011/240191

Ueda M, Asano T, Nakazawa M, Miyatake K, Inouye K (2008) Purification and characterization of novel raw-starch-digesting and cold-adapted alpha-amylases from Eisenia foetida Comp. Biochem Physiol Mol Biol 150(1):125–130

Upasani V, Desai S (1990) Sambhar salt lake chemical composition of the brines and studies on haloalkaliphilic archaebacteria. Arch Microb 154:589–593

Wang Q, Li W, Liu Y, Cao H, Li Z, Guo G (2007) *Bacillus qingdaonensis* sp. nov., a moderately haloalkaliphilic bacterium isolated from a crude sea-salt sample collected near Qingdao in eastern China. Int J Syst Evol Microbiol 57:1143–1147

Wang Q, Hou Y, Xu Z, Miao J, Li G (2008) Optimization of cold-active protease production by the psychrophilic bacterium *Colwellia* sp., NJ341 with response surface methodology. Bioresour Technol 99(6):1926–1931

Watzman H (1997) Left for dead: New Scientist, no. 2068, pp 37–41

Wu G, Wu X, Wang Y, Chi C, Tang Y, Kida K, Wu X, Luan Z (2008) *Halomonas daqingensis* sp. nov., a moderately halophilic bacterium isolated from an oilfield soil. Int J Syst Evol Microbiol 58:2859–2865

Xu X, Wu M, Zhou P, Liu S (2005) *Halobiforma lacisalsi* sp. nov., isolated from a salt lake in China. Int J Syst Evol Microbiol 55:1949–1952

Xu X, Wu Y, Zhou Z, Wang C, Zhou Y, Zhang H, Wang Y, Wu M (2007) *Halomonas saccharevitans* sp. nov., *Halomonas arcis* sp. nov. and *Halomonas subterranea* sp. nov., halophilic bacteria isolated from hypersaline environments of China. Int J Syst Evol Microbiol 57:1619–1624

Xue Y, Fan H, Ventosa A, Grant WD, Jones BE, Cowan DA, Ma Y (2005) *Halalkalicoccus tibetensis* gen. nov., sp. nov., representing a novel genus of haloalkaliphilic archaea. Int J Syst Evol Microbiol 55:2501–2505

Xue Y, Ventosa A, Wang X, Ren P, Zhou P, Ma Y (2008) *Bacillus aidingensis* sp. nov., a moderately halophilic bacterium isolated from Ai-Ding salt lake in China. Int J Syst Evol Microbiol 58:2828–2832

Yoon JH, Kang KH, Park YH (2003) *Halobacillus salinus* sp. nov., isolated from a salt lake on the coast of the East Sea in Korea. Int J Syst Evol Microbiol 53(3):687–693

Yoon J, Kang S, Lee S, Lee M, Oh T (2005a) *Virgibacillus dokdonensis* sp. nov., isolated from a Korean island, Dokdo, located at the edge of the East Sea in Korea. Int J Syst Evol Microbiol 55:1833–1837

Yoon J, Lee C, Oh T (2005b) *Aeromicrobium alkaliterrae* sp. nov., isolated from an alkaline soil, and emended description of the genus *Aeromicrobium*. Int J Syst Evol Microbiol 55:2171–2175

Zhang W, Xue Y, Ma Y, Zhou P, Ventosa A, Grant WD (2002) *Salinicoccus alkaliphilus* sp. nov., a novel alkaliphile and moderate halophile from Baer Soda Lake in Inner Mongolia Autonomous Region, China. Int J Syst Evol Microbiol 52:789–793

Zhang M, Zhao C, Du L, Lu F, Gao C (2008) Expression, purification, and characterization of a thermophilic neutral protease from *Bacillus stearothermophilus* in *Bacillus subtilis*. Sci China Ser C Life Sci J 51(1):52–59

Zhilina TN, Zavarzin GA, Detkova EN, Rainey FA (1996) *Natroniella acetigena* gen. nov. sp. nov., an extremely haloalkaliphilic, homoacetic bacterium: a new member of *Haloanaerobiales*. Curr Microbiol 32(6):320–326

Zvereva EA, Fedorova TV, Kevbrin VV (2006) Cellulase activity of a haloalkaliphilic anaerobic bacterium, strain Z-7026. Extremophiles 10(1):53–60

Index

D.K. Maheshwari, M. Saraf (eds.), *Halophiles*, Sustainable Development and Biodiversity 6, DOI 10.1007/978-3-319-14595-2

Printed by Printforce, the Netherlands